Compendium of Plant Genomes

Whole-genome sequencing is at the cutting edge of life sciences in the new millennium. Since the first genome sequencing of the model plant *Arabidopsis thaliana* in 2000, whole genomes of about 100 plant species have been sequenced and genome sequences of several other plants are in the pipeline. Research publications on these genome initiatives are scattered on dedicated web sites and in journals with all too brief descriptions. The individual volumes elucidate the background history of the national and international genome initiatives; public and private partners involved; strategies and genomic resources and tools utilized; enumeration on the sequences and their assembly; repetitive sequences; gene annotation and genome duplication. In addition, synteny with other sequences, comparison of gene families and most importantly potential of the genome sequence information for gene pool characterization and genetic improvement of crop plants are described.

More information about this series at https://link.springer.com/bookseries/11805

Amanda R. De La Torre
Editor

The Pine Genomes

Springer

Editor
Amanda R. De La Torre
Northern Arizona University
Flagstaff, AZ, USA

ISSN 2199-4781 ISSN 2199-479X (electronic)
Compendium of Plant Genomes
ISBN 978-3-030-93389-0 ISBN 978-3-030-93390-6 (eBook)
https://doi.org/10.1007/978-3-030-93390-6

This Springer imprint is published by the registered company Springer Nature Switzerland AG
The registered company address is: Gewerbestrasse 11, 6330 Cham, Switzerland

This book series is dedicated to my wife Phullara and our children Sourav and Devleena

Chittaranjan Kole

Preface to the Series

Genome sequencing has emerged as the leading discipline in the plant sciences coinciding with the start of the new century. For much of the twentieth century, plant geneticists were only successful in delineating putative chromosomal location, function, and changes in genes indirectly through the use of a number of "markers" physically linked to them. These included visible or morphological, cytological, protein, and molecular or DNA markers. Among them, the first DNA marker, the RFLPs, introduced a revolutionary change in plant genetics and breeding in the mid-1980s, mainly because of their infinite number and thus potential to cover maximum chromosomal regions, phenotypic neutrality, absence of epistasis, and codominant nature. An array of other hybridization-based markers, PCR-based markers, and markers based on both facilitated construction of genetic linkage maps, mapping of genes controlling simply inherited traits, and even gene clusters (QTLs) controlling polygenic traits in a large number of model and crop plants. During this period, a number of new mapping populations beyond F2 were utilized and a number of computer programs were developed for map construction, mapping of genes, and for mapping of polygenic clusters or QTLs. Molecular markers were also used in the studies of evolution and phylogenetic relationship, genetic diversity, DNA fingerprinting, and map-based cloning. Markers tightly linked to the genes were used in crop improvement employing the so-called marker-assisted selection. These strategies of molecular genetic mapping and molecular breeding made a spectacular impact during the last one and a half decades of the twentieth century. But still they remained "indirect" approaches for elucidation and utilization of plant genomes since much of the chromosomes remained unknown and the complete chemical depiction of them was yet to be unraveled.

Physical mapping of genomes was the obvious consequence that facilitated the development of the "genomic resources" including BAC and YAC libraries to develop physical maps in some plant genomes. Subsequently, integrated genetic–physical maps were also developed in many plants. This led to the concept of structural genomics. Later on, emphasis was laid on EST and transcriptome analysis to decipher the function of the active gene sequences leading to another concept defined as functional genomics. The advent of techniques of bacteriophage gene and DNA sequencing in the 1970s was extended to facilitate sequencing of these genomic resources in the last decade of the twentieth century.

As expected, sequencing of chromosomal regions would have led to too much data to store, characterize, and utilize with the-then available computer software could handle. But the development of information technology made the life of biologists easier by leading to a swift and sweet marriage of biology and informatics, and a new subject was born—bioinformatics.

Thus, the evolution of the concepts, strategies, and tools of sequencing and bioinformatics reinforced the subject of genomics—structural and functional. Today, genome sequencing has traveled much beyond biology and involves biophysics, biochemistry, and bioinformatics!

Thanks to the efforts of both public and private agencies, genome sequencing strategies are evolving very fast, leading to cheaper, quicker, and automated techniques right from clone-by-clone and whole-genome shotgun approaches to a succession of second-generation sequencing methods. The development of software of different generations facilitated this genome sequencing. At the same time, newer concepts and strategies were emerging to handle sequencing of the complex genomes, particularly the polyploids.

It became a reality to chemically—and so directly—define plant genomes, popularly called whole-genome sequencing or simply genome sequencing.

The history of plant genome sequencing will always cite the sequencing of the genome of the model plant *Arabidopsis* thaliana in 2000 that was followed by sequencing the genome of the crop and model plant rice in 2002. Since then, the number of sequenced genomes of higher plants has been increasing exponentially, mainly due to the development of cheaper and quicker genomic techniques and, most importantly, the development of collaborative platforms such as national and international consortia involving partners from public and/or private agencies.

As I write this preface for the first volume of the new series "Compendium of Plant Genomes," a net search tells me that complete or nearly complete whole-genome sequencing of 45 crop plants, eight crops and model plants, eight model plants, 15 crop progenitors and relatives, and three basal plants is accomplished, the majority of which are in the public domain. This means that we nowadays know many of our model and crop plants chemically, i.e., directly, and we may depict them and utilize them precisely better than ever. Genome sequencing has covered all groups of crop plants. Hence, information on the precise depiction of plant genomes and the scope of their utilization are growing rapidly every day. However, the information is scattered in research articles and review papers in journals and dedicated Web pages of the consortia and databases. There is no compilation of plant genomes and the opportunity of using the information in sequence-assisted breeding or further genomic studies. This is the underlying rationale for starting this book series, with each volume dedicated to a particular plant.

Plant genome science has emerged as an important subject in academia, and the present compendium of plant genomes will be highly useful to both students and teaching faculties. Most importantly, research scientists involved in genomics research will have access to systematic deliberations on the plant genomes of their interest. Elucidation of plant genomes is of interest not only for the geneticists and breeders, but also for practitioners of an array of plant science disciplines, such as taxonomy, evolution, cytology, physiology, pathology, entomology, nematology, crop production, biochemistry, and obviously bioinformatics. It must be mentioned that information regarding each plant genome is ever-growing. The contents of the volumes of this compendium are, therefore, focusing on the basic aspects of the genomes and their utility. They include information on the academic and/or economic importance of the plants, description of their genomes from a molecular genetic and cytogenetic point of view, and the genomic resources developed. Detailed deliberations focus on the background history of the national and international genome initiatives, public and private partners involved, strategies and genomic resources and tools utilized, enumeration on the sequences and their assembly, repetitive sequences, gene annotation, and genome duplication. In addition, synteny with other sequences, comparison of gene families, and, most importantly, the potential of the genome sequence information for gene pool characterization through genotyping by sequencing (GBS) and genetic improvement of crop plants have been described. As expected, there is a lot of variation of these topics in the volumes based on the information available on the crop, model, or reference plants.

I must confess that as the series editor, it has been a daunting task for me to work on such a huge and broad knowledge base that spans so many diverse plant species. However, pioneering scientists with lifetime experience and expertise on the particular crops did excellent jobs editing the respective volumes. I myself have been a small science worker on plant genomes since the mid-1980s and that provided me the opportunity to personally know several stalwarts of plant genomics from all over the globe. Most, if not all, of the volume editors are my longtime friends and colleagues. It has been highly comfortable and enriching for me to work with them on this book series. To be honest, while working on this series I have been and will remain a student first, a science worker second, and a series editor last. And I must express my gratitude to the volume editors and the chapter authors for providing me the opportunity to work with them on this compendium.

I also wish to mention here my thanks and gratitude to the Springer staff, particularly Dr. Christina Eckey and Dr. Jutta Lindenborn, for the earlier set of volumes and presently Ing. Zuzana Bernhart for all their timely help and support.

I always had to set aside additional hours to edit books beside my professional and personal commitments—hours I could and should have given to my wife, Phullara, and our kids, Sourav and Devleena. I must mention that they not only allowed me the freedom to take away those hours from them but also offered their support in the editing job itself. I am really not sure whether my dedication of this compendium to them will suffice to do justice to their sacrifices for the interest of science and the science community.

New Delhi, India Chittaranjan Kole

Preface

Pines (*Pinus*) are the world's most economically important forest tree species. With more than 100 species, pines are also the most abundant extant group of Gymnosperms. Pines are naturally distributed in the Northern hemisphere, where they inhabit pure or mixed-species forests or are planted for commercial uses. Some species such as *Pinus radiata* are also planted as commercial species in the Southern hemisphere. Efforts to understand their complex biology, functions and evolution were limited by their non-model system attributes (e.g., long generation times, slow growth, difficulty to clone or vegetative propagate) and huge genome sizes (20–40 Gbp) with high percentages (>70%) of repeat sequences, mostly transposable elements. In the last five years, improved and more accessible sequencing and bioinformatic tools have allowed significant changes in the study of the genomics and transcriptomics of pines. Since 2014, four species (*Pinus taeda*, *Pinus lambertiana*, *Pinus pinaster* and *Pinus radiata*) have been sequenced, and numerous transcriptomic resources have been developed.

This book is the first comprehensive compilation of the most up-to-date research in the genomics, transcriptomics and breeding of pine species across Europe, North America and Australia. The twelve chapters in this book aim to cover different aspects in genomic and transcriptomic research mainly focusing on the species with sequenced genomes but also in other pines of ecological and economical importance. In the Chap. 1, recent advances in whole-genome sequencing, transcriptome sequencing and target enrichment of nuclear genes for North American pine species are described. In the absence of chromosome-level reference genomes, studies on the genome architecture have been based on the presence of genetic and linkage maps. Genetic mapping and comparative mapping approaches are reviewed with an emphasis on *P. taeda* in Chap. 2.

Transposable elements are major components of pines and gymnosperm genomes. Although initial studies have revealed important information on their structure, classification and genome organization, many questions remain regarding their role in adaptive responses and genome x environment interactions in long-generation species such as pines. Chapter 3 provides a comprehensive review on the latest discoveries and future research perspectives for the study of transposable elements in plants, with an emphasis on pine species.

Rapid changes in the climate due to increases in temperature and altered precipitation regimes pose a significant challenge for natural and planted populations of pines. Our ability to predict future responses to environmental changes will only come from a thorough understanding of the genomic and transcriptomic basis of abiotic stress. In this book, associations between genotypes and environmental variables are tested across the *P. lambertiana* species' natural distribution (Chap. 4). Plastic responses to low water availability analyzed with transcriptome analysis for *P. pinaster* are reviewed in Chaps. 5 and 9.

Whole-genome sequencing and the development of a transcriptome atlas in *P. pinaster* are fully covered in Chap. 5. In addition, this chapter provides a comprehensive review of some of the most important research in *P. pinaster*, including recent findings in molecular breeding, transgenesis, comparative genomics, gene expression regulation, biotic and abiotic stress and genetic architecture and variation in the species. Perspectives about the impact of these recent discoveries and future research approaches are also discussed.

Most phenotypic traits of commercial importance in pines and plants in general have complex genomic architecture, meaning that a large number of genes are usually involved. Chapter 6 summarizes recent studies on complex traits in *P. taeda*, while Chaps. 7 and 8 focus on the genomics of disease resistance against fungal pathogens causing three major diseases in North American pine species: white pine blister rust, pitch canker and fusiform rust.

Transcriptomic approaches in European pines such as *P. pinaster* and *P. sylvestris* are covered in Chaps. 9 and 10. While Chap. 9 focuses on the transcriptomic, proteomics, metabolomic and genetic transformation used in functional genomic studies in *P. pinaster*; Chap. 10 focuses on the transcriptional and genomic responses to radiation stress in *P. sylvestris* populations from the Chernobyl exclusion zone after the Chernobyl nuclear power accident in the late '80s.

Given the economic importance of many widely distributed pine species, there is wide interest on improving breeding efficiency by shortening breeding cycles in long-generation pines. A big limitation, however, is the little knowledge about the genomics of complex traits in conifer species. The genomic selection was, therefore, developed as a potential solution. Genomic selection has the potential to shorten breeding cycles when compared with conventional (pedigree-based) breeding, reduce the cost of phenotyping and also does not require the identification of causal genes (as in marker-assisted breeding). Chapter 11 reviews recent research advances in genomic selection in *P. sylvestris* in Sweden and *P. radiata* in New Zealand.

Finally, the book concludes by discussing the future needs and applications in pine genomics by proposing a collaborative international advisory committee that organizes and prioritizes species to be sequenced and publicly accessed by the scientific community (Chap. 12). All the recent genomic and transcriptomic resources and studies described in this book have paved the way for understanding the complex biology of this very important group of plants and will help future management, conservation and breeding efforts.

Flagstaff, Arizona, USA
June 2021

Amanda R. De La Torre

Contents

Contributors

Sara Abrahamsson SKOGFORSK (The Forestry Research Institute of Sweden), Sävar, SE, Sweden

Ana Alves BioISI - Biosystems & Integrative Sciences Institute, Faculdade de Ciências, Universidade de Lisboa, Lisboa, PT, Portugal

Ricardo Alía Departamento de Ecología y Genética Forestal, Centro de Investigación Forestal (CIFOR, CSIC), Instituto Nacional de Investigación y Tecnología Agraria y Alimentaria (INIA), Madrid, ES, Spain

Isabel Arrillaga Biotechnology and Biomedicine (BiotecMed) Institute and Plant Biology Department, University of Valencia, Valencia, ES, Spain

Concepción Ávila Facultad de Ciencias, Departamento de Biología Molecular y Bioquímica, Universidad de Málaga, Málaga, Spain

Laurent Bouffier INRAE, Univ. Bordeaux, BIOGECO, Cestas, FR, France

Jeremy T. Brawner Department of Plant Pathology, University of Florida, Gainesville, FL, USA

Katharina B. Budde Department of Forest Genetics and Forest Tree Breeding, Buesgen Institute, Georg-August University of Göttingen, Göttingen, Germany

José Antonio Cabezas Departamento de Ecología y Genética Forestal, Centro de Investigación Forestal (CIFOR, CSIC), Instituto Nacional de Investigación y Tecnología Agraria y Alimentaria (INIA), Madrid, ES, Spain;
Unidad Mixta de Genómica y Ecofisiología Forestal, Instituto Nacional de Investigación y Tecnología Agraria y Alimentaria (INIA)/Universidad Politécnica de Madrid (INIA/UPM), Madrid, ES, Spain

Ainhoa Calleja-Rodríguez SKOGFORSK (The Forestry Research Institute of Sweden), Sävar, SE, Sweden

Francisco R. Cantón Dpto. Biología Molecular y Bioquímica. Facultad de Ciencias Campus de Teatinos S/N, Universidad de Málaga, Málaga, ES, Spain

Vanessa Castro-Rodríguez Facultad de Ciencias, Departamento de Biología Molecular y Bioquímica, Universidad de Málaga, Málaga, Spain

Rafael A. Cañas Facultad de Ciencias, Departamento de Biología Molecular y Bioquímica, Universidad de Málaga, Málaga, ES, Spain

María Teresa Cervera Departamento de Ecología y Genética Forestal, Centro de Investigación Forestal (CIFOR, CSIC), Instituto Nacional de Investigación y Tecnología Agraria y Alimentaria (INIA), Madrid, ES, Spain;
Unidad Mixta de Genómica y Ecofisiología Forestal, Instituto Nacional de Investigación y Tecnología Agraria y Alimentaria (INIA)/Universidad Politécnica de Madrid (INIA/UPM), Madrid, ES, Spain

Francisco M. Cánovas Facultad de Ciencias, Departamento de Biología Molecular y Bioquímica, Universidad de Málaga, Málaga, ES, Spain

John M. Davis School of Forest, Fisheries, and Geomatics Sciences, University of Florida, Gainesville, FL, USA

Amanda R. De La Torre School of Forestry, Northern Arizona University, Flagstaff, Arizona, USA

Fernando N. de la Torre Facultad de Ciencias, Departamento de Biología Molecular y Bioquímica, Universidad de Málaga, Málaga, Spain

Nuria de María Departamento de Ecología y Genética Forestal, Centro de Investigación Forestal (CIFOR, CSIC), Instituto Nacional de Investigación y Tecnología Agraria y Alimentaria (INIA), Madrid, ES, Spain;
Unidad Mixta de Genómica y Ecofisiología Forestal, Instituto Nacional de Investigación y Tecnología Agraria y Alimentaria (INIA)/Universidad Politécnica de Madrid (INIA/UPM), Madrid, ES, Spain

Marina de Miguel INRAE, Univ. Bordeaux, BIOGECO, Cestas, FR, France;
EGFV, Univ. Bordeaux, Bordeaux Sciences Agro, INRAE, ISVV, Villenave d'Ornon, France

Gustavo T. Duarte Max Plank Institute of Molecular Plant Physiology, Potsdam-Golm, Germany;
Belgian Nuclear Research Centre (SCK CEN), Biosphere Impact Studies, Mol, Belgium

Heidi Dungey Scion (New Zealand Forest Research Institute), Whakarewarewa, Rotorua, New Zealand

Carmen Díaz-Sala Departamento de Ciencias de la Vida (Fisiología Vegetal), Universidad de Alcalá, Alcalá de Henares, ES, Spain

Marta Callejas Díaz Departamento de Ecología y Genética Forestal, Centro de Investigación Forestal (CIFOR, CSIC), Instituto Nacional de Investigación y Tecnología Agraria y Alimentaria (INIA), Madrid, ES, Spain

Jorge El-Azaz Dpto. Biología Molecular y Bioquímica. Facultad de Ciencias Campus de Teatinos S/N, Universidad de Málaga, Málaga, ES, Spain

Daniel Ence School of Forest, Fisheries, and Geomatics Sciences, University of Florida, Gainesville, FL, USA

Maria Rosario García-Gil Department of Forest Genetics and Plant Physiology, Umeå Plant Science Centre, Swedish University of Agricultural Sciences, Umeå, SE, Sweden

Stanislav A. Geras'kin Russian Institute of Radiology and Agroecology, Obninsk, Russian Federation

David S. Gernandt Departamento de Botánica, Instituto de Biología, Universidad Nacional Autónoma de México, Ciudad de México, Mexico

Santiago C. González-Martínez INRAE, Univ. Bordeaux, BIOGECO, Cestas, FR, France

Natalie Graham Scion (New Zealand Forest Research Institute), Whakarewarewa, Rotorua, New Zealand

Delphine Grivet Departamento de Ecología y Genética Forestal, Centro de Investigación Forestal (CIFOR, CSIC), Instituto Nacional de Investigación y Tecnología Agraria y Alimentaria (INIA), Madrid, ES, Spain

María Ángeles Guevara Departamento de Ecología y Genética Forestal, Centro de Investigación Forestal (CIFOR, CSIC), Instituto Nacional de Investigación y Tecnología Agraria y Alimentaria (INIA), Madrid, ES, Spain;
Unidad Mixta de Genómica y Ecofisiología Forestal, Instituto Nacional de Investigación y Tecnología Agraria y Alimentaria (INIA)/Universidad Politécnica de Madrid (INIA/UPM), Madrid, ES, Spain

Laura Hernández-Escribano Departamento de Ecología y Genética Forestal, Centro de Investigación Forestal (CIFOR, CSIC), Instituto Nacional de Investigación y Tecnología Agraria y Alimentaria (INIA), Madrid, ES, Spain

Agathe Hurel INRAE, Univ. Bordeaux, BIOGECO, Cestas, FR, France

Ahmed Ismael Scion (New Zealand Forest Research Institute), Whakarewarewa, Rotorua, New Zealand

Jeremy S. Johnson Department of Environmental Studies, Prescott College, Prescott, Arizona, USA

Carmen Jurado-Mañogil Department of Plant Breeding, CEBAS-CSIC, Murcia, Spain

Jaroslav Klápště Scion (New Zealand Forest Research Institute), Whakarewarewa, Rotorua, New Zealand

Jun-Jun Liu Canadian Forest Service, Natural Resources Canada, Victoria, BC, Canada

Carol A. Loopstra Department of Ecology and Conservation Biology, Texas A&M University, College Station, TX, USA

Mengmeng Lu Department of Biological Sciences, University of Calgary, Calgary, Canada

Miriam López-Hinojosa Departamento de Ecología y Genética Forestal, Centro de Investigación Forestal (CIFOR, CSIC), Instituto Nacional de Investigación y Tecnología Agraria y Alimentaria (INIA), Madrid, ES, Spain;
Unidad Mixta de Genómica y Ecofisiología Forestal, Instituto Nacional de Investigación y Tecnología Agraria y Alimentaria (INIA)/Universidad Politécnica de Madrid (INIA/UPM), Madrid, ES, Spain

Juan Majada CETEMAS, Forest and Wood Technology Research Centre, Asturias, ES, Spain

Lorenzo Federico Manjarrez Departamento de Ecología y Genética Forestal, Centro de Investigación Forestal (CIFOR, CSIC), Instituto Nacional de Investigación y Tecnología Agraria y Alimentaria (INIA), Madrid, ES, Spain;
Unidad Mixta de Genómica y Ecofisiología Forestal, Instituto Nacional de Investigación y Tecnología Agraria y Alimentaria (INIA)/Universidad Politécnica de Madrid (INIA/UPM), Madrid, ES, Spain

Pablo Martínez-García Department of Plant Breeding, CEBAS-CSIC, Murcia, Spain

Pedro J. Martínez-García Department of Plant Breeding, CEBAS-CSIC, Murcia, Spain

Jorge Mas-Gómez Department of Plant Breeding, CEBAS-CSIC, Murcia, Spain

Isabel Mendoza-Poudereux Biotechnology and Biomedicine (BiotecMed) Institute and Plant Biology Department, University of Valencia, Valencia, ES, Spain

Célia Miguel iBET, Instituto de Biologia Experimental e Tecnológica, Oeiras, Portugal;
BioISI - Biosystems & Integrative Sciences Institute, Faculdade de Ciências, Universidade de Lisboa, Lisboa, PT, Portugal

Inês Modesto Department of Plant Biotechnology and Bioinformatics and VIB Center for Plant Systems Biology, Ghent University, Ghent, Belgium;
ITQB NOVA, Instituto de Tecnologia Química e Biológica António Xavier, Universidade Nova de Lisboa, Oeiras, PT, Portugal;
BioISI - Biosystems & Integrative Sciences Institute, Faculdade de Ciências, Universidade de Lisboa, Lisboa, PT, Portugal

Marian Morcillo Biotechnology and Biomedicine (BiotecMed) Institute and Plant Biology Department, University of Valencia, Valencia, ES, Spain

David B. Neale University of California-Davis, Davis, CA, USA

C. Dana Nelson USDA Forest Service, Southern Research Station, Southern Institute of Forest Genetics, Saucier, MS, USA;
Forest Health Research and Education Center, Lexington, KY, USA

María Belén Pascual Facultad de Ciencias, Departamento de Biología Molecular y Bioquímica, Universidad de Málaga, Málaga, ES, Spain

Pedro Perdiguero iBET, Instituto de Biologia Experimental e Tecnológica, Oeiras, Portugal;
Centro de Investigación en Sanidad Animal (CISA-INIA), Madrid, ES, Spain

Gary F. Peter School of Forest, Fisheries, and Geomatics Sciences, University of Florida, Gainesville, FL, USA

Alberto Pizarro Departamento de Ciencias de la Vida (Fisiología Vegetal), Universidad de Alcalá, Alcalá de Henares, ES, Spain

Christophe Plomion INRAE, Univ. Bordeaux, BIOGECO, Cestas, FR, France

Tania Quesada School of Forest, Fisheries, and Geomatics Sciences, University of Florida, Gainesville, FL, USA

Annie Raffin INRAE, UEFP, Cestas, FR, France

Rosa Raposo Departamento de Ecología y Genética Forestal, Centro de Investigación Forestal (CIFOR, CSIC), Instituto Nacional de Investigación y Tecnología Agraria y Alimentaria (INIA), Madrid, ES, Spain

Andreia S. Rodrigues ITQB NOVA, Instituto de Tecnologia Química e Biológica António Xavier, Universidade Nova de Lisboa, Oeiras, PT, Portugal;
Swammerdam Institute for Life Sciences, University of Amsterdam, Amsterdam, NL, Netherlands

Dainis E. Rungis Latvian State Forest Research Institute, Salaspils, Latvia

Jcrome Salse INRAE, Univ. Clermont Auvergne, GDEC, Clermont-Ferrand, FR, France

Manoj K. Sekhwal School of Forestry, Northern Arizona University, Flagstaff, Arizona, USA

Richard A. Sniezko USDA Forest Service, Dorena Genetic Resource Center, Cottage Grove, Oregon, USA

Lieven Sterck Department of Plant Biotechnology and Bioinformatics and VIB Center for Plant Systems Biology, Ghent University, Ghent, Belgium

Kristian Stevens Department of Plant Pathology, University of California, Davis, CA, USA;
Department of Evolution and Ecology, University of California-Davis, Davis, CA, USA

Mari Suontama SKOGFORSK (The Forestry Research Institute of Sweden), Sävar, SE, Sweden

Leopoldo Sánchez INRAE, ONF, Orléans, BioForA FR, France

Jean-Francois Trontin BioForBois, FCBA Technological Institute, Wood & Construction Industry Dpt, Cestas, FR, France

Polina Y. Volkova Russian Institute of Radiology and Agroecology, Obninsk, Russian Federation

Angelika F. Voronova Latvian State Forest Research Institute, Salaspils, Latvia

Alejandra Vázquez-Lobo Centro de Investigación en Biodiversidad y Conservación, Universidad Autónoma del Estado de Morelos, Cuernavaca, Morelos, Mexico

María Dolores Vélez Departamento de Ecología y Genética Forestal, Centro de Investigación Forestal (CIFOR, CSIC), Instituto Nacional de Investigación y Tecnología Agraria y Alimentaria (INIA), Madrid, ES, Spain;
Unidad Mixta de Genómica y Ecofisiología Forestal, Instituto Nacional de Investigación y Tecnología Agraria y Alimentaria (INIA)/Universidad Politécnica de Madrid (INIA/UPM), Madrid, ES, Spain

Matthew Weiss Department of Biology, Northern Arizona University, Flagstaff, Arizona, USA

Rafael Zas MBG-CSIC, Pontevedra, ES, Spain

Abbreviations

ABA	Abscisic Acid
AFLP	Amplified Fragment Length Polymorphism
AIF	Apoptosis-inducing factor
AUX	Auxin
BAC	Bacterial Artificial Chromosome
BRs	Brassinosteroids
cDNA	Complementary DNA
ChIP-seq	Chromatin Immunoprecipitation Sequencing
CNV	Copy number variation
COS	Conserved Orthologous Sequences
CP	Candidate or validation Population
CRISPR	Clustered Regularly Interspaced Short Palindromic Repeats
CTK	Cytokinin
DBH	Diameter at Breast Height
DEG	Differentially Expressed Gene
EAA	Environmental Association Analysis
EM	Expectation Maximization
eQTL	Expression quantitative trait locus
ESTP	Expressed Sequenced Tags Polymorphism
EST	Expressed Sequenced Tags
ET	Ethylene
ETI	Effector-Triggered Immunity
FA	Fluctuation Asymmetry
FT	Flowering locus T-like protein
GA	Gibberellin
GBLUP	Genomic-based Best Linear Unbiased Prediction
GBS	Genotyping by Sequencing
gDNA	Global DNA
GEA	Genome-wide Environmental Analysis
GEBV	Genomic Breeding Values
GMO	Genetically Modified Organism
GS	Genomic Selection
GWAS	Genome-wide Association Study
HM	Heavy Metal
HMW	High Molecular Weight
HR	Hypersensitive-like Response

HSPs	Heat-shock proteins
ICRP	International Commission on Radiation Protection
Indel	Insertion/Deletion
IRAP	Inter-retrotransposon amplified polymorphism
ISH	In situ Hybridization
IUCN	International Union for Conservation of Nature
iWUE	Intrinsic Water Use Efficiency
JA	Jasmonic Acid
LA	Linkage Analysis
LC	Liquid Chromatography
LCM	Laser Capture Microdissection
LD	Linkage Disequilibrium
LiDAR	Light Detection and Ranging
lncRNA	Long non-coding RNA
LOD	Logarithm of the Odds
LTR	Long Terminal Repeat
MAPK	Mitogen-activated Protein Kinase
MAS	Marker-Assisted Selection
MGR	Major Gene Resistance
miRNA	MicroRNA
MITE	Miniature Inverted Repeat Transposable Element
MLM	Mixed Linear Model
MNPs	Multiple Nucleotide Polymorphisms
MP	Mate pair
MS	Mass Spectrometry
MWAS	Metagenome-wide Association Study
NBS	Nucleotide Binding Site
NCBI	National Center for Biotechnology Information
NGS	Next Generation Sequencing
NIRs	Near-Infrared Spectroscopy
NMR	Nuclear Magnetic Resonance
NRM	Numerator Relationship Matrix
ONT	Oxford Nanopore Technology
ORF	Open Reading Frame
PAV	Presence/Absence variation
PBLUP	Pedigree-based Best Linear Unbiased Prediction
PCA	Principal Component Analysis
PCD	Programmed Cell Death
PE	Paired end
PEG	Polyethylene Glycol
PPC	Pine Pitch Canker
PR	Pathogenesis-related
PTI	Pattern-Triggered Immunity
PWN	Pine Wood Nematode
QDR	Quantitative Disease Resistance
QTL	Quantitative Trait Loci
QTN	Quantitative Trait Nucleotide

RAPD	Random Amplified Polymorphic DNA
RBIP	Retrotransposon-based Insertional Polymorphism
RDA	Redundancy Analysis
REMAP	Retrotransposon Microsatellite Amplified Polymorphism
RFLP	Restriction Fragment Length Polymorphism
RIVP	Retrotransposon Internal Variation Polymorphism
RNAi	RNA interference
ROS	Reactive Oxygen Species
RRM	Realized Relationship Matrix
SA	Salicylic Acid
SE	Somatic Embryogenesis
SLs	Strigolactones
SNP	Single Nucleotide Polymorphism
sRNA	Small RNA
SSAP	Sequence-Specific Amplification Polymorphism
ssBLUP	Single-Step Best Linear Unbiased Prediction
SSR	Simple Sequence Repeat
SUP	Single-Uredinial Pustule
TE	Transposable Element
TF	Transcription Factor
TIR	Terminal Inverted Repeat
TP	Training Population
TSDs	Target Site Duplications
VOCs	Volatile Organic Compounds
WGD	Whole-genome Duplication
WPBR	White Pine Blister Rust
WUE	Water use efficiency

1 Advances in the Genomic and Transcriptomic Sequencing of North American Pines

Alejandra Vázquez-Lobo,
David S. Gernandt, Pedro J. Martínez-García,
and Amanda R. De La Torre

Abstract

Genetic and evolutionary questions are being addressed in pines using a host of high-throughput sequencing strategies, including whole-genome sequencing, transcriptome sequencing, and target enrichment of nuclear genes. Some of the questions being addressed include the genetic basis of pathogen and drought resistance, differential expression, genetic mapping, phylogeography, and phylogenetics. Pine genomes are enormous, ranging from 20 to 40 Gb. At present, draft genomes are available for only two pine species, *P. taeda* (loblolly pine) and *P. lambertiana* (sugar pine), but most other approximately 80 species of North American pines have been represented in evolutionary studies based on complete plastomes, low-copy nuclear genes, and transcriptomes. A number of online databases have been developed and made publicly available for comparative studies of pines and other conifers.

1.1 Nuclear Genomes

Pine species are naturally distributed in the Northern Hemisphere and are also planted as commercial species in the Southern Hemisphere. Due to their importance for commercial forestry, they are considered the world's most economically important forest species. Efforts to understand their complex biology and evolution were limited by the absence of reference genomes. Pines, as other conifers, are slow-growing, long-lived species and possess enormous genomes (20–40 Gb) with a high number of repeat elements (Wegrzyn et al. 2014). Limitations of short-read sequencing technologies, computational power, and assembly software made the sequencing of pine genomes a daunting task just 10 years ago (De La Torre et al. 2014, 2019). To date, only two North American pine species have been sequenced: *Pinus taeda* and *Pinus lambertiana*.

The first sequenced genome was *Pinus taeda* (loblolly pine) in 2014, the most planted forest

A. Vázquez-Lobo (✉)
Centro de Investigación en Biodiversidad y Conservación, Universidad Autónoma del Estado de Morelos, 62209 Cuernavaca, Morelos, Mexico
e-mail: alejandra.vazquez@uaem.mx

D. S. Gernandt
Departamento de Botánica, Instituto de Biología, Universidad Nacional Autónoma de México, 04510 Ciudad de México, Mexico
e-mail: dgernandt@ib.unam.mx

P. J. Martínez-García
Department of Plant Breeding, CEBAS-CSIC, Murcia, Spain
e-mail: pjmartinez@cebas.csic.es

A. R. De La Torre
Northern Arizona University, 200 E Pine Knoll Dr, Flagstaff, AZ AZ86011, USA
e-mail: Amanda.de-la-torre@nau.edu

A. R. De La Torre (ed.), *The Pine Genomes*, Compendium of Plant Genomes,
https://doi.org/10.1007/978-3-030-93390-6_1

tree in North America and a key species for commercial forestry in the southwestern United States (McKeand et al. 2021). Whole-genome shotgun sequencing was used to analyze the 22 Gb genome and develop the first two genome assemblies (versions 1.0 and 1.01) of the *P. taeda* genome (Wegrzyn et al. 2014; Zimin et al. 2014). Version 1.0 was based on the MaSuRCA (Zimin et al. 2013) assembly of paired-end reads from a haploid female gametophyte (or megagametophyte) and long insert linking read pairs ("super-reads") from diploid needle tissue (Zimin et al. 2014). This version resulted in a draft genome sequence of 20.15 Gb (spanning 23.2 Gbp) with an N50 scaffold size of 66.9 kbp (Zimin et al. 2014), from which 82% was composed of repetitive elements (Neale et al. 2014). Version 1.01 employed scaffolding from independent genome and transcriptome assemblies (Wegrzyn et al. 2014). Structural annotation identified 50,172 gene models with long intron lengths varying from 2.7 to 100 kbp (Neale et al. 2014). Current version available from the TreeGenes database (see below) is version 2.01, which contains 51,751 protein sequences (file Pita.2_01.pep.fa, last accessed in June 2021).

Another economic and ecologically important pine species in North America is sugar pine (*Pinus lambertiana*). *P. lambertiana* populations are severely affected by the exotic fungal pathogen *Cronartium ribicola* (white pine blister rust, WPBR) throughout their natural range (Weiss et al. 2020). Therefore, interest in identifying the genes coding resistance to the disease was a significant motivation to decode its genome. The sequencing and assembly of *P. lambertiana* followed a similar procedure to that used in *P. taeda* (in fact, both species were sequenced by the same group of researchers at University of California-Davis, US). Paired-end libraries for short-read Illumina sequencing were constructed from haploid megagametophyte tissue, error-corrected and later used to construct "super-reads" with MaSuRCA 2.3.0 (Zimin et al. 2013). Mate pairs from diploid tissue libraries were cleaned and filtered and added to the haploid data for genome assembly with SOAPdenovo2 (Luo et al. 2012). Newly developed Pacific Biosciences (PacBio) and Illumina RNA-seq data were used for additional scaffolding steps (Gonzalez-Ibeas et al. 2016). The total length of assembly version 1.0 including all scaffolds and contigs >200 bp was 27.6 Gbp from a 31 Gbp estimated genome size (Stevens et al. 2016). An important contribution of this assembly was the identification of candidate genes for *Cr1* (major gene for WPBR resistance) that could significantly contribute to marker-assisted breeding efforts (Stevens et al. 2016). In a more recent study, long-reads from 10X Genomics (www.10Xgenomics.com) were used to build and improve the assembly, generating an eightfold improvement over the original NG50 scaffold of 247 kb (Crepeau et al. 2017).

1.2 Plastid Genomes and Target Gene Sequencing

An important aim in the generation of bioinformatic resources for pines has been the search for markers for evolutionary studies. Nuclear genomes of pines are characterized by their high levels of gene duplication and high frequency of repeat regions, making it difficult to identify useful nuclear markers for phylogenetic and population genetic analyses, which usually assume orthology relationships of genetic variants. Plant mitochondrial genomes have low substitution rates and relatively high rates of rearrangement and transfer to and from the nucleus. Until recently most DNA-based phylogenetic and population studies in *Pinus* have been based on plastid markers. To take into account coalescent processes, phylogenetic studies of low-copy nuclear genes were adopted (e.g., Syring et al. 2007; Willyard et al. 2007; DeGiorgio et al. 2014).

The first fully sequenced plastid genome of a gymnosperm was that of the hard pine (subgenus *Pinus*) *Pinus thunbergii* (Wakasugi et al. 1994). The plastome of *P. koraiensis*, a soft pine (subgenus Strobus), was made available in public sequence databases a few years later. Adoption of short-read sequencing accelerated the pace of plastome sequencing until presently complete or nearly complete genomes are available for more than 100 pine species, including all species from North America. Plastome sequencing has allowed inference of phylogenies with well-resolved relationships; however, there is clearly discordance between organellar gene trees and nuclear gene trees of pines (e.g., Wang and Wang 2014; Gernandt et al. 2018), and some relationships among species remain uncertain. Due to the size of the pine genome, alternative approaches to whole-genome sequencing were needed. Therefore, efforts were made to identify suitable low-copy nuclear regions for evolutionary studies.

Target enrichment has been used to acquire sequences for most (Neves et al. 2013) or a fraction (Gernandt et al. 2018) of putative low-copy nuclear genes in pines for studies ranging from genetic mapping to phylogenetics. Biotinylated RNA oligonucleotides are used to enrich genomic libraries for specific regions of interest such as exomes (the entire protein coding fraction of the genome), which can then be characterized with massively parallel sequencing (Gnirke et al. 2009). Target enrichment can be combined with genome skimming to also include in the same sequencing runs the high-copy fraction of genomes, particularly nuclear ribosomal DNA and complete plastomes, a strategy called Hyb-Seq (Weitemier et al. 2014). Hyb-Seq can be performed on degraded or historical samples and has a low per-sample price (Hale et al. 2020). Because the method includes flanking sequences of targeted genes or exons, it provides additional information on the history of markers that may have undergone gene duplication or loss. This method has been used to characterize hundreds of genes for phylogenetic studies of the three most species-rich clades of exclusively North American pines, subsects. *Australes*, *Ponderosae*, and *Cembroides* (Gernandt et al. 2018; Montes et al. 2019; Willyard et al. 2021) and to study population genetics and local adaptation (Peláez et al. 2020).

Can evolutionary questions be addressed better by characterizing hundreds or a few thousand nuclear genes and complete plastomes with Hyb-Seq or by characterizing transcriptomes? It has been argued that genes and data derived from genes, in particular, transcriptomes should not be analyzed with coalescence methods because of the likelihood that they have undergone recombination. This is particularly the case for those genes that are divided into exons dispersed more broadly across the genome and represent unlinked/independent estimates of gene tree relationships (Springer and Gatesy 2016).

1.3 Transcriptomic Resources

The characterization of transcriptomes is a basic tool for the annotation of reference genomes, so for the annotation of the *P. taeda* reference genome, transcriptomes of different tissues at different stages of development for the species were generated. This allowed the identification of more than 80,000 transcripts, of which about 45,000 genes were successfully mapped (Wegrzyn et al. 2014). Similarly, for *P. lambertiana*, by characterizing a reference transcriptome from different tissues and taking advantage of different sequencing platforms, close to 30,000 transcripts have been functionally annotated (Gonzalez-Ibeas et al. 2016). As mentioned above, the sugar pine is threatened by the WPBR, as are its white bark pine relatives with similar distributions. Through a comparative analysis of transcriptomes of the western white pine (*P. monticola*), limber pine (*P. flexilis*), white bark pine (*P. albicaulis*), and sugar pine (*P. lambertiana*), signals of positive selection were found in different genes, including candidates to WPBR resistance (Baker et al. 2018). Transcriptome sequences for loblolly pine and sugar pine are available in the TreeGenes database (see below).

Identification of the genetic basis of processes and phenotypes in pines requires a more detailed analysis, considering biotic or abiotic factors, and comparing tissues and species. For example, characterization of transcriptomes of *P. patula* and *P. tecunumanii* from plantations in South Africa (species from Mexico and Central America) using tissues infected with a fungus (*Fusarium circinatum*) and differential expression analyses has allowed the identification of genes involved in the response to pathogens in these species (Visser et al. 2019, 2018, 2015). Similarly, through differential expression analysis (DE) specific genes have been identified for the response to a fungal infection (*Dothistroma septosporum*) in *P. contorta* (Lu et al. 2021). Through DE transcriptomic analyses, genes involved in wood maturation in *P. radiata* (Li et al. 2011) and in resin tapping in *P. elliotti* (de Oliveira Junkes et al. 2019) have also been identified.

The development of bioinformatic resources for pines of arid environments is of special interest, since in many cases these species are the only forest resource in said environments. For species of arid climates in Mexico, investigations have been carried out on *P. pinceana*, through the characterization of transcriptomes of individuals from different populations in the range of distribution of the species (Figueroa-Corona et al. 2021) and for *P. cembroides*, a differential expression analysis has been carried out to identify the changes in gene expression in juvenile and adult leaves (Webster et al. in prep).

While low-copy genes are informative for evolutionary inferences, identification of functional genes requires characterization of the transcriptome and detection of differentially expressed genes. Recent advances have been made in this regard, by obtaining and characterizing new transcriptomes for 107 pine species from megametophytes or young needles for an evolutionary study of the genus *Pinus* (Jin et al. 2021). This study included 66 species distributed in America, of which 31 are mainly distributed in the United States and Canada and 35 are from Mexico, the Caribbean, and Central America.

1.4 Databases for Genomic and Transcriptomic Resources

1.4.1 Plaza

The database and online resource PLAZA (http://bioinformatics.psb.ugent.be/plaza/) version 4.0 were created to allow comparative, evolutionary, and functional genomic analyses among plant species through a user-friendly web interface (Van Bel et al. 2018). PLAZA allows users to browse genomes, gene families, and phylogenetic trees; to find functional information through BLAST; and to explore genome organization through different visualization tools (e.g., Ks graphs, Skyline plots, WGDotplot) based on gene collinearity or synteny information (Proost et al. 2015). PLAZA Gymnosperms includes structural and functional annotation of 16 gymnosperm species, including 777,165 genes clustered in 30,041 multi-species gene families (last accessed in June 2021). In the case of gymnosperm species lacking reference genome sequences, PLAZA uses curated transcriptomic data to identify genes and gene families and make them available for comparative genomics analyses. To date, PLAZA contains data on only three Pinus species: *Pinus taeda*, *Pinus sylvestris,* and *Pinus pinaster* (last accessed in June 2021).

1.4.2 TreeGenes

The Dendrome project and the associated TreeGenes database (https://treegenesdb.org) were created in the early 1990s as a repository to store genetic linkage and Expressed Sequence Tags (ESTs) data with a focus on commercial Pinaceae species (Wegrzyn et al. 2008, 2019; Falk et al. 2018). Over the years, TreeGenes expanded to include curated data in addition to data provided by users (Falk et al. 2018). To accommodate the needs of larger datasets as a result of the high-throughput sequencing, TreeGenes incorporated more efficient models for data storage and later moved to the Tripal framework, a more flexible, efficient, and sustainable platform (Falk et al. 2018). The Tripal Gateway framework supports cross-site query, data transfer, access to analytical pipelines (e.g., Galaxy), and different modules such as Tripal Plant PopGen Submit (TPPS), Tripal Sequence Similarity Search (TSeq), and OrthoQuery (Falk et al. 2018). The TSeq module allows sequence similarity search against genes, TreeGenes UniGenes, proteins, and full genome through NCBI BLASTX, BLASTN, or BLASTP. Genetic, phenotypic, and/or environmental data submission from users can be uploaded through the TPPS, and genetic linkage data can be uploaded through the Genetic Map Submission window in TreeGenes. More recent developments in TreeGenes include CartograTree, a module that allows the integration of genetic, phenotypic, and environmental data for trees with geographic coordinates (Vasquez-Gross et al. 2013). The database expanded to include any forest tree with genomic and transcriptomic resources and now hosts 38 genomes and 3 M transcripts from 2143 species (last accessed June 2021). However, from those 38 genomes only 7 are gymnosperms (*Ginkgo biloba*, *Gnetum montanum*, *Picea abies*, *Picea glauca*, *Pinus lambertiana*, *Pinus taeda*, and *Pseudotsuga menziesii*) and 2 are pine species.

1.4.3 ConGenIE

The Conifer Genome Integrative Explorer (ConGenIE, http://congenie.org) database was created as a subdomain under the umbrella of the Plant Genome Integrative Explorer (http://PlantGenIE.org; Sundell et al. 2015). This database aimed to host the genomic and transcriptomic data developed by the Norway spruce genome project (Nystedt et al. 2013), and was focused on developing tools for functional genomic analyses. In recent years, ConGenIE was expanded and now hosts genome sequences and transcriptomic data of *Picea glauca* x *Picea engelmannii* (assembly WS7711-v1.0), *Picea glauca* (assembly PG29-v1.0), and *Pinus taeda* (v1.0). It also allows the cross-link between species using the other PlantGenIE subdomains such as AtGenIE.org (*Arabidopsis thaliana* Genome Integrative Explorer), and PopGenIE.org (*Populus* Genome Integrative Explorer).

References

Baker EAG, Wegrzyn JL, Sezen UU, Falk T, Maloney PE, Vogler DR, Delfino-Mix A, Jensen C, Mitton J, Wright J, Neale DB (2018) Comparative transcriptomics among four white pine species. G3 Genes Genomes Genet 8(5):1461–1474. https://doi.org/10.1534/g3.118.200257

Crepeau MW, Langley CH, Stevens KA (2017) From pine cones to read clouds: rescaffolding the megagenome of sugar pine (*Pinus lambertiana*). G3 Genes|Genomes|Genet 7(5):1563–1568. https://doi.org/10.1534/g3.117.040055

De La Torre AR, Birol I, Bousquet J, Ingvarsson PK, Jansson S, Jones SJM, Keeling CI, MacKay J, Nilsson O, Ritland K et al (2014) Insights into conifer giga-genomes. Plant Physiol 166:1–9

De La Torre AR, Piot A, Liu B, Wilhite B, Weiss M, Porth I (2019) Functional and morphological evolution in gymnosperms: a portrait of implicated gene families. Invited Contribution to Special Issue in Evol Appl 13(1):210–227

De Oliveira Junkes CF, de Araújo Júnior AT, de Lima JC, de Costa F, Füller T, de Almeida MR, Neis FA, da Silva Rodrigues-Correa KC, Fett JP, Fett-Neto AG (2019) Resin tapping transcriptome in adult slash pine (*Pinus elliottii var. elliottii*). Ind Crops Prod 139 (June):111545. https://doi.org/10.1016/j.indcrop.2019.111545

DeGiorgio M, Syring J, Eckert AJ, Liston A, Cronn R, Neale DB, Rosenberg NA (2014) An empirical evaluation of two-stage species tree inference strategies using a multilocus dataset from North American pines. BMC Evol Biol 14:67

Falk T, Herndon N, Grau E, Buehler S, Richter P, Zaman S, Baker EM, Ramnath R, Ficklin S, Staton M, Feltus FA, Jung S, Main D, Wegrzyn JL (2018) Growing and cultivating the forest genomics database, TreeGenes. Database. https://doi.org/10.1093/database/bay084 Bottom of Form

Figueroa-Corona L, Valerio PD, Wegrzyn J, Piñero D (2021) Transcriptome of weeping pinyon pine, *Pinus pinceana*, shows differences across heterogeneous habitats. Trees. https://doi.org/10.1007/s00468-021-02125-8

Gernandt DS, Aguirre Dugua X, Vázquez-Lobo A, Willyard A, Moreno Letelier A, Pérez de la Rosa JA, Piñero D, Liston A (2018) Multi-locus phylogenetics, lineage sorting, and reticulation in *Pinus* subsection *Australes*. Am J Bot 105:711–725

Gnirke A, Melnikov A, Maguire J, Rogov P, LeProust EM, Brockman W, Fennell T et al (2009) Solution hybrid selection with ultra-long oligonucleotides for massively parallel targeted sequencing. Nat Biotechnol 27:182–189

Gonzalez-Ibeas D, Martínez-García PJ, Famula RA, Delfino-Mix A, Stevens KA et al (2016) Assessing the gene content of the megagenome: sugar pine (*Pinus lambertiana*). G3 (Bethesda) 6:3787–3802

Hale H, Gardner EM, Viruel J, Pokorny L, Johnson MG (2020) Strategies for reducing per-sample costs in target capture sequencing for phylogenomics and population genomics in plants. Appl Plant Sci 8

Jin WT, Gernandt DS, Wehenkel C, Xia XM, Wei XX, Wang XQ (2021) Phylogenomic and ecological analyses reveal the spatiotemporal evolution of global pines. Proc Natl Acad Sci United States of America 118 (20). https://doi.org/10.1073/PNAS.2022302118

Li X, Wu HX, Southerton SG (2011) Transcriptome profiling of wood maturation in *Pinus* radiata identifies differentially expressed genes with implications in juvenile and mature wood variation. Gene 487(1):62–71. https://doi.org/10.1016/j.gene.2011.07.028

Lu M, Feau N, Vidakovic DO, Ukrainetz N, Wong B, Aitken SN, Hamelin RC, Yeaman S (2021) Comparative gene expression analysis reveals mechanism of pinus contorta response to the Fungal Pathogen *Dothistroma septosporum*. Mol Plant Microbe Interact 34(4):397–409. https://doi.org/10.1094/MPMI-10-20-0282-R

Luo R, Liu B, Xie Y, Li Z, Huang W et al (2012) SOAPdenovo2: an empirically improved memory-efficient short-read de novo assembler. Gigascience 1 (1):18

McKeand SE, Payn KG, Heine AJ, Abt RC (2021) Economic significance of continued improvement of loblolly pine genetics and its efficient deployment to landowners in the southern United States. J For 119:62–72. https://doi.org/10.1093/jofore/fvaa044

Montes JR, Peláez P, Willyard A, Moreno-Letelier A, Piñero D, Gernandt DS (2019) Phylogenetics of *Pinus* subsection *Cembroides* Engelm. (Pinaceae) inferred from low-copy nuclear gene sequences. Syst Bot 44:501–518

Neale DB, Wegrzyn JL, Stevens KA, Zimin AV, Puiu D, Crepeau MW, Cardeno C, Koriabine M, Holtz-Morris AE, Liechty JD et al (2014) Decoding the massive genomes of loblolly pine using haploid DNA and novel assembly strategies. Genome Biol 15:R59

Neves LG, Davis JM, Barbazuk WB, Kirst M (2013) Whole-exome targeted sequencing of the uncharacterized pine genome. Plant J 75:146–156

Nystedt B, Street NR, Wetterbom A, Zuccolo A, Lin YC, Scofield DG, Vezzi F, Delhomme N, Giacomello S, Alexeyenko A et al (2013) The Norway spruce genome sequence and conifer genome evolution. Nature 497:579–584

Peláez P, Ortiz-Martínez A, Figueroa-Corona L, Montes JR, Gernandt DS (2020) Population structure, diversifying selection, and local adaptation in *Pinus patula*. Am J Bot 107:1555–1566

Proost S, Van Bel M, Vaneechoutte D, Van de Peer Y, Inzé D, Mueller-Roeber B, Vandepoele K (2015) PLAZA 3.0: an access point for plant comparative genomics. Nucl Acids Res 43(D1):D974–D981. https://doi.org/10.1093/nar/gku986

Springer MS, Gatesy J (2016) The gene tree delusion. Mol Phylogenet Evol 94:1–33

Stevens KA, Wegrzyn JL, Zimin A, Puiu D, Crepeau M, Cardeno C, Paul R, Gonzalez-Ibeas D, Koriabine M, Holtz-Morris AE, Martínez-García PJ, Sezen UU, Marçais G, Jermstad K, McGuire PE, Loopstra CA, Davis JM, Eckert A, de Jong P, Yorke JA, Salzberg SL, Neale DB, Langley CH (2016) Sequence of the sugar pine megagenome. Genetics 204:1613–1626. https://doi.org/10.1534/genetics.116.193227

Sundell D, Mannapperuma C, Netotea S, Delhomme N, Lin Y-C, Sjödin A, Van de Peer Y, Jansson S, Hvidsten TR, Street NR (2015) The plant genome integrative explorer resource: PlantGenIE.org. New Phytol 208:1149–1156. https://doi.org/10.1111/nph.13557

Syring J, Farrell K, Businský R, Cronn R, Liston A (2007) Widespread genealogical nonmonophyly in species of *Pinus* subgenus *Strobus*. Syst Biol 56:163–181

Van Bel M, Diels T, Vancaester E, Kreft L, Botzki A, Van de Peer Y, Coppens F, Vandepoele K (2018) PLAZA 4.0: an integrative resource for functional, evolutionary and comparative plant genomics. Nucl Acids Res 46 (D1):D1190–D1196. https://doi.org/10.1093/nar/gkx1002

Vasquez-Gross HA, Yu JJ, Figueroa B, Gessler DDG, Neale DB, Wegrzyn JL (2013) CartograTree: connecting tree genomes, phenotypes and environment. Mol Ecol Resour 13:528–537. https://doi.org/10.1111/1755-0998.12067

Visser EA, Wegrzyn JL, Myburg AA, Naidoo S (2018) Defence transcriptome assembly and pathogenesis related gene family analysis in *Pinus Tecunumanii* (Low Elevation). BMC Genom 19(1):1–13. https://doi.org/10.1186/s12864-018-5015-0

Visser EA, Wegrzyn JL, Steenkamp ET, Myburg AA, Naidoo S (2019) Dual Rna-Seq analysis of the Pine-Fusarium Circinatum interaction in resistant (*Pinus Tecunumanii*) and susceptible (*Pinus Patula*) hosts. Microorganisms 7(9):7–9. https://doi.org/10.3390/microorganisms7090315

Visser EA, Wegrzyn JL, Steenkmap ET, Myburg AA, Naidoo S (2015) Combined de Novo and genome guided assembly and annotation of the *Pinus* patula juvenile shoot transcriptome. BMC Genom 16(1):1–13. https://doi.org/10.1186/s12864-015-2277-7

Wakasugi T, Tsudzuki J, Ito S, Nakashima K, Tsudzuki T, Sugiura M (1994) Loss of all ndh genes as determined by sequencing the entire chloroplast genome of the black pine Pinus thunbergii. Proc Natl Acad Sci 91:9794–9798

Wang B, Wang X-R (2014) Mitochondrial DNA capture and divergence in *Pinus* provide new insights into the evolution of the genus. Mol Phylogenet Evol 80:20–30

Wegrzyn JL, Lee JM, Tearse BR, Neale DB (2008) TreeGenes: a forest tree genome database. Int J Plant Genom 412875:7

Wegrzyn JL, Liechty JD, Stevens KA, Wu LS, Loopstra CA, Vasquez-Gross AH, Dougherty WM, Lin BY, Zieve JJ, Martínez-García PJ, Holt C, Yandell M, Zimin AV, Yorke YA, Crepeau MW, Puiu D, Salzberg SL, de Jong PJ, Mockaitis K, Main D, Langley CH, Neale DB (2014) Unique features of the loblolly pine (*Pinus taeda L.*) megagenome revealed through sequence annotation. Genetics 196(3):891–909. https://doi.org/10.1534/genetics.113.159996

Wegrzyn JL, Staton MA, Street NR., Main D, Grau E, Herndon N, Buehler S, Falk T, Zaman S, Ramnath R, Richter P, Sun L, Condon B, Almsaeed A, Chen M, Mannapperuma C, Jung S, Ficklin S (2019) Cyberinfrastructure to improve forest health and productivity:

the role of tree databases in connecting genomes, phenomes, and the environment. Front Plant Sci 10: 813. https://www.frontiersin.org/article/, https://doi.org/10.3389/fpls.2019.00813

Weiss M, Sniezko R, Puiu D, Crepeau MW, Stevens K, Salzberg SL, Langley CH, Neale DB, De La Torre AR (2020) Genomic basis of white pine blister rust quantitative disease resistance and its relationship with qualitative resistance. Plant J. https://doi.org/10.1111/tpj.14928

Weitemier K, Straub SCK, Cronn RC, Fishbein M, Schmickl R, McDonnell A, Liston A (2014) Hyb-Seq: combining target enrichment and genome skimming for plant phylogenomics. Appl Plant Sci 2:1400042

Willyard A, Syring J, Gernandt DS, Liston A, Cronn R (2007) Fossil calibration of molecular divergence infers a moderate mutation rate and recent radiations for Pinus. Mol Biol Evol 24:90–101

Willyard A, Gernandt DS, Cooper B, Douglas C, Finch K, Karemera H, Lindberg E, Langer SK, Lefler J, Marquardt P, Pouncey DL (2021) Phylogenomics in the Hard Pines (*Pinus* subsection *Ponderosae;* Pinaceae*)* Confirms Paraphyly in *Pinus* ponderosa, and Places *Pinus jeffreyi* with the California Big Cone Pines. Sys Bot 46:538–561

Zimin A, Marais G, Puiu D, Roberts M, Salzberg S et al (2013) The MaSuRCA genome assembler. Bioinformatics 29:2669–2677

Zimin A, Stevens KA, Crepeau MW, Holtz-Morris A, Koriabine M, Marçais G, Puiu D, Roberts M, Wegrzyn JL, de Jong PJ, Neale DB, Salzberg SL, Yorke JA, Langley CH (2014) Sequencing and assembly of the 22-Gb loblolly pine genome. Genetics 196(3):875–890. https://doi.org/10.1534/genetics.113.159715

Advances in Genetic Mapping in Pines

2

Pedro J. Martínez-García, Alejandra Vázquez-Lobo, Pablo Martínez-García, Jorge Mas-Gómez, Carmen Jurado-Mañogil, and Kristian Stevens

Abstract

This chapter summarizes the history and current status of genetic mapping in pines. We review the genetic mapping approaches, comparative mapping, and the different uses of genetic maps in these species, with a special focus on loblolly pine (*Pinus taeda* L.), one of the first conifer genomes sequenced. Clear advances in the number of recombinants and markers have been observed in the last century. High-density genetic maps are now a reality for these species with massive genome sequences. They provide improved resolution in genetic studies for a better dissection of complex traits. They also improve the quality of these massive genome assemblies and further the goal of a high-quality contiguous physical map. In the present, the implementation of new genomics technologies is allowing new scenarios for genetic mapping in these species.

2.1 Introduction

Genetic mapping—also called linkage mapping—is an important tool that allows researchers to carry out detailed genetic analyses in plant species. This tool is based on the construction of a genetic map, which can be obtained from one or more populations of the same species. The genetic map is based on the idea of linkage, meaning that the closer two genes are to each other on the chromosome, the greater the likelihood they will be inherited together. By following the inheritance patterns, the relative locations of genes and other important features existing in chromosomes can be established. Maps allow the dissection of the architecture underlying complex polygenic traits through the identification of quantitative trait loci (QTLs). At the same time, genetic maps have been used to study genome structure and evolution in different species. One of the main characteristics of the genetic maps, their map resolution (a.k.a. marker density) has improved substantially with the advent of next

P. J. Martínez-García (✉) · P. Martínez-García · J. Mas-Gómez · C. Jurado-Mañogil
Department of Plant Breeding, CEBAS-CSIC, Murcia, Spain
e-mail: pjmartinez@cebas.csic.es

A. Vázquez-Lobo
Centro de Investigación en Biodiversidad y Conservación, Universidad Autónoma del Estado, de Morelos, Cuernavaca, Morelos 62209, México

K. Stevens
Department of Plant Pathology, University of California, Davis 95616, CA, USA

K. Stevens
Department of Evolution and Ecology, University of California-Davis, Davis, CA, USA

A. R. De La Torre (ed.), *The Pine Genomes*, Compendium of Plant Genomes,
https://doi.org/10.1007/978-3-030-93390-6_2

generation sequencing (NGS) technologies. This is more evident in species of the genus *Pinus*, which along with the other gymnosperms have massive and complex genomes. Originally, genetic maps in this species showed very low density and a small number of markers. Today the availability of hundreds of thousands of markers has made this approach more efficient, allowing the improvement of the genome of *Pinus taeda* and other species of *Pinus*. In this chapter, a complete description of the evolution of genetic mapping strategies in these species is presented.

2.2 Genetic Mapping Approaches in *Pinus taeda*

Usually, in most plant species, genetic maps have been obtained from segregating populations derived from crosses between inbred lines. These strategies provide some advantages because segregation originates during the meiosis of a single parent. Thus, allele number in a *locus* is reduced down to two variants in diploid species, being easy to know the linkage phase (Groover et al. 1994; Sewell et al. 1999). However, in obligate outcrossing species such as pines, it is generally not possible to generate those kinds of populations due to several reasons such as significant genetic load (Grattapaglia and Sederoff 1994) or inbreeding depression. In addition, pines species are highly heterozygous with a very long generation time. Despite the commented obstacles and time constraints, a three-generation outbred *Pinus taeda pedigree* was created in early genetic analysis (Devey et al. 1991). In a three-generation pedigree, the segregation in the progeny comes from separated meiosis and crossover in the two parents. Then, it is possible to find up to four alleles per *locus* in the progeny and it is necessary to use an indirect approach to determine the linkage phase, using genotype information of grandparents or segregation ratios in the offspring (Groover et al. 1994; Sewell et al. 1999). This full-sib progeny test (also called outbreed F_2 progeny), in the three-generation pedigree, resembles the typical F_1 or CP (four-way cross-population) obtained in other outbreed species.

In the most common approach in loblolly pine (*Pinus taeda*), and other pines such as *Pinus lambertiana*, after the cross between two parents with a large resulting progeny, the genotype marker information is divided into two sets, each one including the information of meiosis segregation of each parent (Groover et al. 1994; Sewell et al. 1999; Jermstad et al. 2011). These two sets are used to create two distinct maps, which subsequently are joined to form a "sex-averaged map", where multiallelic and codominant markers in both parents are used as reference points to align the two sex-maps (Groover et al. 1994; Sewell et al. 1999). This approach has been extended with the construction of consensus genetic maps, integrated by the information of two or more different crosses. The mapping of several populations improves map resolution, because a higher number of markers provides more coverage (Sewell et al. 1999).

Moreover, a different strategy can be approached in pines, since it is possible to analyze maternal segregation, by genotyping the haploid tissue of the megagametophyte (Neves et al. 2014; Remington et al. 1999). In gymnosperms, the female haploid gametophyte develops as storage reserve tissue for the developing embryo and can be easily dissected from the seed. Although the haploid megagametophyte provides a small amount of DNA, it is enough for reducing the sequence complexity of genome assembly in pines (Zimin et al. 2014; Stevens et al. 2016) and constructing genetic maps based on segregation before fertilization, from the maternal lineage (Remington et al. 1999). This approach provides two advantages: first, a large population to obtain a high confidence level is not needed, and second, it assesses only the maternal meiosis. Consensus linkage maps combining the haploid linkage maps with sex-averaged linkage maps have also been constructed, achieving agreement among the different approaches to create linkage genetic maps (Westbrook et al. 2015a).

The aforementioned approaches have been applied in the genetic mapping of loblolly pine for over 25 years. Although, in essence, the strategies have not changed, marker coverage has grown dramatically. As a consequence of the development of new techniques for massive genotyping in *P. taeda*, map coverage has increased from 75 markers/loci, in the first map obtained by Devey et al. (1994), to 26,021 markers/*loci* in the most recent map obtained by De La Torre et al. (2019) (Table 2.1). The earliest efforts for the construction of genetic maps of *P. taeda* included electromorph-based markers, such as RFLPs, RAPDs, and AFLPs (Devey et al. 1994; Groover et al. 1994; Remington et al. 1999; Sewell et al. 1999). These markers are usually found distributed throughout the genome and are methodologically accessible; however, they have low coverage, and usually, the identity of each *locus* is unknown. Furthermore, with these markers, it is not possible to distinguish dominant genotypes from the heterozygous ones; therefore, they are referred to as dominant markers, and the genotypes in the parental generation should be inferred from the segregation proportion in the progeny. First maps also included *loci* of isozymes (Brown et al. 2001; Krutovsky et al. 2004; Sewell et al. 1999), which have the advantage of being codominant markers with known identity (proteins with enzymatic activity); however, they represented a low coverage and time-consuming laboratory work. Later, the identification of simple sequence repeat or microsatellites (SSR) in *P. taeda* allowed the inclusion of these markers, increasing map density and coverage (Echt et al. 2011).

Table 2.1 Summary and description of the different genetic maps obtained for *Pinus taeda* (loblolly pine) during the last three decades

References	Type of markers	No. of markers	Population (no. of progeny used)	Type of map	LG	cM	Software
Devey et al. (1994)	RFLP Isozymes	75	Base 1 pedigree (95)	Sex-averaged map	20	632	GMENDEL 2.0
Groover et al. (1994)	RFLP	162	qtl1 pedigree (172)	Maternal and paternal map	17/23	520/498	JoinMap
Sewell et al. (1999)	RFLP Isozymes RAPD	357	base1 (95) and qtl1 (172) pedigrees	Consensus	20	1300	MapManager 2.0 MapMaker 2.6.1 JoinMap 1.4
Remington et al. (1999)	AFLP	184	7–56 (93 Haploid seeds) (Megagametophyte Tissue)	Haploid map	12	1528	PGRI MAPMAKER 2.0
Devey et al. (1999)	RFLP SSR	223	base1 (95) and qtl1 (48) pedigrees	Consensus	20	1281	MapManager 2.6.5 MapMaker 1.0 JoinMap 2.0
Brown et al. (2001)	RFLP Isozymes ESTP	235	base1 (95) and qtl1 (172) pedigrees	Consensus	12	1165	MapMaker JoinMap
Temesgen et al. (2001)	ESTP RFLP	104	base1 (95) and qtl1 (172) pedigrees	Consensus	23	1108	MapManager MapMaker JoinMap 1.4

(continued)

Table 2.1 (continued)

References	Type of markers	No. of markers	Population (no. of progeny used)	Type of map	LG	cM	Software
Zhou et al. (2003)	SSR	51	base1 pedigree (118)	Sex-averaged map	15	795	CRI-MAP
Krutovsky et al. (2004)	RFLP Isozymes ESTP	302	base1 (95) and qtl1 (172) pedigrees	Consensus	12	1274	JoinMap
Eckert et al. (2009)	SSR ESTP Isozymes RAPD RFLP SNP	373	base1 (95) and qtl1 (172) pedigrees	Consensus	12	1228	JoinMap
Eckert et al. (2010)	SNP	1635	qtl1 pedigree (172)	Sex-averaged map	12	1898	
Echt et al. (2011)	SSR	429	base1 (97) and qtl1 (170) pedigrees	Consensus	12	1429	JoinMap
Martínez-García et al. (2013)	SSR ESTP Isozymes RAPD RFLP SNP	2466	base (202) and qtl (487) pedigrees	Consensus	12	1476	
Neves et al. (2014)	SNP Indels MNP PAV	2841	10–5 (72 Haploid seeds) (Megagametophyte Tissue)	Haploid map	12	1637.4	JoinMap 3.0
Westbrook et al. (2015b)	SNP PAV	3352	BC1 & 10–5 (345)	Combined	12	1781.61	JoinMap 3.0
Westbrook et al. (2015a)	SNP SSR RFLP ESTP MNP PAV	3856	QTL-BASE1 (267) QTL-BASE2 (689) BC1 (490) 10–5 (72 haploids)	Consensus Map	12	2305.42	JoinMap 4.1 MergeMap LPmerge
de La Torre et al. (2019)	SNP	26,021	QTL (377) BASE (100)	Consensus Map	12	2270.41	ASMap v.0.4 R package JoinMap 5.0

The development of QTL mapping in large genomes requires high-density genetic maps, making especially desirable the use of Single Nucleotide Polymorphisms (SNPs) as markers. Nowadays, SNPs have become the most used markers for desirable characteristics such as codominance, abundance in the genome, and amenability to high-throughput approaches. Moreover, the large genome sizes due to repeated DNA elements in *P. taeda* and other conifers, make the process of screening and development of markers more complex compared to angiosperms (Kinlaw and Neale 1997; Morse et al. 2009; Ritland et al. 2011). This mentioned complexity can be reduced with the use of SNPs.

The discovery of many SNPs through resequencing approaches has allowed the study of the entire functional gene space rather than a small set of candidate genes (Eckert et al. 2009). The first use of SNP markers to build a genetic map for loblolly pine placed only 35 SNPs (Eckert et al. 2009), but subsequent analyses quickly achieved greater coverage. A total of 1,635 SNPs were mapped in the next genetic map (Eckert et al. 2010), a large increase in the number of markers in comparison with previous works. After this work, two SNP assays with 7,000 SNPs each one, were designed and used to complete the first high-density consensus map in loblolly pine, placing 2,466 markers with an average marker density of 0.62 cM/marker (Martínez-García et al. 2013). Simultaneously, an exome sequence capture approach was used to build another genetic map of 2,841 markers with a density of 0.58 cM/marker (Neves et al. 2014). Later, another two maps were generated, through a combination of the mentioned maps, having almost the same density (0.60 cM/marker), but being a consensus map among more populations (Westbrook et al. 2015a, b). Finally, the development of an SNP array of 635 k SNPs from a whole-genome resequencing work (De La Torre et al. 2019b) was used to create the last (to date) ultradense linkage map in *P. taeda* containing 26,021 of those SNPs (De La Torre et al. 2019b).

2.3 Comparative Mapping

Comparative mapping represents a powerful tool in genetic mapping, which is based on the presence of orthologous *loci* among species with a recent common ancestor. Two closely related species will have more syntenic and collinear genomes. Therefore, it is possible to extrapolate information from one organism to another, which is useful in genetic map construction and QTLs inference (Chagné et al. 2003). In addition, comparisons among genomes could bring information about evolutionary relationships (Frary et al. 2008; Krutovsky et al. 2004).

Different types of molecular markers have been used for comparative mapping. In early studies, RFLPs and isozymes were the most commonly used markers (Bonierbale et al. 1988; Tanksley et al. 1988, 1992); later, RAPDs or AFLPs were also applied (Chancerel et al. 2011). However, all these markers present problems in the identification of orthologous *loci* (Komulainen et al. 2003). The availability of fast and wide sequencing analysis has made possible the detection of putatively orthologous SNPs markers through methods such as Expressed Sequence Tag Polymorphism (ESTP) or Conservative Ortholog Sequences (COS). These methods allowed the development of SNPs arrays (Chancerel et al. 2011; Fulton et al. 2002) with higher reliability in ortholog identification (Fulton et al. 2002; Komulainen et al. 2003). Despite the high number of potential markers, not all of them provide relevant information for comparative studies. Informative markers are those which are conserved among species and therefore enable to find orthologous *loci* in target species and are also single-copy *locus* in the genome (paralog *loci* lead to unreliable results). Due to this fact, RFLPs, RAPDs, AFLPs, and SSRs have limitations because they are usually located in non-coding regions which are less conserved among species (Brown et al. 2001; Komulainen et al. 2003) and more likely to be homoplastic. On the other hand, the nature of ESTPs and their COS-derived markers as coding regions make them perfect candidates for conserved sequences (Chagné et al. 2003; Fulton et al. 2002; Komulainen et al. 2003). Therefore, a data mining process must be carried out before mapping (Frary et al. 2008; Fulton et al. 2002; Liewlaksaneeyanawin et al. 2009; Neale and Krutovsky 2004).

Despite 300 MY of evolution (Hertling et al. 2020), conifer genomes seem to be highly conserved among species, showing low diversification rates (Ahuja and Neale 2005; Buschiazzo et al. 2012). Therefore, this group is a perfect target for comparative mapping, since a high degree of genome synteny and collinearity is assumed among the members of this lineage. *P. taeda*, the most studied conifer, represents a great reference for comparative functional and structural genomics (Jermstad et al. 2011; Komulainen et al. 2003; Neale and Krutovsky 2004). The identification of loblolly pine RFLPs *loci* in other 12 conifer species provided new insights into Pinaceae genomic architecture and settled the basis for comparative mapping in the *Pinus* genus (Ahuja et al. 1994). *P. taeda* was the reference species in several of those studies (Brown et al. 2001; Chancerel et al. 2011; Devey et al. 1999; Jermstad et al. 2011; Komulainen et al. 2003; Krutovsky et al. 2004; Liewlaksaneeyanawin et al. 2009; Neale and Krutovsky 2004). Later, the construction of genetic maps of *P. taeda* and *P. radiata* using the same set of RFLPs and SSRs markers allowed the detection of 60 RFLPs and 9 SSRs *loci* which were polymorphic in both species (Devey et al. 1999). As a result, the constructed maps had the specific features to ease comparative studies within the genus, representing a new framework for future comparative analysis.

Through the years, comparative techniques have improved at the same time as genetic maps have been constructed. Consequently, new tools have been developed for a better integrative knowledge. One of the main aspects in comparative mapping is the detection of orthologous *loci* with conserved collinearity among compared species, that can be used as anchor points in genetic maps alignment (Brown et al. 2001). Anchor *loci* were identified by using *P. taeda* ESTPs and identifying orthologous markers in other *Pinus* species. Later, a comparison between *P. taeda* and slash pine (*P. elliottii*) maps revealed high synteny and collinearity, resulting in the construction of a template map with 35 anchor *loci*.

Few years later, the development of COS markers (Fulton et al. 2002) made available a new valuable tool for orthologs identification, making comparative studies much easier. This technique was applied to the Pinaceae family (Liewlaksaneeyanawin et al. 2009), following the methodology described by Fulton et al. (2002). This process uses an EST database to obtain orthologous *loci* through the BLAST tool. As the main aim was to encompass a wide range of the Pinaceae family species, they used loblolly pine and a mixture of spruce EST databases. Although this technique retrieves putative single-copy markers, it demands validation by using alternative methods such as single-banded amplicon detection by PCR (Liewlaksaneeyanawin et al. 2009). Despite having limitations, COS markers remain as a powerful tool for comparative mapping, and later studies have brought to light its usefulness (Chancerel et al. 2011; Jermstad et al. 2011).

The high degree of synteny within the genome of the Pinaceae family has been established through an unprecedented comparative analysis of the genetic maps of the reference loblolly pine and Douglas fir (*Pseudotsuga menziesii*). Genetic maps of these two genera were contrasted for the first time using ESTP and RFLP markers, confirming that it is possible and useful to compare species within the Pinaceae family (Krutovsky et al. 2004).

Genetic mapping has intrinsic properties for comparing genomes, allowing the transfer of information from one species to another and providing supplementary information, accelerating the identification of genomic regions associated with phenotypic traits of interest. As described above, several techniques have been developed across the years to make comparative mapping easier in pine trees. As a result, *P. taeda* has arisen as a great reference for comparative mapping to construct genetic maps in other species. These works have set a precedent for future research to launch breeding programs.

2.4 Uses of *P. taeda* Genetic Mapping

One of the main applications of genetic mapping is the identification of chromosomal regions and/or anonymous *loci* associated with simple and complex phenotypes, such as wood quality in pines. In recent years, the emergence of new sequencing tools has accelerated the identification of genes associated with phenotypic traits using both transcriptomic analysis and genome-wide association studies (GWAS) in many species of angiosperms. However, the large size of the pine genome and the presence of repeated regions have limited the application of these methodologies, and also the genome assembly through massive sequencing. Therefore, the genetic mapping will continue to be an essential tool for the development of functional genomics and to order scaffolds in the assembly of pine genomes.

2.5 QTL Mapping

Identification and selection of traits of interest are the main aspects of breeding programs, and the availability of reference genetic maps has made this work easier for loblolly pine. Locating markers across the genome leads to possible associations with traits and their responsible *loci,* also known as Quantitative Trait Loci (QTL). This is useful especially when the genome sequence is not available as occurs in conifers (Lu et al. 2017; Zimin et al. 2017) and allows marker-assisted breeding. Thereby, constructed genetic maps have helped to identify several QTLs in *P. taeda* (Fig. 2.1).

Loblolly pine is one of the major species used for wood production (Rajan et al. 2020). Due to this fact, most of the QTL identification studies are focused on wood quality-related *loci*. However, complex traits are determined by different factors, so a dissection of the trait is key in the identification of QTLs. Wood quality is a complex, polygenic trait influenced by factors like cell wall composition or wood specific gravity, among others. In this way, specific factors are the main goal in QTL identification rather than the complex "wood quality" trait. For example, phenotyping wood quality has been focused on properties like cell wall molecular content (Sewell et al. 2002), microfibril angle, and volume percentage of latewood (Sewell et al. 2000). Wood specific gravity is the most studied trait, with the identification of 14 QTLs in the loblolly pine genetic map (Devey et al. 1999; Groover et al. 1994; Sewell et al. 2000). All these wood properties-related QTLs appear to have little effect, with individual contributions of 3–8% on the phenotype (Neale and Wheeler 2004). These QTLs are spread all over the genome rather than being concentrated in linkage groups as families (Brown et al. 2003). In addition, since growth is as much important as cell wall properties for the exploitation of wood, several growth-related QTLs were identified. They focused on the height of the tree and stem diameter, straightness, and forking defects (Kaya et al. 1999; Xiong et al. 2016).

As a powerful tool for trait-associated *loci,* QTL mapping has played an important role in *P. taeda* breeding. However, trait-marker relationships underwent some changes with the development of new statistical approaches such as genome-wide association studies (GWAS) made possible by the high-density genotyping of SNP markers. GWAS takes advantage of historical recombination on the basis of linkage

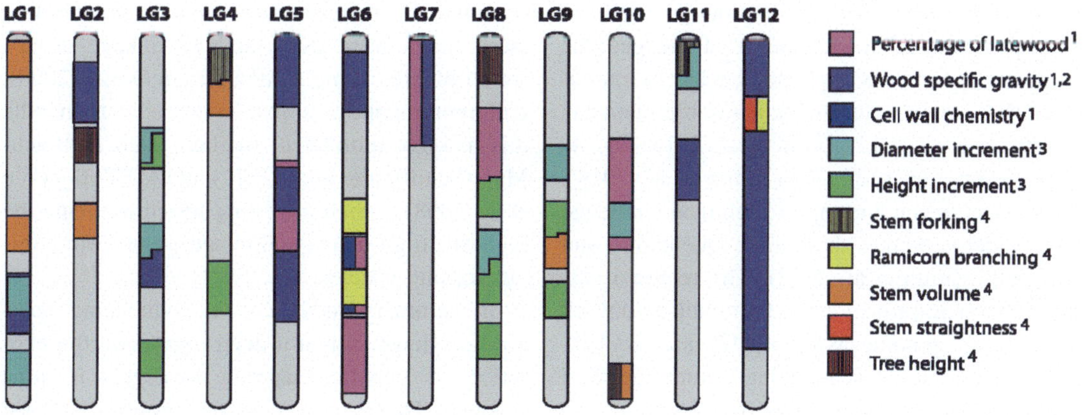

Fig. 2.1 Schematic representation of the location of QTLs for different traits studied in *Pinus taeda* ([1]Brown et al. 2001 [2]Groover et al. 1994; [3]Kaya et al. 1999; [4]Xiong et al. 2016).

disequilibrium, between a genotyped marker and a causal locus, and thus is not limited to a specific genetic background, so controlled pedigrees are no longer needed (González-Martínez et al. 2007). These techniques allowed the association of genetic markers with not only wood property traits (González-Martínez et al. 2007) but also with aridity-related traits (Cumbie et al. 2011; Eckert et al. 2010; González-Martínez et al. 2008), disease resistance (Amerson et al. 2015; Cumbie et al. 2020; Quesada et al. 2010, 2014), metabolite production (Eckert et al. 2012; De La Torre et al. 2019b), and even mycorrhiza traits (Piculell et al. 2019), among others. In the future, a combination of QTL mapping and association studies, together with other types of QTL approaches such as expression quantitative trait locus (eQTL) analysis, will help in elucidating the genetic basis of quantitative trait variation and will allow the identification of genetic variants as Quantitative Trait Nucleotide (QTN).

2.6 Genome Assembly

The genome sequencing of some conifer species, such as loblolly pine, began to be available later than those of angiosperms (Wegrzyn et al. 2013). The initial draft assembly of loblolly pine (0.6) was released and publicly available in 2012 (http://loblolly.ucdavis.edu/bipod/ftp/Genome_Data/genome/pinerefseq/Pita/v0.6/). It contained 18.5 Gbp with an N50 contig size of 800 bp. A first characterization of the genome sequence showed its high percentage of repetitive DNA, mainly due to transposable elements (Wegrzyn et al. 2013; Zimin et al. 2014) and its large genome size (22 Gb) (Zimin et al. 2014). These facts meant a complex technical challenge for whole-genome shotgun sequencing and assembly (Zimin et al. 2014). The released version 0.6 was improved in the first published work of *P. taeda* genome assembly (*P. taeda* v1.0), showing a genome sequencing containing 20.1 Gbp, with an N50 contig size of 8.2 kbp and an N50 scaffold size of 66.9 kbp (Neale et al. 2014). A few years later, a new assembly (*P. taeda v2.0*) was published using PacBio technology, based on long-read single-molecule sequencing. As a result of this novel strategy, the number of scaffolds was reduced. A total of 452,742 scaffolds larger than 5000 bp were obtained, with an N contig size of 25,631 kbp and an N50 scaffold size of 107,821 (Zimin et al. 2017), showing a great improvement in genome contiguity.

Despite the large efforts, scaffolds were not anchored to chromosomes in any of these studies (Neale et al. 2014; Zimin et al. 2017). Indeed, although high-density maps have been obtained (Neves et al. 2014; Martínez-García et al. 2013; De La Torre et al. 2019a, b), to provide a preliminary ordering of scaffolds, the reference genome is still quite fragmented (De La Torre et al. 2019b). Similarly, to other conifers species in which only a low percentage of the genome is ordered, the high-density maps are the main resource to be used as a reference to improve the assembly of the genome (Bernhardsson et al. 2019; Martínez-García et al., 2013; Neves et al., 2014). In an ongoing effort to increase long scale genomic contiguity, DNA from over 500 haploid *P. taeda* megagametophytes from a single mother tree has been extracted (PineRefSeq Consortium 2019). The goal of this effort is to place the total number of scaffolds, 140,452, from genome assembly V2.0, onto a genetic map. After sequencing, SNP calling, and filtering, a total of 90 million high-quality SNPs from a single heterozygous diploid mother was obtained. In a preliminary stage, a small set of these SNPs were used, and 12 linkage groups were obtained using Joinmap (van Ooijen 2006). Different mapping software are able to handle this massive number of markers such as BatchMap (Schiffthaler et al. 2017) and MSTMap (Wu et al. 2008), which are being tested to create the highest high-density linkage map for this outcrossing species.

In summary, advances in sequencing technologies have shown a deep impact on the evolution of genetic mapping strategies in pine species. These advances are providing forest scientists new tools that are allowing them to review historical data, improving the resolution in the efforts to dissect the genomic architecture of important traits, such as traits related to local adaptation to climate change or wood quality traits.

References

Ahuja MR, Devey ME, Groover AT, Jermstad KD, Neale DB (1994) Mapped DNA probes from loblolly pine can be used for restriction fragment length polymorphism mapping in other conifers. Theor Appl Genet 88(3–4):279–282. https://doi.org/10.1007/BF00223632

Ahuja MR, Neale DB (2005) Evolution of genome size in conifers. Silvae Genetica 54(1–6):126–137. https://doi.org/10.1515/sg-2005-0020

Amerson HV, Nelson CD, Kubisiak TL, Kuhlman EG, Garcia SA (2015) Identification of nine pathotype-specific genes conferring resistance to fusiform rust in loblolly pine (*Pinus taeda* L.). Forests 6(8):2739–2761. https://doi.org/10.3390/f6082739

Bernhardsson C, Vidalis A, Wang X, Scofield DG, Schiffthaler B, Baison J, Street NR, Rosario García-Gil M, Ingvarsson PK (2019) An ultra-dense haploid genetic map for evaluating the highly fragmented genome assembly of Norway spruce (*Picea abies*). G3: Genes Genomes Genetics 9(5):1623–1632. https://doi.org/10.1534/g3.118.200840

Bonierbale MW, Plaisted RL, Tanksley SD (1988) RFLP maps based on a common set of clones reveal modes of chromosomal evolution in potato and tomato. Genetics 120(4):1095–1103. https://doi.org/10.1007/BF00053724

Brown GR, Bassoni DL, Gill GP, Fontana JR, Wheeler NC, Megraw RA, Davis MF, Sewell MM, Tuskan GA, Neale DB (2003) Identification of quantitative trait loci influencing wood property traits in Loblolly pine (*Pinus taeda* L.). III. QTL verification and candidate gene mapping. Genetics 164(4):1537 1546

Brown GR, Kadel EE, Bassoni DL, Kiehne KL, Temesgen B, van Buijtenen JP, Sewell MM, Marshall KA, Neale DB (2001) Anchored reference loci in loblolly pine (*Pinus taeda* L.) for integrating pine genomics. Genetics 159(2):799–809

Buschiazzo E, Ritland C, Bohlmann J, Ritland K (2012) Slow but not low: genomic comparisons reveal slower evolutionary rate and higher dN/dS in conifers compared to angiosperms. BMC Evol Biol 12(1):8. https://doi.org/10.1186/1471-2148-12-8

Chagné D, Brown G, Lalanne C, Madur D, Pot D, Neale D, Plomion C (2003) Comparative genome and QTL mapping between maritime and loblolly pines. Mol Breeding 12(3):185–195. https://doi.org/10.1023/A:1026318327911

Chancerel E, Lepoittevin C, Le Provost G, Lin YC, Jaramillo-Correa JP, Eckert AJ, Wegrzyn JL, Zelenika D, Boland A, Frigerio JM, Chaumeil P, Garnier-Géré P, Boury C, Grivet D, González-Martínez SC, Rouzé P, Van de Peer Y, Neale DB, Cervera MT, Plomion C (2011) Development and implementation of a highly-multiplexed SNP array for genetic mapping in maritime pine and comparative mapping with loblolly pine. BMC Genomics 12 (1):368. https://doi.org/10.1186/1471-2164-12-368

Cumbie WP, Eckert AJ, Wegrzyn JL, Whetten R, Neale DB, Goldfarb B (2011) Association genetics of carbon isotope discrimination, height and foliar nitrogen in a natural population of *Pinus taeda* L. Heredity 107(2):105–114. https://doi.org/10.1038/hdy.2010.168

Cumbie WP, Huber DA, Steel VC, Rottmann W, Cannistra C, Pearson L, Cunningham M (2020) Marker associations for fusiform rust resistance in a clonal population of loblolly pine (*Pinus taeda*, L.). Tree Genetics and Genomes 16(6). https://doi.org/10.1007/s11295-020-01478-4

De La Torre AR, Puiu D, Crepeau MW, Stevens K, Salzberg SL, Langley CH, Neale DB (2019) Genomic architecture of complex traits in loblolly pine. New Phytol 221(4):1789–1801. https://doi.org/10.1111/nph.15535

De La Torre AR, Wilhite B, Neale DB (2019) Environmental genome-wide association reveals climate adaptation is shaped by subtle to moderate allele frequency shifts in loblolly pine. Genome Biol Evol 11 (10):2976–2989. https://doi.org/10.1093/gbe/evz220

Devey ME, Fiddler TA, Liu BH, Knapp SJ, Neale DB (1994) An RFLP linkage map for loblolly pine based on a three-generation outbred pedigree. Theor Appl Genet 88(3–4):273–278. https://doi.org/10.1007/BF00223631

Devey ME, Sewell MM, Uren TL, Neale DB (1999) Comparative mapping in loblolly and radiata pine using RFLP and microsatellite markers. Theor Appl Genet 99(3–4):656–662. https://doi.org/10.1007/s001220051281

Devey ME, Jermstad KD, Tauer CG, Neale DB (1991) Inheritance of RFLP loci in a loblolly pine three-generation pedigree. Theoretical and Applied Genetics 83(2):238–242. https://doi.org/10.1007/BF00226257

Echt CS, Saha S, Krutovsky KV, Wimalanathan K, Erpelding JE, Liang C, Nelson CD (2011) An annotated genetic map of loblolly pine based on microsatellite and cDNA markers. BMC Genet 12:1–6. https://doi.org/10.1186/1471-2156-12-17

Eckert AJ, Pande B, Ersoz ES, Wright MH, Rashbrook VK, Nicolet CM, Neale DB (2009) High-throughput genotyping and mapping of single nucleotide polymorphisms in loblolly pine (*Pinus taeda* L.). Tree Genetics Genomes 5(1):225–234. https://doi.org/10.1007/s11295-008-0183-8

Eckert AJ, van Heerwaarden J, Wegrzyn JL, Nelson CD, Ross-Ibarra J, González-Martínez SC, Neale DB (2010) Patterns of population structure and environmental associations to aridity across the range of loblolly pine (*Pinus taeda* L., Pinaceae). Genetics 185(3):969–982. https://doi.org/10.1534/genetics.110.115543

Eckert AJ, Wegrzyn JL, Cumbie WP, Goldfarb B, Huber DA, Tolstikov V, Fiehn O, Neale DB (2012) Association genetics of the loblolly pine (*Pinus taeda*, Pinaceae) metabolome. New Phytol 193(4):890–902. https://doi.org/10.1111/j.1469-8137.2011.03976.x

Frary A, Doganlar S, Ratnaparkhe M (2008) Comparative mapping. Principal and Practices in Plant Genomics

Fulton TM, Van der Hoeven R, Eannetta NT, Tanksley SD (2002) Identification, analysis, and utilization of conserved ortholog set markers for comparative genomics in higher plants. Plant Cell 14 (7):1457–1467. https://doi.org/10.1105/tpc.010479

González-Martínez SC, Huber D, Ersoz E, Davis JM, Neale DB (2008) Association genetics in *Pinus taeda* L II carbon isotope discrimination. Heredity 101 (1):19–26. https://doi.org/10.1038/hdy.2008.21

González-Martínez SC, Wheeler NC, Ersoz E, Nelson CD, Neale DB (2007) Association genetics in *Pinus taeda* L I. Wood property traits. Genetics 175(1):399–409. https://doi.org/10.1534/genetics.106.061127

Grattapaglia D, Sederoff R (1994) Genetic linkage maps of *Eucalyptus grandis* and *Eucalyptus urophylla* using a pseudo-testcross: mapping strategy and RAPD markers. Genetics 137(4):1121–1137

Groover A, Devey M, Fiddler T, Lee J, Megraw R, Mitchel-Olds T, Sherman B, Vujcic S, Williams C, Neale D (1994) Identification of quantitative trait loci influencing wood specific gravity in an outbred pedigree of loblolly pine. Genetics 138(4):1293–1300

Herting J, Stützel T, Klaus KV (2020) The ancestral conifer cone: What did it look like? a modern trait-evolution approach. Int J Plant Sci 181(9):871–886. https://doi.org/10.1086/710489

Jermstad KD, Eckert AJ, Wegrzyn JL, Delfino-Mix A, Davis DA, Burton DC, Neale DB (2011) Comparative mapping in *Pinus*: Sugar pine (*Pinus lambertiana* Dougl.) and loblolly pine (*Pinus taeda* L.). Tree Genetics Genomes 7(3):457–468. https://doi.org/10.1007/s11295-010-0347-1

Kaya Z, Sewell MM, Neale DB (1999) Identification of quantitative trait loci influencing annual height-and diameter-increment growth in loblolly pine (*Pinus taeda* L.). Theor Appl Genetics 98(3–4):586–592. https://doi.org/10.1007/s001220051108

Kinlaw CS, Neale DB (1997) Complex gene families in pine genomes. Trends Plant Sci 2(9):356–359. https://doi.org/10.1016/S1360-1385(97)84624-9

Komulainen P, Brown GR, Mikkonen M, Karhu A, García-Gil MR, O'Malley D, Lee B, Neale DB, Savolainen O (2003) Comparing EST-based genetic maps between Pinus sylvestris and *Pinus taeda*. Theor Appl Genet 107(4):667–678. https://doi.org/10.1007/s00122-003-1312-2

Krutovsky KV, Troggio M, Brown GR, Jermstad KD, Neale DB (2004) Comparative mapping in the pinaceae. Genetics 168(1):447–461. https://doi.org/10.1534/genetics.104.028381

Liewlaksaneeyanawin C, Zhuang J, Tang M, Farzaneh N, Lueng G, Cullis C, Findlay S, Ritland CE, Bohlmann J, Ritland K (2009) Identification of COS markers in the Pinaceae. Tree Genet Genomes 5(1):247–255. https://doi.org/10.1007/s11295-008-0189-2

Lu M, Krutovsky KV, Nelson CD, West JB, Reilly NA, Loopstra CA (2017) Association genetics of growth and adaptive traits in loblolly pine (*Pinus taeda* L.) using whole-exome-discovered polymorphisms. Tree Genetics Genomes 13(3). https://doi.org/10.1007/s11295-017-1140-1

Martínez-García PJ, Stevens KA, Wegrzyn JL, Liechty J, Crepeau M, Langley CH, Neale DB (2013) Combination of multipoint maximum likelihood (MML) and regression mapping algorithms to construct a high-density genetic linkage map for loblolly pine (*Pinus taeda* L.). Tree Genetics Genomes 9(6):1529–1535. https://doi.org/10.1007/s11295-013-0646-4

Morse AM, Peterson DG, Islam-Faridi MN, Smith KE, Magbanua Z, Garcia SA, Kubisiak TL, Amerson HV, Carlson JE, Nelson CD, Davis JM (2009) Evolution of genome size and complexity in *Pinus*. PLoS ONE 4 (2):e4332. https://doi.org/10.1371/journal.pone.0004332

Neale DB, Krutovsky KV (2004) Comparative genetic mapping in trees: the group of conifers. In: Molecular marker systems in plant breeding and crop improvement. Springer, pp 267–277

Neale DB, Wheeler NC (2004) Mapping of quantitative trait loci in loblolly pine and Douglas-fir: a summary. For Genet 11(3–4):173–178

Neale DB, Wegrzyn JL, Stevens KA, Zimin AV, Puiu D, Crepeau MW, Langley CH (2014) Decoding the massive genome of loblolly pine using haploid DNA and novel assembly strategies. Genome Biol 15(3):1–13

Neves LG, Davis JM, Barbazuk WB, Kirst M (2014) A high-density gene map of loblolly pine (*Pinus taeda* L.) based on exome sequence capture genotyping. G3 Genes Genomes Genetics 4(1):29–37. https://doi.org/10.1534/g3.113.008714

Piculell BJ, Martínez-García PJ, Nelson CD, Hoeksema JD (2019) Association mapping of ectomycorrhizal traits in loblolly pine (*Pinus taeda* L.). Mol Ecol 8(8):2088–2099. https://doi.org/10.1111/mec.15013

PinerefSeq Consortium (2019). A hybrid genetic genomic map for *Pinus taeda*. Proceedings of the WFGA Conference. Oral presentation.

Quesada T, Gopal V, Cumbie WP, Eckert AJ, Wegrzyn JL, Neale DB, Goldfarb B, Huber DA, Casella G, Davis JM (2010) Association mapping of quantitative disease resistance in a natural population of loblolly pine (*Pinus taeda* L.). Genetics 186 (2):677–686. https://doi.org/10.1534/genetics.110.117549

Quesada T, Resende MFR, Muñoz P, Wegrzyn JL, Neale DB, Kirst M, Peter GF, Gezan SA, Nelson CD, Davis JM (2014) Mapping fusiform rust resistance genes within a complex mating design of loblolly pine. Forests 5(2):347–362. https://doi.org/10.3390/f5020347

Rajan K, Djioleu A, Kandhola G, Labbé N, Sakon J, Carrier DJ, Kim J-W (2020) Investigating the effects of hemicellulose pre-extraction on the production and characterization of loblolly pine nanocellulose. Cellulose 1–14

Remington DL, Whetten RW, Liu B-H, O'Malley DM (1999) Construction of an AFLP genetic map with nearly complete genome coverage in *Pinus taeda*. Theor Appl Genet 98(8):1279–1292. https://doi.org/10.1007/s001220051194

Ritland K, Krutovsky KV, Tsumura Y, Pelgas B, Isabel N, Bousquet J (2011) Genetic mapping in conifers. Genetics Genomics Breeding Conifers 196:238

Schiffthaler B, Bernhardsson C, Ingvarsson PK, Street NR (2017) BatchMap: a parallel implementation of the OneMap R package for fast computation of F1 linkage maps in outcrossing species. PLoS ONE 12(12): e0189256. https://doi.org/10.1371/journal.pone.0189256

Sewell MM, Bassoni DL, Megraw RA, Wheeler NC, Neale DB (2000) Identification of QTLs influencing wood property traits in loblolly pine (*Pinus taeda* L.). I. Physical wood properties. Theor Appl Genetics 101 (8):1273–1281

Sewell MM, Davis MF, Tuskan GA, Wheeler NC, Elam CC, Bassoni DL, Neale DB (2002) Identification of QTLs influencing wood property traits in loblolly pine (*Pinus taeda* L.). II. Chemical wood properties. Theor Appl Genetics 104(2–3):214–222

Sewell MM, Sherman BK, Neale DB (1999) A consensus map for loblolly pine (*Pinus taeda* L.). I. Construction and integration of individual linkage maps from two outbred three-generation pedigrees. Genetics 151(1):321–330. https://doi.org/10.1093/genetics/151.1.321

Stevens KA, Wegrzyn JL, Zimin A, Puiu D, Crepeau M, Cardeno C, Paul R, Gonzalez-Ibeas D, Koriabine M, Holtz-Morris AE, Martínez-García PJ, Sezen UU, Marçais G, Jermstad K, McGuire PE, Loopstra CA, Davis JM, Eckert AJ, de Jong P, Yorke JA, Salzberg SL, Neale DB, Langley CH (2016) Sequence of the sugar pine megagenome. Genetics 204(4):1613–1626. https://doi.org/10.1534/genetics.116.193227

Tanksley SD, Bernatzky R, Lapitan NL, Prince JP (1988) Conservation of gene repertoire but not gene order in pepper and tomato. Proc Natl Acad Sci 85(17):6419–6423. https://doi.org/10.1073/pnas.85.17.6419

Tanksley SD, Ganal MW, Prince JP, De Vicente MC, Bonierbale MW, Broun P, Fulton TM, Giovannoni JJ, Grandillo S, Martin GB, Messeguer R, Miller JC, Miller L, Paterson AH, Pineda O, Roder MS, Wing RA, Wu W, Young ND (1992) High density molecular linkage maps of the tomato and potato genomes. Genetics 132(4):1141–1160

Temesgen B, Brown GR, Harry DE, Kinlaw CS, Sewell MM, Neale DB (2001) Genetic mapping of expressed sequence tag polymorphism (ESTP) markers in loblolly pine (Pinus taeda L.). Theoretical and Applied Genetics 102(5):664–675. https://doi.org/10.1007/s001220051695

Van Ooijen, J. W. (2006). JoinMap 4, software for the calculation of genetic linkage maps in experimental populations. Kyazma B.V. Wageningen, Netherlands

Wegrzyn JL, Lin BY, Zieve JJ, Dougherty WM, Martínez-García PJ, Koriabine M, Holtz-Morris A, DeJong P, Crepeau M, Langley CH, Puiu D, Salzberg SL, Neale DB, Stevens KA (2013) Insights into the loblolly pine genome: characterization of BAC and fosmid sequences. PLoS ONE 8(9):e72439. https://doi.org/10.1371/journal.pone.0072439

Westbrook JW, Chhatre VE, Wu LS, Chamala S, Neves LG, Muñoz P, Martínez-García PJ, Neale DB, Kirst M, Mockaitis K, Dana Nelson C, Peter GF, Davis JM, Echt CS (2015a) A consensus genetic map for *Pinus taeda* and *Pinus elliottii* and extent of linkage disequilibrium in two genotype-phenotype discovery populations of *Pinus taeda*. G3 Genes Genomes Genetics 5(8):1685–1694. https://doi.org/10.1534/g3.115.019588

Westbrook JW, Walker AR, Neves LG, Munoz P, Resende Jr MFR, Neale DB, Wegrzyn JL, Huber DA, Kirst M, Davis JM, Peter GF (2015b) Discovering candidate genes that regulate resin canal number in *Pinus taeda* stems by integrating genetic analysis across environments, ages, and populations. New Phytol 205(2):627–641. https://doi.org/10.1111/nph.13074

Wu Y, Bhat P, Close TJ, Lonardi S (2008) Efficient and accurate construction of genetic linkage maps from minimum spanning tree of a graph. Plos Genetics 4(10)

Xiong JS, McKeand SE, Isik F, Wegrzyn J, Neale DB, Zeng ZB, da Costa e Silva L, Whetten RW (2016) Quantitative trait loci influencing forking defects in an outbred pedigree of loblolly pine. BMC Genet 17 (1):1–11. https://doi.org/10.1186/s12863-016-0446-6

Zhou Y, Gwaze DP, Reyes-Valdés MH, Bui T, Williams CG (2003) No clustering for linkage map based on low-copy and undermethylated microsatellites. Genome 46(5):809–816. https://doi.org/10.1139/g03-062

Zimin A, Stevens KA, Crepeau MW, Holtz-Morris A, Koriabine M, Marçais G, Puiu D, Roberts M, Wegrzyn JL, de Jong PJ, Neale DB, Salzberg SL, Yorke JA, Langley CH (2014) Sequencing and assembly of the 22-Gb loblolly pine genome. Genetics 196(3):875–890. https://doi.org/10.1534/genetics.113.159715

Zimin AV, Stevens KA, Crepeau MW, Puiu D, Wegrzyn JL, Yorke JA, Langley CH, Neale DB, Salzberg SL (2017) An improved assembly of the loblolly pine mega-genome using long-read single-molecule sequencing. GigaScience 6(1). https://doi.org/10.1093/gigascience/giw016

3 Transposable Elements in Pines

Angelika F. Voronova
and Dainis E. Rungis

Abstract

The importance of mobile genetic elements or transposons in genome evolution and adaptation is being increasingly recognized. Transposons were shown to be involved in gene regulation, shaping chromosome structures, reshuffling coding sequences and generating genetic diversity. Transposable elements constitute the largest fraction of plant genomes and are diverse by structure and distribution. There are still unanswered questions, with regard to their interaction with the environment, regulation mechanisms and convenience of horizontal transfer, but involvement in the epigenetic mechanisms could explain a fraction of missing heritability. Due to their repetitive nature and ubiquity throughout genomes, the study of transposons requires distinctly developed methods and assays. Even with the rapid development of sequencing technologies, genomic assembly and annotation of transposable elements in large genomes are still problematic, and therefore they are routinely excluded from analysis. In this chapter, we aim to summarize the main directions of transposable element research and basic findings in plants with an emphasis on more complex pine genomes providing further perspectives in this area. Although, compared to model species, transposon research in conifers is currently at an initial stage, further studies could uncover additional aspects of their role in genome architecture, adaptive responses and environment–genome interactions.

3.1 Introduction

Mobile Genetic elements or Transposable Elements (TE) are sequences which are able to change their location or proliferate within one genome, or were able in the past. Plant genomes contain a large proportion of TE-derived repeats, most of which are inactive or degraded, and only a small proportion of TE families are able to transpose, most often in response to specific conditions. After the discovery of the first active transposition events in the maize genome by McClintock (1950), it took a long time for the idea of the genome as a static structure to change (Wendel and Wessler 2000; Wessler 2001;

A. F. Voronova · D. E. Rungis (✉)
Latvian State Forest Research Institute, Salaspils, Latvia
e-mail: dainis.rungis@silava.lv

A. F. Voronova
e-mail: angelika.voronova@silava.lv

A. R. De La Torre (ed.), *The Pine Genomes*, Compendium of Plant Genomes,
https://doi.org/10.1007/978-3-030-93390-6_3

Peaston et al. 2004; Grandbastien and Casacuberta 2012). Lately, it has been recognized that TEs are determinants of genome dynamics and act as a tool of evolution (Madlung and Comai 2004; Hawkins et al. 2006; Levin and Moran 2011; Anderson et al. 2019).

The distribution of mobile genetic elements in plant genomes is closely associated with the biology of the species: history of origin and evolution, mating system, generation time, effective population size, as well as with associated genome properties such as ploidy level, recombination frequency and sex-chromosomes (Wessler 2006; Le Rouzic et al. 2007; Levin and Moran 2011; Kejnovský et al. 2012; Li et al. 2015). In addition to polyploidy, enlarged genome size generally reflects higher TE content (Ma et al. 2004; Zhang and Wessler 2004; Hawkins et al. 2006; Ammiraju et al. 2007; Wicker and Keller 2007). Despite being predominantly diploid, conifer genomes are extremely large, reflecting a high repetitive DNA content (Ahuja and Neale 2005; Morse et al. 2009; Kovach et al. 2010; Magbanua et al. 2011). Plant genome diversity studies have also demonstrated the association of TE distribution with harsh microclimatic conditions (Kalendar et al. 2000; Wendel and Wessler 2000; Knight et al. 2005; Voronova and Rungis 2013; Voronova et al. 2017). A full "life cycle" of TE activity includes transcriptional activation interconnected with the methylation state of the genome region, which is followed by the production or recruitment of enzymes necessary for movement, and then transposition, resulting in a new insertion in a different genomic location. TE activity is induced by a range of stress factors, including environmental changes, and could be associated with species adaptation (Wessler 1996; Capy et al. 2000; Chénais et al. 2012; Grandbastien 2015). In addition, recent studies of non-coding RNA and other epigenetic factors have also revealed functional roles for truncated and fixed TE insertions in genomes as a source of regulatory motifs for transcriptional activation could remain after the ability for transposition is lost (Jurka 2008; Feschotte 2008; Hadjiargyrou and Delihas 2013; Galindo-González et al. 2017).

Gymnosperms are a distinct group of plant species of ancient origin (Bowe et al. 2000; Lu et al. 2014), with rare genome duplication events in their evolutionary history (Neale et al. 2014; Li et al. 2015), which is reflected by high genomic synteny between species (Neale and Kinlaw 1997; Komulainen et al. 2003; Lu et al. 2016). However, intergenic regions of conifer genomes are characterized by high diversification of sequences containing numerous families of transposable elements, many of ancient origin and long retention times (Kovach et al. 2010; De La Torre et al. 2014; Zimin et al. 2014). Natural selection has operated continuously on conifer populations, almost without anthropological impact, therefore strong adaptation to local environments is pronounced in these tree species (Wachowiak et al. 2009; Buschiazzo et al. 2012; Chhatre et al. 2013). Studies of mobile genetic elements in pine species genomes could reveal additional aspects of environment–genome interactions, as well as provide practical information for the development of molecular markers for use in breeding and studies of adaptive variation.

3.2 Classification and Structure of Mobile Genetic Elements

TEs have been classified according to their mechanisms of transposition, structure, ability to produce proteins for transposition, and phylogenetic relationships of TE sequences. The first classification system (Finnegan 1989) divided mobile genetic elements into two classes. Class I, or retrotransposons, replicate via an RNA intermediate, resulting in the proliferation of copies of the element within the genome. Class II or DNA transposons use transposase to excise the TE sequence and integrate this sequence into other genomic locations. It was generally believed that all DNA transposons are low-copy elements due to their non-replicative mode of transposition, however, later additional replicative mechanisms for DNA TE were identified (Mendiola et al. 1994; Kapitonov and Jurka 2001). This two-class classification system is still broadly used, however, much greater diversity is found within each TE class, therefore, division into orders, superfamilies, families and subfamilies was later introduced (Wicker et al. 2007) (Fig. 3.1). With the development of high-throughput sequencing technologies, additional TEs and even novel transposition mechanisms were revealed (Cappello et al. 1985; Evgen'ev et al. 1997; Kapitonov and Jurka 2001; Curcio and Derbyshire 2003; Fischer and Suttle 2011; Piégu et al. 2015). Currently, autonomous and non-autonomous elements are recognized within both TE classes, based on the presence or absence of protein-coding domains. Non-autonomous elements do not contain functional protein-coding domains required for transposition, but retain recognition sites used for transposition, and therefore can transpose by using transposes and functional enzymes produced from other TEs (Hartl et al. 1992; Flavell et al. 1994; Feschotte et al. 2002). Following transposition, target site duplications (TSD) are produced by the cell repair machinery, with exception of rare elements which utilize a circular junction mechanism (Cappello et al. 1985). TSDs can be used for the identification of recent transposition events or recently emerged chimeric TE structures (Sabot et al. 2005; Vicient et al. 2005; Yin et al. 2013), however, for old TEs, these sites are degraded and are not always recognizable.

Classes are further divided into orders (Fig. 3.1), however, other researchers have suggested further division into five major classes according to the enzymes encoded by TEs: Long terminal repeat (LTR) retrotransposons, non-LTR retrotransposons, cut-and-paste DNA transposons, rolling-cycle DNA transposons and self-synthesizing DNA transposons (Kapitonov and Jurka 2008). Superfamilies are recognized within each order, which share the similarity of core proteins at the amino acid level and represent ancient evolutionary lineages. Non-autonomous elements such as SINEs, are recognized by their structure and some short conservative motifs (Wicker et al. 2007; Wicker 2012). TE families are species-specific diverged clades of sequences, which are at least 80% similar at the nucleotide sequence level and share at least 80% coverage (Wicker et al. 2007). A broader view, taking into account the biological meaning of classification, which does not always meet the strict 80/80 sequence similarity framework has also been proposed (Wicker 2012). For instance, one TE family should have highly similar regulatory sequences, e.g., LTRs, which utilize similar enzymes for transposition, but contain large insertions in the element body, reflecting divergence of these elements. Functionally and evolutionarily, these should be classified into subfamilies, however, they are often categorized

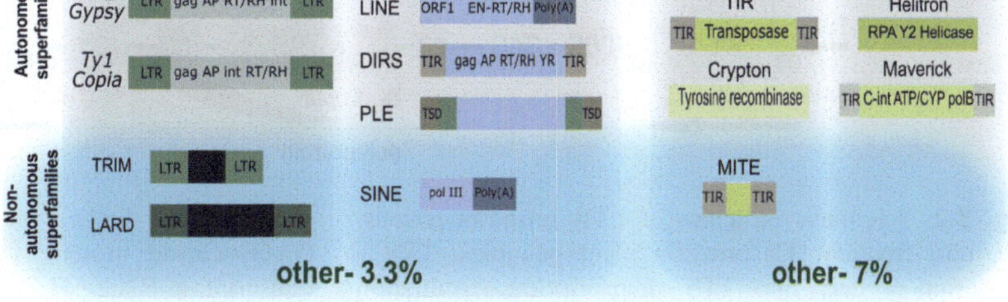

Fig. 3.1 Classification of mobile genetic elements based on Wicker et al. (2007). TE content (as percentage of genome) for Sugar pine (*P. lambertiana*) is indicated (Stevens et al. 2016)

as separate families, thus artificially inflating TE family counts within genomes. In other cases, short non-autonomous elements derived from larger TEs could share similarity in very limited sections such as regulatory sequences, which, however, underlines the similar origin of these TEs and usage of similar enzymes for transposition. According to the curators of the largest TE database *Repbase*, some ancient TE families could share even less than 75% nucleotide sequence similarity (Kapitonov and Jurka 2008). These considerations are important for the classification of TEs in large, diversified genomes such as pine genomes, where some elements could share more than 90% similarity in their regulatory sequences, but contain large insertion/deletion polymorphisms, and would then not meet the 80/80 rule over their entire length for classification as closely related TEs. This can lead to the definition of an excessive number of TE families, disconnected from real biological significance (Voronova et al. 2020).

3.3 Long Terminal Repeats Retrotransposons

Similar to other plant genomes, the largest proportion of pine genomes consists of LTR retrotransposons (RLXs) and derived sequences (Fig. 3.1). These belong to Class I TEs, which resemble retroviruses in their structure and transposition mode. RLXs are further classified into *Pseudoviridae* (or *Ty1*, *Copia*) and *Metaviridae* (or *Ty3*, *Gypsy*) genera which do not contain any true viruses (Capy 2005; Grandbastien 2008; Piégu et al. 2015). Comparative studies of protein domains and structure of LTR retrotransposons revealed common ancient lineages or clades of LTR RE within large data sets of repeats found in plant species, resulting in additional classification categories (Wicker and Keller 2007; Llorens et al. 2011; Neumann et al. 2019). Autonomous LTR retrotransposons are flanked with direct long terminal repeats (LTR), and contain capsid protein GAG and POL protein-coding genes, which encode nucleocapsid proteins and transposition enzymes: protease, reverse transcriptase, RNaseH and integrase. Reverse transcriptase is the most conservative domain of LTR REs, and has been widely used in genome studies or element identification and classification (Kumar and Bennetzen 1999). *Copia* superfamily elements contain an integrase domain before the reverse transcriptase domain in the polyprotein (Fig. 3.2). In contrast to retroviruses, retrotransposons lack genes coding for envelope proteins, therefore, retrotransposons lack infection potential. However, some retrotransposons could have additional envelope-like domains, which are highly divergent and may facilitate transmembrane transport from the nucleus or act as a chaperone for replication (Havecker et al. 2004). Additional Open Reading Frame (ORF) with unknown function or coding for antisense host transcripts is found in some retrotransposons between the pol gene and the 3'LTR (Bureau et al. 1994; Martínez-Izquierdo et al. 1997; Kumekawa et al. 1999; Kalendar et al. 2020).

LTRs are direct repeats found in the RLX termini, which contain regulatory signals for transposition and are highly diverged (Kumar and Bennetzen 1999; Grandbastien and Casacuberta 2012). LTRs of retrotransposons are formed

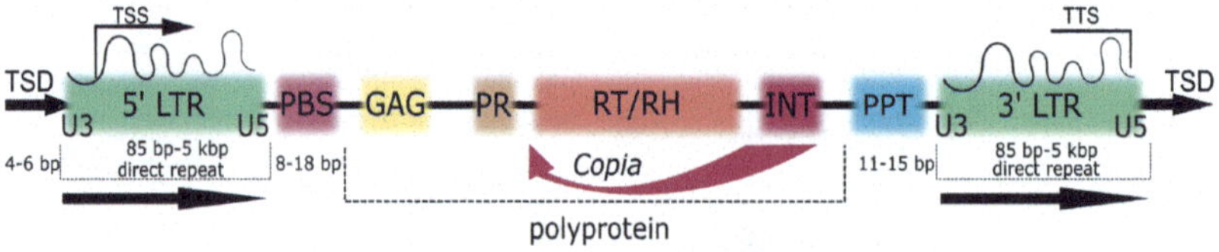

Fig. 3.2 Schematic structure of LTR retrotransposons, the most frequent TE order in plant genomes. TSS—transcription start site; TTS—transcription termination site; U3 and U5—unique 5' and 3' regions of LTRs; TSD—target site duplications; PBS—primer binding site; GAG—capsid protein; PR—protease; RT/RH—reverse transcriptase-RNase-H; INT—integrase; PPT—polypurine tract

during the retrotransposition process and are identical to each other immediately after insertion, therefore, evaluation of the sequence divergence between two LTRs from one TE enables approximation of the time of transposition (San-Miguel et al. 1998). However, recombination events resulting in gene conversion between similar LTRs of one element or between different copies of elements could lead to higher sequence similarity and therefore introduce errors in these estimations (Vitte and Panaud 2003). LTRs are usually 0.1–2 kb in length with some exceptions (Martínez-Izquierdo et al. 1997; Macas and Neumann 2007; Gao et al. 2012b). LTRs and their structural features are used in RLX identification, as they are a main distinguishing feature for non-autonomous LTR retrotransposons (Rho et al. 2007; You et al. 2015). Transcription of LTR retrotransposons starts and ends in the 5' and 3' LTRs, respectively, therefore they contain host transcription initiation and termination sites, e.g., Pol II promoters or TATA-boxes, cis-acting elements, transcription factor binding sites (TFBS), CpG islands and polyadenylation sites (Takeda et al. 1999; Vicient et al. 2005; Tapia et al. 2005; Gao et al. 2012a; Yokosho et al. 2016). LTRs could promote transcription of nearby sequences even if the transposon body is deleted (Klaver and Berkhout 1994; Kumar and Bennetzen 1999; Butelli et al. 2012). Typical LTRs contain unique sequences U3 and U5 with inverted repeats and a central repeat (R) (Fig. 3.2). Downstream, after the 5' LTR, the primer binding site (PBS) is located, which is the tRNA binding site used for reverse transcription priming of the (−) strand (reviewed by Voytas and Boeke 1993; Flavell 1995; Mak and Kleiman 1997).

The PBS motif is the most conservative sequence between different REs, and is used in plant genome studies for molecular marker development or novel LTR RE isolation (Kalendar et al. 2010). tRNA fragments now are classified in a class of small regulatory RNAs which target PBS sites of LTR REs in RNA interference pathways and block reverse transcription of the element, preventing mobility (Martinez 2017; Schorn and Martienssen 2018). Upstream from the 3'LTR, the polypurine tract (PPT) is located, which is used for (+) strand priming (Fig. 3.2).

3.4 DNA Transposons

DNA transposons were among the first mobile genetic elements described in maize (McClintock 1950) and later in prokaryotes (Broda 1978). Generally, DNA TEs are simpler in structure and shorter in size than RLXs. The amount and distribution of DNA transposons in conifer genomes is much lower (Fig. 3.1), however, their impact on well-studied plant protein-coding genes is unambiguous (McClintock 1950; Bureau and Wessler 1994; Casa et al. 2000; Jiang et al. 2003; Liu et al. 2005; Castelletti et al. 2014; Venkatesh and Nandini 2020). Autonomous cut-and-paste DNA transposons (Tc1/mariner, hAT, etc.) contain a single transposase gene that is flanked with terminal inverted repeats (TIRs), and integration sites contain unique TSDs, which remain after excision of the element and serve as a hallmark of recent excision. The transposase gene contains a TIR binding domain, a catalytic domain and a nuclear localization signal, therefore, DNA transposition is independent of cellular proteins or co-factors (Bainton et al. 1991; Lampe et al. 1996). Rolling-circle DNA transposons (e.g., *Helitrons*) were first discovered in plants (Kapitonov and Jurka 2001), and they contain rolling-circle replication initiator and DNA helicase domains (Thomas and Pritham 2015). Due to their transposition mechanism, they do not form TSDs and TIRs (Kapitonov and Jurka 2007). Self-synthesizing DNA transposons have not to date been found in plants (Kapitonov and Jurka 2006; Krupovic and Koonin 2016).

Non-autonomous DNA transposons are usually short sequences, which contain only TIRs, recognized by transposases from autonomous elements. The most abundant type of non-autonomous DNA TEs in plants are Miniature Inverted-repeat Transposable Elements (MITEs) which preferentially reside in gene regions (Bureau and Wessler 1994; Wessler et al. 1995; Zhang et al. 2000; Yang et al. 2007). About 119 potential MITE elements were identified in Loblolly pine (*P. taeda*) BAC sequences (Magbanua et al. 2011). 297 families were found in the Loblolly pine genome using bioinformatic approaches, and sequences were included in the publicly available ConTEdb database (Yi et al. 2018). A recent comparative study of two pine reference genomes (Loblolly pine and Sugar pine) independently discovered *Plater* MITE family whose insertions were predominantly found in gene regions. 74 and 87 *Plater* MITE copies were inserted in the introns of genes, but 191 and 65 copies were preferentially distributed in 0–1 kb flanks Loblolly pine and Sugar pine genes, respectively (Voronova et al. 2020).

3.5 TEs and Genome Organization

The distribution of each TE family in genomes is family- and species-specific (Hawkins et al. 2006; Du et al. 2010; Jedlicka et al. 2020). However, some general tendencies exist. Monocotyledon genomes contain more TEs compared to dicotyledon genomes, due to the recent evolutionary expansion of these TEs (SanMiguel et al. 1998; Vitte and Bennetzen 2006). RLX insertions are more frequent in genomes of outcrossing species (Wright et al. 2001; Tam et al. 2007; Lockton and Gaut 2010), therefore, the mating system was also reported as influencing TE frequencies (Charlesworth and Charlesworth 1995; Wright and Schoen 1999; Morgan 2001).

Within genomes, TE distribution varies between chromosomes, as well as coding and non-coding regions. The patterns of distribution within genomes could be influenced by any insertional preferences of active TEs, as well as recombination, which can also rearrange old and inactive TE sequences. As previously mentioned, DNA transposons or short non-autonomous RLXs are found more frequently in gene regions (Witte et al. 2001; Wei et al. 2009), as well as there are located evolutionary younger insertions (Xu and Du 2014). Differential distribution is most likely influenced by purifying selection, which removes highly deleterious mutations, therefore, most TE insertions are found in neutral non-coding genomic loci, but some insertions could have an impact on gene regulation (Wendel and Wessler 2000; Wessler 2006; Jurka 2008). Best studied LINE-1 retrotransposon insertion preference into relaxed chromatin of tissues-specific genes was reported for human neuronal cells (Muotri et al. 2005; Singer et al. 2010). Somatic mosaicism is even more pronounced in plants and was studied in perennial plants with relation to resistance (Simberloff and Leppanen 2019), therefore mobile genetic elements in conifers could be studied in this aspect in the future.

In angiosperm genomes, the distribution of *Copia* retrotransposons is usually dispersed across chromosomes, while Gypsy LTR REs tend to concentrate in centromeric and telomeric regions (Jiang et al. 1996; Miller et al. 1998; Pereira 2004; Wang et al. 2006; Neumann et al. 2011; Domingues et al. 2012). Dispersed chromosomal distribution of several RLX was observed in conifers, where only one TE family was found to be clustered in telomeres (Friesen et al. 2001). Monophyletic grouping of reverse transcriptase domain sequences suggested a common ancestry with subsequent species-specific proliferation of certain conifer TE lineages (Friesen et al. 2001). Currently, in plant genomes, a number of retrotransposons have been described with prevalence in pericentromeric, subtelomeric and constitutive heterochromatin regions (Miller et al. 1998; Lippman et al. 2004; Dai et al. 2007; Gao et al. 2008; Neumann et al. 2011; Kejnovský et al. 2012; de Castro Nunes et al. 2018). Integrase genes of centromere-associated retrotransposons contain chromodomains, which could regulate specific targeting of element transposition to centromeres (Kordiš 2005; Gao et al. 2008). Comparative studies of plant retrotransposons have established that specific *Gypsy* lineages in plant genomes contain similar chromodomains (Neumann et al. 2011). However, centromeric retrotransposons have not yet been extensively studied in pine genomes (Neumann et al. 2011). Although some homologous lineages were present in the loblolly pine genome, they contained group C chromodomains, which lack conserved aromatic cage residues known to interact with methylated histone H3 lysine 9 (H3K9), which is required for proper chromosome condensation, but their localization was not centromeric (Neumann et al. 2011). This result indicates probable divergence of centromere sequences in gymnosperm species.

In plant genomes with high LTR RE content, clusters and nested repeats are abundantly found in intergenic regions, which are formed after multiple transpositions of TEs into each other (SanMiguel et al. 1996; Li et al. 2004; Gao et al. 2015). Repeat nesting is a serious problem for automated TE discovery in complex plant genomes (SanMiguel et al. 1996; Shirasu et al. 2000; Kronmiller and Wise 2008; Flutre et al. 2012; Gao et al. 2015). Positioning of highly repetitive TEs in the assembly in large plant genomes is challenging using data from widely used short-read sequencing techniques. Repeat nesting was recently studied by evaluation of the age of each LTR pair (Jedlicka et al. 2019). Gypsy elements were found to more frequently form nested repeats, but nesting was associated with palindromic sequences and heterochromatin. Members of one TE family tend to insert into each other more frequently (Pereira 2004; Jedlicka et al. 2019) and nested insertions usually were found in the 3' UTR of disrupted elements (Jedlicka et al. 2019).

The diversity, repetitive nature and high abundance of TEs in pine genomes lead to the situation where the differentiation of full-length elements from partial, chimeric or nested copies is often not possible without resequencing using long read sequencing techniques. Automated detection of repeats in short-read sequencing data sets can lead to inaccurate detection and assembly, therefore, different analysis flows with manual curation is required for accurate annotation (Flutre et al. 2012). Computationally detectable short motifs such as TSDs or TIRs can accumulate point mutations leading to unrecognizability of the insertion boundaries by de novo TE detection applications (Fig. 3.3).

Comparison of reference pine genomes and different versions of genome assemblies suggest persistent problems in TE positioning and inflated TE family counts (Voronova et al. 2020). Furthermore, individual polymorphisms associated with TEs or genes result in large insertions or deletions between different genomes, and species reference genomes should include variants from different genotypes (Tettelin et al. 2005; Morgante et al. 2007; Nguyen et al. 2015; Bayer et al. 2020). In the *Pinus* genera, reference genomes have been sequenced and assembled for several species, but genomic variation between individuals within one species remains largely undiscovered. TE copy number variation among

Fig. 3.3 Example of three true TE families, and resulting inflated TE element count by automated TE identification (not all combinations are indicated)

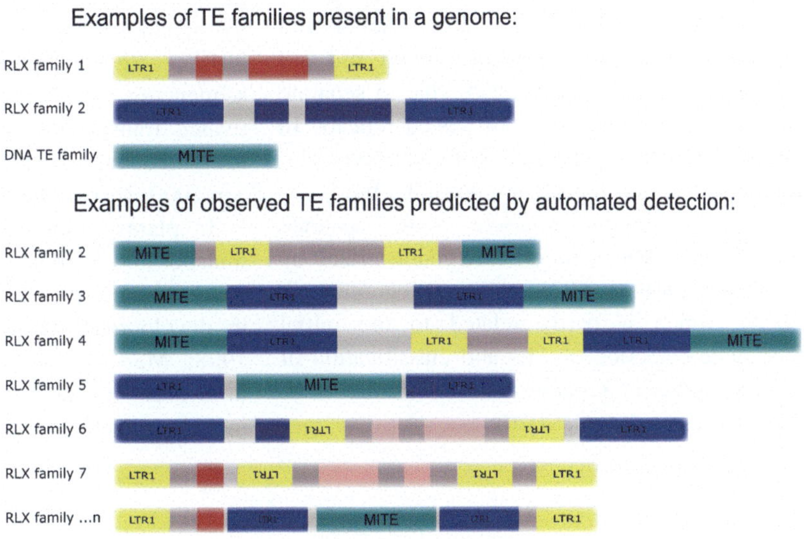

eight individuals of Scots pine (*P. sylvestris)* revealed differences depending on the evolutionary age of RE families: moderately distributed families were more variable between genotypes (Voronova et al. 2017), and this initial study demonstrates the extreme diversity of non-coding portions of the pine genome.

Recombination processes of highly similar sequences in the genome could be involved in TE copy number variation, TE reshuffling and TE-associated gene reshuffling. Unequal homologous recombination between elements or their LTRs could lead to the formation of solo LTRs and genome size reduction (Vicient 1999; Shirasu et al. 2000; Devos et al. 2002; Vitte and Panaud 2003). In conifer genomes, unequal recombination-associated loss of LTR intervening sequences occurs less often than in angiosperms, therefore, full-length TE sequences are more prevalent in conifer genomes (Nystedt et al. 2013; Cossu et al. 2017). TE abundance is related to higher genome-wide methylation, which is elevated in conifer genomes (Saéz-Laguna et al. 2014; Ausin et al. 2016); it may be that extended heterochromatin regions prevent recombination processes (Groth et al. 2007).

Due to the error-prone reverse transcription process, copies from one parental element are more diversified, but template switching by reverse transcriptase could produce chimeric elements flanked by identical TSDs, which are produced during the integration process (Sabot et al. 2005; Vicient et al. 2005; Sabot and Schulman 2007). If such insertions are old and TSDs are mutated, it is not possible to separate such chimeric elements from nested repeats. In large, diversified genomes, both chimeric elements and nested repeats are possible, but accumulated mutations interfere with accurate identification. Pine genomes are characterized by increased amounts of TE families (Wegrzyn et al. 2014), which is partly related to the initial automatic detection of TEs and classification of nested repeats as separate families (Flutre et al. 2012; Voronova et al. 2020).

3.6 Function of Transposable Elements

Since the discovery of TEs, their transposition has been linked to stress conditions and evolutionary change (McClintock 1984; Wessler 1996; Kalendar et al. 2000; Capy et al. 2000; Beguiristain et al. 2001; Kazazian 2004; Wendel et al. 2012; Grandbastien 2015; Baluska et al. 2018; Lee et al. 2019). In normal conditions, these sequences are usually tightly controlled by the host organism using methylation mechanisms (Mlura et al. 2001; Casacuberta and Santiago 2003; Kato et al. 2003; Lippman et al. 2003; Slotkin and Martienssen 2007). For a long period, transposable elements were considered to be "junk DNA" and selfish elements (Orgel and Crick 1980). The functional role of transposable element insertions is less evident, and resulting phenotypic changes are less studied compared with the effect of protein-coding gene mutations and polymorphism (Zhou et al. 2012; Zhao et al. 2018; Lee et al. 2019). Information from various plant species and genes where TE-derived insertions are linked to phenotypic alterations is accumulating (Fray and Grierson 1993; Kashkush et al. 2003; Kobayashi et al. 2004; Studer et al. 2011; Chu et al. 2011; Butelli et al. 2012; Tsuchiya and Eulgem 2013; Lai et al. 2019). Multiple types of regulative non-coding RNAs originate from TE sequences, nested elements and their relicts remaining after purifying selection (Wang et al. 2009; Hadjiargyrou and Delihas 2013; Qin et al. 2015; Lee et al. 2019). Recent progress in transcriptome sequencing has also revealed functions for inactivated transposable elements, which could be reshuffled by recombination processes (Bennetzen 2000; Devos et al. 2002).

In plants, transcription and transposition of transposable elements are associated with stress conditions, meristematic tissues and certain stages in development (Bureau et al. 1994; Wessler 1998; Martínez and Slotkin 2012). Reported influences of TE insertions in gene regions include gene interruption by transposition;

alteration of gene expression levels via providing additional transcription initiation signals or downregulation by methylation; exon shuffling and alternate splicing; initiation of antisense transcription; production of non-coding RNA or providing target sites for these; providing additional poly-A signals that affect transcript stability and transport from nucleus (Fig. 3.4). Using these regulatory influences, multiple insertions of similar TEs could result in the recruitment of dynamic gene networks (Varagona et al. 1992; Takeda et al. 1999; Xiao et al. 2008; Feschotte 2008; Rebollo et al. 2012; Grandbastien 2015; Mita and Boeke 2016).

It is obvious that it is not possible to find one specific function for all existing diversity of

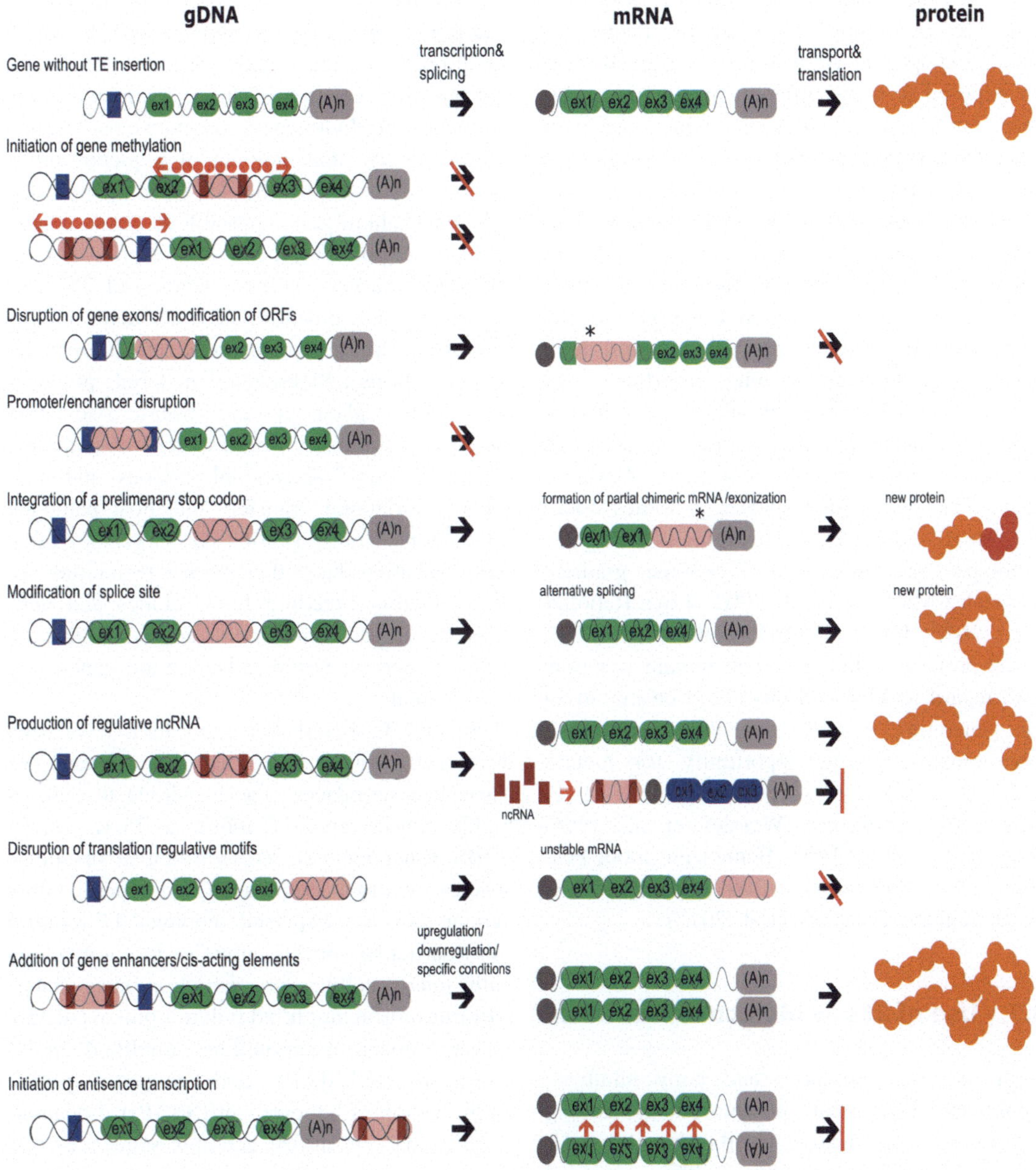

Fig. 3.4 Examples of the potential regulatory impact of TE insertions on gene transcription and translation

mobile genetic elements within one genome or in different species. Furthermore, different insertions of similar sequences belonging to one family could have (or not have) different functional roles depending on their localization in the genome, age, interplay with other factors, etc. In the rice genome, transcription of approximately 33% of stress-responsive genes is associated with TEs (Krom et al. 2008). A similar proportion of genes in other plant species may be associated with TEs and could contribute to species- or population-specific adaptation. Long-lived trees are exposed to many different pathogens, herbivores, environmental clues, etc., which vary over time. The extreme diversity of TEs in pine genomes could reflect the long and ongoing adaptation processes of each species or population to local environments (Voronova et al. 2017, 2020). Some ancient and old TE families and their insertions may have lost their function over the years and therefore behave like neutral loci, slowly degrading. Some other insertions may have lost their transposition activity, but underwent fixation in populations due to selective advantages under specific conditions. Advantageous TE loci in pine species genomes could have small additive effects resulting in a range of genotypes resulting in a phenotype gradient (González-Martínez et al. 2002; Loya-Rebollar et al. 2013; Lu et al. 2019; Hall et al. 2021). Some portion of the genomes remain dynamic with regard to TEs, and this could be important for generating genetic diversity in changing environments, giving opportunity for natural selection to select genotypes adapted to newly established conditions (Wessler et al. 1995; Grandbastien et al. 1997; Bennetzen 2000; Feschotte 2008; Baucom et al. 2009; Belyayev et al. 2010; Galindo-González et al. 2017).

3.7 Use of TEs as Molecular Markers

High sequence similarity and copy numbers, significant insertional polymorphism among individuals, conservative motifs shared by TE families and the distribution of TEs across all chromosomes and plant species make TEs extremely convenient sequences for molecular marker development (Purugganan and Wessler 1995; Waugh et al. 1997; Peterson et al. 2002; Schulman et al. 2004; Kalendar and Schulman 2006; Schulman 2007; Kalendar et al. 2011; Monden and Tahara 2015; Bhat et al. 2020). TE-derived molecular markers reflect restructuration of comparatively large sequence segments from hundreds of bp to several kbp. Retrotransposon-derived insertions are irreversible in subsequent generations, in contrast to SNPs and SSRs, which could revert to their initial variants (homoplasy), complicating the use of these markers for the evaluation of phylogenetic relationships. Using a comparatively small amount of sequence information, for example, the sequence of one distributed TE family, it is possible to obtain a large amount of polymorphism data. Application of TE-based markers includes studies of TE activation and transposition, TE-associated genome stability in different conditions, genetic diversity of populations and breeding material, phylogenetic studies, phenotype and disease association studies, evolutionary studies, genetic mapping, differentiation of somaclonal varieties, and studies of horizontal transfer and host–pathogen interactions. Transposon-based markers are often more sensitive than other marker techniques for differentiating breeding lines, clones and varieties developed using mutagenesis, which are differentiated phenotypically, but are genetically very similar.

Several TE-based marker systems have been developed, which are based on different TE-associated sequences. These include the use of highly conservative TE motifs as PCR primers (iPBS, Kalendar et al. 2010), which are useful for species without sequenced genomes. Other genetic tools use dispersed annotated TE families from particular species, alone or in combination with microsatellites or restriction sites (inter-retrotransposon amplified polymorphism (IRAP), retrotransposon-microsatellite amplified polymorphism (REMAP) and sequence-specific amplification polymorphism (SSAP)). Location-specific markers utilize genomic sequences surrounding known TEs (retrotransposon-based insertional polymorphism (RBIP)). Each approach has advantages and disadvantages—iPBS markers are easy and cheap to apply, but reproducibility can be low, similar to other non-specific methods such as RAPDs (Halldén et al. 1996; Bussell et al. 2005). The second category of markers (IRAP, REMAP, SSAP) produce large amounts of data, as usually motifs from species-specific distributed RLX are used, in addition, the markers are dominant and require knowledge about species-specific TE sequences. The third type of markers (RBIP) usually rely on detailed investigations, require more effort for development and implementation, but they are precise and codominant (allow resolution of TE insertion containing or empty loci). These markers could be associated with specific locations or genes, and can be used to investigate phylogenetic relationships, as well as gene regulation and related phenotypes (Flavell et al. 1998; Butelli et al. 2012; Sampath and Yang 2014; Venkatesh and Nandini 2020). Methods utilized for the resolution of amplified fragments vary from gel electrophoresis and hybridization assays to capillary electrophoresis and high-throughput sequencing techniques (Monden and Tahara 2015; Kalendar et al. 2019).

IRAP, REMAP and SSAP methods are PCR-based techniques broadly used in plant diversity studies, which analyze amplified DNA fragment length differences between several genomic features. IRAP amplifies fragments between several nested RLX, REMAP between RLX and microsatellites, and SSAP between RLX and restriction sites (Kalendar et al. 2011). PCR-based techniques are limited by amplification fragment length that can be produced in defined PCR conditions and polymerase characteristics. Results are dependent on differences in primer annealing due to SNP mutations in primer annealing regions, TE insertions or deletions, as well as DNA quality. IRAP markers use conservative LTR sequences from highly distributed RLX families within particular plant species. Fragments can be produced using a single primer that amplifies the genomic region between two RLXs of one family inserted in opposite orientations close enough to each other to produce amplification products. If a combination of two primers from different RLX families is used, fragments are amplified in a competitive manner, meaning that not all fragments from single primer amplifications will be present in the amplification with a combination of the same primers (Kalendar and Schulman 2014). Successful TE family and marker selection with subsequent high-quality resolution methods allow visualization of multiple fragments, therefore producing high information content (Kalendar and Schulman 2006, 2014). The SSAP method is similar to AFLP and uses PCR primers complementary to TEs, which enables higher reproducibility than the AFLP method (Waugh et al. 1997; Leigh et al. 2003). Retrotransposon Internal Variation Polymorphism (RIVP) utilizes amplification of internal RE domains (Purugganan and Wessler 1995).

Retrotransposon-based markers have not been widely used for pine population studies, which is related to a low level of genome and pine-specific TE research, as well as high genetic diversity of pines and large genomes. IRAP markers were developed for Masson pine (*Pinus massoniana*) using a genome walking technique, and were used to assess genetic diversity in breeding lines (Fan et al. 2014). IRAP markers were developed for Scots pine using RLX isolation from transcribed sequences, and high diversity within Scots pine populations was revealed and IRAP markers were able to differentiate subpopulations growing in differing conditions based on different frequencies of RLX bands (Voronova and Rungis 2013).

3.8 TE Investigations in Pine Species

Given the preponderance of TE investigations in other plant species, as well as the contrasts with the large genomes and life cycles of conifers, investigations of TE diversity and distribution in distinct and ancient plant lineages such as gymnosperms are of high scientific interest. Conifers are characterized by large outcrossing populations, with distributions mainly determined by natural selection forces, and are therefore highly adapted to local environments (Savolainen and Pyhäjärvi 2007; Wachowiak et al. 2009). At the level of taxa, conifers display lower phenotypic and gene nucleotide diversity and reduced recombination rates due to longer generation times (Heuertz et al. 2006; Pyhäjärvi et al. 2007; Chen et al. 2010). In addition, large population sizes and intensive gene flow restrict fixation of new mutations in populations (Savolainen and Pyhajarvi 2007). Therefore, conifers are characterized by high genetic diversity within populations, low nucleotide substitution rates and low phenotypic variation between species. DNA sequences of pine species that diverged 10 million years ago differ by only 5% at neutral sites (Savolainen and Pyhajarvi 2007). While low linkage disequilibrium (LD) levels are widely reported in gymnosperms (Neale and Savolainen 2004), non-coding regions may have more extended regions of LD (Moritsuka et al. 2012). However, the large genome size of conifers delayed sequencing attempts until 2013, when the first spruce genome sequence was announced (Nystedt et al. 2013). Reassociation studies revealed a high repetitive content of conifer genomes, with 70–75% consisting of repetitive DNA (Miksche and Hotta 1973; Rake et al. 1980; Peterson et al. 2002). Later, genome sequence homology analyses confirmed that approximately 65% of the Loblolly pine genome is composed of repetitive REs, while de novo estimates were around 79.3% (Wegrzyn et al. 2014). Overall repetitive content of other pine species was estimated to be variable, for example, for Scots pine, it was estimated at 52% (Nystedt et al. 2013). A higher content of interspersed repeats was reported in Sugar pine (79%) compared with Loblolly pine (74%) (Stevens et al. 2016). These are approximate estimates which depend on the analysis pipeline utilized, as higher repetitive content will be estimated with lower stringency parameters, etc.

Attempts to characterize mobile elements in conifers before genomic sequences were available were made by amplification of reverse transcriptase domains using degenerate primers. This enabled the identification of retrotransposons (Table 3.1) and allowed for comparative phylogenetic studies of TEs from different plant species (Voytas et al. 1992; Flavell et al. 1992; Friesen et al. 2001; Fan et al. 2013).

Conserved *Gypsy* and *Copia* orders were identified, however, diverged TE elements remained undiscovered using these methods. Hybridization with isolated short reverse transcriptase fragments detected high TE copy numbers and a dispersed distribution across chromosomes. Similar to angiosperms, *Copia* elements were absent from centromere and telomere regions (Kamm et al. 1996; Brandes et al. 1997; Friesen et al. 2001). Cross-hybridization of isolated RTs suggested a monophyletic distribution of TEs among pine species and even more distantly related gymnosperms (Kamm et al. 1996; Kossack and Kinlaw 1999). Bacterial artificial chromosome (BAC) analyses of loblolly pine revealed all orders of TEs, a high content of LTR REs, particularly from the Gypsy superfamily, and a high proportion of pine-specific TE families, most of which were only weakly similar to angiosperm TEs (Morse et al. 2009; Kovach et al. 2010; Magbanua et al. 2011; Wegrzyn et al. 2013). Sequencing of several conifer genomes confirmed these previous investigations and uncovered an enormous diversity of pine TEs (Neale et al. 2014; Wegrzyn et al. 2014; Zimin et al. 2014, 2017; Stevens et al. 2016). The 100 most highly represented REs account for less than 20% of the loblolly pine genome, therefore, at least 40–60% of the repetitive interspersed fraction could be composed of low-copy number elements (Wegrzyn et al. 2014). In general, this could account for the increased diversity and degradation levels of RE sequences in conifers, compared to angiosperm genomes (Kovach et al. 2010; Magbanua et al. 2011; Nystedt et al. 2013; Wegrzyn et al. 2014; Zimin et al. 2014). Additionally, large genome sizes and repeat nesting could produce overestimations of TE family content identified by automated assembly and annotation pipelines (Voronova et al. 2020). The ratio of solo LTRs to full-length elements in gymnosperm genomes is considerably lower than in angiosperms (1:9 in Norway spruce versus 1:1 in *A. thaliana* or 16:1 for BARE-1 in barley),

Table 3.1 TEs identified in *Pinus* species

TE name	Classification	Species were first identified	References	Method
TPE1	RLX *Copia*	Slash pine (*P. elliottii)*	Kamm et al. (1996)	Shared DNA cloning and sequencing
IFG	RLX *Gypsy*	Monterey pine (*P. radiata)*	Kossack and Kinlaw (1999)	RFLP fragment cloning and sequencing
PpRT1	RLX *Gypsy*	Maritime pine (*P. pinaster)*	Rocheta et al. (2007)	Genome walking
PpRT2,3,4	RLX *Gypsy*	Maritime pine	Miguel et al. (2008)	PCR amplification
Gymny	RLX *Gypsy*	Loblolly pine	Morse et al. (2009)	BAC sequencing
PMRT1-58	RLX *Copia*	Masson pine	Fan et al. (2013)	PCR amplification of RTs
REPM	RLX *Gypsy*	Masson pine	Fan et al. (2013)	PCR amplification of RTs
Copia-17-PTa	RLX *Copia*	Scots pine	Voronova and Rungis (2013)	iPBS amplification
Ouachita, Bastrop, Ozark, Appalachian, Angelina, Talladega	RLX *Gypsy*	Loblolly pine	Wegrzyn et al. (2013)	BAC and fosmid sequences
Piedmont	RLX unknown	Loblolly pine	Wegrzyn et al. (2013)	BAC and fosmid sequences
Congaree, Cumberland, Pinewoods	RLX *Copia*	Loblolly pine	Wegrzyn et al. (2013)	BAC and fosmid sequences
Silava-PTa	RLX *Copia*	Scots pine	Voronova and Rungis (2013)	iPBS amplification
PIER db	TE families from automatic detection	Loblolly pine	Neale et al. (2014, Wegrzyn et al. (2014)	Whole genome NGS sequencing
PARTC	RLX *Copia*	Loblolly pine, Norway spruce (*Picea abies*)	Zuccolo et al. (2015)	Sequence analysis
MITE Plater, DNA TE Irbe	DNA TE MITE	Loblolly pine Sugar pine	Voronova et al. (2020)	Reference genome analysis
RLX Daugava	RLX Copia	Loblolly pine, Sugar pine	Voronova et al. (2020)	Reference genome analysis

which indicates a lower rate of RE removal from gymnosperm genomes (Nystedt et al. 2013). The ratio of partial elements to full-length elements is 1.2:1 in Loblolly pine, 3.2:1 in Siberian fir (*Abies sibirica*) and the average is 2:1 in most conifers (Nystedt et al. 2013; Wegrzyn et al. 2014), suggesting a significant proportion of full-length elements, in contrast to angiosperm genomes.

The *Gypsy* RLX *IFG* is one of the most studied TEs in pines. First isolated from Monterey pine, this family is also present in Sugar pine (Kossack and Kinlaw 1999). A related RLX,

PpRT1 with a unique 1.7 kb insertion, was identified in Maritime pine, Loblolly pine and Swiss stone pine (*P. cembra)* (Rocheta et al. 2007; Magbanua et al. 2011). Despite its relatively short length (approx. 4kbp), IFG is frequently found in pine genomes and therefore the occupation area of IFG is one of the highest among TEs in some pine species. Hybridization studies revealed that about 5.8% of the Loblolly pine genome is occupied by the IFG family (Magbanua et al. 2011), while a shotgun whole genome sequencing study estimated occupancy of 0.4% (Wegrzyn et al. 2014), and copy number estimation using absolute quantification real-time PCR suggested 0.8% occupancy for this family (Voronova et al. 2017). All these methods have their own drawbacks, however, considering the significant intraspecies variation of RLX sequences, estimates for species could be only approximations. Even small variation in the relative proportion of these elements can be significant, as 1% of the Loblolly pine genome is equivalent to 432 Mbp of sequence, which is comparable with genome sizes of some mosses and algae. The evolutionary history of IFG is ancient, as it was later also identified in angiosperms, from Cork oak (*Quercus suber*) under the name *Corky* (gb|EU862277.1|), and IFG homologous sequences have been identified in the grape genome (Godinho et al. 2012). In grape, some IFG insertions have accumulated mutations, but some contained identical LTRs, indicative of recent transposition (Godinho et al. 2012). Comparison of two distantly related reference pine species revealed three homologous genes with IFG insertions, all of which are protein kinases, indicating that these insertions probably took place in a common ancestor (Voronova et al. 2020). Despite the fact that some IFG inserts contain a chromodomain, insertions were identified in the vicinity and within introns of 14 genes in the grape genome (Godinho et al. 2012). In the Loblolly pine and Sugar pine genomes, 14–18 genes contained IFG parts in 0–1 kb gene flanks, while 99–317 genes contained IFG insertions in gene introns, respectively (Voronova et al. 2020). Comparison of the plant RLX classified as *IFG7* to the *Reina* clade of chromoviruses demonstrated vertical transmission throughout the plant kingdom (Neumann et al. 2019), implying the absence of horizontal transfer. The *IFG-7* family is diverged in the gymnosperm families *Cupressaceae*, *Taxaceae*, *Podocarpaceae*, *Araucariaceae* and *Ephedrales*, while ginkgo (*Ginkgo biloba*) contains the RLX *Alisey*, which has high nucleotide sequence similarity with IFG, and together with the common distribution of other RLXs, suggests a closer phylogenetic relationship of ginkgo with pine species (Kossack and Kinlaw 1999; Voronova et al. 2017). Although the IFG family is an ancient and widely distributed family in pine genomes, some pine species contain elevated IFG RLX content. For example, two pine species inhabiting extreme environments were found to contain elevated copies of IFG in their genomes. Macedonian pine (*Pinus peuce)* (120,290 copies) is distributed at high elevations, and Italian stone pine (*Pinus pinea)* (223,966 copies) has a very southern distribution in Syria and Libya, hinting that the expansion of this RE may be associated with adaptation to extreme conditions. These species also have the largest genome sizes of the pine species included in the study, however, the IFG-7a_PTa family also had the highest genome occupancy in the Italian stone pine (1.62%) and the Macedonian pine genomes (0.97%). High copy numbers may reflect transposition bursts in these species, while the association of increased copy numbers throughout the whole genome and increased copy numbers within protein-coding gene regions or insertions within genes involved in adaptive responses in these species is not yet known, and requires further investigation.

Another widely studied RLX is *Congaree*, a *Copia* superfamily element 15 kbp in length and which has the highest occupancy compared to other retroelements in the Loblolly pine genome (Wegrzyn et al. 2014). *Congaree* is prevalent in four pine species: Weymouth pine (*P. strobus*), Scots pine, Loblolly pine and Lodgepole pine (*P. contorta*), displaying the highest genome occupancy proportion among RLXs (1.45–2.28%) (Voronova et al. 2017). The *Congaree* family is diverged in pine species and could contain several subfamilies, as several closely related RLXs with diversified LTRs under different names were revealed by sequence comparison (*Ouachita*, *Bastrop*, *RLC_515*). Despite many genomic copies, this element was not among the most distributed TEs in gene regions of Loblolly pine. However, the relative *Congaree* copy number variation among Scots pine individuals was higher compared to the relatively older *IFG* family. RLX *Copia* element *Pinewoods* (Wegrzyn et al. 2014) belongs to the homologous TE family *PARTC* isolated from Norway spruce (Zuccolo et al. 2015) as full-length nucleotide sequence identity between elements is up to 85%, but the description of elements in different species usually results in new naming of TE families. *Pinewoods* elements in Loblolly pine and Scots pine have 91% nucleotide sequence identity. This family of elements is also distributed in other gymnosperm species. For example, *Pinewoods* was found to be 82% similar to *pGity5* from ginkgo, which again underlines the slower rate of TE differentiation in gymnosperms genera compared to angiosperms.

Occupancy of RLX could significantly vary between pine species genomes, as was observed for TE proliferation in other plant species or lineages. For example, the Scots pine genome contains an elevated fraction of *Angelina* and *Riga-4* (R4) RLXs, which were previously identified as being differentially expressed in response to stress conditions (Voronova et al. 2014; Voronova 2019). The relative proportion of the most distributed RLX was also variable across pine species (Fig. 3.5). The reason for such differences is not currently understood, and should be investigated in the future. It is important to note that not all mentioned pine species have genome sequences available, therefore, they may also contain other proliferated TEs as in the case of Scots pine or some TE families with more divergent sequences.

RLX distribution is associated with the phylogeny of plant species and could uncover some additional insights into the evolutionary history of pine species, as unlike protein-coding genes, TE insertions are highly variable and diverse. For example, amplification product polymorphism of RLX families revealed differentiation within species belonging to the *Contortae* subsection of pines, which is supported by previous studies of phylogeny based on nuclear and plastid genomes (Gernandt et al. 2005; Parks et al. 2012; Voronova et al. 2017). It has been suggested that the subsection *Contortae* was derived relatively recently, and has been placed in the section *Pinus* by analysis of nuclear and plastid sequences (Eckert and Hall 2006; Palmé et al. 2009). However, some ambiguity about the phylogenetic placement of this subsection remains, due to highly variable sites identified by whole plastid genome analysis (Parks et al. 2009). In addition, the division of *Contortae* between northern and southern glacial refugia may explain the divergence of species within this subsection (Wheeler and Guries 1982; Millar 1998). Considering the monophyly of pine species, investigations of particular RLX insertions

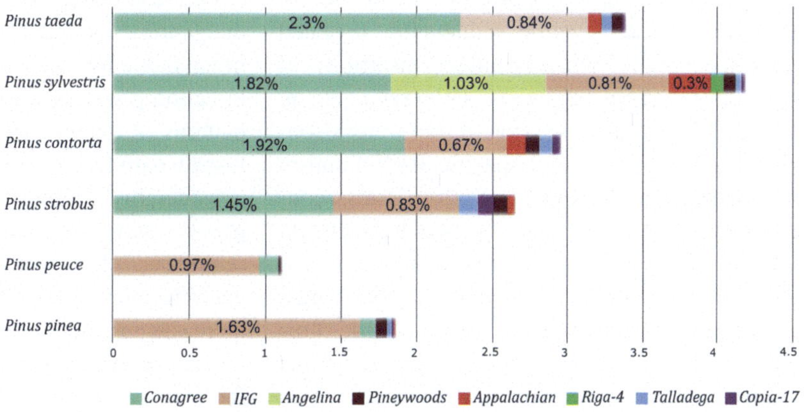

Fig. 3.5 Occupation area (%) of LTR retrotransposon families considering species genome size, element length and estimated average copy number (Voronova et al. 2017)

could help to reveal additional details of pine species distribution and origin. A highly distributed and ancient RLX family such as *IFG* could be a good candidate for phylogenetic studies.

Evidence of possible transposition (or somatic recombination) was demonstrated in Scots pine clones by analyzing IRAP fragments in ramets growing in different plantations (Voronova and Rungis 2013). Scots pine-specific IRAP markers, unlike neutral SSR markers, distinguished Scots pine subpopulations growing in different hydrological regimes (Voronova and Rungis 2013). The frequency of retrotransposon insertions was elevated in pine individuals growing in dry conditions, and these results were later extended by analysis of copy number variation of specific RLXs in these trees (Voronova, unpublished data). Similar associations of RLX insertion frequencies with adverse conditions were demonstrated for wild barley populations growing in the Evolution Canyon microsite, Lower Nahal Oren, Mount Carmel, Israel (Kalendar et al. 2000).

The association of transposable elements with stress responses and genome evolution promoted detailed studies of TE activity in pine genomes. The *PpRT* reverse transcriptase domain was first identified in expressed sequence tag (EST) databases, indicating that this element was expressed (Miguel et al. 2008). However, most of the other identified elements in pine species contained frameshift mutations in the polyprotein ORFs (Brandes et al. 1997; Kossack and Kinlaw 1999), larger rearrangements and nesting, therefore, most pine TEs were considered not to be transpositionally active (Wegrzyn et al. 2013). However, whole genome sequencing revealed some potentially active younger insertions of low-copy number elements (Wegrzyn et al. 2014). Differential expression of retrotransposon-associated fragments and specific RLX families was detected in response to various stress conditions in Scots pine ramets like heat shock, salicylic acid and abscisic acid treatment, infestation with insects and inoculation with fungal pathogens (Voronova et al. 2011, 2014; Voronova 2019). Pine whole transcriptome sequencing studies investigating stress responses or developmental stages also reveal significant upregulation of TEs (Elbl et al. 2015; Liu et al. 2015; Cañas et al. 2017), but usually these sequences were not further analyzed. Detection of transcription of elements is not indicative of transposition of TEs. As previously described, several levels of TE regulation can be defined: including transcription of TEs utilizing host genome enzymes, which depends on the methylation status of the genomic region. Truncated or nested element transcription could be co-expressed with genes in pine genomes, where non-coding regions like introns are longer and contain various numbers of TE insertions (Ahuja and Neale 2005; Wegrzyn et al. 2014). Detailed analysis is complicated by the absence of accurate annotation of pine TEs, high-quality reference genomes and gene annotations which continue to be improved (Zimin et al. 2017). In addition, multiple copy numbers of highly similar TEs prevent the precise localization of elements in large genomes using short-read assemblies, therefore, additional methods or information are needed for TE discovery and analyses (Novák et al. 2013; Nelson et al. 2017; Goerner-Potvin and Bourque 2018).

Expression of mobile genetic elements solely as a result of random chromatin relaxation of TE-containing genome regions was disputed by a study of comparative RLX expression in response to different fungal pathogens (Voronova 2019). Random expression of the studied RLXs frequently found in pine genomes would result in the proportional expression of the most distributed retrotransposon families. Nevertheless, no direct connection between RLX family copy number in the genome and relative levels of RLX transcriptional induction was revealed in different stages of infection progress. Comparison of the levels and patterns of expression after inoculation with two fungal pathogens indicated that responses were considerably different both in stress-responsive genes and in non-coding retrotransposon sequences. Infection by a more aggressive necrotrophic pathogen resulted in insignificant increases in studied retrotransposon transcription levels compared to rapid increases of the same RLX families after infection by a conditionally biotrophic fungus (Voronova 2019). It was therefore suggested that transcription of pine TEs may be associated with transcription of protein-coding genes, which have considerably longer introns in the pine genome (Ahuja and Neale 2005; Wegrzyn et al. 2014). There is growing evidence of the role of non-coding RNA in host–pathogen interplay through different RNA interference mechanisms (Weiberg and Jin 2015; Wang et al. 2017; Poretti et al. 2020). TE sequences can be acquired by pathogens via horizontal gene transfer (Panaud 2016) and pathogenic strategies could rapidly evolve utilizing this mechanism. Based on further TE-associated gene analyses in pine reference genomes, it was suggested that genes with diverse TE content in the non-coding regions could provide an adaptive advantage by facilitating rapid transcriptional responses in diverse stress conditions (Voronova et al. 2020). Segmental transcriptional activation of stress-responsive genes and associated TEs could lead to the previously observed retrotransposon transcription patterns in response to fungal pathogens. Most studied retrotransposon families like *IFG*, *Pinewoods* and *Congaree* with a wide distribution throughout the whole genome were not among the families that were distributed in gene introns or flanks, confirming previous observations from expression studies.

A comparative study of interspersed repeats in Loblolly pine and Sugar pine gene non-coding regions (introns and flanking regions) revealed several additional pine TE families, which were preferentially distributed in genes. Insertions of the newly identified DNA TE *Plater,* belonging to the non-autonomous MITE elements, were found to be significantly enriched in 0–2 kb gene flanking regions, while most other TE diversity was observed in gene introns (Voronova et al. 2020) (Fig. 3.6).

Important TFBS were found in TEs distributed in gene regions. For example, *Plater* MITE contains ten ARR1 binding sites which are involved in response to cytokinins in other plant species (Mok 1994; Argueso et al. 2012; Hwang et al. 2012). In addition, TATA-box, DOF, W-box, GT and MYB binding sites found in the *Plater* consensus sequence are usually associated with transcription initiation and stress response (Sakai et al. 2001; Yanagisawa 2004; Taniguchi et al. 2007; Eulgem and Somssich 2007). No genes containing *Plater* MITE insertions within both non-coding regions (introns and flanks) were identified in single genes in both pine species, suggesting probable impact of localization of *Plater* MITE insertions on gene function and elimination of MITE insertions from unfavorable positions by natural selection.

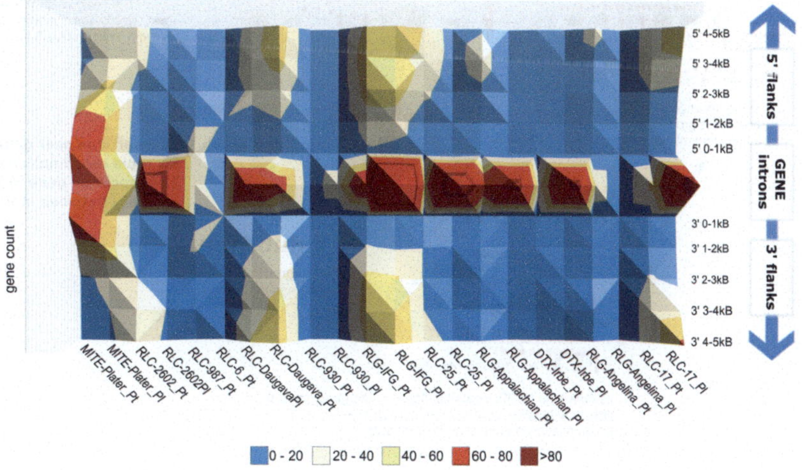

Fig. 3.6 Comparison of TE distribution in gene non-coding regions of high-quality genes from the Sugar pine genome v.1.01 (PILA) and filtered annotated gene set of Loblolly pine v.2.0 (PITA), reproduced from Voronova et al. (2020)

The newly described *Irbe* DNA TE had not been identified previously using automated TE identification and annotation pipelines, as it is an ancient element with degraded TSDs (Voronova et al. 2020). However, the transposase coding region could be recognized, as well as the distribution of the highly similar *Irbe* DNA TE sequences within gene introns. Interestingly, the *Irbe* element contains sequence homology to the mature microRNA sly-miR9472-3p from a drought-tolerant tomato line (Candar-Cakir et al. 2016; Liu et al. 2017), suggesting that a stress-responsive gene network could be regulated by this sequence. Some microRNAs are highly conserved among plants, including angiosperms and gymnosperms, suggesting common regulatory pathways and mechanisms (Chen 2008; Sun 2012; Chávez Montes et al. 2014; You et al. 2017; Galdino et al. 2019; Krivmane et al. 2020). TEs are considered a natural source of many microRNA and other non-coding RNAs (Piriyapongsa and Jordan 2008; Li et al. 2011; Qin et al. 2015). Longer LTR retrotransposons were also found in the pine gene introns, for example, the *Copia* LTR RE *Daugava* contains AG-repeats in the LTRs, which is homologous to a light-responsive TFBS in plants (Liu et al. 2013). Gene networks formed by transposable elements could be highly advantageous, as such changes in non-coding regions could be rapid, specific for given tissues or stresses and combinatorial, each insertion adding a small effect. Furthermore, pine genomes contain gene duplications and families, and a change in only one gene from a gene family could be less harmful to the host.

Advantageous flexible stress-responsive gene networks could be formed by TE insertions in proximity or in the introns of genes (Fig. 3.7). A high diversity of different TE families was found in important stress-responsive genes, which could play an important role in the regulation of transcriptional responses in a broad range of stress conditions. Genes having the highest numbers of TE insertions within their introns were found to be annotated with many co-occurring GO terms indicating involvement in a range of cellular processes. For example, in Loblolly pine potassium channel coding gene exhibit the largest number of intronic TE insertions, members of protein kinase coding genes, transcription factors and cytochrome genes were found frequently among genes containing TE

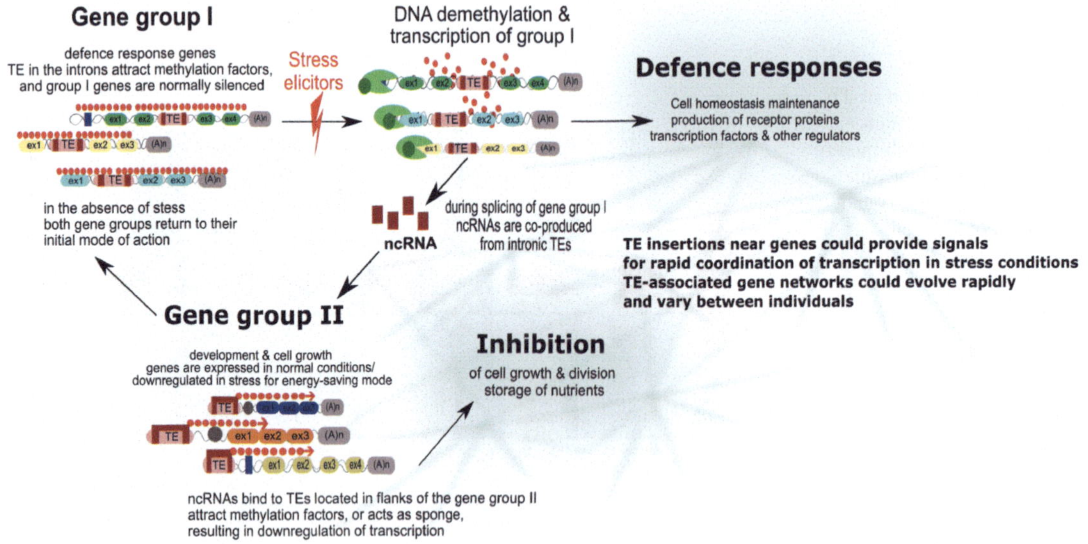

Fig. 3.7 Example of a suggested TE-associated gene networking mechanism based on *Plater* MITE distribution in the Loblolly pine genome, based on Voronova et al. (2020)

insertions in both Loblolly pine and Sugar pine (Voronova et al. 2020). Expression of plant genes with large introns is enhanced in a broader range of tissues (Camiolo et al. 2009; Das and Bansal 2019). Additionally, in both pine species, observed GC content of introns containing TEs was on average lower compared to nearby exons (39% vs. 44%), similar to reports from other plant species (Mizuno and Kanehisa 1994; Singh et al. 2016; Voronova et al. 2020). Lower GC content in introns indicates a more accessible conformation of DNA for various regulatory factors which are components of TEs (Schwartz et al. 2009; Gelfman and Ast 2013; Ullah et al. 2018). In addition, some pathogens could acquire host TE-associated genetic features to downregulate the transcription of host genes containing TEs by recently discovered RNA-interacting mechanisms, leading to competition and rapid diversification of TE sequences. Evolutionary change via recombination, transposition and/or fast degradation of RE sequences result in high genetic diversity and could induce adaptability of populations that undergo continuous and variable natural selection pressures. Further investigations in this area will broaden the understanding about plant–environment interactions.

3.9 Conclusions and Future Perspectives

The importance of transposable elements in adaptive responses is increasingly being recognized, however, the processes and mechanisms are only beginning to be studied in pine genomes. High-quality genome sequences and annotation are the basis for systematic TE investigations. Given that transposable elements are the largest and most variable elements within genomes, reference genome sequences from one or a few individuals are probably not sufficient to capture the diversity found within species or populations. In addition, transposable element sequences can have many different functions and effects on their host genome depending on their location, age and properties. As they form a large proportion of genomes, they could influence genome architecture and integrity merely by their presence. They can influence gene regulation either by disrupting gene structure, introducing chromatin remodeling factors, modifying expression by producing non-coding RNA or introducing additional regulatory motifs. These functions and influences are particularly relevant in pine species. Pines species have a long evolutionary history, are widely distributed and are highly adapted to a diverse range of habitats. They also exhibit a high degree of phenotypic plasticity. As pines, and conifers in general, possess such famously large genomes, a large proportion of which consist of transposable elements, the study of the role of these TEs in pines is of significant scientific interest.

To date, studies of TEs in pines have shown that many overall properties are shared with other conifer genera, as well as other plant species. This includes their genomic distribution, prevalent TE classes, diversity and number of families. However, comparative studies indicate that expansion of many families has occurred after speciation, leading to distinct differences in the distribution of TEs between species. It is well established that TEs are expressed and transposed in response to stress conditions and could also be transmitted via horizontal transfer, but the adaptive significance of this has been less well understood. Similar to observations in other plant species, recent studies in pine have shown transcriptional activation of TEs in response to a range of stress conditions. The overall genomic distribution of highly prevalent TE families does not always reflect the relative representation of TE sequences in the transcriptome. Comparative investigation of gene non-coding regions revealed elevated amounts of TEs in pine gene introns, which considering the saturation of TEs with stress-responsive motifs and potential production of regulative non-coding RNA could modulate a more rapid and coordinated reorientation of transcription patterns in stress conditions. TE-associated gene networking could be highly advantageous for long-lived trees as these networks are able to form evolution more rapidly than adaptive changes in protein-coding sequences. TEs can facilitate the reshuffling or reorganization of non-coding sequences of genes without disrupting evolutionarily valuable protein-coding sequences. The presence of gene families allows diversification of genes by TE insertions without impacting host fitness. Transposition enables adaptation to changes in the combination of stress factors or unprecedented types of stress by the interconnection of different gene networks or the inclusion/exclusion of genes in networks. Furthermore, such a system can be evolutionarily effective, allowing random TE-associated changes in species producing large amounts of pollen and seeds, which then undergo natural selection for the most adaptive genotypes. General functional patterns are shared between species, but adaptation to local environments has developed after speciation, probably reflected by species or population-specific patterns of TE distribution.

Advances in sequencing technologies will undoubtedly improve the quality and quantity of sequenced pine genomes. However, the extreme diversity of TEs and their interactions (e.g., recombination, nesting, mutation, transposition and even possible horizontal transfer) will present challenges to researchers beyond the technical tasks of obtaining high-quality genome assemblies. Future TE research will benefit from more complete pine-specific gene annotations, investigation of gene families and intraspecies diversity. Investigation of the diversity found within species and populations will lead to more comprehensive insights into the role of TEs on adaptive responses in evolutionarily ancient, long-lived, widespread and highly adaptable keystone species such as pines.

References

Ahuja MR, Neale DB (2005) Evolution of genome size in animals. Silvae Genet 54:126–137. https://doi.org/10.1515/sg-2005-0020

Ammiraju JSS, Zuccolo A, Yu Y et al (2007) Evolutionary dynamics of an ancient retrotransposon family provides insights into evolution of genome size in the genus Oryza. Plant J 52:342–351. https://doi.org/10.1111/j.1365-313X.2007.03242.x

Anderson SN, Stitzer MC, Brohammer AB et al (2019) Transposable elements contribute to dynamic genome content in maize. Plant J 100:1052–1065. https://doi.org/10.1111/tpj.14489

Argueso CT, Ferreira FJ, Epple P et al (2012) Two-component elements mediate interactions between cytokinin and salicylic acid in plant immunity. PLoS Genet 8:e1002448. https://doi.org/10.1371/journal.pgen.1002448

Ausin I, Feng S, Yu C et al (2016) DNA methylome of the 20-gigabase Norway spruce genome. Proc Natl Acad Sci U S A 113:E8106–E8113. https://doi.org/10.1073/pnas.1618019113

Bainton R, Gamas P, Craig NL (1991) Tn7 transposition in vitro proceeds through an excised transposon intermediate generated by staggered breaks in DNA. Cell 65:805–816. https://doi.org/10.1016/0092-8674(91)90388-F

Baluska F, Gagliano M, Witzany G (2018) Memory and learning in plants. Springer International Publishing, Cham

Baucom RS, Estill JC, Leebens-Mack J, Bennetzen JL (2009) Natural selection on gene function drives the evolution of LTR retrotransposon families in the rice genome. Genome Res 19:243–254. https://doi.org/10.1101/gr.083360.108

Bayer PE, Golicz AA, Scheben A et al (2020) Plant pan-genomes are the new reference. Nat Plants 6:914–920. https://doi.org/10.1038/s41477-020-0733-0

Beguiristain T, Grandbastien MA, Puigdomènech P, Casacuberta JM (2001) Three Tnt1 subfamilies show different stress-associated patterns of expression in tobacco. Consequences for retrotransposon control and evolution in plants. Plant Physiol 127:212–221. https://doi.org/10.1104/pp.127.1.212

Belyayev A, Kalendar R, Brodsky L et al (2010) Transposable elements in a marginal plant population: temporal fluctuations provide new insights into genome evolution of wild diploid wheat. Mob DNA 1:6. https://doi.org/10.1186/1759-8753-1-6

Bennetzen JL (2000) Transposable element contributions to plant gene and genome evolution. Plant Mol Biol 42:251–269. https://doi.org/10.1023/A:1006344508454

Bhat RS, Shirasawa K, Monden Y, et al (2020) Developing transposable element marker system for molecular breeding. In: Methods in molecular biology. Humana Press Inc., pp 233–251

Bowe LM, Coat G, dePamphilis CW (2000) Phylogeny of seed plants based on all three genomic compartments: extant gymnosperms are monophyletic and Gnetales' closest relatives are conifers. Proc Natl Acad Sci 97:4092–4097. https://doi.org/10.1073/pnas.97.8.4092

Brandes A, Heslop-Harrison JS, Kamm A et al (1997) Comparative analysis of the chromosomal and genomic organization of Ty1-copia-like retrotransposons in pteridophytes, gymnosperms and angiosperms. Plant Mol Biol 33:11–21. https://doi.org/10.1023/A:1005797222148

Broda P (1978) DNA insertion elements, plasmids and episomes. FEBS Lett 93:380–381. https://doi.org/10.1016/0014-5793(78)81148-x

Bureau TE, Wessler SR (1994) Stowaway: a new family of inverted repeat elements associated with the genes of both monocotyledonous and dicotyledonous plants. Plant Cell 6:907–916. https://doi.org/10.1105/tpc.6.6.907

Bureau TE, White SE, Wessler SR (1994) Transduction of a cellular gene by a plant retroelement. Cell 77:479–480

Buschiazzo E, Ritland C, Bohlmann J, Ritland K (2012) Slow but not low: genomic comparisons reveal slower evolutionary rate and higher dN/dS in conifers compared to angiosperms. BMC Evol Biol 12:8. https://doi.org/10.1186/1471-2148-12-8

Bussell JD, Waycott M, Chappill JA (2005) Arbitrarily amplified DNA markers as characters for phylogenetic inference. Perspect Plant Ecol Evol Syst 7:3–26. https://doi.org/10.1016/j.ppees.2004.07.001

Butelli E, Licciardello C, Zhang Y et al (2012) Retrotransposons control fruit-specific, cold-dependent accumulation of anthocyanins in blood oranges. Plant Cell 24:1242–1255. https://doi.org/10.1105/tpc.111.095232

Camiolo S, Rau D, Porceddu A (2009) Mutational biases and selective forces shaping the structure of *Arabidopsis* genes. PLoS ONE 4:e6356. https://doi.org/10.1371/journal.pone.0006356

Cañas RA, Li Z, Pascual MB et al (2017) The gene expression landscape of pine seedling tissues. Plant J 91:1064–1087. https://doi.org/10.1111/tpj.13617

Candar-Cakir B, Arican E, Zhang B (2016) Small RNA and degradome deep sequencing reveals drought-and tissue-specific micrornas and their important roles in drought-sensitive and drought-tolerant tomato genotypes. Plant Biotechnol J 14:1727–1746. https://doi.org/10.1111/pbi.12533

Cappello J, Handelsman K, Lodish HF (1985) Sequence of Dictyostelium DIRS-1: an apparent retrotransposon with inverted terminal repeats and an internal circle junction sequence. Cell 43:105–115. https://doi.org/10.1016/0092-8674(85)90016-9

Capy P (2005) Classification and nomenclature of retrotransposable elements. Cytogenet Genome Res 110:457–461. https://doi.org/10.1159/000084978

Capy P, Gasperi G, Biémont C, Bazin C (2000) Stress and transposable elements: co-evolution or useful parasites? Heredity (edinb) 85:101–106. https://doi.org/10.1046/j.1365-2540.2000.00751.x

Casa AM, Brouwer C, Nagel A et al (2000) The MITE family Heartbreaker (Hbr): molecular markers in maize. Proc Natl Acad Sci 97:10083–10089. https://doi.org/10.1073/pnas.97.18.10083

Casacuberta JM, Santiago N (2003) Plant LTR-retrotransposons and MITEs: control of transposition and impact on the evolution of plant genes and genomes. Gene 311:1–11. https://doi.org/10.1016/S0378-1119(03)00557-2

Castelletti S, Tuberosa R, Pindo M, Salvi S (2014) A MITE transposon insertion is associated with differential methylation at the maize flowering time QTL vgt1. G3 Genes Genomes Genet 4:805–812. https://doi.org/10.1534/g3.114.010686

Charlesworth D, Charlesworth B (1995) Quantitative genetics in plants: the effect of the brreeding system on genetic variability. Evolution (N Y) 49:911–920. https://doi.org/10.1111/j.1558-5646.1995.tb02326.x

Chávez Montes RA, De Fátima R-Cárdenas F, De Paoli E et al (2014) Sample sequencing of vascular plants demonstrates widespread conservation and divergence of microRNAs. Nat Commun 5:1–15. https://doi.org/10.1038/ncomms4722

Chen X (2008) MicroRNA metabolism in plants. Curr Top Microbiol Immunol 320:117–136

Chen J, Källman T, Gyllenstrand N, Lascoux M (2010) New insights on the speciation history and nucleotide diversity of three boreal spruce species and a Tertiaryrelict. Heredity (Edinb) 104:3–14. https://doi.org/10.1038/hdy.2009.88

Chénais B, Caruso A, Hiard S, Casse N (2012) The impact of transposable elements on eukaryotic genomes: from genome size increase to genetic adaptation to stressful environments. Gene 509:7–15

Chhatre VE, Byram TD, Neale DB et al (2013) Genetic structure and association mapping of adaptive and selective traits in the east Texas loblolly pine (*Pinus taeda* L.) breeding populations. Tree Genet Genomes 9:1161–1178. https://doi.org/10.1007/s11295-013-0624-x

Chu C-G, Tan CT, Yu G-T et al (2011) A novel retrotransposon inserted in the dominant Vrn-B1 allele confers spring growth habit in tetraploid wheat (*Triticum turgidum* L.). G3 (Bethesda) 1:637–45. https://doi.org/10.1534/g3.111.001131

Cossu RM, Casola C, Giacomello S et al (2017) LTR retrotransposons show low levels of unequal recombination and high rates of intraelement gene conversion in large plant genomes GBE. Genome Biol Evol 9:3449–3462. https://doi.org/10.1093/gbe/evx260

Curcio MJ, Derbyshire KM (2003) The outs and ins of transposition: from MU to kangaroo. Nat Rev Mol Cell Biol 4:865–877

Dai J, Xie W, Brady TL et al (2007) Phosphorylation regulates integration of the yeast Ty5 retrotransposon into heterochromatin. Mol Cell 27:289–299. https://doi.org/10.1016/j.molcel.2007.06.010

Das S, Bansal M (2019) Variation of gene expression in plants is influenced by gene architecture and structural properties of promoters. PLoS ONE 14:e0212678. https://doi.org/10.1371/journal.pone.0212678

de Castro NR, Orozco-Arias S, Crouzillat D et al (2018) Structure and distribution of centromeric retrotransposons at diploid and allotetraploid coffea centromeric and pericentromeric regions. Front Plant Sci 9:175. https://doi.org/10.3389/fpls.2018.00175

De La Torre AR, Birol I, Bousquet J et al (2014) Insights into conifer giga-genomes. Plant Physiol 166:1724–1732. https://doi.org/10.1104/pp.114.248708

Devos KM, Brown JKM, Bennetzen JL (2002) Genome size reduction through illegitimate recombination counteracts genome expansion in *Arabidopsis*. Genome Res 12:1075–1079. https://doi.org/10.1101/gr.132102

Domingues DS, Cruz GMQ, Metcalfe CJ et al (2012) Analysis of plant LTR-retrotransposons at the fine-scale family level reveals individual molecular patterns. BMC Genomics 13:137. https://doi.org/10.1186/1471-2164-13-137

Du J, Tian Z, Hans CS et al (2010) Evolutionary conservation, diversity and specificity of LTR-retrotransposons in flowering plants: insights from genome-wide analysis and multi-specific comparison. Plant J 63:584–598. https://doi.org/10.1111/j.1365-313X.2010.04263.x

Eckert AJ, Hall BD (2006) Phylogeny, historical biogeography, and patterns of diversification for Pinus (Pinaceae): phylogenetic tests of fossil-based hypotheses. Mol Phylogenet Evol 40:166–182. https://doi.org/10.1016/j.ympev.2006.03.009

Elbl P, Lira BS, Andrade SCS et al (2015) Comparative transcriptome analysis of early somatic embryo formation and seed development in Brazilian pine, Araucaria angustifolia (Bertol.) Kuntze. Plant Cell Tissue Organ Cult 120:903–915. https://doi.org/10.1007/s11240-014-0523-3

Eulgem T, Somssich IE (2007) Networks of WRKY transcription factors in defense signaling. Curr Opin Plant Biol 10:366–371. https://doi.org/10.1016/j.pbi.2007.04.020

Evgen'ev MB, Zelentsova H, Shostak N, et al (1997) Penelope, a new family of transposable elements and its possible role in hybrid dysgenesis in Drosophila virilis. Proc Natl Acad Sci U S A 94:196–201. https://doi.org/10.1073/pnas.94.1.196

Fan F, Cui B, Zhang T et al (2014) LTR-retrotransposon activation, IRAP marker development and its potential in genetic diversity assessment of masson pine (*Pinus massoniana*). Tree Genet Genomes 10:213–222. https://doi.org/10.1007/s11295-013-0677-x

Fan F, Wen X, Ding G, Cui B (2013) Isolation, identification, and characterization of genomic LTR retrotransposon sequences from masson pine (*Pinus massoniana*). Tree Genet Genomes 9:1237–1246. https://doi.org/10.1007/s11295-013-0631-y

Feschotte C (2008) Transposable elements and the evolution of regulatory networks. Nat Rev Genet 9:397–405

Feschotte C, Jiang N, Wessler SR (2002) Plant transposable elements: where genetics meets genomics. Nat Rev Genet 3:329–341. https://doi.org/10.1038/nrg793

Finnegan DJ (1989) Eukaryotic transposable elements and genome evolution. Trends Genet 5:103–107. https://doi.org/10.1016/0168-9525(89)90039-5

Fischer MG, Suttle CA (2011) A virophage at the origin of large DNA transposons. Science 332:231–234. https://doi.org/10.1126/science.1199412

Flavell AJ (1995) Retroelements, reverse transcriptase and evolution. Comp Biochem Physiol Part B Biochem Mol Biol 110:3–15. https://doi.org/10.1016/0305-0491(94)00122-B

Flavell AJ, Dunbar E, Anderson R et al (1992) Ty1-copia group retrotransposons are ubiquitous and heterogeneous in higher plants. Nucleic Acids Res 20:3639–3644. https://doi.org/10.1093/nar/20.14.3639

Flavell AJ, Knox MR, Pearce SR, Ellis THN (1998) Retrotransposon-based insertion polymorphisms (RBIP) for high throughput marker analysis. Plant J 16:643–650. https://doi.org/10.1046/j.1365-313X.1998.00334.x

Flavell AJ, Pearce SR, Kumar A (1994) Plant transposable elements and the genome. Curr Opin Genet Dev 4:838–844. https://doi.org/10.1016/0959-437X(94)90068-X

Flutre T, Permal E, Quesneville H (2012) Transposable element annotation in completely sequenced eukaryote genomes. In: Grandbastien MA, Casacuberta J (eds) Plant transposable elements. Springer, Heidelberg, pp 17–39

Fray RG, Grierson D (1993) Identification and genetic analysis of normal and mutant phytoene synthase genes of tomato by sequencing, complementation and co-suppression. Plant Mol Biol 22:589–602. https://doi.org/10.1007/BF00047400

Friesen N, Brandes A, Heslop-Harrison JS (2001) Diversity, origin, and distribution of retrotransposons (gypsy and copia) in conifers. Mol Biol Evol 18:1176–1188. https://doi.org/10.1093/oxfordjournals.molbev.a003905

Galdino JH, Eguiluz M, Guzman F, Margis R (2019) Novel and conserved miRNAs among Brazilian pine and other gymnosperms. Front Genet 10:222. https://doi.org/10.3389/fgene.2019.00222

Galindo-González L, Mhiri C, Deyholos MK, Grandbastien MA (2017) LTR-retrotransposons in plants: engines of evolution. Gene 626:14–25. https://doi.org/10.1016/j.gene.2017.04.051

Gao C, Xiao M, Jiang L et al (2012) Characterization of transcriptional activation and inserted-into-gene preference of various transposable elements in the Brassica species. Mol Biol Rep 39:7513–7523. https://doi.org/10.1007/s11033-012-1585-0

Gao D, Jiang N, Wing RA et al (2015) Transposons play an important role in the evolution and diversification of centromeres among closely related species. Front Plant Sci 6:216. https://doi.org/10.3389/fpls.2015.00216

Gao D, Jimenez-Lopez JC, Iwata A et al (2012) Functional and structural divergence of an unusual LTR retrotransposon family in plants. PLoS ONE 7:1–12. https://doi.org/10.1371/journal.pone.0048595

Gao X, Hou Y, Ebina H et al (2008) Chromodomains direct integration of retrotransposons to heterochromatin. Genome Res 18:359–369. https://doi.org/10.1101/gr.7146408

Gelfman S, Ast G (2013) When epigenetics meets alternative splicing: the roles of DNA methylation and GC architecture. Epigenomics 5:351–353. https://doi.org/10.2217/epi.13.32

Gernandt DS, Geada López G, Ortiz García S, Liston A (2005) Phylogeny and classification of *Pinus*. Taxon 54:29–42. https://doi.org/10.2307/25065300

Godinho S, Paulo OS, Morais-CECÍLIO L, Rocheta M (2012) A new gypsy-like retroelement family in vitis vinifera. Vitis J Grapevine Res 51:65–72. https://doi.org/10.1186/1471-2164-9-469

Goerner-Potvin P, Bourque G (2018) Computational tools to unmask transposable elements. Nat Rev Genet 19:688–704. https://doi.org/10.1038/s41576-018-0050-x

González-Martínez SC, Alía R, Gil L (2002) Population genetic structure in a Mediterranean pine (*Pinus pinaster* Ait.): a comparison of allozyme markers and quantitative traits. Heredity (edinb) 89:199–206. https://doi.org/10.1038/sj.hdy.6800114

Grandbastien M-A, Lucas H, Morel J-B et al (1997) The expression of the tobacco Tnt1 retrotransposon is linked to plant defense responses. Genetica 100:241–252. https://doi.org/10.1023/A:1018302216927

Grandbastien MA (2008) Retrotransposons of plants. In: Mahy BWJ, Marc HV, Van Regenmortel (ed) Encyclopedia of virology. Elsevier, pp 428–436

Grandbastien MA (2015) LTR retrotransposons, handy hitchhikers of plant regulation and stress response. Biochim Biophys Acta Gene Regul Mech 1849:403–416. https://doi.org/10.1016/j.bbagrm.2014.07.017

Grandbastien MA, Casacuberta JM (2012) Plant transposable elements. Springer, Berlin

Groth A, Rocha W, Verreault A, Almouzni G (2007) Chromatin challenges during DNA replication and repair. Cell 128:721–733

Hadjiargyrou M, Delihas N (2013) The intertwining of transposable elements and non-coding RNAs. Int J Mol Sci 14:13307–13328. https://doi.org/10.3390/ijms140713307

Hall D, Olsson J, Zhao W et al (2021) Divergent patterns between phenotypic and genetic variation in Scots pine. Plant Commun 2:100139. https://doi.org/10.1016/j.xplc.2020.100139

Halldén C, Hansen M, Nilsson NO et al (1996) Competition as a source of errors in RAPD analysis. Theor Appl Genet 93:1185–1192. https://doi.org/10.1007/BF00223449

Hartl DL, Lozovskaya ER, Lawrence JG (1992) Nonautonomous transposable elements in prokaryotes and eukaryotes. Genetica 86:47–53. https://doi.org/10.1007/BF00133710

Havecker et al., 2004.Havecker ER, Gao X, Voytas DF (2004) The diversity of LTR retrotransposons. Genome Biol 5. https://doi.org/10.1186/gb-2004-5-6-225

Hawkins JS, Kim HR, Nason JD et al (2006) Differential lineage-specific amplification of transposable elements is responsible for genome size variation in Gossypium. Genome Res 16:1252–1261. https://doi.org/10.1101/gr.5282906

Heuertz M, De Paoli E, Källman T et al (2006) Multilocus patterns of nucleotide diversity, linkage disequilibrium and demographic history of Norway spruce [*Piceaabies* (L.) Karst]. Genetics 174:2095–2105. https://doi.org/10.1534/genetics.106.065102

Hwang I, Sheen J, Müller B (2012) Cytokinin signaling networks. Annu Rev Plant Biol 63:353–380. https://doi.org/10.1146/annurev-arplant-042811-105503

Jedlicka P, Lexa M, Kejnovsky E (2020) What can long terminal repeats tell us about the age of LTR retrotransposons, gene conversion and ectopic recombination? Front Plant Sci 11:644. https://doi.org/10.3389/fpls.2020.00644

Jedlicka P, Lexa M, Vanat I et al (2019) Nested plant LTR retrotransposons target specific regions of other elements, while all LTR retrotransposons often target palindromes and nucleosome-occupied regions: in silico study. Mob DNA 10:1–14. https://doi.org/10.1186/s13100-019-0186-z

Jiang J, Nasuda S, Dong F et al (1996) A conserved repetitive DNA element located in the centromeres of cereal chromosomes. Proc Natl Acad Sci U S A 93:14210–14213. https://doi.org/10.1073/pnas.93.24.14210

Jiang N, Bao Z, Zhang X et al (2003) An active DNA transposon family in rice. Nature 421:163–167. https://doi.org/10.1038/nature01214

Jurka J (2008) Conserved eukaryotic transposable elements and the evolution of gene regulation. Cell Mol Life Sci 65:201–204

Kalendar R, Amenov A, Daniyarov A (2019) Use of retrotransposon-derived genetic markers to analyse genomic variability in plants. Funct Plant Biol 46:15–29

Kalendar R, Antonius K, Smýkal P, Schulman AH (2010) iPBS: a universal method for DNA fingerprinting and retrotransposon isolation. Theor Appl Genet 121:1419–1430. https://doi.org/10.1007/s00122-010-1398-2

Kalendar R, Bachmair A, Vicient CM, Casacuberta JM (2020) Additional ORFs in plant LTR-retrotransposons. Addit ORFs Plant LTR-Retrotransposons Front Plant Sci 11:555. https://doi.org/10.3389/fpls.2020.00555

Kalendar R, Flavell AJ, Ellis THN et al (2011) Analysis of plant diversity with retrotransposon-based molecular markers. Heredity (edinb) 106:520–530. https://doi.org/10.1038/hdy.2010.93

Kalendar R, Schulman AH (2006) IRAP and REMAP for retrotransposon-based genotyping and fingerprinting. Nat Protoc 1:2478–2484. https://doi.org/10.1038/nprot.2006.377

Kalendar R, Schulman AH (2014) Transposon-based tagging: IRAP, REMAP, and iPBS. Methods Mol Biol 1115:233–255. https://doi.org/10.1007/978-1-62703-767-9_12

Kalendar R, Tanskanen J, Immonen S et al (2000) Genome evolution of wild barley (Hordeum spontaneum) by BARE-1 retrotransposon dynamics in response to sharp microclimatic divergence. Proc Natl Acad Sci 97:6603–6607. https://doi.org/10.1073/pnas.110587497

Kamm A, Doudrick RL, Heslop-Harrison JS, Schmidt T (1996) The genomic and physical organization of Ty1-copia-like sequences as a component of large genomes in *Pinus elliottii* var. *elliottii* and other gymnosperms. Proc Natl Acad Sci U S A 93:2708–2713. https://doi.org/10.1073/pnas.93.7.2708

Kapitonov VV, Jurka J (2001) Rolling-circle transposons in eukaryotes. Proc Natl Acad Sci U S A 98:8714–8719. https://doi.org/10.1073/pnas.151269298

Kapitonov VV, Jurka J (2008) A universal classification of eukaryotic transposable elements implemented in Repbase. Nat Rev Genet 9:411–412

Kapitonov VV, Jurka J (2007) Helitrons on a roll: eukaryotic rolling-circle transposons. Trends Genet 23:521–529

Kapitonov VV, Jurka J (2006) Self-synthesizing DNA transposons in eukaryotes. Proc Natl Acad Sci U S A 103:4540–4545. https://doi.org/10.1073/pnas.0600833103

Kashkush K, Feldman M, Levy AA (2003) Transcriptional activation of retrotransposons alters the expression of adjacent genes in wheat. Nat Genet 33:102–106. https://doi.org/10.1038/ng1063

Kato M, Miura A, Bender J et al (2003) Role of CG and non-CG methylation in immobilization of transposons in Arabidopsis. Curr Biol 13:421–426. https://doi.org/10.1016/S0960-9822(03)00106-4

Kazazian HH (2004) Mobile elements: drivers of genome evolution. Science 303:1626–1632. https://doi.org/10.1126/science.1089670

Kejnovský E, Hawkins J, Feschotte C (2012) Plant transposable elements : biology and evolution 2

Klaver B, Berkhout B (1994) Comparison of 5' and 3' long terminal repeat promoter function in human immunodeficiency virus. J Virol 68:3830–3840

Knight CA, Molinari NA, Petrov DA (2005) The large genome constraint hypothesis: evolution, ecology and phenotype. Ann Bot 95:177–190. https://doi.org/10.1093/aob/mci011

Kobayashi S, Goto-Yamamoto N, Hirochika H (2004) Retrotransposon-induced mutations in grape skin color. Science 304:982. https://doi.org/10.1126/science.1095011

Komulainen P, Brown GR, Mikkonen M et al (2003) Comparing EST-based genetic maps between *Pinus sylvestris* and *Pinus taeda*. Theor Appl Genet 107:667–678. https://doi.org/10.1007/s00122-003-1312-2

Kordiš D (2005) A genomic perspective on the chromodomain-containing retrotransposons: chromoviruses. Gene 347:161–173

Kossack DS, Kinlaw CS (1999) IFG, a gypsy-like retrotransposon in *Pinus* (Pinaceae), has an extensive history in pines. Plant Mol Biol 39:417–426. https://doi.org/10.1023/A:1006115732620

Kovach A, Wegrzyn J, Parra G et al (2010) The *Pinus taeda* genome is characterized by diverse and highly diverged repetitive sequences. BMC Genomics 11:420. https://doi.org/10.1186/1471-2164-11-420

Krivmane B, Šņepste I, Šķipars V et al (2020) Identification and in silico characterization of novel and conserved microRNAs in methyl jasmonate-stimulated scots pine (*Pinus sylvestris* L.) needles. Forests 11:384. https://doi.org/10.3390/f11040384

Krom N, Recla J, Ramakrishna W (2008) Analysis of genes associated with retrotransposons in the rice genome. Genetica 134:297–310. https://doi.org/10.1007/s10709-007-9237-3

Kronmiller BA, Wise RP (2008) TEnest: automated chronological annotation and visualization of nested plant transposable elements. Plant Physiol 146:45–59. https://doi.org/10.1104/pp.107.110353

Krupovic M, Koonin EV (2016) Self-synthesizing transposons: unexpected key players in the evolution of viruses and defense systems. Curr Opin Microbiol 31:25–33

Kumar A, Bennetzen JL (1999) Plant retrotransposons. Annu Rev Genet 33:479–532. https://doi.org/10.1146/annurev.genet.33.1.479

Kumekawa N, Ohtsubo H, Horiuchi T, Ohtsubo E (1999) Identification and characterization of novel retrotransposons of the gypsy type in rice. Mol Gen Genet 260:593–602. https://doi.org/10.1007/s004380050933

Lai Y, Cuzick A, Lu XM et al (2019) The arabidopsis RRM domain protein EDM 3 mediates race-specific disease resistance by controlling H3K9me2-dependent alternative polyadenylation of *RPP 7* immune receptor transcripts. Plant J 97:646–660. https://doi.org/10.1111/tpj.14148

Lampe DJ, Churchill MEA, Robertson HM (1996) A purified mariner transposase is sufficient to mediate transposition in vitro. EMBO J 15:5470–5479. https://doi.org/10.1002/j.1460-2075.1996.tb00930.x

Lockton S, Gaut BS (2010) The evolution of transposable elements in natural populations of self-fertilizing Arabidopsis thaliana and its outcrossing relativeArabidopsis lyrata. BMC Evol Biol 10:24–28. https://doi.org/10.1186/1471-2148-10-10

Le Rouzic A, Boutin TS, Capy P (2007) Long-term evolution of transposable elements. Proc Natl Acad Sci 104:19375–19380. https://doi.org/10.1073/pnas.0705238104

Lee H, Zhang Z, Krause HM (2019) Long noncoding RNAs and repetitive elements: junk or intimate evolutionary partners? Trends Genet 35:892–902. https://doi.org/10.1016/j.tig.2019.09.006

Leigh F, Kalendar R, Lea V et al (2003) Comparison of the utility of barley retrotransposon families for genetic analysis by molecular marker techniques. Mol Genet Genomics 269:464–474. https://doi.org/10.1007/s00438-003-0850-2

Levin HL, Moran JV (2011) Dynamic interactions between transposable elements and their hosts. Nat Rev Genet 12:615–627. https://doi.org/10.1038/nrg3030

Li W, Zhang P, Fellers JP et al (2004) Sequence composition, organization, and evolution of the core Triticeae genome. Plant J 40:500–511. https://doi.org/10.1111/j.1365-313X.2004.02228.x

Li Y, Li C, Xia J, Jin Y (2011) Domestication of transposable elements into MicroRNA genes in plants. PLoS ONE 6:e19212. https://doi.org/10.1371/journal.pone.0019212

Li Z, Baniaga AE, Sessa EB et al (2015) Early genome duplications in conifers and other seed plants. Sci Adv 1:1–8. https://doi.org/10.1126/sciadv.1501084

Lippman Z, Gendrel A-V, Black M et al (2004) Role of transposable elements in heterochromatin and epigenetic control. Nature 430:471–476. https://doi.org/10.1038/nature02651

Lippman Z, May B, Yordan C et al (2003) Distinct mechanisms determine transposon inheritance and methylation via small interfering RNA and histone modification. PLoS Biol 1:e67. https://doi.org/10.1371/journal.pbio.0000067

Liu B, Shan XH, Liu ZL et al (2005) Mobilization of the active MITE transposons mPing and Pong in rice by introgression from wild rice (Zizania latifolia Griseb.). Mol Biol Evol 22:976–990

Liu JJ, Sturrock RN, Sniezko RA et al (2015) Transcriptome analysis of the white pine blister rust pathogen Cronartium ribicola: De novo assembly, expression profiling, and identification of candidate effectors. BMC Genomics 16:1–16. https://doi.org/10.1186/s12864-015-1861-1

Liu M, Yu H, Zhao G et al (2017) Profiling of drought-responsive microRNA and mRNA in tomato using high-throughput sequencing. BMC Genomics 18:1–18. https://doi.org/10.1186/s12864-017-3869-1

Liu Y, Yin J, Xiao M et al (2013) Characterization of structure, divergence and regulation patterns of plant promoters. J Mol Biol Res 3:23–36. https://doi.org/10.5539/jmbr.v3n1p23

Llorens C, Futami R, Covelli L et al (2011) The gypsy database (GyDB) of mobile genetic elements: release 2.0. Nucleic Acids Res 39. https://doi.org/10.1093/nar/gkq1061

Loya-Rebollar E, Sáenz-Romero C, Lindig-Cisneros RA et al (2013) Clinal variation in *Pinus* hartwegii populations and its application for adaptation to climate change. Silvae Genet 62:86–95. https://doi.org/10.1515/sg-2013-0011

Lu M, Krutovsky K V., Nelson CD, et al (2016) Exome genotyping, linkage disequilibrium and population structure in loblolly pine (*Pinus taeda* L.). BMC Genomics 17:730. https://doi.org/10.1186/s12864-016-3081-8

Lu M, Loopstra CA, Krutovsky KV (2019) Detecting the genetic basis of local adaptation in loblolly pine (*Pinus taeda* L.) using whole exome-wide genotyping and an integrative landscape genomics analysis approach. Ecol Evol 9:6798–6809. https://doi.org/10.1002/ece3.5225

Lu Y, Ran JH, Guo DM et al (2014) Phylogeny and divergence times of gymnosperms inferred from single-copy nuclear genes. PLoS ONE 9:e107679. https://doi.org/10.1371/journal.pone.0107679

Ma J, Devos KM, Bennetzen JL (2004) Analyses of LTR-retrotransposon structures reveal recent and rapid genomic DNA loss in rice. Genome Res 14:860–869. https://doi.org/10.1101/gr.1466204

Macas J, Neumann P (2007) Ogre elements—a distinct group of plant Ty3/gypsy-like retrotransposons. Gene 390:108–116. https://doi.org/10.1016/j.gene.2006.08.007

Madlung A, Comai L (2004) The effect of stress on genome regulation and structure. Ann Bot 94:481–495. https://doi.org/10.1093/aob/mch172

Magbanua Z V., Ozkan S, Bartlett BD et al (2011) Adventures in the enormous: a 1.8 million clone BAC library for the 21.7 Gb genome of loblolly pine. PLoS One 6. https://doi.org/10.1371/journal.pone.0016214

Mak J, Kleiman L (1997) Primer tRNAs for reverse transcription. J Virol 71:8087–8095

Martínez-Izquierdo JA, García-Martínez J, Vicient CM (1997) What makes Grande1 retrotransposon different? Genetica 100:15–28. https://doi.org/10.1023/a:1018332218319

Martinez G (2017) To cite this article: German Martinez (2017) tRNAs as primers and inhibitors of retrotransposons. Mob Genet Elements 7:1–6. https://doi.org/10.1080/2159256X.2017.1393490

Martínez G, Slotkin RK (2012) Developmental relaxation of transposable element silencing in plants: functional or byproduct? Curr Opin Plant Biol 15:496–502. https://doi.org/10.1016/j.pbi.2012.09.001

McClintock B (1950) The origin and behavior of mutable loci in maize. Proc Natl Acad Sci U S A 36:344–355. https://doi.org/10.1073/pnas.36.6.344

McClintock B (1984) The significance of responses of the genome to challenge. Science 226:792–801. https://doi.org/10.1126/science.15739260

Mendiola MV, Bernales I, de la Cruz F (1994) Differential roles of the transposon termini in IS91 transposition. Proc Natl Acad Sci U S A 91:1922–1926. https://doi.org/10.1073/pnas.91.5.1922

Miguel C, Simões M, Oliveira MM, Rocheta M (2008) Envelope-like retrotransposons in the plant kingdom: evidence of their presence in gymnosperms (*Pinus pinaster*). J Mol Evol 67:517–525. https://doi.org/10.1007/s00239-008-9168-3

Miksche JP, Hotta Y (1973) DNA Base composition and repetitious DNA in several conifers. Chromosoma 41:29–36. https://doi.org/10.1007/BF00284072

Millar CI (1998) Early evolution of pines. In: Richardson DM (ed) Ecology and biogeography of *Pinus*. Cambridge University Press, pp 69–91

Miller JT, Dong F, Jackson SA et al (1998) Retrotransposon-related DNA sequences in the centromeres of grass chromosomes. Genetics 150:1615–1623

Mita P, Boeke JD (2016) How retrotransposons shape genome regulation. Curr Opin Genet Dev 37:90–100. https://doi.org/10.1016/j.gde.2016.01.001

Mizuno M, Kanehisa M (1994) Distribution profiles of GC content around the translation initiation site in different species. FEBS Lett 352:7–10. https://doi.org/10.1016/0014-5793(94)00898-1

Mlura A, Yonebayashi S, Watanabe K et al (2001) Mobilization of transposons by a mutation abolishing full DNA methylation in Arabidopsis. Nature 411:212–214. https://doi.org/10.1038/35075612

Mok MC (1994) Cytokinin and plant development—an overview. In: Mok, DWS, Mok MC (ed) Cytokinins chemistry, activity, and function. CRC Press, Ann Arbor, Michigan, pp 155–166

Monden Y, Tahara M (2015) Plant transposable elements and their application to genetic analysis via high-throughput sequencing platform. Hortic J 84:283–294. https://doi.org/10.2503/hortj.MI-IR02

Morgan MT (2001) Consequences of life history for inbreeding depression and mating system evolution in plants. Proc R Soc B Biol Sci 268:1817–1824.https://doi.org/10.1098/rspb.2001.1741

Morgante M, De Paoli E, Radovic S (2007) Transposable elements and the plant pan-genomes. Curr Opin Plant Biol 10:149–155. https://doi.org/10.1016/j.pbi.2007.02.001

Moritsuka E, Hisataka Y, Tamura M et al (2012) Extended linkage disequilibrium in noncoding regions in a conifer, cryptomeria japonica. Genetics 190:1145–1148. https://doi.org/10.1534/genetics.111.136697

Morse AM, Peterson DG, Islam-Faridi MN et al (2009) Evolution of genome size and complexity in *Pinus*. PLoS ONE 4:1–11. https://doi.org/10.1371/journal.pone.0004332

Muotri AR, Chu VT, Marchetto MCN et al (2005) Somatic mosaicism in neuronal precursor cells mediated by L1 retrotransposition. Nature 435:903–910. https://doi.org/10.1038/nature03663

Neale DB, Kinlaw C (1997) Complex gene families in pine genomes. Trends Plant Sci 2:32–35

Neale DB, Savolainen O (2004) Association genetics of complex traits in conifers. Trends Plant Sci 9:325–330. https://doi.org/10.1016/j.tplants.2004.05.006

Neale DB, Wegrzyn JL, Stevens KA et al (2014) Decoding the massive genome of loblolly pine using haploid DNA and novel assembly strategies. Genome Biol 15:1–13. https://doi.org/10.1186/gb-2014-15-3-r59

Nelson MG, Linheiro RS, Bergman CM (2017) McClintock: an integrated pipeline for detecting transposable element insertions in whole-genome shotgun sequencing data. G3 Genes Genomes Genet 7:2763–2778. https://doi.org/10.1534/g3.117.043893

Neumann P, Navrátilová A, Koblížková A et al (2011) Plant centromeric retrotransposons: a structural and cytogenetic perspective. Mob DNA 2:1–16. https://doi.org/10.1186/1759-8753-2-4

Neumann P, Novák P, Hoštáková N, MacAs J (2019) Systematic survey of plant LTR-retrotransposons elucidates phylogenetic relationships of their polyprotein domains and provides a reference for element classification. Mob DNA 10:1

Nguyen N, Hickey G, Zerbino DR et al (2015) Building a pan-genome reference for a population. J Comput Biol 22:387–401. https://doi.org/10.1089/cmb.2014.0146

Novák P, Neumann P, Pech J et al (2013) RepeatExplorer: a Galaxy-based web server for genome-wide characterization of eukaryotic repetitive elements from next-generation sequence reads. Bioinformatics 29:792–793. https://doi.org/10.1093/bioinformatics/btt054

Nystedt B, Street NR, Wetterbom A et al (2013) The Norway spruce genome sequence and conifer genome evolution. Nature 497:579

Orgel LE, Crick FHC (1980) Selfish DNA: the ultimate parasite. Nature 284:604–607. https://doi.org/10.1038/284604a0

Palmé AE, Pyhäjärvi T, Wachowiak W, Savolainen O (2009) Selection on nuclear genes in a *Pinus* phylogeny. Mol Biol Evol 26:893–905. https://doi.org/10.1093/molbev/msp010

Panaud O (2016) Horizontal transfers of transposable elements in eukaryotes: the flying genes. C R Biol 339:296–299. https://doi.org/10.1016/j.crvi.2016.04.013

Parks M, Cronn R, Liston A (2012) Separating the wheat from the chaff: mitigating the effects of noise in a plastome phylogenomic data set from *Pinus* L. (Pinaceae). BMC Evol Biol 12:1. https://doi.org/10.1186/1471-2148-12-100

Parks M, Cronn R, Liston A (2009) Increasing phylogenetic resolution at low taxonomic levels using massively parallel sequencing of chloroplast genomes. BMC Biol 7:1–17. https://doi.org/10.1186/1741-7007-7-84

Peaston AE, Evsikov AV, Graber JH et al (2004) Retrotransposons regulate host genes in mouse oocytes and preimplantation embryos human genomes (International Human Genome Sequencing Consortium, 2001; Mouse Genome Sequencing Consortium, 2002). Almost all mammalian TEs are retro. Dev Cell 7:597–606

Pereira V (2004) Insertion bias and purifying selection of retrotransposons in the *Arabidopsis thaliana* genome. Genome Biol 5:R79. https://doi.org/10.1186/gb-2004-5-10-r79

Peterson DG, Schulze SR, Sciara EB et al (2002) Integration of cot analysis, DNA cloning, and high-throughput sequencing facilitates genome characterization and gene discovery. Genome Res 12:795–807. https://doi.org/10.1101/gr.226102

Piégu B, Bire S, Arensburger P, Bigot Y (2015) A survey of transposable element classification systems—a call for a fundamental update to meet the challenge of their diversity and complexity. Mol Phylogenet Evol 86:90–109

Piriyapongsa J, Jordan IK (2008) Dual coding of siRNAs and miRNAs by plant transposable elements. RNA 14:814–821. https://doi.org/10.1261/rna.916708

Poretti M, Praz CR, Meile L et al (2020) Domestication of high-copy transposons underlays the wheat small RNA response to an obligate pathogen. Mol Biol Evol 37:839–848. https://doi.org/10.1093/molbev/msz272

Purugganan MD, Wessler SR (1995) Transposon signatures: species-specific molecular markers that utilize a class of multiple-copy nuclear DNA. Mol Ecol 4:265–270. https://doi.org/10.1111/j.1365-294X.1995.tb00218.x

Pyhäjärvi T, García-Gil MR, Knürr T et al (2007) Demographic history has influenced nucleotide diversity in European Pinus sylvestris populations. Genetics177:1713–1724. https://doi.org/10.1534/genetics.107.077099

Qin S, Jin P, Zhou X et al (2015) The role of transposable elements in the origin and evolution of microRNAs in human. PLoS ONE 10:e0131365. https://doi.org/10.1371/journal.pone.0131365

Rake AV, Miksche JP, Hall RB, Hansen KM (1980) DNA Reassociation kinetics of four conifers. Can J Genet Cytol 22:69–79. https://doi.org/10.1139/g80-010

Rebollo R, Romanish MT, Mager DL (2012) Transposable elements: an abundant and natural source of regulatory sequences for host genes. Annu Rev Genet 46:21–42. https://doi.org/10.1146/annurev-genet-110711-155621

Rho M, Choi J-HH, Kim S et al (2007) De novo identification of LTR retrotransposons in eukaryotic genomes. BMC Genomics 8:1–16. https://doi.org/10.1186/1471-2164-8-90

Rocheta M, Cordeiro J, Oliveira M, Miguel C (2007) PpRT1: the first complete gypsy-like retrotransposon isolated in *Pinus pinaster*. Planta 225:551–562. https://doi.org/10.1007/s00425-006-0370-5

Sabot F, Schulman AH (2007) Template switching can create complex LTR retrotransposon insertions in Triticeae genomes. BMC Genomics 8:5–9. https://doi.org/10.1186/1471-2164-8-247

Sabot F, Sourdille P, Bernard M (2005) Advent of a new retrotransposon structure: the long form of the Veju elements. Genetica 125:325–332. https://doi.org/10.1007/s10709-005-7926-3

Saéz-Laguna E, Guevara MÁ, Diáz LM et al (2014) Epigenetic variability in the genetically uniform forest tree species *Pinus pinea* L. PLoS ONE 9:e103145. https://doi.org/10.1371/journal.pone.0103145

Sakai H, Honma T, Takashi A et al (2001) ARR1, a transcription factor for genes immediately responsive to cytokinins. Science 294:1519–1521. https://doi.org/10.1126/science.1065201

Sampath P, Yang T-J (2014) Miniature inverted-repeat transposable elements (MITEs) as valuable genomic resources for the evolution and breeding of brassica crops. Plant Breed Biotechnol 2:322–333. https://doi.org/10.9787/pbb.2014.2.4.322

SanMiguel P, Gaut BS, Tikhonov A et al (1998) The paleontology of intergene retrotransposons of maize. Nat Genet 20:43–45. https://doi.org/10.1038/1695

SanMiguel P, Tikhonov A, Jin Y-K et al (1996) Nested retrotransposons in the intergenic regions of the maize genome. Science 274:765–768. https://doi.org/10.1126/science.274.5288.765

Savolainen O, Pyhäjärvi T (2007) Genomic diversity in forest trees. Curr Opin Plant Biol 10:162–167. https://doi.org/10.1016/j.pbi.2007.01.011

Schorn AJ, Martienssen R (2018) Tie-break: host and retrotransposons play tRNA. Trends Cell Biol 28:793–806

Schulman AH (2007) Molecular markers to assess genetic diversity. Euphytica 158:313–321. https://doi.org/10.1007/s10681-006-9282-5

Schulman AH, Flavell AJ, Ellis THN (2004) The application of LTR retrotransposons as molecular markers in plants. Methods Mol Biol 260:145–173. https://doi.org/10.1385/1-59259-755-6:145

Schwartz S, Meshorer E, Ast G (2009) Chromatin organization marks exon-intron structure. Nat Struct Mol Biol 16:990–995. https://doi.org/10.1038/nsmb.1659

Shirasu K, Schulman AH, Lahaye T, Schulze-Lefert P (2000) A contiguous 66-kb barley DNA sequence provides evidence for reversible genome expansion. Genome Res 10:908–915. https://doi.org/10.1101/gr.10.7.908

Simberloff D, Leppanen C (2019) Plant somatic mutations in nature conferring insect and herbicide resistance. Pest Manag Sci 75:14–17. https://doi.org/10.1002/ps.5157

Singer T, McConnell MJ, Marchetto MCN et al (2010) LINE-1 retrotransposons: mediators of somatic variation in neuronal genomes? Trends Neurosci 33:345–354. https://doi.org/10.1016/j.tins.2010.04.001

Singh R, Ming R, Yu Q (2016) Comparative analysis of GC content variations in plant genomes. Trop Plant Biol 9:136–149. https://doi.org/10.1007/s12042-016-9165-4

Slotkin RK, Martienssen R (2007) Transposable elements and the epigenetic regulation of the genome. Nat Rev Genet 8:272–285. https://doi.org/10.1038/nrg2072

Stevens KA, Wegrzyn JL, Zimin A et al (2016) Sequence of the sugar pine megagenome. Genetics 204:1613–1626. https://doi.org/10.1534/genetics.116.193227

Studer A, Zhao Q, Ross-Ibarra J, Doebley J (2011) Identification of a functional transposon insertion in the maize domestication gene tb1. Nat Genet 43:1160–1163. https://doi.org/10.1038/ng.942

Sun G (2012) MicroRNAs and their diverse functions in plants. Plant Mol Biol 80:17–36. https://doi.org/10.1007/s11103-011-9817-6

Takeda S, Sugimoto K, Otsuki H, Hirochika H (1999) A 13-bp cis-regulatory element in the LTR promoter of the tobacco retrotransposon Tto1 is involved in responsiveness to tissue culture, wounding, methyl jasmonate and fungal elicitors. Plant J 18:383–393. https://doi.org/10.1046/j.1365-313X.1999.00460.x

Tam SM, Causse M, Garchery C, et al (2007) The distribution of copia-type retrotransposons and the evolutionary history of tomato and related wild species. J EvolBiol 20:1056–1072. https://doi.org/10.1111/j.1420-9101.2007.01293.x

Taniguchi M, Sasaki N, Tsuge T et al (2007) ARR1 directly activates cytokinin response genes that encode proteins with diverse regulatory functions. Plant Cell Physiol 48:263–277. https://doi.org/10.1093/pcp/pcl063

Tapia G, Verdugo I, Yañez M et al (2005) Involvement of ethylene in stress-induced expression of the TLC1.1 retrotransposon from Lycopersicon chilense Dun. Plant Physiol 138:2075–2086. https://doi.org/10.1104/pp.105.059766

Tettelin H, Masignani V, Cieslewicz MJ et al (2005) Genome analysis of multiple pathogenic isolates of Streptococcus agalactiae: implications for the microbial "pan-genome." Proc Natl Acad Sci U S A 102:13950–13955. https://doi.org/10.1073/pnas.0506758102

Thomas J, Pritham EJ (2015) Helitrons, the eukaryotic rolling-circle transposable elements. In: Mobile DNA III. American Society of Microbiology, pp 893–926

Tsuchiya T, Eulgem T (2013) An alternative polyadenylation mechanism coopted to the Arabidopsis RPP7 gene through intronic retrotransposon domestication.

Proc Natl Acad Sci 110:E3535–E3543. https://doi.org/10.1073/pnas.1312545110

Ullah F, Hamilton M, Reddy ASN, Ben-Hur A (2018) Exploring the relationship between intron retention and chromatin accessibility in plants. BMC Genomics 19:21. https://doi.org/10.1186/s12864-017-4393-z

Varagona MJ, Purugganan M, Wessler SR (1992) Alternative splicing induced by insertion of retrotransposons into the maize waxy gene. Plant Cell 4:811–820. https://doi.org/10.1105/tpc.4.7.811

Venkatesh, Nandini B (2020) Miniature inverted-repeat transposable elements (MITEs), derived insertional polymorphism as a tool of marker systems for molecular plant breeding. Mol Biol Rep 47

Vicient CM (1999) Retrotransposon BARE-1 and its role in genome evolution in the genus hordeum. Plant Cell Online 11:1769–1784. https://doi.org/10.1105/tpc.11.9.1769

Vicient CM, Kalendar R, Schulman AH (2005) Variability, recombination, and mosaic evolution of the barley BARE-1 retrotransposon. J Mol Evol 61:275–291. https://doi.org/10.1007/s00239-004-0168-7

Vitte C, Bennetzen JL (2006) Analysis of retrotransposon structural diversity uncovers properties and propensities in angiosperm genome evolution. Proc Natl Acad Sci 103:17638–17643. https://doi.org/10.1073/pnas.0605618103

Vitte C, Panaud O (2003) Formation of solo-LTRs through unequal homologous recombination counterbalances amplifications of LTR retrotransposons in rice Oryza sativa L. Mol Biol Evol 20:528–540. https://doi.org/10.1093/molbev/msg055

Voronova A (2019) Retrotransposon expression in response to in vitro inoculation with two fungal pathogens of Scots pine (*Pinus sylvestris* L.). BMC Res Notes 12:243. https://doi.org/10.1186/s13104-019-4275-3

Voronova A, Belevich V, Jansons A, Rungis D (2014) Stress-induced transcriptional activation of retrotransposon-like sequences in the Scots pine (*Pinus sylvestris* L.) genome. Tree Genet Genomes 10:937–951. https://doi.org/10.1007/s11295-014-0733-1

Voronova A, Belevich V, Korica A, Rungis D (2017) Retrotransposon distribution and copy number variation in gymnosperm genomes. Tree Genet Genomes 13:1–23. https://doi.org/10.1007/s11295-017-1165-5

Voronova A, Jansons Ā, Ruņģis D (2011) Expression of retrotransposon-like sequences in Scots pine (*Pinus sylvestris*) in response to heat stress. Exp Biol 9:121–127

Voronova A, Rendón-Anaya M, Ingvarsson P et al (2020) Comparative study of pine reference genomes reveals transposable element interconnected gene networks. Genes (basel) 11:1216. https://doi.org/10.3390/genes11101216

Voronova A, Rungis D (2013) Development and characterisation of irap markers from expressed retrotransposon-like sequences in *Pinus sylvestris* L. Proc Latv Acad Sci Sect B Nat Exact, Appl Sci 67:485. https://doi.org/10.2478/prolas-2013-0082

Voytas DF, Boeke JD (1993) Yeast retrotransposons and tRNAs. Trends Genet 9:421–427. https://doi.org/10.1016/0168-9525(93)90105-Q

Voytas DF, Cummings MP, Koniczny A et al (1992) Copia-like retrotransposons are ubiquitous among plants. Proc Natl Acad Sci 89:7124–7128. https://doi.org/10.1073/pnas.89.15.7124

Wachowiak W, Balk PA, Savolainen O (2009) Search for nucleotide diversity patterns of local adaptation in dehydrins and other cold-related candidate genes in Scots pine (*Pinus sylvestris* L.). Tree Genet Genomes 5:117–132. https://doi.org/10.1007/s11295-008-0188-3

Wang D, Qu Z, Yang L et al (2017) Transposable elements (TEs) contribute to stress-related long intergenic noncoding RNAs in plants. Plant J 90:133–146. https://doi.org/10.1111/tpj.13481

Wang Y, Tang X, Cheng Z et al (2006) Euchromatin and pericentromeric heterochromatin: comparative composition in the tomato genome. Genetics 172:2529–2540. https://doi.org/10.1534/genetics.106.055772

Wang Z, Gerstein M, Snyder M (2009) RNA-Seq: a revolutionary tool for transcriptomics. Nat Rev Genet 10:57–63. https://doi.org/10.1038/nrg2484

Waugh R, McLean K, Flavell AJ et al (1997) Genetic distribution of Bare-1-like retrotransposable elements in the barley genome revealed by sequence-specific amplification polymorphisms (S-SAP). Mol Gen Genet 253:687–694. https://doi.org/10.1007/s004380050372

Wegrzyn JL, Liechty JD, Stevens KA et al (2014) Unique features of the loblolly pine (*Pinus taeda* L.) megagenome revealed through sequence annotation-term breeding programs and large-scale. https://doi.org/10.1534/genetics.113.159996

Wegrzyn JL, Lin BY, Zieve JJ et al (2013) Insights into the loblolly pine genome: characterization of BAC and fosmid sequences. PLoS ONE 8:e72439. https://doi.org/10.1371/journal.pone.0072439

Wei F, Stein JC, Liang C et al (2009) Detailed analysis of a contiguous 22-Mb region of the maize genome. PLoS Genet 5:e1000728. https://doi.org/10.1371/journal.pgen.1000728

Weiberg A, Jin H (2015) Small RNAs—the secret agents in the plant-pathogen interactions. Curr Opin Plant Biol 26:87–94. https://doi.org/10.1016/j.pbi.2015.05.033

Wendel JF, Greilhuber J, Doležel J, Leitch IJ (2012) Plant genome diversity volume 1: plant genomes, their residents, and their evolutionary dynamics. Plant Genome Divers Vol 1 Plant Genomes, their Resid their Evol Dyn 1–279. https://doi.org/10.1007/978-3-7091-1130-7

Wendel JF, Wessler SR (2000) Retrotransposon-mediated genome evolution on a local ecological scale. Proc Natl Acad Sci U S A 97:6250–6252. https://doi.org/10.1073/pnas.97.12.6250

Wessler SR (1998) Transposable elements and the evolution of gene expression. Symp Soc Exp Biol 51:115–122

Wessler SR (2006) Transposable elements and the evolution of eukaryotic genomes. Proc Natl Acad Sci U S A 103:17600–17601
Wessler SR (1996) Plant retrotransposons: turned on by stress. Curr Biol 6:959–961. https://doi.org/10.1016/S0960-9822(02)00638-3
Wessler SR (2001) Plant transposable elements a hard act to follow. Plant Physio 125:149–151
Wessler SR, Bureau TE, White SE (1995) LTR-retrotransposons and MITEs: important players in the evolution of plant genomes. Curr Opin Genet Dev 5:814–821
Wheeler NC, Guries RP (1982) Biogeography of lodgepole pine. Can J Bot 60:1805–1814. https://doi.org/10.1139/b82-227
Wicker T (2012) So many repeats and so little time: how to classify transposable elements. In: Grandbastien MA, Casacuberta J (eds) Plant transposable elements. Springer, Berlin, pp 1–15
Wicker T, Keller B (2007) Genome-wide comparative analysis of copia retrotransposons in Triticeae, rice, and Arabidopsis reveals conserved ancient evolutionary lineages and distinct dynamics of individual copia families. Genome Res 17:1072–1081. https://doi.org/10.1101/gr.6214107
Wicker T, Sabot F, Hua-Van A et al (2007) A unified classification system for eukaryotic transposable elements. Nat Rev Genet 8:973–982. https://doi.org/10.1039/b921331g
Witte C-P, Le QH, Bureau T, Kumar A (2001) Terminal-repeat retrotransposons in miniature (TRIM) are involved in restructuring plant genomes. Proc Natl Acad Sci 98:13778–13783. https://doi.org/10.1073/pnas.241341898
Wright SI, Schoen DJ (1999) Transposon dynamics and the breeding system. Genetica 107:139–148. https://doi.org/10.1023/a:1003953126700
Wright SI, Le QH, Schoen DJ, Bureau TE (2001) Population dynamics of an Ac-like transposable element in self- and cross-pollinating arabidopsis. Genetics158:1279–1288. https://doi.org/10.1093/genetics/158.3.1279
Xiao H, Jiang N, Schaffner E et al (2008) A retrotransposon-mediated gene duplication underlies morphological variation of tomato fruit. Science 319:1527–1530. https://doi.org/10.1126/science.1153040
Xu Y, Du J (2014) Young but not relatively old retrotransposons are preferentially located in gene-rich euchromatic regions in tomato (*Solanum lycopersicum*) plants. Plant J 80:582–591. https://doi.org/10.1111/tpj.12656
Yanagisawa S (2004) Dof domain proteins: plant-specific transcription factors associated with diverse phenomena unique to plants. Plant Cell Physiol 45:386–391. https://doi.org/10.1093/pcp/pch055
Yang G, Zhang F, Hancock CN, Wessler SR (2007) Transposition of the rice miniature inverted repeat transposable element mPing in *Arabidopsis thaliana*. Proc Natl Acad Sci 104:10962–10967. https://doi.org/10.1073/pnas.0702080104
Yi F, Ling J, Xiao Y et al (2018) ConTEdb: a comprehensive database of transposable elements in conifers. Database (oxford) 2018:1–7. https://doi.org/10.1093/database/bay131
Yin H, Liu J, Xu Y et al (2013) TARE1, a mutated Copia-Like LTR retrotransposon followed by recent massive amplification in tomato. PLoS ONE 8:e68587. https://doi.org/10.1371/journal.pone.0068587
Yokosho K, Yamaji N, Fujii-Kashino M, Ma JF (2016) Retrotransposon-mediated aluminum tolerance through enhanced expression of the citrate transporter OsFRDL4. Plant Physiol 172:2327–2336. https://doi.org/10.1104/pp.16.01214
You C, Cui J, Wang H et al (2017) Conservation and divergence of small RNA pathways and microRNAs in land plants. Genome Biol 18:1–19. https://doi.org/10.1186/s13059-017-1291-2
You FM, Cloutier S, Shan Y, Ragupathy R (2015) LTR annotator: automated identification and annotation of LTR retrotransposons in plant genomes. Int J Biosci Biochem Bioinforma 5:165–174. https://doi.org/10.17706/ijbbb.2015.5.3.165-174
Zhang Q, Arbuckle J, Wessler SR (2000) Recent, extensive, and preferential insertion of members of the miniature inverted-repeat transposable element family heartbreaker into genic regions of maize. Proc Natl Acad Sci U S A 97:1160–1165. https://doi.org/10.1073/pnas.97.3.1160
Zhang X, Wessler SR (2004) Genome-wide comparative analysis of the transposable elements in the related species *Arabidopsis thaliana* and *Brassica oleracea*. Proc Natl Acad Sci 101:5589–5594. https://doi.org/10.1073/pnas.0401243101
Zhao X, Li J, Lian B et al (2018) Global identification of *Arabidopsis* lncRNAs reveals the regulation of MAF4 by a natural antisense RNA. Nat Commun 9:5056. https://doi.org/10.1038/s41467-018-07500-7
Zhou H, Liu Q, Li J et al (2012) Photoperiod- and thermo-sensitive genic male sterility in rice are caused by a point mutation in a novel noncoding RNA that produces a small RNA. Cell Res 22:649–660. https://doi.org/10.1038/cr.2012.28
Zimin A, Stevens KA, Crepeau MW et al (2014) Sequencing and assembly of the 22-gb loblolly pine genome. Genetics 196:875–890. https://doi.org/10.1534/genetics.113.159715
Zimin AV, Stevens KA, Crepeau MW et al (2017) An improved assembly of the loblolly pine mega-genome using long-read single-molecule sequencing. Gigascience 6:1–4. https://doi.org/10.1093/gigascience/giw016
Zuccolo A, Scofield DG, De Paoli E, Morgante M (2015) The Ty1-copia LTR retroelement family PARTC is highly conserved in conifers over 200MY of evolution. Gene 568:89–99. https://doi.org/10.1016/j.gene.2015.05.028

4 Genomics of Climate Adaptation in *Pinus Lambertiana*

Matthew Weiss, Manoj K. Sekhwal, David B. Neale, and Amanda R. De La Torre

Abstract

The largest pine tree in the world, sugar pine, also has one the largest genomes ever sequenced in any plant species. The reference genome, transcriptome, and SNP markers developed for the species have been used to support breeding disease resistant trees and answer questions related to genome obesity. These resources also create an opportunity to explore associations between genetic variation and environmental variables. By combining ordination and association techniques such as PCA, MLM, and RDA, we identified markers and genes associated with environmental variation across the Sugar Pine species range. Most of the genes identified by our analysis were associated with precipitation though temperature and continentality were also found to be associated with putatively adaptive genes. The results of the PCA and environmental correlations demonstrated explicit groupings among the environmental variables. Functional annotations for these genes were primarily related to signal transduction and disease resistance, but annotations related to biotic and abiotic stress were also identified. Results further provide insight into the geographic pattern of environmentally correlated genetic variation in the species. These findings may provide important insights to guide management strategies looking to maintain the species through ongoing changes in climate and fire regimes.

4.1 Introduction

Sugar pine is an economically and ecologically important species that is naturally distributed from Baja California (Mexico) to Oregon, with a latitudinal range of 30–43 N degrees, a longitudinal range of 115–124 W degrees, and an elevational range of 0–10,000 ft. Sugar pine is the tallest of its genus reaching heights of 51–63 m (Kinloch and Scheuner 1990). This feature gives the species ecological and cultural significance (Morin et al. 2015). This massive tree has an equally massive genome at 31 billion base pairs which, like all members of the genus, are spread across 12 chromosomes (Saylor 1961; Stevens et al. 2016). A consensus map, originally developed for the species, included 19 linkage groups all of which mapped to the 12 linkage groups from the loblolly reference genome (Jermstad et al. 2011). It is the first *Strobus* pine to have a

M. Weiss
Department of Biology, Northern Arizona University, Flagstaff, Arizona, USA

M. K. Sekhwal · A. R. De La Torre (✉)
School of Forestry, Northern Arizona University, Flagstaff, Arizona, USA
e-mail: Amanda.de-la-torre@nau.edu

D. B. Neale
University of California, Davis, CA, USA

A. R. De La Torre (ed.), *The Pine Genomes*, Compendium of Plant Genomes,
https://doi.org/10.1007/978-3-030-93390-6_4

published reference genome (Stevens et al. 2016; Crepeau et al. 2017) and transcriptome (Gonzalez-Ibeas et al. 2016). It also has multiple field site resources such as progeny trials and a two-generations full-sib cross designed for QTL mapping (Jermstad et al. 2011; Vazquez-Lobo et al. 2017). Additionally, the tree has economic importance as a timber species with high quality and dimensional stability when compared with other softwoods (Kinloch and Scheuner 1990).

Sugar pine currently faces numerous threats to its continued survival and management. The most significant threat facing sugar pine is white pine blister rust, the disease caused by the fungal pathogen *Cronaritum ribicola* (Kinloch 2003). This invasive pathogen affects all five needle or strobus pines of which sugar pine is one the most susceptible (Kinloch and Scheuner 1990). Along with western white pine, sugar pine has breeding programs that measure slow rust reactions and quantitative resistance due to the existence of avirulent strains of *Cronartium ribicola* capable of overcoming major gene resistance in these species (Kinloch et al. 2007). These breeding programs resulted in numerous resources to assess sugar pine genotypes across the species range for associations with disease resistance (Jermstad et al. 2011). While a greater understanding of quantitative disease resistance has been the original purpose for developing these resources, sugar pine also faces numerous environmental threats. There is some genetic evidence for a recent bottleneck in populations near Lake Tahoe, which may be explained by historical logging and fire suppression. As such, sugar pine populations may already be reduced from historic levels (Maloney et al. 2011).

Sugar pine conservation also has increased relevance in light of a changing climate. As a long-lived tree species, it may be unable to migrate or rapidly adapt in response to a changing climate (Aitken et al. 2008). A recent extreme weather event in the form of drought was correlated with high levels of mortality for sugar pine (Pile et al. 2018). Such extreme weather events are expected to continue to increase due to climate change (Bellprat et al. 2019). Furthermore, despite accounting for only 1.4% of individuals, large trees such as white fir and sugar pine accounted for 49.4% of the biomass in a mixed conifer forest sampled in Yosemite National Park. It is suggested that such large trees may be disproportionately important for carbon sequestration (Lutz et al. 2012). Since the species is threatened by both biotic and abiotic variables, preserving the species may require an understanding of the relationship between not just genetics and disease resistance, but genetics and environment. The available resources for this species include a large amount of environmental and genetic data for trees across the species range. The genetics of traits that affect responses to climate, such as drought tolerance, frost tolerance, and phenology, as well some important aspects of disease resistance, are frequently polygenic (Howe et al. 2000; González-Martínez et al. 2008; Poland et al. 2009; Eckert et al. 2010). Gaining a greater knowledge of the underlying genetics of these adaptations will be necessary for policy and management of the species into the future.

Understanding the underlying neutral genetic variation gives additional context for any adaptive variation we detect. Previous studies have examined population structure in the species using an analysis of chloroplast haplotypes across the complete species range (Liston et al. 2007). Another study examined a subset of the species, representing populations from the southern Sierra Nevada to the Oregon border, by using both PCoA of populations and an individual-based method using a Bayesian cluster analysis of genetic markers (Vangestel et al. 2016). In this chapter, we seek to expand on previous efforts using individual-based methods to understand population structure across the entire study area which covers the entire species range outside of Baja California. We will use the results of our population structure analysis to provide further context to our genotype–environment association in order to (1) determine which environmental variables are most strongly associated with genetic variation and (2) identify genes and gene families associated with environmental variation.

The study of the genomics of climate adaptation in conifers presents unique challenges and possibilities that are thus far understudied relative to other systems (Prunier et al. 2016). Large genome sizes have slowed the development of reference genomes relative to that of other plant species (Nystedt et al. 2013; Soltis and Soltis 2016). These massive genome sizes combined with extensive regions of repetitive elements require larger numbers of markers and samples in order to sample reasonable fractions of genome-wide variation (Lamara et al. 2016; Hall et al. 2016). Many examinations of population structure, such as F_{st} outlier analysis, require predefined populations. The continuous distribution of many conifer species, and high levels of gene flow due to wind pollination and outcrossing, make predefining populations difficult and largely irrelevant to the biology of the species (Martins et al. 2016). The recent development of individual-based methods for the detection of loci under environmental selection gives us an opportunity to better assess environmental associations in these species (Rellstab et al. 2015).

Genome-wide methods to detect loci associated with environmental variation can analyze one variable (univariate GEA) or multiple variables at a time (multivariate GEA). Both methods are individual-based and allow the control of population structure while testing for environmental associations (Rellstab et al. 2015). The univariate GEA tests for pairwise associations between loci and individual environmental predictors. Testing each environmental predictor separately drastically raises the number of tests performed resulting in an increase in type I error, a variety of multiple testing corrections has been used to solve this problem, many of which result in some degree of increase in type II error (Hayes 2013). Multivariate methods such as the redundancy analysis (RDA), not only eliminate the multiple testing issue by analyzing many loci and predictors simultaneously, but also more accurately portray the current understanding of the genetic architecture of complex traits where many loci are seen to contribute to a phenotype (Hall et al. 2016; Fahrenkrog et al. 2017).

4.2 Methods

4.2.1 Genome-Wide SNP Data

Single Nucleotide Polymorphisms (SNPs) used for our analysis were originally developed for a genome-wide association study of disease resistance in sugar pine (Weiss et al. 2020). Megagametophyte tissue from seeds of 1371 trees representing the entire species range excepting Baja California were obtained from the Placerville gene bank in California. Maternal genotypes for each tree were inferred through pooling DNA from eight to ten megagametophytes. Two platforms were used to genotype SNPs from these samples, a 600 K Affymetrix array (Thermo Scientific, Inc.) and an 80 k Illumina Infinium array (Illumina, Inc). After filtering for quality control, 125,238 SNPs of these SNPs from 1015 individuals were retained for further analysis.

4.2.2 Population Structure

Population structure was determined by using a Bayesian cluster analysis with fastSTRUCTURE (Raj et al. 2014). This analysis was conducted using 10 independent runs of $K = 2 - 10$. Each run used 80–90 iterations with an average of 88 iterations per run. The optimal value of K, representing the number of genetic lineages, was selected using the program chooseK.py (Raj et al. 2014). Ten replicates of each cluster analysis were aligned and visualized using CLUMPP (Jakobsson and Rosenberg 2007). Input files for both Adegenet and fastSTRUCTURE were created using Plink v 1.07 (Purcell et al. 2007).

Additionally, a principal component analysis (PCA) was used to determine population structure with the Adegenet R package (Jombart 2008). Individuals were assigned to clusters using a k-means clustering algorithm to a PCA of SNP variation across individuals. K-means has been found to be superior to fastSTRUCTURE in both computational speed and power in the analysis of large numbers of loci and is free from assumptions regarding linkage disequilibrium and Hardy–Weinberg equilibrium (Jombart et al. 2010; Stift et al. 2019). The number of clusters was selected using the silhouette algorithm in the R package factoextra testing 1–10 clusters with 100 bootstraps.

4.2.3 Climate Data

Individuals with both geographic and genetic information were selected for a set of 739 individuals and 125,238 SNPs. Geographic information in the form of latitude and longitude was obtained from seed tree locations. Individuals were excluded when geographic information did not match the sugar pine species range or the national forest of origin resulting in a loss of four likely erroneous records and retention of 735 trees for further analysis. Climate normals from 1961–1990 for each individual using ClimateNA_v5.60 software package (Wang et al. 2016) based on latitude and longitude information for each tree. All directly calculated and derived annual variables were obtained for a total of 23 climate variables. Correlations between all climate variables and principal components were tested in R v3.5.3 using ggcorrplot (). To further visualize the relationship between environmental variables, a principal component analysis was conducted on our data set including all 23 climate variables as well as Latitude and Longitude using the princomp () function in R v3.5.3.

4.2.4 Univariate GEA

Associations between SNPs and environmental variables were tested using TASSEL v.5 (Bradbury et al. 2007). Climate data for 735 trees was combined with the first five principal components of a PCA for climate data to create a set of 28 environmental observations tested for association. SNPs with maf ≤ 0.05 were retained for a total of 125,238 SNPs. Population structure was controlled for using the first five principal components obtained from a PCA analysis of the SNP data. Relatedness was accounted for using a kinship matrix. These components were incorporated into a mixed linear model

$$y = X\beta + \boldsymbol{Zu} + \boldsymbol{e}$$

with y representing the environmental observations, β representing fixed random effects of genotype data with kinship and population structure acting as covariates, additive effects represented by u and residuals represented by e. Correlation between minor allele frequency and proportion heterozygous was calculated for all SNPs (n = 125,238) using Pearson's correlation coefficient.

4.2.5 Redundancy Analysis

Redundancy analysis was conducted for the detection of single nucleotide polymorphisms (SNPs) that are environmental outliers based on methods outlined by Forester et al. (2018). The environmental variables used as predictors for the RDA were derived from ClimateWNA (Wang et al. 2016). Environmental variables were then plotted with a heatmap in order to detect highly correlated variables in the R vegan package (Oksanen and Simpson 2009) (Fig. 4.1). Highly correlated variables ($R^2 > 0.6$) were removed, leaving mean annual temperature, temperature difference (TD), mean annual precipitation, annual heat moisture index, relative humidity, and mean annual radiation for posterior analysis. The RDA retained 6 constrained axes (one for each of the predictors). Unconstrained axes were calculated for each of the constrained axes and used to calculate an adjusted R^2 which describes the proportion of variance explained by the constrained ordination axis. Inertia plots were used to select the three most informative axes for use in the analysis (Fig. 4.4). Individual trees and SNPs were plotted into this constrained ordination. Loadings for SNPs on each of the three retained axes were extracted and candidate SNPs were determined as those that were 3.5 standard deviations from the mean (two-tailed p-value $\sim$0.0005) in order to reduce the false positive rate. Candidate SNPs were plotted on the constrained ordination, color

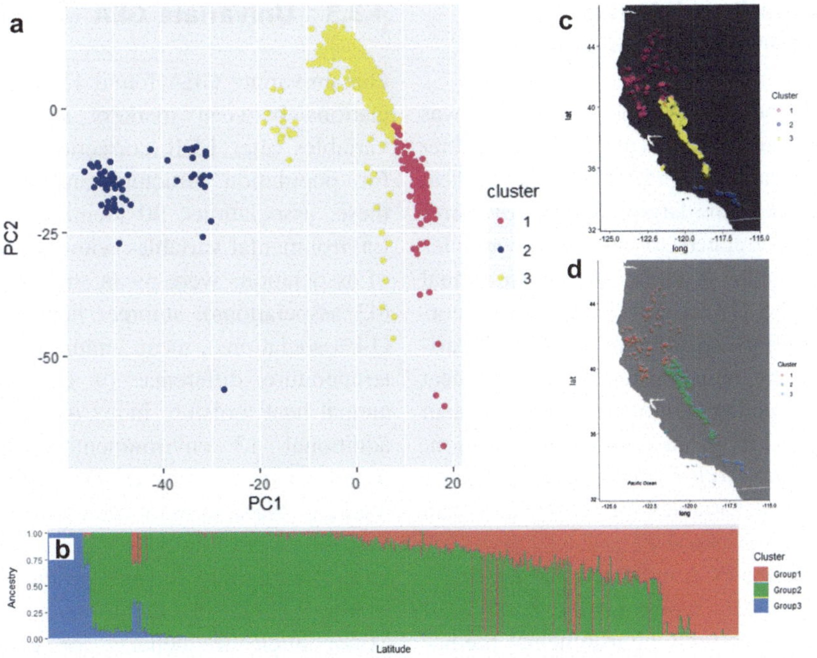

Fig. 4.1 The results of the population structure analysis showing **a** A PCA of genetic variation with colors representing groups as defined by a K-means clustering algorithm. **c** The same trees plotted geographically by cluster as determined by the PCA analysis. **b** A bar plot of percent of ancestry as determined by fastSTRUCTURE with individuals organized by latitude with the lowest latitude to the left. **d** The same individuals plotted on the map with color representing the group representing a plurality of ancestry for any one individual

coded by the environmental trait with which they were associated (Fig. 4.3).

4.2.6 Functional Gene Annotations

The genes associated with candidate SNPs were annotated using different approaches such as pfam (Finn et al. 2014) and blastp (Johnson et al. 2008), BlastKOALA (Kanehisa et al. 2016b). The Pfam was ran using the HMMER (Finn et al. 2011) at default parameters with an e-value of 1.0 to search proteins families. The blastp was ran at an expected threshold of −0.05; matrix-BLOSUM 62; database- non-redundant protein sequence (nr) to search the similar hits. The BlastKOALA at KEGG (Kanehisa et al. 2016a) was performed for protein pathways and annotation. The identical matching genes were chosen for identifying annotation and KEGG pathways.

4.3 Results

4.3.1 Population Structure

Results from our fastSTRUCTURE analysis give novel insights into the population structure of sugar pine. Previous studies had not sampled the Transverse Range, trees from which represented a unique ancestry group in our analysis ($N = 85$), This group represented the most geographically isolated group we sampled though a small number of individuals were found to exist sympatrically with an individual from cluster two in Big Sur. One group was shown to exist in the Sierra Nevada representing a majority of our sampled individuals ($N = 570$). A final group was seen in The Cascade Range of northern California and southern Oregon representing the remainder of our sampled individuals.

4.3.2 Correlations Between Environmental Variables

Every environmental variable we tested was significantly correlated to at least one other environmental variable (Fig. 4.2). Summer heat moisture index had the largest number of correlations being correlated to all other variables except mean annual precipitation. Mean annual radiation had the fewest correlations being correlated with only five other environmental variables: annual heat moisture index, summer heat moisture index, climate moisture deficit, mean annual precipitation, mean summer precipitation. Latitude was found to have fewer correlations to the environmental variables for our species than longitude (Figs. 4.3 and 4.4).

Results from the PCA indicated that the first 5 principal components represented over 89% of variation of the measured environmental predictors. These principal components were retained for subsequent analysis. The results of the PCA and environmental correlations demonstrated clear groupings among our environmental variables relating to broader environmental concepts. The first PC described 61.2% of environmental variation and was related to variables associated with temperature with mean annual temperature, degree days above 0 °C, degree days above 18 °C, mean coldest month temperature, extreme minimum temperature, and extreme maximum temperature showing strong negative correlations to PC1 while degree days below zero, degree days below 18, beginning of frost-free period and precipitation as snow showing strong positive correlations to PC1. PC2 was strongly correlated to variables involved in precipitation with mean annual precipitation and mean summer precipitation, showing strong positive correlations to PC2, while summer heat moisture deficit, annual heat moisture deficit, and climate moisture deficit were seen to be negatively correlated to this principal component. PC3 had its strongest correlation to temperature difference which was our only environmental predictor measuring continentality.

4.3.3 Univariate GEA

Our univariate GEA found 132 significant associations between markers and environmental variables after FDR correction and controlling for population structure and relatedness. Of these, associations, 90 identified unique SNPs. Environmental variables with the largest number of associations were mean summer precipitation (43 associations), summer heat moisture index (14 associations), mean annual precipitation and temperature difference (9 associations each), annual heat moisture index (6 associations). An additional 13 environmental variables were found to have marker associations with 27 markers. Markers associated with environmental variables were found on the majority of linkage groups with only linkage groups 2, 4, and 10 lacking environmentally associated markers. Minor allele frequencies for identified SNPs ranged from 0.050 to 0.334 with an average of 0.098. The proportion of heterozygotes ranged from 0.007 to 0.616 with an average of 0.143.

4.3.4 RDA

The adjusted R^2 for our RDA was 0.015 suggesting that 1.5% of our genome-wide variation is explained by environmental variables. The analysis revealed 949 candidate markers involved with two environmental variables. Mean annual precipitation had 569 associated markers, while temperature difference (maximum annual temperature–minimum annual temperature) had 380 associations. Mean annual precipitation was seen to be correlated with mean summer precipitation and climatic moisture deficit indicating that these environmental variables may also be involved in selection on candidate markers identified as being involved with mean annual precipitation. Mean summer precipitation was seen to have a larger number of marker associations as indicated by the univariate analysis, and as such may be the main driver of selection for markers in the multivariate analysis

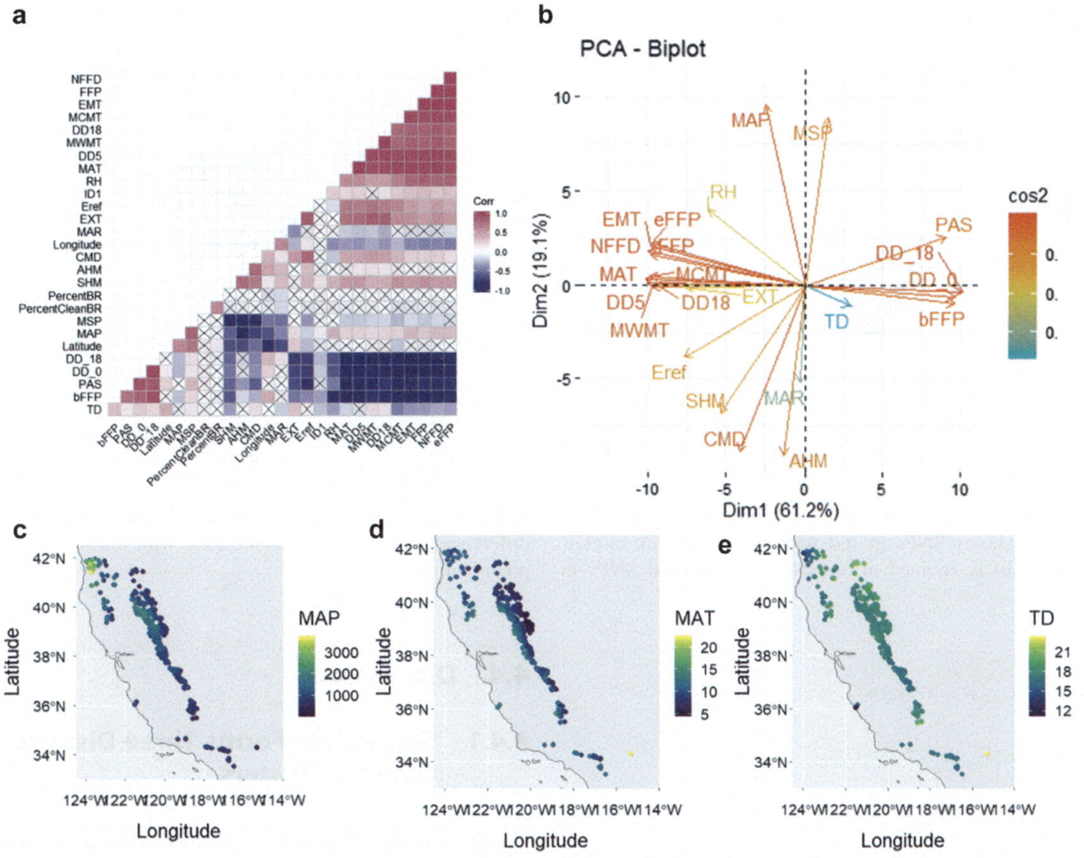

Fig. 4.2 Figure (**a**) is a correlation matrix of all environmental predictors used in our analysis as well as latitude and longitude. Cells with an [X] were not significantly correlated. Figure (**b**) is a PCA of the environmental space generated from the environmental predictors measured at each site. Figures describing the environmental variation in (**c**) mean annual precipitation, **d** mean annual temperature and **e** continentality for all maternal genotypes sampled in this study

found to be involved with mean annual precipitation. Temperature difference was not seen to be strongly correlated with other environmental variables.

Of the SNPs identified in the RDA, 13 SNPs had also been identified in the univariate GEA (Table 4.1). The RDA identified nine of these SNPs as involved in mean annual precipitation and four of these SNPs as involved in temperature difference. Minor allele frequencies for these SNPs ranged from 0.050 to 0. 174 with an average of 0.081. Proportion of heterozygotes ranged from 0 to 0.078 with an average of 0.030. Of SNPs located on scaffolds with linkage information, only linkage 3 SNPs on linkage group 7 and a single SNP on linkage group 6 were identified. Two SNPs identified genes both of which had annotations. PILA_26583 contained LRR, NACHT, and WD40 domains and an eggnog description indicating a resistance protein. PILA_27631 associated with MAP, MSP, and PC_2 environmental variables (univariate GEA) and MAP (RDA) was identified as a transcription factor. PILA_06301 associated with MSP (univariate GEA) and MAP (RDA) environmental variables was identified as Phytolongin Phyl1.2-

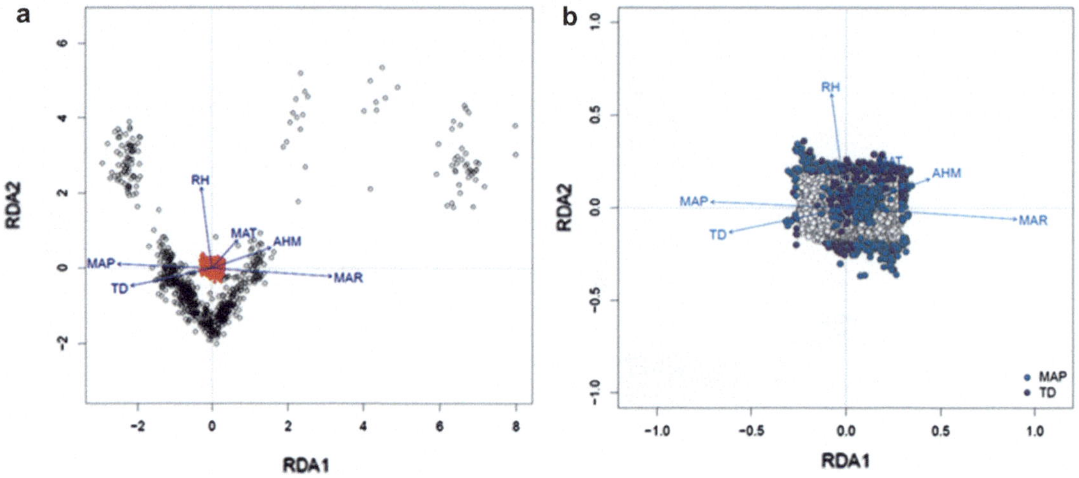

Fig. 4.3 A plot of the redundancy analysis. Figure (**a**) shows SNPs in red with trees as black circles. Figure (**b**) is zoomed in on SNPs with neutral SNPs in white, purple SNPs are associated with temperature difference, and blue are associated with mean annual precipitation

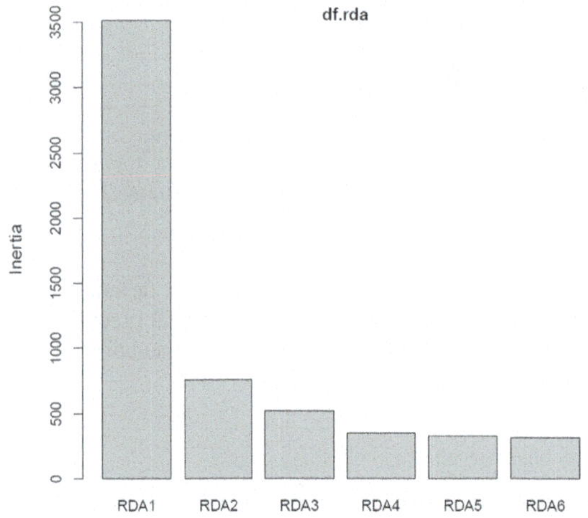

Fig. 4.4 Inertia plot of axes from the RDA

like. PILA_20458 associated with PC_4 (univariate GEA) and TD (RDA) environmental variables was identified as LRR receptor-like serine/threonine-protein kinase (Table 4.1). PILA_01563 gene associated with TD and MAP (RDA) environmental variables was involved in MAPK signaling pathway and PILA_26583 gene associated with TD and PC_4 (RDA and univariate) was involved in circadian rhythm (Tables 4.2 and 4.3).

4.4 Discussion

4.4.1 Sugar Pine Forms Three Distinct Genetic Clusters

To our knowledge, our evaluation of sugar pine population structure is the most geographically complete analysis of genome-wide population structure conducted in the species to date. Our study expands upon previous efforts (Vangestel et al. 2016) by including populations in the transverse range, a mountain range in southern California, effectively covering the entire species range excepting an isolated population in northern Baja California. The results indicated three distinct population clusters: One in northern Californian and Central Oregon, one in the Sierra Nevada, and a third in the transverse range. The first two populations confirmed a previously identified contact zone between the Cascade and Sierra Nevada mountain ranges in northern California. The third cluster in the transverse range was unidentified by any previous genetic analysis. A previous marker-based study of neutral genetic variation did not include individuals from this range (Vangestel et al. 2016), while a

Table 4.1 Table showing SNPs that were significant for environmental associations in both the univariate GEA and RDA. Univariate GEA and RDA columns indicate the environmental predictor(s) found to be associated with each analysis. MAF indicates the minor allele frequency for each locus. Proportion Missing is the percentage of missing data at each locus. Proportion heterozygous is the proportion of individuals that were heterozygous for a given loci

Marker	CDS/Gene	Univariate GEA	RDA	MAF	Proportion missing	Proportion heterozygous	GenBank annotation	KEGG pathway
AX-175388677	NA	TD	TD	0.07224	0.03155	0.04816	NA	NA
AX-175438560	NA	eFFP	TD, MAP	0.05718	0.01646	0.07531	NA	NA
AX-175448126	NA	SHM, DD_0, DD18, NFFD, eFFP, FFP, PAS, EMT, RH, PC_1	MAP	0.09488	0.0096	0.00693	NA	NA
AX-175536842	NA	SHM	MAP, TD	0.06653	0	0.02606	NA	NA
AX-175594929	NA	PC_4	MAP	0.05402	0.0096	0.05263	NA	NA
AX-175606265	NA	PC_4	MAP	0.10015	0.06859	0.10898	NA	NA
AX-175760597	NA	MAT, MWMT, DD5, DD18, NFFD, eFFP, FFP, EMT, RH, PC1	TD	0.06285	0.01783	0.06704	NA	NA
AX-175779067	PILA_27631	MAP, MSP, PC_2	MAP	0.17422	0.03155	0.18414	Transcription factor (MYB30)	Transcription factors
AX-175797892	NA	MAR	MAP	0.05441	0.00412	0.10606	NA	NA
AX-175853764	NA	eFFP	MAP	0.06206	0.01646	0.06276	NA	NA
seq-rs37799-SP	PILA_06301	MSP	MAP	0.07955	0.03429	0.10795	Phytolongin Phyl1.2-like	Membrane trafficking
seq-rs4561-SP	NA	MAP, MSP, MAR	MAP	0.05891	0.06859	0.04713	NA	NA
seq-rs9898-SP	PILA_20458	PC_4	TD	0.12723	0.07819	0.18006	LRR receptor-like serine/threonine-protein kinase	NA

Table 4.2 Table showing SNPs that were associated with two or more uncorrelated environmental predictors. The column titled "Analysis" specifies the type of GEA analysis that identified the association where "both" signifies that associations were uncovered in both the RDA and univariate GEA. Predictor indicates the predictors associated to the SNP. Gene indicates the gene located by the associated SNP. Protein family indicates the protein family annotation for the gene. GenBank description indicates the annotation using NCBI-blast. KEGG pathway illustrates the pathway of the corresponding genes

Analysis	Marker	Predictors	Gene	Protein family	GenBank annotation	Pfam annotation	KEGG pathway
RDA	AX-175438560	TD, MAP	NA	NA	NA	NA	NA
RDA	AX-175733789	TD, MAP	NA	NA	NA	NA	NA
RDA	AX-175611554	TD, MAP	NA	NA	NA	NA	NA
RDA	seq-rs42290-SP	TD, MAP	PILA_01563	Inhibitor_I29, GRAN, SIGNAL, TRANS	Mitogen-activated protein kinase YODA-like	Protein kinase domain	MAPK signaling pathway
RDA	AX-175424188	TD, MAP	PILA_01563	Inhibitor_I29, GRAN, SIGNAL, TRANS	Mitogen-activated protein kinase YODA-like	Protein kinase domain	MAPK signaling pathway
RDA	AX-175970382	TD, MAP	PILA_01563	Inhibitor_I29, GRAN, SIGNAL, TRANS	Mitogen-activated protein kinase YODA-like	Protein kinase domain	MAPK signaling pathway
RDA	AX-175536842	TD, MAP	NA	NA	NA	NA	NA
RDA	AX-175474959	TD, MAP	NA	NA	NA	NA	NA
RDA	AX-175818904	TD, MAP	NA	NA	NA	NA	NA
Univariate	seq-rs33205-SP	MAP, MSP, TD	PILA_11488	Enolase_C, Enolase_N	Myosin-binding protein 3-like	Zein-binding	Chromosome and associated proteins
Univariate	seq-rs20069-SP	MAP, MSP, TD	PILA_00972	NB-ARC, AAA	Pentatricopeptide repeat-containing protein	PPR repeat family	
Both	seq-rs9898-SP	TD, PC_4	PILA_26583	NB-ARC, WD40, NACHT, AAA	Protein SPA1-RELATED 3-like isoform	WD domain, G-beta repeat	Circadian rhythm
Both	AX-175438560	eFFP, TD, MAP	NA	NA	NA	NA	NA
Both	AX-175448126	MAP, FFP, EMT	NA	NA	NA	NA	NA
Both	AX-175536842	SHM, MAP, TD	NA	NA	NA	NA	NA
Both	AX-175853764	eFFP, MAP	NA	NA	NA	NA	NA

Table 4.3 Key to abbreviations to the 23 derived and directly calculated annual climate variables taken from ClimateWNA

Abbr	Variable	Derived/directly calculated
AHM	annual heat-moisture index (MAT + 10)/(MAP/1000))	Directly calculated
bFFP	the day of the year on which FFP begins	Directly calculated
CMD	Hargreaves climatic moisture deficit (mm)	Directly calculated
DD < 0	Degree days below 0 °C, chilling degree days	Directly calculated
DD < 18	Degree days below 18 °C, heating degree days	Directly calculated
DD > 18	Degree days above 18 °C, cooling degree days	Directly calculated
DD > 5	Degree days above 5 °C, growing degree days	Directly calculated
eFFP	The day of the year on which FFP ends	Directly calculated
EMT	Extreme minimum temperature over 30 years	Directly calculated
Eref	Hargreaves reference evaporation (mm)	Derived
EXT	Extreme maximum temperature over 30 years	Derived
FFP	Frost-free period	Derived
MAP	Mean annual precipitation (mm),	Derived
MAR	Mean annual solar radiation (MJ m^{-2} d^{-1})	Derived
MAT	Mean annual temperature (°C),	Derived
MCMT	Mean coldest month temperature (°C),	Derived
MSP	May to September precipitation (mm),	Derived
MWMT	Mean warmest month temperature (°C),	Derived
NFFD	The number of frost-free days	Derived
PAS	Precipitation as snow (mm) between August in previous year and July in current year	Derived
RH	Mean annual relative humidity (%)	Derived
SHM	summer heat-moisture index ((MWMT)/(MSP/1000))	Derived
TD	Temperature difference between MWMT and MCMT, or continentality (°C),	Derived

previous study of chloroplast haplotypes grouped this population in the transverse range with the population we identified in the Sierra Nevada into a single population (Liston et al. 2007). Remarkably, both our PCA and fastSTRUCTURE analyses indicated greater differentiation between our transverse range population and the Sierra Nevada population, than that of the Sierra Nevada population and the Cascade Range population. Since chloroplast DNA is inherited paternally, this may be an indication that gene flow from seeds, but not pollen, experiences a significant barrier caused by the divide between the transverse range and the southern Sierra Nevada. Furthermore, many individuals in the southern Sierra Nevada have some percentage of ancestry from the transverse range population, while no individuals in the transverse range are seen with ancestry from the Sierra Nevada population indicating a potential directionality of pollen gene flow consistent with dominant wind patterns for the region.

4.4.2 Loci Under Selection for Environmental Variables are Driven by Precipitation

Despite the fact that the majority of environmental variation in our study area was defined by temperature, the majority of genetic associations from both analyses were for environmental predictors relating to precipitation or temperature difference. Associations between markers and mean annual precipitation accounted for 60% of all associations identified by the RDA. The remaining 40% of associations from the RDA were involved in temperature difference, a measure of continentality. Results from the univariate GEA also showed a high proportion of associations with precipitation-related environmental variables with 43% of identified associations related to mean annual precipitation, mean summer precipitation or the second principal component of the PCA of environmental variation. Temperature-related environmental variables accounted for 16% of associations. These associations were with ten variables correlated to mean annual temperature and PC1 which comprised mean annual temperature, degree days below zero, degree days above 18, degree days above 5, end of frost-free period, frost-free period, number of frost-free days, extreme minimum temperature, mean warmest monthly temperature and PC1 itself. Variables that were correlated to both mean annual temperature and mean annual precipitation such as climatic moisture deficit, annual heat moisture index, and summer heat moisture index accounted for 17% of associations.

A possible explanation for the limited role of temperature in local adaptation may be higher gene flow between populations of sugar pine growing in environments with different temperatures. Regions with extreme values for precipitation and continentality exist in distinct and relatively isolated regions of the species range (Fig. 4.2a and c), increasing the possibility of an accumulation of alleles adapted to these environments. For example, the entire southern portion of the species range exists in low precipitation environments with the highest precipitation environment occurring in the northwest extreme of the species range. The southern populations likely have few opportunities to exchange alleles with trees growing in the moist northwest of the range. Values for temperature, however, vary throughout the species range with many regions with disparate values in close proximity (Fig. 4.2b). Furthermore, along the Sierra Nevada, where temperature appears to be driven by elevation, many populations in high-temperature environments exist mere miles to the west of low-temperature environments. The prevailing west to east wind would all but ensure gene flow between these two environments. This increased gene flow due to proximity between regions with disparate environments would inhibit local adaptation for anyone environmental characteristic.

4.4.3 SNPs Associated with Multiple Environmental Predictors Identify Genes Involved in Signal Transduction

A total of 17 unique SNPs had significant associations with two or more uncorrelated environmental predictors (Table 4.2). These SNPs were located in four unique genes. Our RDA identified 12 SNPs that were associated with both mean annual precipitation and temperature difference. Annotations for these SNPs indicated that these associations accounted for 1.3% of all associations identified by the RDA. Only three of these SNPs were from annotated regions of the sugar pine genome v.1.5. These three SNPs were all collocated on a gene, PILA_01563, involved in signal transduction. Regulatory gene products, such as calcium-dependent protein kinases (CDPKs), mitogen-activated protein kinases (MAPKs), bZIP, MYB, and WRKY can cause changes in plant morphology or physiology by regulating signal transduction pathways or acting as transcription factors to regulate the expression of downstream genes. Subsequently, they can enable plants to successfully survive in the arid environment (Yang et al. 2021).

From the univariate GEA, 14 SNPs were identified that had associations with 2 or more environmental predictors. Only two of these SNPs were associated with environmental predictors that were uncorrelated, seq-rs20069-sp and seq-rs33205. The first of these SNPs identified a gene, PILA_11488, with an enolase domain. Enolases have been reported to be upregulated in maize and rice when plants are undergoing salt and drought stress (Rabello et al. 2008; Hu et al. 2011). The remaining gene PILA_00972, as well as a gene identified by both analyses, PILA_26583, contained NB-ARC domains which are commonly seen in association with LRR domains as part of resistance proteins and contain signal transduction domains (McHale et al. 2006). It has been well reported that resistance (R) proteins in plants are involved in pathogen recognition and following activation of innate immune responses. Most R proteins contain a central nucleotide-binding domain (NB-ARC) (van Ooijen et al. 2008). Additionally, PILA_26583 contained a WD40 domain, also known to be involved in signal transduction. Genes involved in signal transduction may play a role in abiotic stress acclimatization. Signal transduction has been reported to be upregulated in stress acclimated tobacco leaves (Vranová et al. 2002); and plays an important role in response to cold, salinity, and drought by connecting sensing and response pathways (Huang et al. 2012). As such, signal transduction may be an important common denominator for many types of abiotic stress. In addition, the signaling pathways are interconnected to the defense network, in which numerous transcription factors, biochemical pathways, and phytohormones are involved in plant defense against pathogens (Liu and Lam 2019). Ku et al. (2018) comprehensively evaluated how the crosstalks among different signaling pathways coordinate and balance the defense signaling in response to biotic and abiotic stresses.

In our study, gene PILA_20458 associated with PC_4 (univariate GEA) and TD (RDA) was identified as LRR receptor-like serine/threonine-protein kinase (Table 4.1). The most known R proteins in plants are the nucleotide-binding (NB) site leucine-rich repeat (LRR) that play important roles in plant defense responses to various pathogens. Xu et al. found that the *NBS-LRR* gene (*ZmNBS25*) in maize enhances disease resistance in rice and Arabidopsis (Xu et al. 2018). In this study, we identified gene PILA_01563 associated with TD, MAP environmental factors were involved in MAPK signaling pathway. It has been reported that calcium-dependent signaling and mitogen-activating protein kinases (MAPKs) act during abiotic stress. For instance, in *Populus euphratica*, the calcium-dependent protein kinase 10 (CPK10) is expressed under drought and frost and activates both drought- and frost-responsive genes to induce stress tolerance (Chen et al. 2013).

In the ordination analysis (univariate and RDA), identified gene PILA_26583 associated with TD, PC_4 environmental variables was involved in circadian rhythm. A WD40 type protein (*OsRACK1A*) in rice was regulated by circadian clocks and plays an important role in the salt stress response (Zhang et al. 2018). Moreover, a total of 743 WD40 proteins were identified in wheat which showed the specific characteristics at the reproductive developmental stage, and in responses to stresses, including cold, heat, drought, and powdery mildew infection pathogen (Hu et al. 2018).

4.5 Conclusions

Our analysis detected multiple genes that are putatively involved in adaptation to environmental variation across the species range of sugar pine. While an ordination-based analysis was able to detect a larger number of significant genes and loci, the association-based analysis detected associations with a greater number of environmental variables including temperature, a variable that was not strongly correlated to any of the variables detected in our ordination-based techniques. A small proportion of these genes were detected by both analyses. Additionally, some genes were identified as being correlated or associated with two or more uncorrelated environmental variables, suggesting that these genes may have pleiotropic effects on climate adaptations. In light of increasing environmental challenges caused by climate change, such genetic resources that allow us to detect environmental adaptation will be increasingly valuable for breeding and monitoring long-lived conifer species such as sugar pine into the future. As the threats that plague this valuable species increase, so does our ability to study the underlying ability of these species to tolerate and adapt to these threats. As such, this study should serve as a baseline for greater advances in our understanding of the relationship between environment and genetics in the species.

Acknowledgements This project was supported by the U. S. Department of Agriculture/National Institute of Food and Agriculture [McIntire Stennis project 1020440] awarded to A.D.L.T; by the U.S. Department of Agriculture/National Institute of Food and Agriculture (award # 2017-67013-26214) awarded to A.D.L.T and D.B.N at the University of California-Davis; and by the NAU School of Forestry new faculty start-up funds awarded to A.D.L.T.

References

Aitken SN, Yeaman S, Holliday JA, Wang T, Curtis-McLane S (2008) Adaptation, migration or extirpation: climate change outcomes for tree populations. Evol Appl 1(1):95–111

Bellprat O, Guemas V, Doblas-Reyes F, Donat MG (2019) Towards reliable extreme weather and climate event attribution. Nat Commun 10(1):1–7

Bradbury PJ, Zhang Z, Kroon DE, Casstevens TM, Ramdoss Y, Buckler ES (2007) TASSEL: software for association mapping of complex traits in diverse samples. Bioinform Appl 23:2633–2635

Chen JH, Xue B, Xia XL, Yin WL (2013) A novel calcium-dependent protein kinase gene from *Populus euphratica*, confers both drought and cold stress tolerance. Biochem Biophys Res Commun 441:630–636

Crepeau MW, Langley CH, Stevens KA (2017) From pine cones to read clouds: Rescaffolding the megagenome of sugar pine (Pinus lambertiana). G3: Genes Genomes Genet 7(5):1563–1568

Eckert AJ, Bower AD, GonzÁlez-MartÍnez SC, Wegrzyn JL, Coop G, Neale DB (2010) Back to nature: ecological genomics of loblolly pine (Pinus taeda, Pinaceae). Mol Ecol 19(17):3789–3805

Fahrenkrog AM, Neves LG, Resende MFR, Dervinis C, Davenport R, Barbazuk WB, Kirst M (2017) Population genomics of the eastern cottonwood (Populus deltoides). Ecol Evol 7:9426–9440

Finn RD, Bateman A, Clements J, Coggill P, Eberhardt RY, Eddy SR, Heger A, Hetherington K, Holm L, Mistry J, Sonnhammer EL, Tate J, Punta M (2014) Pfam: the protein families database. Nucleic Acids Res 42:D222-230

Finn RD, Clements J, Eddy SR (2011) HMMER web server: interactive sequence similarity searching. Nucleic Acids Res 39:W29-37

Forester BR, Lasky JR, Wagner HH, Urban DL (2018) Comparing methods for detecting multilocus adaptation with multivariate genotype-environment associations. Mol Ecol 27:2215–2233

González-Martínez SC, Huber D, Ersoz E, Davis JM, Neale DB (2008) Association genetics in Pinus taeda L. II. Carbon isotope discrimination. Heredity, 101 (1):19–26

Hall D, Hallingbäck HR, Wu HX (2016) Estimation of number and size of QTL effects in forest tree traits. Tree Genet Genomes 12:110

Hayes B (2013) Overview of Statistical Methods for Genome-Wide Association Studies (GWAS). Humana Press, Totowa, NJ, pp 149–169

Howe GT, Saruul P, Davis J, Chen THH (2000) Quantitative genetics of bud phenology, frost damage, and winter survival in an F2 family of hybrid poplars. Theor Appl Genet 101(4):632–642

Hu R, Xiao J, Gu T, Yu X, Zhang Y, Chang J, Yang G, He G (2018) Genome-wide identification and analysis of WD40 proteins in wheat (*Triticum aestivum* L.). BMC Genomics 19:803–816

Hu X, Lu M, Li C, Liu T, Wang W, Wu J, Tai F, Li X, Zhang J (2011) Differential expression of proteins in maize roots in response to abscisic acid and drought. Acta Physiol Plant 33:2437–2446

Huang GT, Ma SL, Bai LP, Zhang L, Ma H, Jia P, Liu J, Zhong M, Guo ZF (2012) Signal transduction during cold, salt, and drought stresses in plants. Mol Biol Rep 39:969–987

Jakobsson M, Rosenberg NA (2007) CLUMPP: a cluster matching and permutation program for dealing with label switching and multimodality in analysis of population structure. Bioinformatics 23:1801–1806

Jermstad KD, Eckert AJ, Wegrzyn JL, Delfino-Mix A, Davis DA, Burton DC, Neale DB (2011) Comparative mapping in Pinus: sugar pine (*Pinus lambertiana* Dougl.) and loblolly pine (*Pinus taeda* L.). Tree Genetics Genomes 7(3):457–468

Johnson M, Zaretskaya I, Raytselis Y, Merezhuk Y, McGinnis S, Madden TL (2008) NCBI BLAST: a better web interface. Nucleic Acids Res 36:W5-9

Jombart T (2008) Adegenet: A R package for the multivariate analysis of genetic markers. Bioinformatics 24:1403–1405

Jombart T, Devillard S, Balloux F (2010) Discriminant analysis of principal components: a new method for the analysis of genetically structured populations. BMC Genet. 11(1):94. https://doi.org/10.1186/1471-2156-11-94

Kanehisa M, Sato Y, Kawashima M, Furumichi M, Tanabe M (2016) KEGG as a reference resource for gene and protein annotation. Nucleic Acids Res 44: D457–D462

Kanehisa M, Sato Y, Morishima K (2016) BlastKOALA and GhostKOALA: KEGG tools for functional characterization of genome and metagenome sequences. J Mol Biol 428:726–731

Kinloch BB (2003) White pine blister rust in North America: past and prognosis. Phytopathology 93 (8):1044–1047

Kinloch BB, Davis DA, Burton D (2007) Resistance and virulence interactions between two white pine species and blister rust in a 30-year field trial. Tree Genet Genomes 4(1):65–74

Ku YS, Sintaha M, Cheung MY, Lam HM (2018) Plant hormone signaling crosstalks between biotic and abiotic stress responses. Int J Mol Sci 19:3206–3241

Lamara M, Raherison E, Lenz P, Beaulieu J, Bousquet J, Mackay J (2016) Genetic architecture of wood properties based on association analysis and co-expression networks in white spruce. New Phytol 210:240–255

Liston A, Parker-Defeniks M, Syring JV, Willyard A, Cronn R (2007) Interspecific phylogenetic analysis enhances intraspecific phylogeographical inference: a case study in Pinus lambertiana. Mol Ecol 16:3926–3937

Lutz JA, Larson AJ, Swanson ME, Freund JA (2012) Ecological importance of large-diameter trees in a temperate mixed-conifer forest. PLoS ONE 7(5)

Liu JZ, Lam HM (2019) Signal transduction pathways in plants for resistance against pathogens. Int J Mol Sci 20:2335–2341

Maloney PE, Vogler DR, Eckert AJ, Jensen CE, Neale DB (2011) Population biology of sugar pine (Pinus lambertiana Dougl.) with reference to historical disturbances in the Lake Tahoe Basin: Implications for restoration. Forest Ecol Manage 262(5):770–779

Martins H, Caye K, Luu K, Blum MGB, François O (2016) Identifying outlier loci in admixed and in continuous populations using ancestral population differentiation statistics

McHale L, Tan X, Koehl P, Michelmore RW (2006) Plant NBS-LRR proteins: adaptable guards. Genome Biol 7:212

Morin NR, Brouillet L, Levin GA (2015) Flora of North America North of Mexico. Rodriguesia 66(4):973–981

Nystedt B, Street N, Wetterbom A, Zuccolo A, Nature YL et al (2013) The Norway spruce genome sequence and conifer genome evolution. Nature 497(7451):579–584

Oksanen J, Simpson GL (2009) The vegan package boreal avian modelling project view project theory in ethnobotany view project

Pile LS, Meyer MD, Rojas R, Roe O (2018) Characterizing tree mortality after extreme drought and insect outbreaks in the Southern Sierra Nevada, vol 444

Poland JA, Balint-Kurti PJ, Wisser RJ, Pratt RC, Nelson RJ (2009) Shades of gray: the world of quantitative disease resistance. Trends Plant Sci 14(1):21–29. Elsevier Current Trends.

Prunier J, Verta J-P, MacKay JJ (2016) Conifer genomics and adaptation: at the crossroads of genetic diversity and genome function. New Phytol 209:44–62

Purcell S, Neale B, Todd-Brown K, Thomas L, Ferreira MAR, Bender D, Maller J, Sklar P, de Bakker PIW, Daly MJ et al (2007) PLINK: a tool set for whole-genome association and population-based linkage analyses. Am J Human Genetics 81:559–575

Rabello AR, Guimarães CM, Rangel PHN, da Silva FR, Seixas D, de Souza E, Brasileiro ACM, Spehar CR, Ferreira ME, Mehta Â (2008) Identification of drought-responsive genes in roots of upland rice (Oryza sativa L). BMC Genomics 9:1–13

Raj A, Stephens M, Pritchard JK (2014) fastSTRUCTURE: variational inference of population structure in large SNP data sets. Genetics 197:573–589

Rellstab C, Gugerli F, Eckert AJ, Hancock AM, Holderegger R (2015) A practical guide to environmental association analysis in landscape genomics. Mol Ecol 24:4348–4370

Saylor LC (1961) A karyotypic analysis of selected species of Pinus. Master's Thesis North Carolina State University. Genetica 10:77–84

Soltis PS, Soltis DE (2016) Ancient WGD events as drivers of key innovations in angiosperms. Curr Opin Plant Biol 30:159–165

Stevens KA, Wegrzyn JL, Zimin A (2016) Sequence of the sugar pine megagenome. Genetics 204:1613–1626

Stift M, Kolář F, Meirmans PG (2019) Structure is more robust than other clustering methods in simulated mixed-ploidy populations. Heredity 123:429–441

Vangestel C, Vázquez-Lobo A, Martínez-García PJ, Calic I, Wegrzyn JL, Neale DB (2016) Patterns of neutral and adaptive genetic diversity across the natural range of sugar pine (Pinus lambertiana Dougl.). Tree Genetics Genomes 12:51

van Ooijen G, Mayr G, Kasiem MM, Albrecht M, Cornelissen BJ, Takken FL (2008) Structure-function analysis of the NB-ARC domain of plant disease resistance proteins. J Exp Bot 59:1383–1397

Vranová E, Atichartpongkul S, Villarroel R, Van Montagu M, Inzé D, Van Camp W (2002) Comprehensive analysis of gene expression in Nicotiana tabacum leaves acclimated to oxidative stress. Proc Natl Acad Sci USA 99:10870–10875

Wang T, Hamann A, Spittlehouse D, Carroll C (2016) Locally downscaled and spatially customizable climate data for historical and future periods for North America (I Álvarez, Ed.). PLOS ONE 11:e0156720.

Weiss M, Sniezko RA, Puiu D, Crepeau MW, Stevens K, Salzberg SL, Langley CH, Neale DB, De La Torre AR (2020) Genomic basis of white pine blister rust quantitative disease resistance and its relationship with qualitative resistance. Plant J tpj.14928

Xu Y, Liu F, Zhu S, Li X (2018) The maize NBS-LRR gene ZmNBS25 enhances disease resistance in rice and Arabidopsis. Front Plant Sci 9:1033

Yang XY, Lu MQ, Wang YF, Wang YR, Liu ZJ, Chen S (2021) Response mechanism of plants to drought stress. Horticulturae 7:50–86

Zhang D, Wang Y, Shen J, Yin J, Li D, Gao Y, Xu W, Liang J (2018) OsRACK1A, encodes a circadian clock-regulated WD40 protein, negatively affect salt tolerance in rice. Rice 11:45–60

5 Maritime Pine Genomics in Focus

Lieven Sterck, Nuria de María, Rafael A. Cañas, Marina de Miguel, Pedro Perdiguero, Annie Raffin, Katharina B. Budde, Miriam López-Hinojosa, Francisco R. Cantón, Andreia S. Rodrigues, Marian Morcillo, Agathe Hurel, María Dolores Vélez, Fernando N. de la Torre, Inês Modesto, Lorenzo Federico Manjarrez, María Belén Pascual, Ana Alves, Isabel Mendoza-Poudereux, Marta Callejas Díaz, Alberto Pizarro, Jorge El-Azaz, Laura Hernández-Escribano, María Ángeles Guevara, Juan Majada, Jerome Salse, Delphine Grivet, Laurent Bouffier, Rosa Raposo, Amanda R. De La Torre, Rafael Zas, José Antonio Cabezas, Concepción Ávila, Jean-Francois Trontin, Leopoldo Sánchez, Ricardo Alía, Isabel Arrillaga, Santiago C. González-Martínez, Célia Miguel, Francisco M. Cánovas, Christophe Plomion, Carmen Díaz-Sala, and María Teresa Cervera

A. R. De La Torre (ed.), *The Pine Genomes*, Compendium of Plant Genomes,
https://doi.org/10.1007/978-3-030-93390-6_5

L. Sterck · I. Modesto
Department of Plant Biotechnology and Bioinformatics and VIB Center for Plant Systems Biology, Ghent University, 9052 Ghent, Belgium
e-mail: lieven.sterck@psb.vib-ugent.be

I. Modesto
e-mail: ismodesto@itqb.unl.pt

N. de María · M. López-Hinojosa · M. D. Vélez · L. F. Manjarrez · M. C. Díaz · L. Hernández-Escribano · M. Á. Guevara · D. Grivet · R. Raposo · J. A. Cabezas · R. Alía · M. T. Cervera (✉)
Departamento de Ecología y Genética Forestal, Centro de Investigación Forestal (CIFOR, CSIC), Instituto Nacional de Investigación y Tecnología Agraria y Alimentaria (INIA), Madrid, ES, Spain
e-mail: cervera@inia.es

N. de María
e-mail: ndemaria@inia.es

M. López-Hinojosa
e-mail: miriam.lopez@inia.es

M. D. Vélez
e-mail: velez.mdolores@inia.es

L. F. Manjarrez
e-mail: lorenzo.federico@inia.es

M. C. Díaz
e-mail: callejas.marta@inia.es

M. Á. Guevara
e-mail: aguevara@inia.es

D. Grivet
e-mail: dgrivet@inia.es

R. Raposo
e-mail: raposo@inia.es

J. A. Cabezas
e-mail: cabezas.joseantonio@inia.es

R. Alía
e-mail: alia@inia.es

N. de María · M. López-Hinojosa · M. D. Vélez · L. F. Manjarrez · M. Á. Guevara · J. A. Cabezas · M. T. Cervera
Unidad Mixta de Genómica y Ecofisiología Forestal, Instituto Nacional de Investigación y Tecnología Agraria y Alimentaria (INIA)/Universidad Politécnica de Madrid (INIA/UPM), Madrid, ES, Spain

R. A. Cañas · F. R. Cantón · F. N. de la Torre · M. B. Pascual · J. El-Azaz · C. Ávila · F. M. Cánovas
Dpto. Biología Molecular y Bioquímica. Facultad de Ciencias Campus de Teatinos S/N, Universidad de Málaga, Málaga, ES, Spain
e-mail: racanas@uma.es

F. R. Cantón
e-mail: frcanton@uma.es

F. N. de la Torre
e-mail: fdelatorre@uma.es

M. B. Pascual
e-mail: bpascual@uma.es

J. El-Azaz
e-mail: jelazaz@alu.uma.es

C. Ávila
e-mail: cavila@uma.es

F. M. Cánovas
e-mail: canovas@uma.es

M. de Miguel · A. Hurel · L. Bouffier · S. C. González-Martínez · C. Plomion
INRAE, Univ. Bordeaux, BIOGECO, 33610 Cestas, FR, France
e-mail: marina.de-miguel@inrae.fr

L. Bouffier
e-mail: laurent.bouffier@inrae.fr

S. C. González-Martínez
e-mail: santiago.gonzalez-martinez@inrae.fr

C. Plomion
e-mail: christophe.plomion@inrae.fr

M. de Miguel
EGFV, Univ. Bordeaux, Bordeaux Sciences Agro, INRAE, ISVV, 33882 Villenave d'Ornon, France

P. Perdiguero · C. Miguel
iBET, Instituto de Biologia Experimental e Tecnológica, Apartado 12, 2781-901 Oeiras, Portugal
e-mail: perdiguero.pedro@inia.es

C. Miguel
e-mail: cmmiguel@fc.ul.pt

P. Perdiguero
Centro de Investigación en Sanidad Animal (CISA-INIA), Madrid, ES, Spain

A. Raffin
INRAE, UEFP, 33610 Cestas, FR, France
e-mail: annie.raffin@inrae.fr

K. B. Budde
Department of Forest Genetics and Forest Tree Breeding, Buesgen Institute, Georg-August University of Göttingen, Göttingen, Germany
e-mail: k.budde@uni-goettingen.de

A. S. Rodrigues · I. Modesto
ITQB NOVA, Instituto de Tecnologia Química e Biológica António Xavier, Universidade Nova de Lisboa, Av. República, 2780-157 Oeiras, PT, Portugal

Abstract

The advent of next-generation genome sequencing technologies has allowed approaching the sequencing and analysis of large and complex conifer genomes. Maritime pine (*Pinus pinaster* Ait.) is an economically and ecologically important conifer species widely distributed in South-West Europe, which shows a significant genetic and adaptive variability. This chapter takes on the task of reviewing the insights into the maritime pine genome sequencing breakthrough and its impact on downstream analysis. Maritime pine genome sequencing and assembly approaches are described along with the impact of related tools. A section of the state-of-the-art research on comparative, functional, structural, and translational genomics aimed at dissecting the genetic basis and the specific regulation of biological processes underlying the expression of traits of interest in maritime pine and other conifers is also described. Perspectives about the impact of these tools as well as additional research approaches are discussed.

5.1 Introduction

Maritime pine (*Pinus pinaster* Ait.) is a widely distributed Mediterranean conifer in South-West Europe. It is found in a broad range of ecosystems including continental as well as Atlantic and Mediterranean coastal forests and woodlands. It also reaches North-West Africa where it forms fragmented populations (Abad Viñas et al. 2016). This native species from the Mediterranean basin (Carrión et al. 2000; Rubiales et al. 2009) shows a high level of neutral variation, which is structured in differentiated gene pools related to various glacial refugia (Bucci et al. 2007). The current patched distribution of maritime pine has been shaped by a long historical interaction of these gene pools with natural and anthropogenic disturbances (Carrión et al. 2000;

A. S. Rodrigues
Swammerdam Institute for Life Sciences, University of Amsterdam, Postbus 1210, 1000 BE Amsterdam, NL, Netherlands

M. Morcillo · I. Mendoza-Poudereux · I. Arrillaga
Biotechnology and Biomedicine (BiotecMed) Institute and Plant Biology Department, University of Valencia, Valencia, ES, Spain
e-mail: angeles.morcillo@uv.es

I. Mendoza-Poudereux
e-mail: Isabel.Mendoza@uv.es

I. Arrillaga
e-mail: isabel.arrillaga@uv.es

I. Modesto · A. Alves · C. Miguel
BioISI - Biosystems & Integrative Sciences Institute, Faculdade de Ciências, Universidade de Lisboa, Campo Grande, 1749-016 Lisboa, PT, Portugal
e-mail: acalves@fc.ul.pt

A. Pizarro · C. Díaz-Sala
Departamento de Ciencias de la Vida (Fisiología Vegetal), Universidad de Alcalá, Alcalá de Henares, ES, Spain
e-mail: alberto.pizarro@uah.es

C. Díaz-Sala
e-mail: carmen.diazsala@uah.es

J. Majada
CETEMAS, Forest and Wood Technology Research Centre, B° Pumarabule, Carbayín s/n, 33936 Asturias, ES, Spain
e-mail: jmajada@cetemas.es

J. Salse
INRAE, Univ. Clermont Auvergne, GDEC, 63100 Clermont-Ferrand, FR, France
e-mail: jsalse@clermont.inra.fr

A. R. De La Torre
Northern Arizona University, Flagstaff AZ86011, USA
e-mail: amanda.de-la-torre@nau.edu

R. Zas
MBG-CSIC, Apdo. 28, 36080 Pontevedra, ES, Spain
e-mail: rzas@mbg.csic.es

J.-F. Trontin
BioForBois, FCBA Technological Institute, Wood & Construction Industry Dpt, 71 Route d'Arcachon, Pierroton, 33610 Cestas, FR, France
e-mail: jean-francois.trontin@fcba.fr

L. Sánchez
INRAE, ONF, 45075 Orléans, BioForA FR, France
e-mail: leopoldo.sanchez-rodriguez@inrae.fr

González-Martínez et al. 2007), resulting in highly differentiated populations locally adapted to different conditions (Grivet et al. 2017). However, climate change has increased the recurrence of severe droughts as well as the frequency of heat-induced wildfire, which are major disturbances that are threatening Mediterranean forests and plantations, affecting their dynamics and survival (Batllori et al. 2017; Ruffault et al. 2020).

Maritime pine has also played a relevant socio-economic role in human welfare in western Mediterranean countries. This species has been used as a source of wood and resin. Other subtractive activities have been diversified, including the production of biomass and bioproducts. Maritime pine has also played an important role as a source of high-value non-market services associated with climate regulation and landscape enhancement (e.g., carbon storage, nutrient cycling, water regulation, soil protection, and maintenance of biodiversity), as well as recreation (Abad Viñas et al. 2016).

When compared with other pine species, such as mountain or boreo-alpine pines, Mediterranean pines, including maritime pine, show faster growth rates, larger root systems, and faster rooting through the soil depth at lower temperature and soil moisture (Andivia et al. 2019). Different strategies for stem reorientation also occur between maritime pine and loblolly pine, a species grown in mesic environments. Maritime pine shows a fast-primary response to leaning angles, while a more efficient secondary reorientation was observed in stems of loblolly pine. The amount of cell wall may influence the amount of water for the same xylem area. Lower amount of cell wall in maritime pine than in loblolly pine would result in a higher water uptake for the same xylem area in maritime pine compared to loblolly pine (Ba et al. 2010). In addition, sensitivity of maritime pine secondary growth to climate conditions increases wood heterogeneity regarding the width of growth rings and element dimension (Rozas et al. 2011). Other physiological traits related to plant hydraulic features, such as vulnerability and response to xylem cavitation and embolism or the influence of leaf morphology, water relations and hydraulic traits on the appropriate leaf water supply, which ensures the maintenance of active photosynthesis and productivity, might also contribute to explain the geographical segregation of pine species in Southern Europe (Nardini et al. 2014). *Pinus pinaster* populations across their natural distribution show different resilience strategies to drought; this variation is also observed at individual level within population (Sánchez-Salguero et al. 2018; Feinard-Duranceau et al. 2018). In addition, the successful adaptation of maritime pine to pathogen attack depends on the species, populations, and genetic variation within populations (Elvira-Recuenco et al. 2014). Common mechanisms of response to the pathogen infection, likely representing strategies shared by *Pinus* species, have been identified between species. However, differences have also been found. The response of the sensitive maritime pine to Pine Wood Nematode (PWN) infection shows a more complex pattern of gene expression than resistant species, which could be related with the high susceptibility to PWN of this species (Gaspar et al. 2020). The complementarity and/or expression of these physiological traits at inter- and intraspecific level might explain not only species distribution but also the different growth strategies to cope with abiotic and biotic stress factors across its broad distribution range.

Conifer species have very large and complex genomes (Mackay et al. 2012). The use of high throughput technologies has made it possible to approach sequencing and analysis of these giga-genomes, proving a better knowledge of their structure and functioning. In addition, it has enabled the study of the biology of these long-lived organisms. *Pinus pinaster* has a diploid genome of 28.322 Gbp/C distributed in 12 chromosomes (Zonneveld 2012). Integration of genetic, genomic and transcriptomic-based information, provided by different research teams, are available in several public databases such as Gymno PLAZA (https://bioinformatics.psb.ugent.be/plaza/versions/gymno-plaza/organism/view/Pinus+pinaster), TreeGenes (https://treegenesdb.org/) and SustainPineDB (http://

www.scbi.uma.es/sustainpinedb/). The availability of these databases has allowed the development of a collection of omics-based tools, which are reviewed in this chapter, along with their applications. In addition, phenotypic information is available from segregating progenies and a clonal common garden network (CLONAPIN), which was installed in five different locations representing the natural environmental range of maritime pine (de Miguel et al. 2020). This resource will enable molecular dissection of traits of interest, disentangling factors potentially affecting them.

Considering the maritime pine economic and ecological value as well as its biological features, including its genetic and adaptive variability (Eveno et al. 2008; Grivet et al. 2017), this species has been selected as a model conifer species in South-Western Europe. In this chapter, we review the application of new research technologies as well as the development of genomic tools in this species, that provide (i) a foundation for invigorating progress toward understanding its genome structure, function, and evolution, (ii) new ways to genetically improve its resilience to biotic and abiotic threats while maintaining its productivity to bring out both economic and ecological returns, and (iii) better genetic resources management, conservation, and exploration to face future challenges.

5.2 New Tools

5.2.1 Genome Sequencing

Compared to angiosperms, the availability of assembled and annotated genomes lags behind in gymnosperms, mainly due to their enormous genome size (20–30 Gb) and their highly repetitive nature (Ahuja and Neale 2005; De La Torre et al. 2014; Ojeda et al. 2019). The use of haploid tissue, that offers a significant reduction in sequence complexity, facilitates higher contiguity in their genome assembly (Zimin et al. 2014). Thus, the simplified assembly provided by a haploid genome was the basis of genome sequencing programs for several species, including *Pinus pinaster* (Arrillaga et al. 2014; Neale et al. 2014; Stevens et al. 2016; Shimizu et al. 2017; Li et al. 2020; Scott et al. 2020).

5.2.1.1 Haploid Tissue Generation

Haploid tissue might be achieved in vivo (in seeds) or in vitro by gametophyte culture. Recently, genome editing has been used to generate paternal haploids in wheat (Lv et al. 2020). In maritime pine, template for genome sequencing was obtained from a cell line (L5) derived from megagametophyte cultures (Arrillaga et al. 2014). Briefly, in mid-September, cones were collected from open pollinated maritime pine Oria 6 genotype (Oria, Almeria, Spain), which was selected for its adaptability to extreme drought conditions. Isolated megagametophyte was cultured on a modified Litvay′s medium (Arrillaga et al. 2014). Calli derived from these cultures were transferred every three weeks to the same medium and analyzed for haploid status. The haploid status of the cell line was first determined by chromosome counting, flow cytometry, and seven polymorphic microsatellites (Arrillaga et al. 2014). The selected haploid L5 line was maintained for six months by a three-wesubculture frequency onto the same medium.

Long-term in vitro culture may induce instability in the regenerants, affecting ploidy level or leading to important chromosome rearrangements (Larkin and Scowcroft 1981). Therefore, checking the integrity of the template DNA for the de novo whole sequencing of the genome from the haploid line is an important task. An indirect evaluation of chromosome number can be obtained using flow cytometry, but this does not usually allow to detect aneuploidy involving the presence of only one extra or missing chromosome. Karyological analyses allow the detection of all modifications in chromosome number. However, it is sometimes difficult to obtain enough cells containing metaphases with chromosomes sufficiently contracted and separated to provide a reliable chromosome count (Arrillaga et al. 2014). Chromosome integrity can be checked also using molecular markers, which was the strategy used to validate the suitability of the Oria 6 L5 cell line used to sequence maritime pine genome (Cabezas et al. 2016). The presence of all chromosomes in haploid state and the absence of deletions involving the loss of chromosome arms were confirmed by studying the allelic composition at a microsatellite and 23 SNP loci that were (i) heterozygous in Oria 6 genotype; and (ii) covered all chromosome arms, based on their location on Oria 6 genetic map (de Miguel et al. 2014).

5.2.1.2 High Molecular Weight Genomic DNA Isolation

Third-generation sequencing technologies, which yield read lengths higher than 10 kb and are based on the sequencing of a single DNA molecule, require the use of high-quality, high molecular weight (HMW) genomic DNA (HMW gDNA) for library preparation. Different protocols for HMW gDNA extraction have been optimized and commercial kits are available. These protocols include two processes: nuclei isolation and gDNA extraction. Nuclei isolation is required to concentrate sequencing efforts on HMW gDNA due to the high proportion of chloroplast DNA in the plant cell (Peterson et al. 2000; Zhang et al. 2012). Some of these protocols were originally designed for HMW gDNA extraction from conifers, such as the protocols developed for preparing bacterial artificial chromosome (BAC) libraries, in which purified nuclei are embedded in agarose (matrix plugs) to minimize mechanical shearing of HMW gDNA (Bautista et al. 2007; Zimin et al. 2014). Other protocols have been recently developed to extract HMW gDNA for third-generation sequencing technologies, which are less time-consuming. These protocols have been successfully applied to extract gDNA for nanopore sequencing of *Sequoiadendron giganteum* (giant sequoia) and *Sequoia sempervirens* (coast redwood) genomes (Workman et al. 2018; Scott et al. 2020).

5.2.1.3 Next Generation Sequencing Platforms

The use of next-generation sequencing technologies (NGS) for the sequencing of other large conifer genomes, such as Norway spruce (Nystedt et al. 2013); white spruce (Birol et al. 2013); and loblolly pine (Neale et al. 2014; Zimin et al. 2014) genomes, paved the way for the sequencing of the maritime pine genome. Since then, additional conifer genomes have been sequenced: *Pinus lambertiana* (Stevens et al. 2016), *Pseudotsuga menziesii* var. menziesii (Neale et al. 2017), *Abies alba* (Mosca et al. 2019), *Larix sibirica* (Kuzmin et al. 2019), and *Sequoiadendron giganteum* (Scott et al. 2020).

Paired end (PE) and mate pair (MP) genomic libraries of different sizes, and shotgun sequencing using a combination of 454 and Illumina platforms, have been used for the genome sequencing. The output sequencing data generated true PE reads with two end tags separated by genomic fragments of increasing sizes to constitute a collection of different libraries. One of the main drawbacks of the 454 and Illumina platforms is the size of the reads that are produced. This is a significant disadvantage for structural genomics approaches since bioinformatics tools do not allow generating high-quality de novo assemblies for large and complex genomes containing high numbers of repetitive elements (De La Torre et al. 2014; Sohn and Nam 2018). The ability to generate long reads using Nanopore or PacBio sequencing has allowed to span large repeat regions, and, therefore assemble fragmented regions during the assembly process.

5.2.1.4 Characterization of BAC Clones and Genome Sequencing Effort

The large conifer genomes are characterized by the high number of repetitive sequences mostly composed of transposons and retrotransposons as well as the accumulation of non-coding regions and extensive gene duplication (Mackay et al. 2012; De La Torre et al. 2014). The shotgun sequencing approach was complemented by the sequencing of isolated bacterial artificial chromosomes (BAC clones) combined with the targeting and sequencing of gene-rich regions in the genome to establish de novo gene structures without a reference genome.

Probes for 866 maritime pine transcripts (Seoane-Zonjic et al. 2016) were designed for sequence gene capture from genomic DNA. Haploid DNA extracted from the same megagametophyte callus (line 5 from the Oria 6 genotype) was isolated, fractionated, and bounded to adapters for 454 sequencing hybridized to 120-mer probes derived from the corresponding transcriptome sequences. The captured DNA was sequenced using the FLX-Titanium platform. The gene models were constructed using GeneAssembler, a new bioinformatics pipeline that reconstructed over the 82% of the gene structures (Seoane-Zonjic et al. 2016). The approach allowed recovering a similar number of exon structures in pine genes without a reference than in genes from other species in which the pipeline was used to validate the system (Fig. 5.1).

A high proportion (85%) of the captured gene models contained sequences of promoter regulatory regions. In parallel, the screening of a *Pinus pinaster* BAC library was performed to isolate clones containing genes whose cDNA sequence was already available. Gene models derived from the gene capture approach were compared with the genomic sequences derived from the BAC clones. The approach was particularly efficient to capture the gene structures of relevant gene families in the maritime pine genome, when they were composed of a few members. This experimental approach proved to be useful for establishing exon/intron boundaries in unknown gene structures, to reconstruct incomplete genes, and to obtain promoter sequences that can be used for transcriptional regulatory studies.

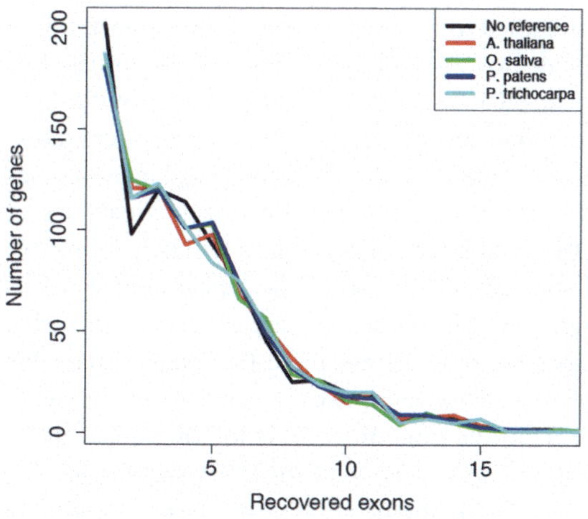

Fig. 5.1 Exon distribution in the generated gene models. *Arabidopsis thaliana* (red), *Oryza sativa* (green), *Physcomitrella patens* (dark blue), *Populus trichocarpa* (light blue), or no reference (black) were used as references

The de novo sequencing of the *Pinus pinaster* genome is based on the combination of different NGS technologies that all use the same single megagametophyte as a template. A genome draft was built based on sequences obtained from 5 single-end libraries, 16 MP libraries of 3.5 Kb, 7 MP libraries of 6 Kb, 4 MP libraries of 8 Kb and 18 MP libraries of 10 Kb using 454 (Roche 454 GS-FLX sequencing) as well as 38 PE libraries and 10 MP libraries of 3 Kb and 5 Kb using illumina—Hiseq2000 sequencing. In addition, a combination of gene capture and BAC clone sequencing was performed. The bulk of these sequences represented more than 65X coverage of the pine genome (Fig. 5.2). Twelve GridIon flowcells of Oxford Nanopore Technology (ONT) data were additionally generated, adding approximately another 5× coverage.

5.2.1.5 Genome Assembly and Annotation

The availability of a near fully sequenced genome sequence is one of the cornerstones of modern-day molecular biology for any species being under study. Moreover, to label a species as a model organism, access to its genome is a prerequisite. With the advances in DNA sequencing technologies over the past decades, generating a genome sequence is no longer the daunting task it used to be for the early model organisms (e.g., *Arabidopsis* and *Homo sapiens*). Nowadays, a range of different approaches to accommodate this task are available. Short read technologies (Illumina; Bentley et al. 2008), that are capable of generating immense amounts of data with high base accuracy at a fairly low cost, as well as long read technologies (PacBio, Eid et al. 2009 and Oxford Nanopore Technology, Deamer et al. 2016), although with lower base accuracy to long range scaffolding techniques (e.g., 10× genomics). Using these different technologies to the best of their abilities makes unraveling genome

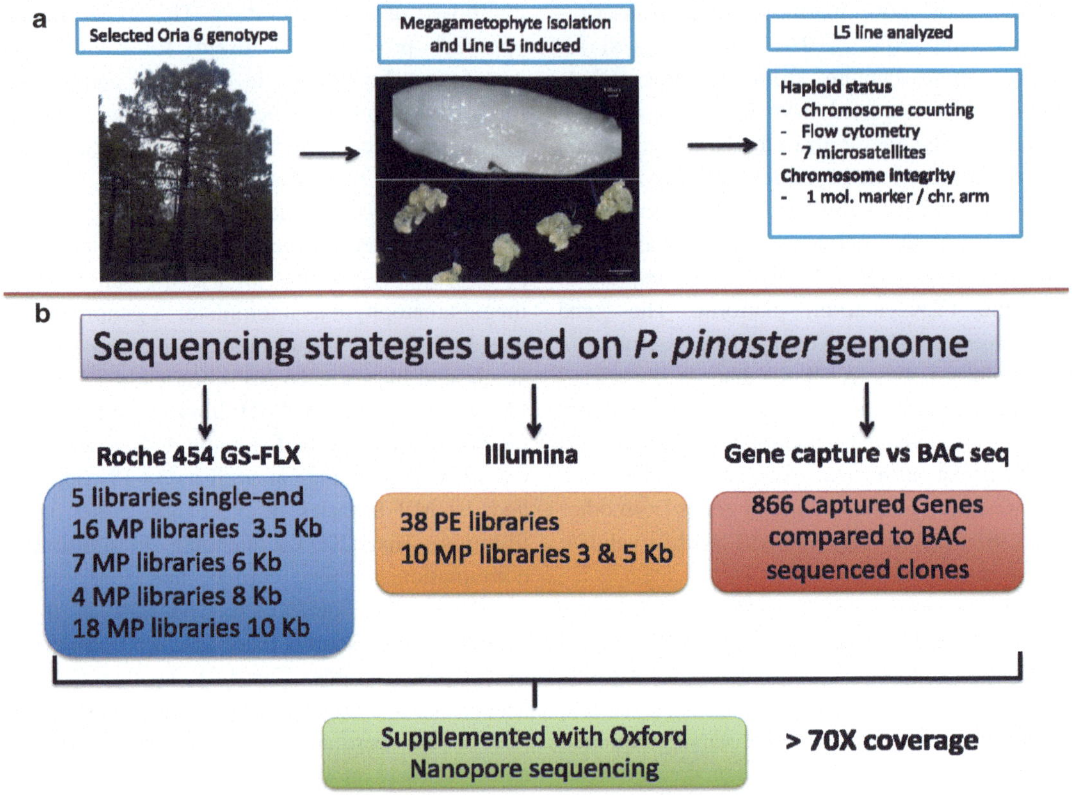

Fig. 5.2 **a** Workflow of the generation of *Pinus pinaster* DNA for genome sequencing, **b** Diagram summarizing the sequencing strategies used for *Pinus pinaster* genome sequencing. Two NGS platforms were used, 454 FLX and Illumina, that combined PE paired-end and MP mate-paired libraries. In addition, a complementary strategy of gene capture and BAC clones sequencing was followed. Third-generation S sequencing technologies were used to complete the assembly

sequences feasible for most organisms. Unfortunately, conifer genomes are somewhat the exception. Due to the characteristics of their genomes such as extremely large sizes, often >20 gigabases, high levels of heterozygosity and complex genome structure composed of many and large repeat regions (De La Torre et al. 2014), genome assembly of conifer genomes remains a challenging undertaking. Nevertheless, several (draft) genomic sequences for various conifer species have been released to the public over the past years (Nystedt et al. 2013; Birol et al. 2013; Zimin et al. 2014; Neale et al. 2017). The genome of the maritime pine will soon be added to that list. Like most of its predecessors, assembling this genome will make use of a hybrid approach, where different sequencing technologies are being combined to overcome most of the issues limiting the de novo assembly of very large genomes (Lee et al. 2019). Short read data will provide high coverage of the genome allowing the generation of an initial draft genome, long read technologies (for *Pinus pinaster* several dozen gigabases of ONT data has been generated) will help to resolve a substantial fraction of the repeat content of its genome. All this is not very different from what has been done to assemble the already available conifer genomes, however, what makes the case of maritime pine different is the biological material that served as input for DNA sequencing.

To tackle the heterozygosity issue, a common problem in de novo genome assembly, the community opted to go for a strategy where all sequencing data is being generated from a single haploid tissue (Sect. 5.2.1.1). The process of de novo genome assembly essentially consists of puzzling all the generated reads back together to form the original biological molecules from where they originally derived. This sounds straightforward but considering the amount of data that needs to be handled for large genomes, such as those of conifers, it is still a challenging undertaking. In practice, this task is performed by software tools. While several software options exist for genomes of moderate size, just a few are available for extremely large genomes. The ABySS software (Simpson et al. 2009) was used for the assembly of the maritime pine genome mainly because of its proven record of accomplishment in assembling large genomes (Birol et al. 2013; Jackman et al. 2017), and for its excellent capability to be deployed on an HPC system offering the required compute power and memory resources required for such large projects. The initial draft genome version based on the Illumina short read data was performed with ABySS, with optimized parameter settings. The next step will be to include the long-read data (ONT) to scaffold the initial draft, improving the contiguity. The last step will entail using genetic maps to assign and order the scaffolds into pseudo-chromosomes.

Even though having the assembled genome sequence is a major step in any genome project, it is still a raw resource. It is, therefore, of limited value for many biologists and for several downstream analyses. Thus, it is key to transform this raw information to knowledge. Genome annotation allows identifying and describing the genetic features present on the genomic sequences (Stein 2001; Yandell and Ence 2012). This mainly involves identifying the location and structure of coding regions (genes), but also many other genetic elements such as non-coding transcripts, repeat regions, and many more. Over the past decades, many tools and approaches have been developed and successfully applied in several genome projects. Though each one has its own advantages and disadvantages, it is generally accepted that the ones that are able to meaningfully integrate several sources of evidence, ranging from intrinsic data over protein similarity to transcript data, perform the best. Keeping that in mind, the EuGene (Foissac et al. 2008) tool was used to annotate the *Pinus pinaster* genome assembly. EuGene can be specifically optimized to perform to the best of its capabilities for a given species (Tuskan et al. 2006; Tomato Genome Consortium 2012; Olsen et al. 2016) and as such will allow to provide a high-quality genome annotation results for the benefit of the conifer scientific community.

5.2.2 Transcriptome Atlas of *Pinus Pinaster*

5.2.2.1 The Biological Relevance of Gene Expression Analysis in Cell and Tissues of Maritime Pine

Conifers are long-living organisms with a complex anatomy. The precise analysis of the transcriptional activity in these specialized cells and tissues is of paramount importance to understand how key molecular and biochemical processes are regulated and how they evolved in this group of economically and ecologically important group of woody plants. A traditional approach to study the precise distribution of individual transcripts in different cell types and tissues has been in situ hybridization (ISH), a technique in which intact RNA included in fixed paraffin-embedded tissue sections is analyzed using specific probes derived from the primary sequence of previously characterized genes. In the last decade, novel and powerful methods have been developed to spatially resolve transcriptome-wide expression and to make large sets of data available to the research community through appropriate bioinformatic platforms. Although the application of these emerging technologies in conifers is still at the initial stages of development, the current compiled data offers a high potential for the understanding of basic biological processes, how they are regulated and how the functioning of different cell types is spatially integrated in the whole organism.

For many years, microarray analysis and direct RNA sequencing of samples containing a large number of cell types, tissues and entire organs only provided valuable information from the bulk of the total RNA population, with dilution of transcript levels for potentially highly expressed genes in specific cell types and tissues. Laser capture microdissection (LCM) of biological organ slices under the microscope enables the efficient selection of specific populations of cells. Furthermore, a combination of LCM and NGS technologies represents a powerful tool to resolve the entire transcriptome of specific cell types and tissues. This experimental approach is particularly well suited for enriching the pre-existing transcriptome in low copy genes as well as for the identification of rare transcripts involved in specialized functions (Morozova et al. 2009). Transcriptome sequencing also permits the identification of transcriptional networks and key regulators in these specific cells and tissues. One path to enhance functional diversity in a genome involves the proliferation of genes into multigene families. The size and diversity of gene families is often reflecting the biology of responses in different plant species, but it is necessary to know their distribution in cell types and their compartmentation to fully understand the nature of these responses. A detailed transcriptome atlas has been achieved for model plant species such as *Arabidopsis* and rice. However, extrapolation of functional knowledge to non-model species is a complicated task. For example, long-lived plant species, such as trees, exhibit specific developmental processes which biological understanding cannot be fully addressed using the available information in model plants. Thus, studies on the transcriptome of cells and tissues are required for the exploration of transcriptome dynamics in trees exhibiting contrasted phenotypes for adaptive traits.

5.2.2.2 In Situ Spatial Distribution of Gene Expression

ISH is a suitable experimental approach to examine the distribution of specific gene transcripts in different cell types and tissues. This sophisticated and extremely laborious technique has been used in a variety of studies to get insights into the function of many conifers structural and regulatory genes during development, and in response to environmental cues (Pérez-Rodríguez et al. 2006; Laajanen et al. 2007; Craven-Bartle et al. 2013; Rupps et al. 2016). While ISH is a powerful tool for studying the spatial expression patterns of genes in plants, its ability to reliably quantify gene expression levels is quite limited. To overcome this limitation, protocols for LCM of specific cell and tissue samples represent a suitable alternative. Furthermore, a combination of ISH and LCM followed by quantitative determination of transcript levels has been used to understand the metabolic functions of individual members of a gene family (Castro-Rodríguez et al. 2015). Another limitation inherent to ISH is that the experimental approach is generally low throughput. Alternatively, the combination of LCM and NGS can provide a high throughput analysis of specific cells and tissues facilitating transcriptome-wide expression studies (Cañas et al. 2017).

5.2.2.3 Transcriptome Sequencing

Transcriptomic studies in conifers have been of great value to understand basic biological functions and global responses to environmental stimuli. Large transcriptome resources have been generated in economical and ecologically important conifers species such as spruces (Ralph et al. 2008; Rigault et al. 2011) and pines (Canales et al. 2014; Baker et al. 2018). These resources have been used in functional genomics approaches for the identification of genes involved in a wide range of processes such as nutrient acquisition, wood formation, embryo development and also to study the molecular responses of trees to a variety of biotic and abiotic stresses.

In maritime pine, a comprehensive characterization of the transcriptome was performed using a combination of two different NGS platforms, 454 and Illumina (Canales et al. 2014). The de novo assembly of the maritime pine transcriptome provided a large catalogue of expressed genes in this conifer species, permitting the annotation of protein coding genes and facilitating, in a great extent, the assembly of the genome.

5.2.2.4 Laser Microdissection Technology for Maritime Pine Transcriptome Analysis

LCM has been used to study the localization of specific transcripts, polypeptides, and metabolites in plants (Yi et al. 2012; Pattison et al. 2015; Zhu et al. 2016). In conifers, LCM has been used as a tool to specifically explore the content of secondary metabolites in vascular tissues and the expression of genes involved in the metabolism of terpenoids (Abbott et al. 2010; Hamberger et al. 2011; Jyske et al. 2015). However, much more limited information is available on the application of LCM in combination with NGS to localize gene expression in conifers (Cañas et al. 2014, 2017; Celedon et al. 2017). The procedure implies the following sequential steps: (i) Cryosectioning of target tissues/cells; (ii) RNA isolation and cDNA synthesis; (iii) cDNA amplification; (iv) library construction; (v) transcriptome sequencing. A brief description of these steps is indicated below.

Normally, methods for LCM in conifers need optimization depending on the tissue that is used as starting material. Protocols for laser capture microdissection of different maritime pine cell types have been established (Cañas et al. 2014). Initially, two different approaches were used: LCM from paraffin-embedded tissues and LCM from flash frozen samples. However, the quality of isolated RNA from paraffin-embedded tissue was poor for transcriptome sequencing and prevented the reproducibility of results. In contrast, cryosections from unfixed frozen tissues produced the best preparations of RNA. Performing sections at low temperatures greatly favor the integrity of the isolated RNA.

The amount of isolated total RNA from LCM samples depended primarily on the number of sections used. However, the amount of total RNA, ranging between 15 and 60 ng, was insufficient for the construction of cDNA libraries and transcriptome sequencing (Cañas et al. 2014). Due to low RNA yield, it was then necessary to synthesize and amplify cDNA from all tissue-type samples. An adapted protocol for Conifer RNA Amplification (CRA+) was developed to perform complementary DNA synthesis and amplification from tiny amounts of total RNA isolated from LCM samples (Further details on this protocol can be found in Cañas et al. 2014). Using the (CRA+) protocol, non-specific amplifications are prevented and read length and read number are considerably enhanced. This technical advance facilitates global gene expression studies in individual tissues of conifers and may also be applied to other plant species. cDNA libraries were generated using the adapted protocols from multiple tissues and transcriptome sequencing was performed using NGS. An illustration of the workflow followed to generate spatial transcriptomics in maritime pine is shown in Fig. 5.3.

5.2.2.5 Bioinformatics Analyses

NGS reads were pre-processed and used for transcriptome assembly of the different tissue/cell types. The transcriptomic sequences are available at ConGenIE.org (Street 2019). The alignment of whole set of transcripts to the reference transcriptome allowed the identification of transcripts and quantification of their relative abundance in the LCM samples. A summary of the results obtained is shown in Fig. 5.4. Visualization of the spatial distribution of specific transcripts can also be examined at ConGenIE.org (Cañas et al. 2017, 2019).

5.2.3 Vegetative Propagation

Asexual reproduction through vegetative propagation (cloning) has long been recognized as a powerful tool to rejuvenate and incidentally get access to true-to-type deployment in plantation forests of selected, 'tested' tree varieties for desirable traits (Franclet et al. 1987). Vegetative propagation also provides ideal support for research purposes (clonal experimental design), evaluation (clonal test, reverse genetics), variety design (genetic engineering), sanitation (meristem culture), and preservation (clonal field or cryopreserved archives) of genetic resources from native or improved stocks (Ahuja 2017).

Compared to classical forestry approaches based on sexual reproduction, clonal forestry

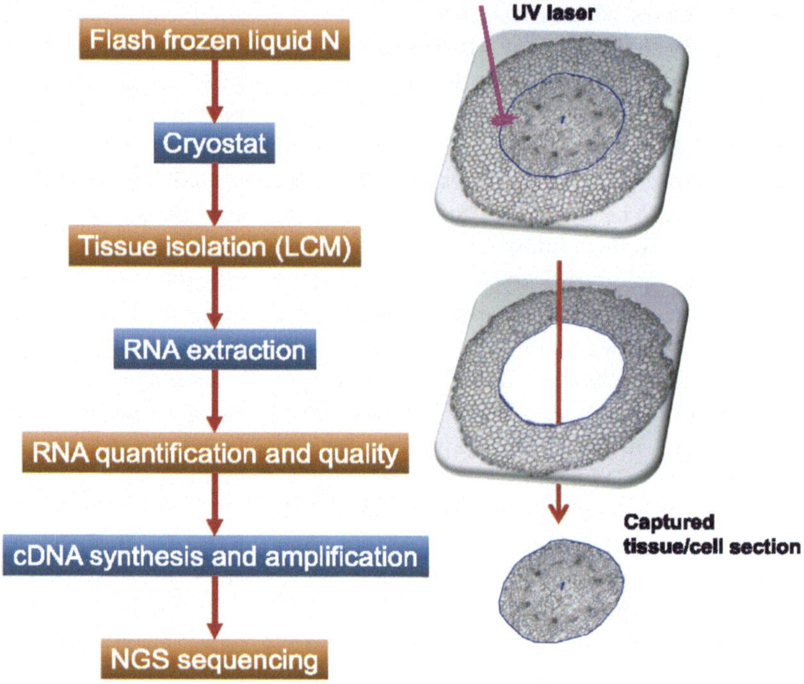

Fig. 5.3 Schematic representation of the workflow followed for LCM and transcriptome sequencing in maritime pine

allows utilization of the whole available genetic variance, both additive and non-additive (Park et al. 2016). It is no longer necessary to wait for sexual maturity of variety progenitors, which typically occurs after long juvenile and then adult vegetative phases in forest trees. Breeding progress can therefore be achieved more efficiently and with more flexibility than seed orchards to cope with the climate crisis and associated environmental and socio-economic cues. This is especially true if vegetative propagation can be combined with early, predictive selection (during the juvenile phase) of valuable genotypes through genome-wide molecular profiling (Park et al. 2016; Ding et al. 2019). Breakthrough technologies for both vegetative propagation and genomic selection are currently in development in conifers, including in maritime pine (Klimaszewska et al. 2016; Plomion et al. 2016b).

Significant efforts have been made to introduce vegetative propagation technologies in maritime pine breeding programs since the 1970s (Franclet et al. 1987). Grafting is currently the routine method for establishing clonal tests, field archives, and seed orchards with selected trees. The rootstock may, however, interfere or even prevent proper evaluation of traits of interest. Grafting is also too expensive for deploying tested clones. Clonal propagation of maritime pine was therefore soon envisaged by stem rooted cuttings (Chaperon et al. 1991), minicuttings (Majada et al. 2011), micropropagation (microcuttings) through axillary or adventitious budding (Dumas and Monteuuis 1995; Tereso et al. 2006a; Álvarez et al. 2009; De Diego et al. 2011) or meristem micrografting (Dumas et al. 1989) followed by micropropagation of rejuvenated micrografts (Trontin et al. 2004). Although field demonstration of micropropagated clones could be achieved (Trontin et al. 2004), both low multiplication rate and maturation issues of propagated material (Monteuuis and Dumas 1992; De Diego et al. 2011) prevented mass propagation of tested clones for commercial application. Currently, rooted macro, mini or micro-stem cuttings can only be used to propagate juvenile maritime pine material, typically seedlings.

The production of cuttings or minicuttings has been optimized for practical application. Main

Fig. 5.4 Spatial transcriptomics in maritime pine. **a** Distribution of transcripts in maritime pine LCM samples. **b** Transcriptional map of unique transcripts

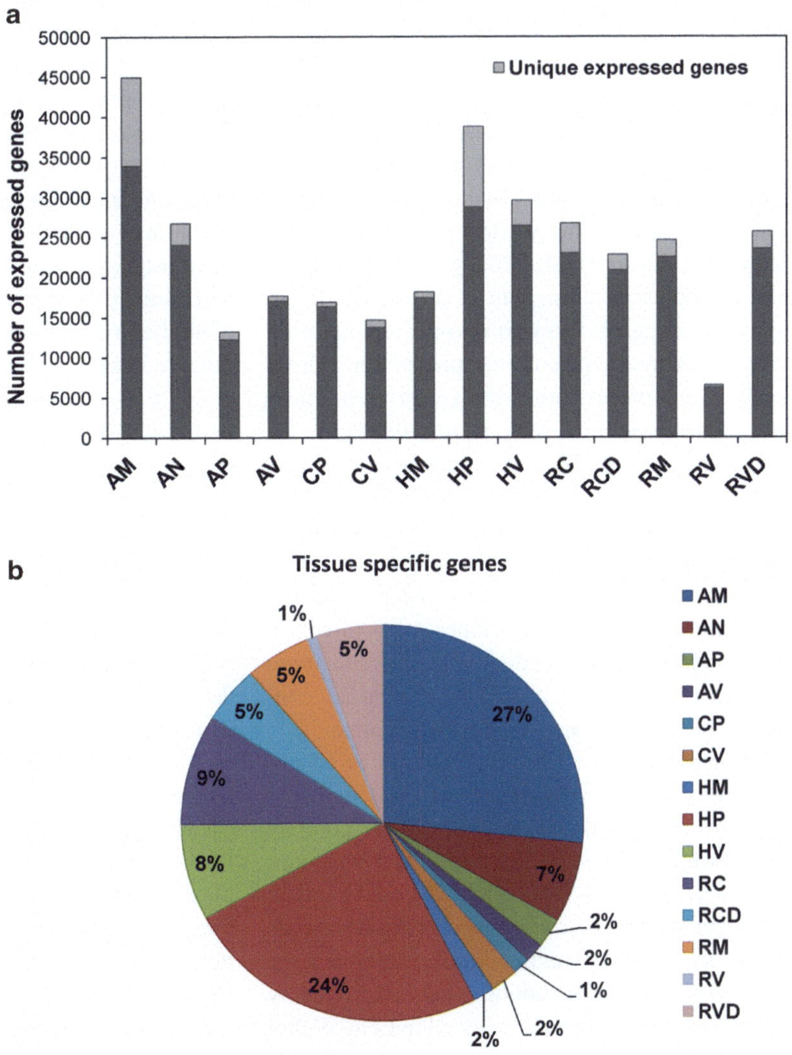

factors affecting the rooting ability of (mini)cuttings (Riov et al. 2020) include (i) the genetic background, (ii) developmental stage and management of mother plants, (iii) shoot collection season, and (iv) auxin treatment. Compared to traditional rooted cuttings, minicuttings offer greater operational, technical, economic, environmental, and quality benefits. Minicutting management and induction treatments significantly influenced rooting ability (Majada et al. 2011). Efficient management of pine mini-hedges could result in both high production yield of rooted minicuttings (4,447–8,581/m^2/year depending on family) and adequate nutritional status. Similar results were reported by Assis et al. (2004) in other pine species (9,600/m^2/year). Nitrogen and carbon concentration in needle tissues were considered as good indicators of rooting success in maritime pine. Strong interaction between family and fertilization level was reported by Martínez-Alonso et al. (2012) suggesting differences in nutrient uptake or assimilation. Moreover, Rowe et al. (2002) showed that nitrogen concentration increased in all families tested as a function of nitrogen supply. The selection of elite genotypes with good rooting ability can therefore be considered for intensive clonal forestry (Shepherd et al. 2005;

Gravel-Grenier et al. 2011). Information about how many cycles of hedging can be made per year for pines remains scarce (Rowe et al. 2002; Sharma and Verma 2011). Up to three cutting cycles per year could be considered in loblolly pine (Rowe et al. 2002) with highest number of 9 cm orthotropic shoots obtained in spring, followed by summer and winter. In maritime pine, Martínez-Alonso et al. (2012) also advised to collect shoots in spring but at an earlier developmental stage (3–5 cm orthotropic shoots) for a ninefold increase in cuttings production. Obviously, variation in rooting ability between family and clones is still an issue in maritime pine for propagating improved forest regeneration stocks. Nevertheless, the technique could be implemented to produce a clonal reference collection named CLONAPIN (www.trees4future.eu) covering the entire genetic diversity of the species, with more than 1,000 clones structured in provenances, families and clones. This collection has been established as clonal garden in contrasted sites in Portugal, Spain and France.

In search for an effective system for propagating mature trees, the possibility of inducing somatic embryogenesis from seeds in conifers (demonstrated for more than 35 years) was considered a paradigm shift (Klimaszewska et al. 2016). First results were reported in maritime pine in the late 1980s (Hughes-Jarlet 1989) and the process has been since continuously improved by various teams in Europe (reviewed in Lelu-Walter et al. 2016; Trontin et al. 2016c see also Llebrés et al. 2018; Arrillaga et al. 2019). As other micropropagation techniques, somatic embryogenesis currently allows only multiplication of very juvenile material (the seed embryo). Somatic embryogenesis from vegetative explants of juvenile to mature trees is still challenging in maritime and other pine species (Trontin et al. 2016a) although demonstrated in spruce (see Varis et al. 2018). Nevertheless, it is the possibility for easy and efficient long-term cryopreservation in liquid nitrogen of initiated embryogenic masses from zygotic embryos (Lelu-Walter et al. 2006; Álvarez et al. 2012) that enables 'retroactive' propagation of mature trees. Once evaluated in the field, clones have lost their capacity for vegetative propagation because of aging but can be regrown and propagated from the juvenile cryopreserved embryogenic stock. Somatic embryogenesis in maritime pine is now sufficiently refined to envisage upscaling and automatization developments towards cost-effective production of somatic seedlings and implementation in clonal forestry (Trontin et al. 2019b). Field trials of somatic trees have been established in France for over 20 years and already demonstrated that they could complete the juvenile and adult vegetative and reproductive phases (Lelu-Walter et al. 2016; Trontin et al. 2016c). However, somatic seedlings exhibited reduced growth compared to standard seedlings during the first 2–3 years growth in field conditions (Trontin et al. 2016c). It was related with insufficient 'maturity' of somatic embryos (Morel et al. 2014b) associated to unbalanced arginine metabolism (Llebrés et al. 2018) resulting in inadequate accumulation and mobilization of protein reserves during the late stages of embryo development and germination. Current research is therefore focused on optimization of both genotype capture and production efficiency of high-quality, vigorous somatic embryos and plants (Trontin et al. 2016c, 2019a; Arrillaga et al. 2019). Recent outcomes revealed further levels of complexity for improving embryo development. Phenotype plasticity could be expressed in conifers (developmental plasticity, tolerance to abiotic stress such as drought) as a result of environmental stimulus (e.g., temperature) during early steps of somatic embryogenesis (Maury et al. 2019a; Castander-Olarieta et al. 2020; Trontin et al. 2020). Temperature priming effects during somatic embryogenesis are already under study in maritime pine (Arrillaga et al. 2019; Trontin et al. 2019b; Pérez-Oliver et al. 2021). Both developmental plasticity and 'memory' of stress may involve complex molecular regulation (Trontin et al. 2016b) including epigenetic mechanisms in a complex crosstalk with plant hormones (Maury et al. 2019b). "Priming" maritime pine megagametophytes at high temperatures upregulated a *WRKY*11-like transcription factor in embryogenic masses as well as increased cytokinin content and improved plant adaptation to heat stress (Pérez-Oliver et al. 2021). One component of the epigenetic complex of regulation of gene expression, the small RNA world (especially micro RNAs), was already shown to be involved in proper embryo development in maritime pine with significant differences detected between zygotic and somatic embryos (Rodrigues et al. 2019).

The possibility for direct vegetative propagation of seed embryos (through somatic embryogenesis) and seedlings (through macro, mini, and microcuttings) attracted much interest for bulk propagation of best families (crosses between elite parents) in conifers (Rosvall et al. 2019) including maritime pine (Majada et al. 2011; Martínez-Alonso et al. 2012). This is especially useful when the genetic value of the clones to be multiplied is still unknown. In addition, multivarietal forestry (Park et al. 2016), defined as the deployment of tested clones in plantation forestry (also referred to as clonal forestry, Rosvall et al. 2019) would be applicable by combining somatic embryogenesis with cryopreservation (Trontin et al. 2019b). A significant collection of cryopreserved embryogenic lines from diverse elite families of the French breeding program is already established (e.g., FCBA collection: over 2000 lines, Trontin et al. 2016c). Multivarietal forestry would allow to optimize genetic gain at selected level of genetic diversity in plantation forestry through the combination of best varieties (for specific traits) in clonal mixtures. That strategy has clear advantages over seed-based forestry for efficient deployment and continuous upgrading of productive varieties that are more resilient in the context of climate change. In this perspective, various initiatives are underway worldwide in conifers including for maritime pine within the MULTIFOREVER project framework (Trontin et al. 2019b).

5.2.4 Transgenesis

Both standard genetic approaches for dissection of gene function and conventional breeding are significantly hindered in perennial trees by delayed adult reproductive phase, high genetic load, and strong inbreeding depression. Hence, the process of forest tree domestication is time-consuming, expensive and still highly challenging in the context of rapid environmental and socio-economic changes related with the climate crisis (Plomion et al. 2016b).

As an important component of forest biotechnology, tree genetic engineering has received much attention since the eighties (Parsons et al. 1986) because phenotype implications of gene insertion or modification (dominant mutations) can be readily studied in primary transformants, i.e., in the absence of genetic recombination (Flachowsky et al. 2009). Furthermore, new trait delivery in target species can be envisaged not only with foreign or even synthetic genes but also with intact or modified versions of native genes. The possibility for accelerated integration of desired gene alleles into model genotypes (for genetic analysis) or elite germplasm (for genetic improvement) would be a paradigm shift in forestry.

In maritime pine, a suitable method for transient expression studies of transgenes (especially promoter sequences) in protoplasts was first developed by Gómez-Maldonado et al. (2001). Soon after, stable transformation methods based on selectable and reporter marker genes were reported by Trontin et al. (2002) after either microprojectile bombardment (biolistic) or co-cultivation of embryogenic tissue with *Agrobacterium tumefaciens*. Transgenic cell lines, somatic embryo, and plants could be obtained from single embryogenic cells within 12 months through somatic embryogenesis (Trontin et al. 2016c). The availability of an effective vegetative propagation system such as somatic embryogenesis was revealed as an essential prerequisite in conifers for efficient production of genetically modified trees from transgenic cells. The possibility for easy cryopreservation of embryogenic tissues further provided a convenient way for long-term securing of clonal archives of transgenic resources (Tereso et al. 2006c; Klimaszewska et al. 2016). The biolistic method yielded complex patterns of transgene integration such as multiple loci and amplification of tandem repeats (Trontin et al. 2007). In contrast, *Agrobacterium*-mediated transformation promoted low copy or single-insertion event of transgene that are less prone to unintended effects such as co-suppression or gene disruption (Tereso et al. 2006c; Trontin et al. 2007). This approach was further refined by independent teams in Portugal (Tereso et al. 2003, 2006b, c), France (Trontin et al. 2007; El-Azaz et al. 2020 Lelu-Walter et al. unpublished results), Spain (Alvarez and Ordás 2013), and Germany (Hassani 2015). Key factors to be considered for successful production of transgenic maritime pine include (i) genetic background and physiology of target embryogenic line (Trontin et al. 2002, 2007; Tereso et al. 2006c); (ii) *Agrobacterium* strain (Trontin et al. 2007; Alvarez and Ordás 2013); (iii) co-cultivation method of embryogenic cells and agrobacteria (Trontin et al. 2002, 2009; Tereso et al. 2006c); (iv) virulence inducers (Trontin et al. 2007, Lelu-Walter et al. unpublished results); (v) antibiotics for controlling agrobacteria (Trontin et al. 2002; Tereso et al. 2006b, c); as well as (vi) selection procedure and associated selectable marker genes such as *nptII* (kanamycin selection, Tereso et al. 2006b; Alvarez and Ordás 2013), *hpt* (hygromycin selection, Trontin et al. 2002; Tereso et al. 2006c) and *bar* (phosphinothricin selection, Trontin et al. 2007; El-Azaz et al. 2020). Both kanamycin (Tereso et al. 2006b; Alvarez and Ordás 2013) and hygromycin (Trontin et al 2007) were reported to affect the regeneration of transgenic lines or somatic embryos. Phosphinothrycin selection (0.1–1.0 mg l^{-1}) is therefore the current preferred method for producing transgenic somatic plants (Trontin et al. 2007; El-Azaz et al. 2020).

Genetic transformation of maritime pine is currently restricted to a few receptive genotypes. This is a severe drawback for implementing transgenic resources in breeding programs. In contrast, the possibility for reverse genetics, defined as ectopic (candidate) gene overexpression or silencing in 'model' genotypes has become a useful tool in research for functional dissection of traits of interest in forest trees such as wood-related or other traits with delayed evaluation (Mauriat et al. 2014). In maritime pine, a protocol for *Agrobacterium*-mediated genetic transformation of one fairly responsive genotype developed by FCBA and INRAE (Trontin JF, Lelu-Walter MA, unpublished results) was disseminated in 2010 to several teams in Europe and further refined in the frame of the multinational project Sustainpine (Hassani et al. 2013a; Trontin et al. 2013, 2017; Carneros et al. 2014). Reverse genetics studies of 36 candidate genes involved in wood formation, carbon and nitrogen metabolisms, stress tolerance and embryogenesis could be implemented by the Sustainpine consortium using strategies for both constitutive overexpression and silencing, especially RNA interference (RNAi). A significant cryopreserved collection of transgenic lines could be established at different laboratories (ca. 400 lines, Trontin et al. 2017) and further enlarged in the frame of the European project ProCoGen (Abarca et al. 2015; Trontin et al. 2015a). Partial results were obtained for candidate genes involved in wood formation (Trontin et al. 2007, 2015b; Mendoza-Poudereux et al. 2014; Abarca et al. 2015; El-Azaz et al. 2020), nitrogen metabolism and other growth-related pathways (Hassani et al. 2013b; Mendoza-Poudereux et al. 2014; Hassani 2015; Ávila et al. 2016), and embryogenesis (Hassani et al. 2013b; Hassani 2015). In the exemplary application of reverse genetics in maritime pine reported recently by El-Azaz et al. (2020), constitutive deregulation of the expression of a *Pinus pinaster MYB8* transcription factor gene (*PpMYB8*) in somatic plants was revealed to alter plant growth and development and to affect lignin deposition. Specifically, RNAi-induced silencing of *PpMYB8* resulted in altered formation of secondary cell wall with significant decreases in both lignin content (by ca. 30%) and H:G ratio of derived monomers of lignin. Interestingly it was demonstrated that *PpMYB8* interacts directly with promoter regulatory elements of members of the arogenate dehydratase gene family involved in the phenylalanine biosynthesis pathway (Sect. 5.3.2.2).

Despite significant efforts for about 20 years, transgenesis in maritime pine remains a long and difficult task requiring substantial expertise in

Agrobacterium-mediated transformation and regeneration of transgenic plants through somatic embryogenesis. As the technology currently involves random transgene insertion and constitutive expression in the plant genome, a significant number of independent transgenic lines must be obtained and thoroughly screened (ideally through genome profiling) for minimal off-target and positional transgene effects (Trontin et al. 2015b). As in other long-lived trees, highly efficient, cost-effective directed mutagenesis would therefore be most welcome in maritime pine for both functional genomics and precision breeding in the context of rapid climate change. The advent of genome edition technologies such as the renowned CRISPR/cas9 system now opens new horizons for trees (Ahuja 2017). Demonstrations are already accumulating in angiosperm trees such as poplar (Fan et al. 2015; Azeez and Busov 2021) and eucalypt (Dai et al. 2020). Promising results are now announced in *Pinus radiata* (C. Poovaiah unpublished results) suggesting that embryogenic cells would be again a convenient support for genome editing in conifers, as expected (Ahuja 2017). Targeted mutagenesis would provide huge opportunities for shortened breeding cycles in long-lived trees, especially as a value-added component of multivarietal forestry integrating genomic selection and somatic embryogenesis (Trontin et al. 2007; Ding et al. 2019). European Union regulations and their implementation at national level may, however, continue to limit the application of these innovative technologies in southwest Europe, where maritime pine is of major importance. The European Court of Justice recently ruled (July 2018) that genome edited plants must be considered as GMOs and comply with Directive 2001/18/EC, one of the most restrictive worldwide.

5.2.5 Molecular Breeding

Breeding large, long-living perennials, like conifers, is a lengthy process with cycles spanning decades. These slow breeding times differentiate conifers from other useful plant species subject to genetic improvement. Reasons lay foremost to the time required to phenotypically assess the candidate trees before selection. On top of the extra time, mature conifers at ages of phenotypic assessment also require space for their normal development, which adds to the difficulties when testing in common gardens. The successful disruptiveness brought by genomic selection (GS) in livestock and crops, notably by unchaining the evaluation of candidates from their phenotypic assessment without affecting selection precision, has profound implications in forest tree breeding. Conifers, by their constitutive features, are potentially an *ideal* host for such an innovation (Muranty et al. 2014). While the scarcity of genomic resources has been a handicap to embrace GS right away for conifers like maritime pine, the accumulated wealth of resources being reviewed here along with recent developments (Plomion et al. 2016a) opens up the possibility of operational GS implementations. Moreover, current challenges posed by global change accentuate the need of developing evaluation strategies that can promptly respond to new selection requirements, like for emerging pests (see Sect. 5.3.3.3) or tolerances to abiotic stresses (see Sect. 5.3.3.2). Additionally, the raising prospects of a new bioeconomy gravitating around wood production have further fueled the demand for newly improved forest tree varieties.

Maritime pine reforestation efforts benefit from one of the most advanced tree breeding programs across Europe, in terms of size, number of cycles of selection, economic impact, and degree of integration of involved stakeholders. Such a system requires at the basis an efficient multiple criteria evaluation scheme, which needs to be backed by a solid knowledge on available genetic variation at multidimensional levels (see Sect. 5.3.5) and on the underlying architectures for the traits under selection (see Sect. 5.3.4). This involves a costly updating on multiple and recurrent assessments, which has fueled the quest for cost-efficient evaluation approaches, even before the first GS proof-of-concepts were engaged for the species.

The use of near-infrared spectroscopy (NIRs) is one of those cost-efficient evaluation methods used notably for traits with a costly assessment, like those related to wood quality components (Lepoittevin et al. 2011). In this latter study, it is shown in a maritime pine breeding pool that the approach yields good quality proxies for wood chemistry traits, essential for the screening of wood resources for the transformation industry. Moreover, such NIRs predictors as well as their underlying traits did not show antagonistic correlations with growth assessments, opening the way to simultaneous improvements. Although selecting only on the basis of NIR evaluations can be a competitive alternative to GS (Rincent et al. 2018), the latter adds the possibility of screening simultaneously for the causal factors to the evaluation capabilities.

When it comes to early assessments for selection, another evident strategy is the use of genetic markers to track the presence of relevant polymorphisms for the trait of interest in the selected population, in what has been often coined as marker-assisted selection (MAS) (Fernando and Grossman 1989). Although there have been notable advancements in mapping relevant quantitative trait loci (QTLs) for several traits in the species, including growth traits (Bartholomé et al. 2016) and traits related to water-use efficiency (Marguerit et al. 2014) (further references on mapping in Sect. 5.3.4), these have not truly materialized into an operational MAS scheme for genetic improvement. Such a lack of breeding implementations is often found in other domesticated species with mapping resources (Muranty et al. 2014), owing typically to the insufficient capture of relevant QTLs or enough genetic variation. Nonetheless, detected QTLs and mapping information have been extensively used when selecting polymorphisms for new genotyping arrays (Plomion et al. 2016a), which subsequently paved the way for breeding and population genetics applications.

Although tracking QTLs with molecular markers is a prevalent objective, other parsimonious uses of markers have made their way more easily into operational breeding. In that sense, marker-based pedigree reconstruction occupies a relevant place in the maritime pine breeding program (Plomion et al. 2016a in Sect. 2.10). This owes to the fact that succeeding mating designs is a difficult task requiring time and effort, mainly due to inconsistent flowering that delay or prevent fulfillment of planned designs. In that context, using a pollen mixture (i.e., polymix) could maximize parental contributions on available female flowering without the operational constraints of controlled bi-parental crosses (for concept, see Lambeth et al. 2001). The drawback comes with the fact that resulting maternal descendants are a mixture of undistinguishable sib cohorts, preventing any reliable forward selection. Fingerprinting is then used to infer paternities and recover the missing pedigree. In that sense, the benefits of a marker-aided polymix scheme compared to the uncorrected pedigree might not be substantial when evaluating maternal contributors, according to the study involving part of the first generation of breeding parents of the maritime pine breeding program (Vidal et al. 2015). However, when such marker-aided polymix scheme is applied to select among the cohorts of descendants, i.e., forward selection mode, the benefits become clearer in the context of maritime pine breeding (Vidal et al. 2017). Indeed, marked-aided pedigree reconstruction allows for a partial disentangling of the maternal sib cohorts, giving access to extra between-families genetic variation. However, these previous schemes are still subject to strategic choices in the operational settings. A recent simulation study (Bouffier et al. 2019) comparing marked-aided polymix to a controlled bi-parental cross scheme shows that the former could more rapidly attain a given level of genetic gain, as it is not constrained by the need of fulfilling the planned mating design. In any case, one aspect that should not be omitted here is the fact that such parsimonious applications allowed for a gradual transition towards GS.

While previous uses required very low densities involving less than hundred multiplexed SNPs, the development of higher density arrays topping to 12 K SNPs (Chancerel et al. 2013; Plomion et al. 2016a) opened the way to the first proof-of-concept GS applications in maritime pine. Indeed, higher marker densities increase the likelihood of tagging close to causal mutations, assuring a stronger and more persistent association between the tagging markers and the underlying QTLs at population level (Muranty et al. 2014). As a result of stronger associations, or tighter linkage disequilibrium (LD), markers are able to resolve higher effects from neighboring QTLs, benefiting the prediction outcome of a genomic evaluation. However, conifers tend to present very rapid decays in LD patterns, and maritime pine is no exception (Plomion et al. 2014), with only a marginal number of pair-wise combinations of positions (less than 0.5%) attaining levels for r^2 (squared correlation coefficient for genotypic states) equal or higher than 0,1 (intra-chromosomal average of 0,01). The consequence is that dense marker coverage is required to efficiently capture most of the causal effects. The first proof-of-concept GS study involving the French maritime pine breeding population (Isik et al. 2016) was based on 2500 SNP informative markers, which certainly represents a lower bound in coverage given the inferred pattern of LD and the genome sizes. In spite of that, results showed reasonable prediction abilities (correlation between predictions and observations), lower than those obtained from a much extensive progeny testing in the breeding program, but still of comparable levels to those of other coetaneous studies in loblolly pine (Zapata-Valenzuela et al. 2012) and in white spruces (Beaulieu et al. 2014). A second GS study (Bartholomé et al. 2016b) raised the number of informative markers to more than 4400 SNPs, while considering a dataset representing 3 subsequent generations of the breeding program, from founders to second generation of selections, with the aim of assessing GS potentiality within and across generations. Results showed much higher prediction abilities than those in previous maritime pine GS study, although new levels were still below those attained by the pedigree-based counterpart. One of the main outcomes, however, was the fact that when training on the two ancestral generations and predicting the latest descendant generation, prediction abilities were at their best for GS, therefore offering a credible shortening of the breeding cycle, and challenging the lengthy progeny testing scheme in terms of gain per time units.

It would be reasonable to think that these first proof-of-concept studies operated at the lower bounds of meaningful marker coverage for GS. On average, their marker coverages were 2.4 SNPs per cM, which according to simulation work with tree breeding in mind (Grattapaglia 2014) is the minimum to equal pedigree-based performances. This highlights the need of new genotyping tools able to provide a higher number of informative markers. Ongoing collaborative research initiatives such as B4EST (http://b4est.eu/) have already produced new genotyping arrays able to combine high numbers of informative markers and low running costs for maritime pine (data not yet published). Great levels of information were obtained from a preliminary step of marker screening, performed with a single use ad hoc high-density SNP array (over 133,000 SNPs) on a diversity panel of 120 representative maritime pine individuals. Only a subset of the best quality markers (13,000) was used for producing the final SNP array. Lowering the development and running cost of the tool was possible by widening the portfolio of current and prospective demands, as the 50 K array comprises four important tree species, including maritime pine.

Other efforts have to be run in parallel to those of mobilizing genomic resources for breeding. These are notably a rethinking of the breeding schemes to accommodate operational GS, in terms of mating designs, phenotyping efforts, and management of genetic diversity.

In addition to their uses for management and evaluation of the breeding populations, genomic resources and genotyping are also very important in upstream (genetic resources) and downstream (production population) breeding steps. Maritime pine gene pools identified with high throughput genotyping by Jaramillo-correa et al. (2015) are now considered to better define in situ genetic resources to be conserved for potential infusion in the breeding population. On the downstream side of production populations, genotyping studies bring powerful information on pollen contamination and mating structure in seed orchards (Bouffier et al. 2017) to optimize the location and design.

5.3 Enabling New Discoveries

5.3.1 Genome Structure Through the Lens of Comparative Genomics

5.3.1.1 Insights from Comparative Mapping Between Conifers

Conifers have specific genome features that have slowed down knowledge on genome structure with respect to their congeners, the angiosperms. While the number of sequenced angiosperm genomes would soon reach the hundreds, there are less than a dozen sequenced gymnosperm genomes. Several features have complicated attempts to sequence the genomes of this group of plants. For instance, conifers have extremely large genomes (>10 Gbp) which pose a significant challenge for short-read sequencing and assembly (Kovach et al. 2010; Mackay et al. 2012; De La Torre et al. 2014). In addition, conifer genomes present large numbers of repetitive elements, which might take up to 86% of the size of the genome (Wan et al. 2018). Consequently, initial draft genome sequences remain highly fragmented (Nystedt et al. 2013; Neale et al. 2014, 2017; Warren et al. 2015; Stevens et al. 2016; Guan et al. 2016; Mosca et al. 2019). More recently, advancements in sequencing technologies have led to longer and more accurate sequencing reads, that when used in combination with chromosome conformation capture libraries, have led to more contiguous conifer genome drafts sequences (Scott et al. 2020).

In the absence of completely contiguous (chromosome-scale) reference genomes in conifers, comparative mapping has relied on high-density linkage maps in order to compare genomes from different lineages (Pelgas et al. 2005, 2006; Pavy et al. 2012, 2017; Liu et al. 2019; De La Torre et al. 2019). This strategy allowed to improve our knowledge on conifer macro-structure evolution, but it also had some limitations. Linkage maps usually sample only a small portion of the gene space in the species' genome (because not all markers segregate in the sampled individuals, and because the number of markers rarely samples the whole genome). In addition, linkage map studies usually include a relatively small group of individuals due to the complexity and long time required to grow several generations of conifers. This also leads to a reduced number of segregating markers, and therefore lower density maps which lack the accuracy needed to unravel fine microstructure rearrangements between species. Despite the limitations, linkage maps have advanced our understanding of genome structure in conifer species. We summarize below what we learned so far from this scale of analysis.

5.3.1.2 Remarkable Macro-Synteny and Macro-Collinearity Conservation Within *Pinaceae* but not Between *Pinaceae* and *Cupressaceae*

Although genome size in conifers ranges from 10 to 50 pg/2C and the haploid number of chromosomes can vary from 9 to 19, the general pattern of karyotypes is highly conserved across species and genera, with most species having 11 or 12 chromosomes (Wang and Ran 2014). Polyploidy is also exceptional in gymnosperms, with only two species from the Cupressaceae known to be polyploids, the hexaploid *Sequoia semperviens* (2n = 66) and the tetraploid *Juniperus chinensis* (2n = 44). Most of the comparative studies of genome structure in conifers have been performed within the Pinaceae, the largest family of conifers. The common conclusion from these works was a remarkable level of interspecific and intergeneric synteny and macro-collinearity conservation for the Pinaceae genomes (Krutovsky et al. 2004; Pelgas et al. 2006; Pavy et al. 2012; Liu et al. 2019). The lack of chromosomal rearrangements within

Pinaceae, comprising genera that diverged around 150 Mya ago (Morse et al. 2009), together with the low rates of speciation observed in these taxa, suggested a slow evolution rate in conifers (Pavy et al. 2012; De La Torre et al. 2017) compared to angiosperms.

Nevertheless, the genome macro-structure stasis observed within the Pinaceae is not maintained when the comparison has been extended to other conifer families. De Miguel et al. (2015) compared a 6 k Pinaceae (n = 12) composite linkage map, using *Pinus pinaster* unigene sequences as reference (Canales et al. 2014), with a *Cryptomeria japonica* (n = 11) linkage map (Moriguchi et al. 2012). This study revealed intense chromosomal shuffling between both botanical families. The authors also proposed an evolutionary model based on 20 contiguous ancestral regions that would have shaped the modern karyotype structure of both conifer families through a different number of chromosomal fusions (de Miguel et al. 2015). Intense chromosomal rearrangements have been well described in angiosperms lineages (Murat et al. 2017) but in conifers, in spite of the efforts to construct ultra-high-density linkage maps using a map-merging strategy (de Miguel et al. 2015; Bernhardsson et al. 2019; Liu et al. 2019), we are still far from a complete coverage of their genomes. In this sense, further studies on genome structure, function, and evolution including different conifer families are needed in order to decipher whether evolutionary mechanisms identified in angiosperm genome evolution have also played a role in the evolution of conifers.

5.3.1.3 Whole Genome Duplication in Conifers: An Open Question

The frequent and rapid chromosomal rearrangements identified in the study of angiosperm genome structure evolution have been usually accompanied by whole-genome duplications (WGDs), a widespread mechanism that contributes to the acquisition of new genes and species diversity (Ren et al. 2018). One ancient WGD event is known to have occurred before the angiosperm-gymnosperm split around 350 MY ago (Jiao et al. 2011). However, whether other WGD events have occurred during the evolutionary history of conifers is still a matter of debate in the scientific community. Nystedt et al. (2013) did not find any evidence of recent WGD in *Picea abies* and advocated an intense activity of transposable elements as the main mechanism of genome size expansion in conifers (Stevens et al. 2016). On the contrary, (Li et al. 2015) reported three WGD for gymnosperms: one specific of the Gnetales clade and two in the ancestry of major conifer clades, Pinaceae and Cupressophytes. Zwaenepoel and Van de Peer (2019) developed a probabilistic method for phylogenomic reconciliation-based WGD inference and contested the results from Li et al. (2015) and Guan et al. (2016), as they did not find evidence of WGD for the Pinaceae nor for the Ginkgo clades. Nevertheless, two independent studies found traces of a WGD for *Ginkgo biloba* (Guan et al. 2016; Roodt et al. 2017) although the authors dated this event at different ages, that is, before or after the separation with the Cycads clade. The opposite results delivered from these studies highlight the challenge to detect traces of very ancient WGD in plants. A reliable answer to the question of the existence of WGD during the evolution of Pines and other conifer lineages may rely on the development of new methodological approaches to infer ancient WGD (Zwaenepoel and Van de Peer 2019).

5.3.1.4 Functional Comparative Genomics

While the knowledge derived from comparative genomics using only DNA sequences provides a rather static view on an organism, a more dynamic view could result from doing comparative genomics on the level of genes, gene families, gene regulation and gene and protein interactions. Analysis of the gene content and how it is regulated is a powerful tool in elucidating biological aspects that explain the potential unique features that define an organism. In the specific case of *Pinus pinaster*, one of the most important traits is drought resilience (Madrigal-González et al. 2017; Caminero et al. 2018). A combination of different functional comparative techniques could be used to elucidate the genetic basis of drought resilience. Analyzing gene family structure and more specifically investigating the expansion and/or contraction of certain gene families with, for instance, CAFE (De Bie et al. 2006) is a starting point. However, the molecular basis of traits is usually not as apparent as gene family size differences and thus more fine-grained analyses are required. Observed differences in phenotypes are frequently explained by a difference in transcription regulation rather than gene content (Romero et al. 2012). Through analyzing gene expression patterns, making use of RNA-seq data, it will be possible to have a detailed view on transcription levels of genes (Behringer et al. 2015). Differences in expression between orthologous genes from closely related species could then be correlated with traits of interest. Once this has been established, the next step is to elucidate the molecular basis of these expression differences. To do so, gene promoter analysis is conducted. By comparing the promoter region of orthologous genes, one can look for absence or presence or combinations of transcription factor binding sites (promoter motifs) directly influencing the corresponding gene transcription (Singh 1998).

5.3.2 Gene Expression Regulation: Analysis of Small RNA and TF

5.3.2.1 Gene Expression Regulation: Analysis of Small RNA

Gene expression regulation by small RNAs (sRNAs) has started to be uncovered in maritime pine through expression profiling of these small molecules of RNA with well-recognized roles in plant development and stress responses (Rodrigues et al. 2019; Perdiguero et al. 2020). Small RNAs are categorized into classes based on their mode of biogenesis and functions, being the microRNAs (miRNAs), most commonly with 21–22-nt, the best-characterized group of sRNAs in plants and other organisms, acting at the post-transcriptional silencing (PTS) level through target transcript cleavage or translation inhibition (Chen 2009). Although the miRNAs are the most well-characterized sRNAs, they form a quite complex group and the term isomiRome has been coined to name the diverse regulatory sequences derived both from the classical precursor sequences and from differential precursor processing to originate RC-miRNAs, nat-miRNAs, isomiRNAs, and miRNA-like sequences (Zhang et al. 2010; Shao et al. 2012; Neilsen et al. 2012; Guo and Chen 2014). Another major sRNA class is formed by the short-interfering RNAs (siRNAs) which, unlike the *MIR* gene-derived miRNAs, arise from the splicing of long double-stranded RNAs into 21–24-nt siRNAs by DCL2-4 enzymes, and may act at the PTS or at the transcriptional level.

Although early reports suggested that conifers were unable to produce significant amounts of 24-nt sRNA sequences (Dolgosheina et al. 2008; Chávez Montes et al. 2014), their expression in conifers is now well established. Recent maritime pine sRNA data also confirm the significant presence of 24-nt sRNAs in this species, but large differences in sRNA expression profile among tissues are obvious. Up to date, maritime pine sRNA sequencing has covered a wide range of vegetative tissues including roots, stems and needles, several zygotic embryo developmental stages, somatic embryos, megagametophytes, and responses to abiotic (drought) and biotic (pine wood nematode) stresses.

Data obtained from embryo and megagametophytes has shown that 21-nt and 24-nt sRNAs are predominant in these tissues (Rodrigues et al. 2019). Small RNA data from developing embryos revealed a strong presence of non-redundant 24-nt sequences, especially in somatic embryos, but many of them are eventually discarded by the bioinformatics analysis due to their very low expression levels. Nevertheless, the presence of the 24-nt sequences is more evident in embryos than in vegetative tissues, highlighting changes in the sRNAs population in the course of maritime pine life cycle.

Maritime pine has been the first conifer to have its isomiRome comprehensively analyzed in vegetative organs (Perdiguero et al. 2020) and, from the 13,441 identified sequences, only a little over ¼ were canonical miRNAs, emphasizing the large diversity of miRNA-associated sequences for which little knowledge is available. In this case, also a differential distribution according to organ/tissue was detected, with roots presenting the larger fraction of isomiRNA or miRNA-like sequences and the higher abundance of 24-nt sequences, which was negatively affected by drought stress, being roots the most responsive type of tissue upon water deficit (Perdiguero et al. 2020).

Up to this date, a total of 22,606 sRNA sequences associated with miRNA biogenesis have been identified in maritime pine sRNA datasets. From these, 117 miRNAs were identified in at least one sRNA library in studies of embryos and megagametophytes (Rodrigues et al. 2019), roots, needles and stems under contrasting water availability (Perdiguero et al. 2020), and stems inoculated with the parasitic nematode *Bursaphelenchus xylophillus* (Modesto et al. unpublished results). Within this group, a reduced subset of 25 sequences (15 canonical/star miRNAs and 10 isomiRNAs) is present in all the analysed libraries. According to the isomiRome annotation, this core subset is represented mainly by sequences 21-22nt long, associated with 15 conserved families (*MIR156*, *MIR162*, *MIR166*, *MIR167*, *MIR168*, *MIR482*, *MIR946*, *MIR947*, *MIR951*, *MIR1312*, *MIR3704*, *MIR11431*, *MIR11532*, *MIR11546* and *MIR11551*). From the 117 miRNAs shared among the three studies, a total of 92 miRNA sequences represents a dynamic miRNA set which varies according to tissue, developmental stage or environmental conditions, and includes 42 sequences corresponding to 38 novel miRNA families identified in maritime pine. It should be noted that the sRNA studies in maritime pine have relied on the use of the reference transcriptome, but mostly on the genome of the closely related species *Pinus taeda* for the identification and annotation of sRNAs. It is, therefore, expected that with the availability of an improved genome assembly of maritime pine it will be possible to further update and refine the available sRNA analyses. It will be also possible to find which loci are responsible for the production of the different types of precursors resulting in different sized sRNAs. Nonetheless, according to the available literature, maritime pine sRNA landscape seems to be generally in line with other reports on sRNAs in conifers where a range of tissues, developmental stages, and environmental responses have been covered (Yakovlev et al. 2010; Wan et al. 2012; Zhang et al. 2013; Niu et al. 2015; Fei et al. 2016; Nakamura et al. 2019; Krivmane et al. 2020). While only a few studies have focused on siRNAs, most of them have identified the miRNA population. Still, a very low number of conifer miRNAs from only four species are deposited in miRbase (www.mirbase.org, release 22.1) when comparing to angiosperm species, being *Picea abies* the conifer species with more sRNA sequence information available, for which 600 mature miRNAs are deposited.

A crucial but challenging task aiming to characterize sRNA functions is the identification of their target genes. For miRNAs, several software tools are available for target prediction but usually a large number of false targets are predicted emphasizing the need for target validation strategies. In maritime pine, validation of miRNA targets has been performed through degradome sequencing by Perdiguero et al. (2020). In this study, 1,592 putative significant interactions involving 957 different maritime pine transcripts and 1,349 isomiRome sequences were predicted in root tissues, suggesting a complex isomiRome-mediated regulation. By identifying the targets of differentially expressed miRNAs, a set of genes were identified which may be relevant in response to drought such as genes encoding cellulose synthase or cinnamyl alcohol dehydrogenase, involved in xylogenesis, and sucrose synthase, directly involved in sugar accumulation, senescence-associated proteins and disease resistance proteins, among others. The identified interactions are important targets to focus on, in order to improve adaptation of pines to environmental stresses. Nevertheless, unequivocal evidence of interaction can only be obtained by other approaches which can be difficult and time-consuming to perform. In maritime pine, the use of a co-transient expression system with the luciferase reporter system (Martinho et al. 2015) has allowed to overexpress a pine miR160 isoform in *Arabidopsis* protoplasts and validate its interaction with a predicted target transcript (Alves et al. unpublished results). The use of luciferase-based sensors for quantifying miRNA activity that respond specifically to both endogenous and overexpressed miRNAs and target mimics can thus be a viable way to confirm in vivo interaction of selected miRNAs and their target transcripts. Future studies on this subject will help clarify the molecular pathways mediated by sRNAs in pines.

5.3.2.2 Gene Expression Regulation: Analysis of Transcription Factors

Transcription factors determine where and when plant genes are transcribed and finally establish the levels of the metabolic end-products that define what processes take place. It is not until mid-2014 when a reference transcriptome for a Mediterranean pine (*Pinus pinaster*) was available (Canales et al. 2014). This genomic resource enabled the access to thousands of transcription factors sequences, to study their distribution in families classified by their DNA-binding domain and make comparisons with other plant transcriptomes. Furthermore, this specific resource facilitated the starting of a more in-depth study of the transcription factors in Mediterranean pine. Before this achievement transcriptional studies were limited to the characterization of individual transcription factors regulating some specific structural genes. Here, we summarize the state of knowledge in the gene regulation of secondary cell wall biosynthesis.

It is well established that members of the large Myb family in plants are involved in the regulation of lignin biosynthesis. Previous studies performed in *Pinus pinaster* showed that some transcription factors belonging to this family: Myb1, Myb4, and Myb8 are of paramount importance in the control of the spatial/temporal expression of *GS1b* (glutamine synthetase) gene. This gene product among other functions is involved in the re-assimilation of the tremendous amounts of ammonium released in the metabolism of phenylpropanoids (Gómez-Maldonado et al. 2004). Although GS1b is expressed in the whole plant, transcript levels are mainly associated with vascular tissues (Avila et al. 2001). The gene is target of Myb8 that acts as activator of its expression. Early studies of transcriptional regulation showed that the gene is target of Myb8 that acts as activator of its expression. Later, additional studies demonstrated that Myb8 is also involved in the co-regulation of other two genes encoding enzymes involved in the process of channeling carbon for lignin biosynthesis: prephenate aminotransferase (PAT) and phenylalanine amonia-lyase (PAL) (Craven-Bartle et al. 2013). Recently, it has been described that Myb8 is also involved in the regulation of the expression of two members of large arogenate dehydrogenase (ADT) gene family of *Pinus pinaster*, ADT-A and ADT-D (El-Azaz et al. 2020). This enzyme catalyzes the last step in biosynthesis of phenylalanine. The amino acid is the precursor of a lot of secondary products in plants included lignin. The ADT family in conifers is composed by 9–10 members, while only a small family of six genes is encoded by the Arabidopsis genome. These findings are likely reflecting a higher demand of precursors for the biosynthesis of secondary metabolites in conifers (El-Azaz et al. 2016). It has been found that both *ADT-A* and *ADT-D* genes are highly expressed in compression wood in maritime pine, a special kind of wood with higher lignin content. Myb 8 has proven to be a positive regulator of the expression of both genes and RNAi-*PpMy8* silenced plants showed a marked decrease on the relative abundance of transcripts for these two genes. The study of the transcriptional network in these silenced plants has allowed to identify another component of the regulatory network PpHY5, an ortholog of the corresponding HY5 in *Arabidopsis*. This transcription factor has been found to stimulate the accumulation of phenylalanine derivatives and the repression of genes involved in secondary cell wall formation (Nguyen et al. 2015; Czemmel et al. 2017) suggesting its role as repressor of the expression of ADT genes in this species.

The study of the Myb family in other conifer species (Bedon et al. 2010) has described the composition of the R2R3 Myb subfamily. The phylogenetic studies displayed that subgroup 4 of the R2R3 Myb, the subgroup in which Myb8 is included in *Picea glauca* and *Pinus taeda* was expanded after the division of gymnosperms and angiosperms. This situation is similar to what is found in the *ADT* gene family, with nine members in maritime pine, and three to six in angiosperms. The presence of additional gene copies of structural and regulatory genes in pathways of great importance for the tree such as the production of wood in conifers suggests an optimization to control specific expression patterns in processes that are crucial for the tree such as lignin biosynthesis. It is important to keep in mind that this process channels as much as 30% of the carbon fixed in photosynthesis (Haslam 1993).

Other transcription factors well characterized to be involved in plant-cell secondary wall formation are some members of NAC family (Yamaguchi and Demura 2010). NAC transcription factors are plant-specific and integrate a big family in many plants with more than 100 members encoded by the genome of the model plant *Arabidopsis thaliana* (Ooka 2003; Nakashima et al. 2012). The maritime pine reference transcriptome enables the identification of 37 NAC genes classified in two groups based in their DNA-binding domain motifs (Pascual et al. 2015). The NAC proteins are involved in multiple plant responses such as development, senescence, response to abiotic and biotic stress, nutrient distribution and cell wall synthesis. The biosynthesis of phenylpropanoids and more specifically the biosynthesis of lignin are very complex pathways that involve the coordinated gene expression of genes encoding enzymes located in different cell compartments and in many plants has proven to be hierarchically structured. In *Arabidopsis* some NAC genes have demonstrated to be on the top of this pyramid with capacity to act as key regulators of the secondary cell wall formation as NST1, SND1, or VND6 (Mitsuda et al. 2007). In maritime pine, the molecular and regulatory properties of the NAC transcription factor PpNAC1, a putative orthologue of NST1 of *Arabidopsis,* have been studied in detail (Pascual et al. 2018). The functional analysis of RNA1 lines of maritime pine with silenced expression of this NAC factor strongly suggests that PpNAC1 is a master regulator phenylalanine biosynthesis and consequently lignin deposition in maritime pine (Pascual et al. 2018). PpNAC1 is mainly localized in secondary xylem of pine adult trees as well as in compression wood where the content of lignin is increased. When *PpNAC1* gene was silenced in pine, the vascular pattern was seriously affected with plants with thick hypocotyls and poor growth. The vascular morphology showed a disorganized stem with altered vascular radial pattern and expanded phloem. The silencing of PpNAC1 downregulated the expression of genes involved in secondary metabolism and in secondary wall formation. The gene is able to control its own expression as well as the expression of some Myb genes that are located downstream in the lignin regulatory network such as Myb 4 and Myb 8. These results partially elucidated the cascade of regulators that could be operative in the control of lignin deposition in maritime pine. As PpNAC1 is a potential ortholog of Arabidopsis NST1, these outcomes also show a certain degree of conservation in the regulation of lignin synthesis. This allows us to speculate that the regulation of this process appeared early in the evolution of seed plants, and has been conserved after the angiosperm-gymnosperm split.

5.3.3 Genes Related to Functions

5.3.3.1 Developmental Plasticity and Adult Cell Reprogramming: The Case of Adventitious Regeneration

Plant tissues have extensive regenerative capacity. Entire plants can be developed from single cells or small cuttings (Xu and Huang 2014). In plants, the possibility to regenerate roots, shoots, or embryos directly from cells other than root or shoot meristems, lateral root initials, or zygotes has been known for years and has been exploited in horticulture, agriculture, and forestry

(Bonga 2016). However, little is known about the mechanisms that enable a somatic differentiated cell to switch its fate into a pluripotent or totipotent cell developed into a root, shoot or embryo, or into repaired damaged tissues (Díaz-Sala 2014). Therefore, determining the way by which cells reset their fate is crucial to understand developmental plasticity (Pizarro and Díaz-Sala 2019; Vilasboa et al. 2019; Díaz-Sala 2020). Knowledge of the molecular and cellular mechanisms involved in recalcitrance of forest tree species for regeneration, which is expressed under a wide range of genetic and physiological conditions, is scarce (Díaz-Sala 2019). It may help explain differences between genotypes, tissues, time of excision, or age. Somatic embryogenesis has been extensively used as biotechnological strategy for maritime pine breeding programs (Lelu-Walter et al. 2016). However, difficulties for the induction and maturation phases have limited the extensive use of this technology. Transcriptome and proteome analyses of zygotic and somatic embryos have been performed in maritime pine to improve somatic embryogenesis for vegetative propagation of conifer species (de Vega-Bartol et al. 2013; Morel et al. 2014a, b; Trontin et al. 2016b; Rodrigues et al. 2018). Functional categories and enrichment analysis of differentially expressed transcripts showed that the molecular regulation from early to late embryogenesis is associated with spatio-temporal modulation of auxin-, gibberellin-, and abscisic acid-mediated responses. Modification of the auxin flow, carbohydrate transport and metabolism, including storage, stress-related, late embryogenesis abundant and energy metabolism proteins, and putative orthologs of genes required for meristem formation and function have been identified during pine embryogenesis. Transcription factors belonging to the WUSCHEL-related HOMEOBOX families associated with the capacity for somatic embryogenesis have been described in maritime pine. The expression domain of the *Pinus pinaster WUSCHEL-related homeobox PpWOXX*, the new WUSHEL-clade member identified in conifers, during somatic embryogenesis and in shoot apex is similar to that described for the *Pinus pinaster WUSCHEL PpWUS*, indicating this gene may play a role in shoot apical meristem maintenance (Alvarez et al. 2018). On the other hand, Arrillaga et al. (2019) described a decrease of the *Pinus pinaster LEAFY COTYLEDON PpLEC1* and *WUSCHEL-related homeobox PpWOX2* gene expression over the course of embryo development in embryogenic lines showing a spiky morphotype with high maturation capability. In addition, the spiky morphotype had a steady increase in auxin and abscisic acid during embryo maturation, whereas the content of both hormones only peaked at the beginning of the maturation in the smooth morphotype. Epigenetic regulation associated with DNA methylation, chromatin remodeling, transposons or sRNA pathways activating or rechanneling a developmental cell memory is an important mechanism regulating different stages of zygotic or somatic embryo development (Klimaszewska et al. 2009; Miguel and Marum 2011; de Vega-Bartol et al. 2013; Rodrigues et al. 2018, 2019). Several interrelated pathways are involved in the plasticity of plant cells for adventitious regeneration in forest tree species. Stress, auxin, and information carried by stem cell genes are common pathways associated with the competence of cells to initiate adventitious morphogenic programs. The role of epigenetic regulation modifying at different levels developmental cell fate is an important mechanism regulating de novo regeneration. Molecular dissection of the mechanisms underlying regeneration capacity and identification of the genes expressed in common interrelated pathways regulating competence for adventitious organogenesis will lead to the identification of an expressional signature characterizing specific levels of regulation, developmental stages, age, position, timing, type of tissues and clones or genetic traits associated with competence for regeneration.

5.3.3.2 Abiotic Stress Response: The Case of Drought Response

Drought is one of the most relevant climatic disturbances affecting the Mediterranean region, reducing growth and increasing tree mortality.

Conifers are "archetypical stress tolerators" (Brodribb et al. 2012). They have developed a variety of strategies to face drought, which include biochemical, physiological, and morphological changes that differ depending on the intensity and duration (Moran et al. 2017; Estravis-Barcala et al. 2020). Isohydric species, as *Pinus pinaster,* close stomata to maintain water potential following abscisic acid (ABA) signaling. Stomatal control of transpiration and decrease in hydraulic capacity are closely related to drought conditions. Tracheid and parenchyma-based xylem and smaller leaf area also reduce water loss, while root morphology (Corcuera et al. 2012) and biomass partitioning are involved in efficient water acquisition (Aranda et al. 2010). Additionally, increased synthesis of molecules that act as osmotic regulators and increased allocation to nonstructural carbohydrates are other adjustments in response to drought (Moran et al. 2017).

Population analyses on the drought-avoiding species *Pinus pinaster* have showed intraspecific variation in drought response, with more resilient saplings than other co-occurring pine species, such as *P. nigra* and *P. sylvestris* (Andivia et al. 2020). Maritime pines from populations with different water availability show differences in functional traits such as conductance, carbon allocation, plant growth, root morphology, and water-use efficiency (Brendel et al. 2002; Nguyen-Queyrens et al. 2002; Nguyen-Queyrens and Bouchet-Lannat 2003; Corcuera et al. 2010, 2012; Sánchez-Gómez et al. 2010; Aranda et al. 2010; Lamy et al. 2011; de Miguel et al. 2012; Gaspar et al. 2013; Sánchez-Salguero et al. 2018; Feinard-Duranceau et al. 2018).

Understanding the mechanisms underlying drought response and tolerance is crucial to predict the impact of recurrent and intense droughts on maritime pine dynamics and its survival under currently ongoing climatic constraints. With this aim, different experimental systems have been designed to induce drought stress, including the use of chemicals such as polyethylene glycol (PEG), (Dubos et al. 2003; Perdiguero et al. 2012b) and imposition of water deprivation (Perdiguero et al. 2012a, 2013; de Miguel et al. 2014; de María et al. 2020). The combination of rapid technical advances and computational developments has allowed a significant increase in the scope of studies. Research focus changed from the expression analysis of a few genes at a time e.g., cDNA-AFLP (Dubos and Plomion 2003; Dubos et al. 2003), to the analysis of hundreds of genes by differential screening of a subtracted library using cDNA microarray (Perdiguero et al. 2012b), and later to thousands of genes by oligonucleotide microarrays (PINARRAY3; Cañas et al. 2015). Today, it is common to study the whole transcriptome in one go using RNA-seq (de Miguel et al. 2014; Cañas et al. 2015; de María et al. 2020; López-Hinojosa et al. 2021). Additionally, a few gene families have been characterized, such as those encoding dehydrins (Perdiguero et al. 2012a, 2014; Velasco-Conde et al. 2012), SnRK kinase family (Colina et al. 2020), and NAC transcription factors (Pascual et al. 2015), identifying several members involved in drought response. Also, studies focused on biological processes acting on drought response, such as cuticle biosynthesis, have led to the characterization of a set of candidate genes (Le Provost et al. 2013).

Most of the maritime pine gene expression studies have analyzed drought response of seedlings (Dubos and Plomion 2003; Dubos et al. 2003), saplings (Le Provost et al. 2013) and trees younger than 5 years old (Perdiguero et al. 2012b; Velasco-Conde et al. 2012; de Miguel et al. 2014; Cañas et al. 2015; de María et al. 2020; López-Hinojosa et al. 2021), which may provide partial information, as it may lack genes or even processes that play relevant roles in mature trees. In addition, due to the progressive recalcitrance to vegetative propagation by cuttings up to 5 years old, only a few studies have been carried out with ramets from the analyzed genotypes (Velasco-Conde et al. 2012; de Miguel et al. 2014; de María et al. 2020; López-Hinojosa et al. 2021) which can be used as biological replicates.

QTL mapping using high-density linkage maps with gene-based markers identified candidate genes in response to drought in maritime pine. De Miguel et al. (2014) used this strategy to study water-use efficiency using an F_1 progeny with genotypes propagated vegetatively to improve the reliability of the phenotypic estimates pointing out 58 genes that could be involved in drought tolerance. Among them, two MYB transcription factors and a histone chaperone, as well as an aquaporin and several genes related with the oxidative stress and the regulation of stomatal aperture. Association mapping (also known as linkage disequilibrium mapping) makes use of natural populations or panels in order to record more recombination events (Cardon and Bell 2001). This renders a very powerful statistical approach and contributes to higher resolution when looking for regions associated with traits. However, it also requires an extensive knowledge of SNPs within the genome. To date, association mapping has been applied in maritime pine to study growth and wood-related traits, fire adaptation and disease susceptibility (Eveno et al. 2008; Lepoittevin et al. 2012; Budde et al. 2014; Cabezas et al. 2015; Bartholomé et al. 2016; Hurel et al. 2021) as further described in Sect. 5.3.4.2 but not drought response. However, the forthcoming availability of the 4TREE Axiom assay developed within the B4EST EU funded project (https://b4est.eu/), a genotyping chip with 20 k SNPs with known positions on *Pinus pinaster* genome, opens the door for the development of new and more powerful genome-wide association studies (GWAS) for the identification of additional candidate genes.

Conifer transcriptome studies have to face the high percentage of genes that, showing high level of homology with other conifer species, cannot be annotated due to the low level or lack of homology with angiosperms. This is likely due to their divergence approximately 350 Myr ago (Zimmer et al. 2007). The analysis of conifer and angiosperms gene families has revealed that they have followed different evolutionary trajectories (reviewed by Mackay et al. 2012). Notwithstanding this difficulty, genes involved in a plethora of processes have been found associated with maritime pine response to drought, among them, genes involved in stress perception and signaling, such as calcium-dependent signaling and protein kinases (Dubos and Plomion 2003; Perdiguero et al. 2012b, 2013; de María et al. 2020). ABA-independent and ABA-dependent regulatory systems act in plant response to drought (Takahashi et al. 2018). Water deficit signals move from roots to leaves, inducing expression of genes involved in ABA biosynthesis (Kuromori et al. 2018), promoting ABA accumulation, which is involved in stomatal control. In addition to ABA, other phytohormones as ethylene (ET), auxin (AUX), jasmonic acid (JA), salicylic acid (SA), gibberellin (GA), cytokinin (CTK), brassinosteroids (BRs), and strigolactones (SLs) are involved in drought response, cross-talking between them to implement plant survival (Ullah et al. 2018). Also, genes involved in phytohormone metabolism, signaling, and transport have been identified in maritime pine (Perdiguero et al. 2012b, 2013; de María et al. 2020). Among them, drought-tolerant pines showed higher accumulation of the different components of the PYR/PYL/RCAR ABA receptor (López-Hinojosa et al. 2021), which in the presence of high levels of ABA, inactivates the repressor PP2C, releasing SnRK2 kinase, that activates transcription factors (TFs) involved in ABA-depended gene expression (Umezawa et al. 2010). Members of TF families MYB, WRKY, bZIP, AP2/ERF (that regulate ABA-responsive gene expression), DREB (an AP2/DREB-type TF), and NAC (involved in the regulation of ABA-independent gene expression), are key players in water stress signaling (Joshi et al. 2016). Several members of these TFs have been found differentially expressed in maritime pines grown under water scarcity, as well as in trees showing contrasting responses to drought (Perdiguero et al. 2012b, 2013; de Miguel et al. 2014, 2016; Cañas et al. 2015; Pascual et al. 2015; de María et al. 2020; López-Hinojosa et al. 2021).

Under drought conditions, maritime pine shows activation of pathways involved in maintenance of cellular homeostasis from dehydration, accumulating e.g., LEA/dehydrins, HSPs/chaperones and protective molecules such as soluble sugars and proline, as well as pathways involved in control and repair of drought-caused damages (activating ROS detoxification system). Additionally, genes involved in a broad variety of processes have been identified differentially expressed in maritime pines facing drought (Velasco-Conde et al. 2012; de Miguel et al. 2014; Perdiguero et al. 2014; Cañas et al. 2015; de María et al. 2020; López-Hinojosa et al. 2021). Among them, genes involved in secondary metabolism, with expression profiles highly associated with their Atlantic or Mediterranean origin (Cañas et al. 2015; López-Hinojosa et al. 2021), as well as genes involved in transport of water, ions, osmolites and hormones, with accumulation of aquaporins, primary transporters that use ATP hydrolysis, such as ABC transporters, and secondary transporters, triggered by ion gradients, such as nitrate transporters in drought-tolerant trees (Perdiguero et al. 2012b, 2013; Cañas et al. 2015; de María et al. 2020). Also, a significant number of genes related to the Calvin and Benson cycle and regulation of photosystem II have been found highly expressed in drought-sensitive trees showing low efficiency in adaptive acclimation response to drought (de Miguel et al. 2014; de María et al. 2020). In addition, genes involved in drought adaptation in maritime pine are related to cell wall modification and lignin biosynthesis; lipid biosynthesis, a major component of cell membranes that also acts as a source for signaling molecules; carbohydrate metabolism including glycolysis and starch and sucrose metabolism (Perdiguero et al. 2013; Cañas et al. 2015; de María et al. 2020; López-Hinojosa et al. 2021). The comparative analysis of drought response between drought-tolerant and drought-sensitive pines revealed drought-responding genes constitutively expressed under nonlimiting water conditions in drought-tolerant trees. A subset of genes involved in the same processes was found to be induced in sensitive individuals subjected to drought (de María et al. 2020). In addition, trees from xeric and mesic maritime pine populations show variation in their abilities to sense environmental conditions to adjust their growth cycles, which among others include differential expression of genes involved in photoperiod control (Cañas et al. 2015). On-going RNA-seq analyses aim to study communication between organs using grafts that combine constructions with rootstock and scion genotypes showing contrasting response to drought (López-Hinojosa and Cervera, unpublished results).

5.3.3.3 Biotic Stress Response: Coping with Pests (PWN), Pathogens (PPC), and Herbivores

Maritime pine trees are easy to find due to their large biomass, and are accessible sources of food and/or a place to breed for a wide array of organisms. Maritime pine has evolved a battery of physical (e.g., bark, resin canals, stone cells, cuticular wax) and chemical defences (e.g., terpenes, phenolics, alkaloids) to deter the damage or minimize the impact that such organisms cause on tree fitness (Vázquez-González et al. 2021). These defences may act constitutively or may be activated after damage recognition (i.e., induced defenses) (Sampedro et al. 2011). Despite the clear adaptive value of resistance mechanisms preserving tree survival and reproduction, most resistance mechanisms show a huge intraspecific genetic variability, both among and within populations (Meijón et al. 2016; López-Goldar et al. 2019; Vázquez-González et al. 2019). Several non-exclusive factors, including but not limited to trade-offs with other life functions, gene flow, environmental heterogeneity and the need to protect from multiple stressors, explain the maintenance of this large variation in spite of their benefits on pine fitness (Vázquez-González et al. 2021).

Maritime pine has co-lived with native pests and diseases since its origin more than 150 million years ago. Despite the pressure imposed by these antagonists, the species has been able to survive and persist up to date, likely due to its effective resistance mechanisms. However, nonprecedent biotic challenges associated with global change are now imposing alarming risks for pine forest sustainability (Wingfield et al. 2015). Global warming and increasing extreme climatic episodes are altering the biology and ecology of both the host and the aggressors, disrupting the natural equilibrium between them. But, more importantly, the raise of global trade associated with global change is drastically increasing the introduction of alien invasive pests and pathogens (Seidl et al. 2017). The introduction of exotic species is particularly alarming, since maritime pine has no specific defenses against them and may face severe losses in population size (Naidoo et al. 2019).

Prevention, mitigation, and management of maritime pine forests health is a main challenge for scientists and managers. This challenge, however, is not easy to overcome. The low profitability of pine forest systems and their environmental context prevent the use of intensive phytosanitary tools, making the search of alternative environmentally friendly and cost-effective ways to preserve forest health an urgent need. Fortunately, the huge intraspecific genetic variation in resistance mechanisms existing within the species provides us with an extremely powerful tool to fight this challenge. Evidence is accumulating that resistance of maritime pine to pests and diseases, even to exotic organisms, is under genetic control (Table 5.1). Resistant genotypes can be identified and selected for breeding purposes (Sniezko and Koch 2017; Woodcock et al. 2018).

Breeding for resistance emerges as a key strategy to preserve forest health in the current context of health decay associated with global change but deployment of resistant genotypes is a hard and time-consuming process (Woodcock et al. 2018). Developing genomic tools to alleviate screening and selection processes would be highly desirable but requires a big knowledge on the molecular processes involved in host-aggressor interactions. Recent efforts made in *Pinus pinaster* in this sense have focused mainly on the two more alarming pathogens currently threatening maritime pine forests in Europe: the pinewood nematode and the Pitch canker.

Pine wilt disease (PWD) is caused by one of such introduced pests, posing a serious threat for maritime pine, a species highly susceptible to this disease (Evans et al. 1996). Despite the efforts to contain PWD since its introduction in the late 1990s, the disease has rapidly spread through Portugal and reached Spain (Mota et al. 1999). PWD is caused by the parasitic nematode *Bursaphelenchus xylophilus*, or pinewood nematode (PWN), that enters the stem tissues of adult trees through the insect vector *Monochamus galloprovincialis* while feeding on the bark. After entering the plant, PWN spreads throughout the stem, feeding on plant cells and causing severe tissue damage. Eventually, water transport is compromised in the xylem, leading to tree wilting and death (Jones et al. 2008; Vicente et al. 2012).

PWD resistant individuals have been identified in Maritime pine (Menéndez-Gutiérrez et al. 2017a, b; Carrasquinho et al. 2018), as well as in other susceptible pine species (Toda and Kurinobu 2002; Xu et al. 2012). Given that the survival trait after PWN inoculation is heritable (Carrasquinho et al. 2018), the implementation of breeding programs to select for resistance can be a valuable tool to control PWD spreading in the Iberian Peninsula. At the same time, it is important to understand how this resistance is achieved and, to this aim, transcriptomic approaches comparing resistant and susceptible maritime pine trees are likely to provide important insights into the defense response. A recent study using this comparative approach to highlight genes and pathways involved in PWN resistance has shown the contribution of cell wall lignification and the activation of the jasmonic acid defense pathway (Modesto et al. 2021). Previous reports on PWN response using uncharacterized plants described a higher number of differentially expressed genes at early timepoints after inoculation (6 h + 24 h post-infection) than at later timepoints (48 hpi and 7 days post-infection) (Gaspar et al. 2017). In the same study, genes involved in response to oxidative stress (e.g., GPX), secondary metabolism and jasmonate synthesis (e.g., OPR3, acyl-CoA oxidase) were shown to be upregulated after inoculation, revealing the importance of these mechanisms in maritime pine response to PWN. The phenylpropanoid biosynthesis pathway was also highly induced, including genes that encode for enzymes involved in lignin synthesis (*peroxidases* and *laccases*), suggesting that cell wall reinforcement may interfere with the ability of

Table 5.1 Literature reporting intraspecific genetic variation in susceptibility of maritime pine to biotic threats

Biotic threat	Genetic level	Outcome	References
Hylobius abietis	Families	Additive variation among 40 half-sibs in susceptibility	Zas et al. (2005)
	Populations	Significant variation in susceptiblity among 10 pine populations	López-Goldar et al. (2018)
	Populations	Significant variation in susceptiblity among 3 pine populations	Suárez-Vidal et al. (2017)
Matsucoccus feytaudi	Populations	Significant variation in density of nymphs among 8 pine populations	Di Matteo and Voltas (2016)
	Populations	Significant variation in insect density and symptoms among 8 pine populations	Schveste and Hugheto (1986)
	Populations	Significant variation in symtoms among 25 pine populations	Harfouche et al. (1995)
Dyorictria sylvestrella	Families	Significant differences in resistance among full-sibs	Jactel et al. (1999)
	Families	Significant differences in resistance among full-sibs	Kleinhentz et al. (1998)
Bursaphelenchus xylophilus	Families	Significant differences in resistance among half-sibs. Selection of resistant families	Carrasquinho et al. (2018)
	Families	Significant differences in resistance among half-sibs	Menéndez-Gutiérrez et al. (2017a)
	Populations	Significant variation in resistance among 9 pine populations	Zas et al. (2015)
	Populations	Significant variation among in resistance among pine populations	Menéndez-Gutierrez et al. (2017b)
Fusarium circinatum	Populations	Significant differences in resistance among 10 populations	Elvira-Recuenco et al. (2014)
	Families	Significant differences in resistance among families within populations	Elvira-Recuenco et al. (2014)
	Genotypes	Significant differences in resistance among clonaly-replicated genotypes within families	Elvira-Recuenco et al. (2014)
	Families	Significant differences among half-sib families	Vivas et al. (2012)
	Populations	Significant differences in resistance among pine populations	Iturritxa et al. (2012)
Diplodia sapinea	Populations	Significant variation in susceptibility among range wide populations	Hurel et al. (2021)
Armillaria ostoyae	Families	Significant differences among half-sib families	Solla et al. (2011)
	Families	Significant differences among half-sib families	Zas et al. (2007)
	Populations	Significant variation in susceptibility among range wide populations	Hurel et al. (2021)
Malampsora pinitorqua	Families	Significant and large additive genetic variance	Baradat and Desprez-Loustau (1997)
	Populations	Provenances from Italy and Landes did not differ in susceptibility but populations from Morocco were highly susceptible	Desprez-Loustau and Baradat (1991)

PWN to infest the stem tissues. At 7 days post-infection, *pinosylvin synthase* gene, involved in the synthesis of the compound pinosylvin having a potential nematicidal effect (Kodan et al. 2002), was found upregulated but it has been speculated that the induction of its synthesis in a late stage probably has a small impact on PWN numbers (Gaspar et al. 2017). Although several genes involved in terpenoid biosynthesis were also upregulated in the early stages of PWN infection (Gaspar et al. 2017), this does not seem to translate into a higher production of terpene compounds (Rodrigues et al. 2017). The induction of terpene synthesis has been observed after mechanical wounding (Rodrigues et al. 2017) and feeding by the PWN insect vector *M. galloprovincialis* (Gonçalves et al. 2020). It seems, therefore, that although the synthesis of terpenes was not induced by PWN, higher levels of these compounds upon the moment of PWN entering the plant may impact its infection success. Despite the induction by PWN inoculation of several genes relevant for plant defense, maritime pine response seems to be, in most cases, insufficient or inadequate to stop PWN infestation of the plant tissues. Gaspar et al. (2017) attributed susceptibility to a late transcriptional response, as observed in other susceptible species (Shin et al. 2009).

Published and ongoing molecular studies for the detection of QTLs and candidate genes involved in resistance will be crucial to assist the selection of resistant maritime pine individuals for breeding purposes.

Maritime pine is also affected by Pine Pitch Canker (PPC), a disease caused by *Fusarium circinatum* Nirenberg and O'Donnell. This pathogen is a fungus introduced to Northern Spain in 2005 (Landeras et al. 2005) and now established in the Atlantic area of Spain and Portugal (EPPO 2011). The disease causes important damages affecting tree growth and wood quality, and it represents an important threat to stand pines and nurseries worldwide (Wingfield et al. 2008). The characteristic disease symptoms are sunken cankers with abundant resin production in branches and stem that progressively girdle the wood causing yellowing of needles and leading to defoliation and dieback of branches. Today, PPC is observed in the Iberian Peninsula on *Pinus radiata* and *Pinus pinaster*. In general terms, *P. radiata* is more susceptible than *Pinus pinaster* (Iturritxa et al. 2013), which is considered a moderately resistant species. Maritime pine shows high genetic variation among provenances, families, and genotypes (Table 5.1), with relatively high narrow-sense heritability estimates ($h^2 = 0.43 - 0.58$ depending on the resistance trait measured) (Elvira-Recuenco et al. 2014). These values suggest this species has a high capacity of response to selection for disease resistance that could be successfully used in breeding programs.

As in PWN disease, a transcriptomic approach was used to understand the molecular mechanisms involved in maritime pine resistance to PPC. The gene expression profiles of maritime pine and the pathogen were determined during the infection process at 3, 5, and 10 days post-inoculation (dpi). At 3 dpi, before the fungus has penetrated the host tissue, highly upregulated genes are those related with early pathogen recognition and activation of defence responses (Hernández-Escribano et al. 2020), events contributing to explain the moderate resistance shown by maritime pine. At 5 and 10 dpi, when the fungus has penetrated and colonized the host tissue, highly expressed genes are related to phytohormone signaling, regulation of ROS, oxidative stress, positive regulation of cell death and signal transduction (Hernández-Escribano et al. 2020), all processes related with pattern-triggered immunity (PTI) activated in response to the pathogen (Boller and He, 2009). Genes of pathogenesis-related proteins, which include chitinases, were upregulated since 3dpi, corroborating early activation of plant defence. As part of the effector-triggered immunity (ETI), plant resistance genes play a crucial role activating downstream defence response (Jones and Dangl 2006). The most highly upregulated gene in maritime pine at all dpi was annotated as a disease resistance gene. BLASTP alignment against the non-redundant NCBI database resulted in best hits corresponding to disease R proteins of *Picea sitchensis* (unknown protein), *Pinus lambertiana* (CC-NBS-LRR like), *P. monticola* (CC-NBS-LRR), and *P. taeda* (NBS-LRR), with identities ranging from 46 to 64% (Hernández-Escribano 2019). The involvement that this resistance gene may have in response to infection needs to be explored since it is a good candidate to be used in maritime pine breeding programs.

Hernández-Escribano et al. (2020) found activation of complex phytohormone signaling pathways at early stages of infection, that involves crosstalk between salicylic acid, jasmonic acid, ethylene, and possibly auxins. Species in *Fusarium fujikori* complex (like *F. circinatum*) are able to synthesize phytohormones that contribute to plant disease (Bömke and Tudzynski 2009; Tsavkelova et al. 2012). They hypothesize the key steps where the pathogen could be manipulating *Pinus pinaster* phytohormone homeostasis to its own benefit, contributing to host susceptibility. They propose that *F. circinatum* prevents salicylic acid biosynthesis from the chorismate pathway by the synthesis of isochorismatase family hydrolase (ICSH) genes, perturbs ethylene homeostasis in the host by expression of genes related to ethylene biosynthesis, and could be blocking jasmonic acid signaling by COI1 suppression. This hypothesis requires testing in *F. circinatum* mutants to confirm the mechanisms underlying this regulation.

Future work in selection of candidate genes that contribute to *Pinus pinaster* resistance, as well as functional studies with *F. circinatum* mutants will enable the acceleration of breeding strategies. Indeed, the big progress done in NGS technologies, and the combination of "omics" approaches will help to elucidate the underlying mechanisms of plant-pathogen interactions.

5.3.4 Genetic Architecture

A major goal in genetics is to decipher the genetic basis of phenotypic variation. In trees, this knowledge is important to guide efforts towards molecular breeding (Grattapaglia et al. 2018), but also to draw conclusions on the kind of natural selection shaping trait variation and consequently the adaptive capacity of populations (Fady et al. 2016). Since the first advances of molecular marker technologies in the early 1990s, maritime pine has been a model species in conifers, at the forefront of research, for studying the genetic architecture of quantitative traits, i.e., number, chromosomal location, allelic effects of QTLs, degree of dominance-epistasis and gene function underlying trait variation. In this section, we illustrate the main achievements from linkage analysis (LA) and association mapping linkage disequilibrium-based approaches (LD), two methods used to detect statistically significant associations between genetic markers and phenotypes.

What must be reminded is that the relatively low heritability of the studied traits combined with the low sample sizes of the studied populations (up to few hundreds), in addition to single tree estimates (without replicate to reduce environmental noises), only allowed detecting a small number of genomic regions showing the strongest effects on trait variation. Given the limited sizes of the mapping populations used in QTL mapping studies, it should also be noticed that reported effect size estimates were substantially inflated (Beavis 1998; Xu 2003). In addition, the severely right-skewed distribution of allelic effects suggested the studied traits to be under polygenic control, a trend largely shared with other plants (Ingvarsson and Street 2011). Noticeably, it is only recently that the degree of polygenicity could be estimated thanks to novel multi-loci detection methods. Last, only one study attempted to compare LA and LD mapping approaches (Bartholomé et al. 2016). This study highlighted their complementarity for deciphering the genetic architecture of quantitative traits in maritime pine.

5.3.4.1 Linkage Mapping Using Pedigreed Populations

Complex traits were first dissected through QTL mapping, which relies on the segregation of alleles in families of known structure, generally two- or three-generation outbred pedigrees. Former QTL studies focused on important productive traits, such as growth, stem form and wood quality (Plomion et al. 1996a; Chagné et al. 2003; Markussen et al. 2003; Pot et al. 2006; Bartholomé et al. 2016b), plant defense-related traits, such as the production of terpenes (Plomion et al. 1996b) and water-use efficiency (WUE), the ratio between net CO_2 assimilation rate to stomatal conductance for water vapor (Brendel et al. 2002; de Miguel et al. 2014; Marguerit et al. 2014). In some studies, coincidences between mapped candidate genes and QTLs were detected. This is the case, for example, for KORRIGAN, a gene involved in the hemicellulose/cellulose biosynthesis, and QTLs for hemicellulose and fiber characteristics (Pot et al. 2006). However, first-generation linkage maps were not densely populated enough with gene-based markers, which hampered the detection of positional candidate genes. While most QTLs displayed rather low effects, a notable exception was for stable carbon isotope composition ($\delta^{13}C$), an integrator of intrinsic WUE, with a major QTL accounting for 67% of the observed phenotypic variance (Marguerit et al. 2014). The use of a unique F2 pedigree with contrasting alleles from different provenances (Landes vs. Corsica) certainly contributed to the detection of this major gene. Current QTL mapping experiments are focusing on the understanding of the genetic basis of tree responses to water availability. A first long-term investigation started in 2012 and is still ongoing in 2021 with 250 trees equipped with micro-dendrometers measuring (with automated data acquisition systems) daily fluctuations in stem diameter (Lagraulet 2015). By combining environmental data obtained at the same scale to that of the "pulse" of trees (i.e., every five minutes along a diurnal cycle) this study will make it possible to study the genetic architecture of the temporal and environmental dynamics of phenotypic plasticity in water supply, water demand and net growth as illustrated in eucalyptus by Bartholomé et al. (2020). Other unpublished QTL studies have been performed based on the same 500 F2 trees for (i) cavitation resistance, an important trait for evaluating the ability of trees to survive and recover from severe drought periods, and (ii) secondary growth plasticity with the analysis of early and late wood alternation for wood density and radial growth, which provides new insights into the genetic control of phenotypic plasticity (i.e., GxE interaction) at the genomic level.

5.3.4.2 Association Mapping Studies

Association mapping involves the use of more or less structured populations to determine the relationship between observed phenotypes and DNA variation. In maritime pine, as in many forest tree species, only a handful of studies reported associations between SNPs (Single Nucleotide Polymorphisms) and trait variability. A decade ago, GWAS was only performed using SNPs and InDels (Insertion-Deletion polymorphisms) obtained from sequenced amplicons within known candidate genes (CG), i.e., functional or expressional CG detected by transcriptomics and proteomics approaches. In a seminal study, Lepoittevin et al. (2012) used 384 SNPs (Illumina GoldenGate technology) from 286 genes and identified two loci associated with growth and wood cellulose content measured on trees of the French breeding populations. Then, using an in situ collection of eastern Iberian trees, Budde et al. (2014) were able to predict 29% of the phenotypic variation in a fire adaptive trait (proportion of serotinous cones) based on 17 SNPs. From the analysis of a common garden experiment, Cabezas et al. (2015) revealed four SNPs in KORRIGAN, a gene for which QTLs had been previously detected (see above), associated with early growth performance. Similarly, Bartholomé et al. (2016) studied 661 founders of the first and second generations French breeding program and detected four loci for stem straightness and three loci for height growth among 2,498 SNPs corresponding to 1,652 gene loci. A more recent study reported associations between SNPs and traits related to biotic interaction, namely susceptibility to two fungal pathogens (*Diplodia sapinea* and *Armillaria ostoyae*), and growth phenology using a clonal test (up to 535 genotypes) genotyped at 5 k loci (Hurel et al. 2021). A total of seven SNPs were found to be associated with pathogen susceptibility under an additive model, but no major effect alleles were detected. Larger GWAS analyses are on-going based on the same clonal collection, as well as an unstructured association population from Corsica (over one thousand trees phenotyped for height, survival, growth phenology, and iWUE), genotyped with the 4TREE Axiom assay developed within the B4EST EU funded project (https://b4est.eu/).

5.3.4.3 Polygenic Association Methods

A new paradigm shift proposing a polygenic basis of quantitative traits (Pritchard et al. 2010; Wisser et al. 2019) was recently applied to maritime pine. In their study, de Miguel et al. (2020) applied polygenic association methods to a clonal trial (vegetatively propagated by cuttings and planted in five sites with contrasted environments) for a series of adaptive traits (height growth, survival, bud phenology, biotic stress resistance, and functional traits). They found a considerable level of polygenicity, of a similar magnitude as in humans (Zeng et al. 2018), and that was stable across environments. They attributed this result partially to negative selection, an evolutionary force that would remove large-effect variants because of their deleterious effects, while small-effect variants would remain unaffected favoring the polygenic basis of the studied traits (O'Connor et al. 2019).

5.3.5 Genetic Variation as a Fuel for Adaptation

Adaptive evolutionary change for a given trait in a population requires different factors to take place. First, it needs an association between the trait and fitness; second, the heritable transmission of the trait across generations; and third, a selective pressure acting in the population. The breeder's equation relates these factors and the response to selection (Walsh and Lynch 2018). How natural selection works also depends on additional factors. For example, Chevin et al. (2010, 2012) proposed that even in a simple evolutionary model, we need to take into account the population phenotypic plasticity and the environmental sensitivity of selection (i.e., the change in the optimum phenotype with the environment) to estimate the critical rate of environmental change beyond which a population must decline and go extinct. In addition, the level of genetic variation in a population is key, as selection cannot act in its absence.

Distinct populations of a species may have different population genetic characteristics resulting in populations with different optimum values for a given trait due to adaptation to the local conditions (Alberto et al. 2013). In *Pinus pinaster,* a species with a complex demographic history and population genetic structure, a wealth of results shows long-term evolutionary change of the populations with some main drivers and traits (and genes) under selection—an outcome that is in accordance with knowledge from other forest tree species. One part of these results comes from comparison of quantitative and molecular genetic differentiation among populations (Q_{ST} vs. F_{ST} approach), showing the effect of directional selection for many growth and adaptive traits (González-Martínez et al. 2002; Lamy et al. 2011; Rodríguez-Quilón et al. 2016) but uniform selection for hydraulic traits (Lamy et al. 2011). In addition, clear signals of local adaptation have been reported, based on the existence of significant climate-trait correlations in *Pinus pinaster* and other Mediterranean pines (Alía et al. 1997; Climent et al. 2008). These correlations are translated at the molecular (gene) level, as shown by significant genotype-environment correlations using different types of markers (González-Martínez et al. 2001; Gómez et al. 2005; Grivet et al. 2011; Jaramillo-correa et al. 2015). Often, these correlations have shown a more important role of temperature-related variables than precipitation ones as drivers of local adaptation in *Pinus pinaster.*

However, short-term evolutionary changes, which may be more relevant in the face of rapid climate change, have been only rarely addressed. Here, we describe different factors affecting the adaptive response of *Pinus pinaster* to natural selection, and also to artificial selection, as an example of the potential for evolutionary change in the species.

5.3.5.1 Response to Selection

Pinus pinaster shows a large level of intra-population genetic diversity, both for molecular markers and quantitative traits. The latter is measured by the additive genetic variance, which according to the breeder's equation is the fuel for future adaptation, as it determines the possibility of selection (both natural and artificial). Genetic studies in *Pinus pinaster* have shown medium to high values of heritability, and therefore an important response to selection is expected for different traits, including growth-related (both growth and growth phenology) traits, reproductive traits, and abiotic (related to frost and drought tolerance) and biotic (related to pest and disease tolerance) response traits (Sampedro et al. 2009, 2010; Aranda et al. 2010; Lepoittevin et al. 2011; Gaspar et al. 2013; Elvira-Recuenco et al. 2014; Zas et al. 2015; Hurel et al. 2021).

The importance of this variation is exemplified by the genetic gain obtained in advance breeding programs of the species that can be around 10–15% of the trait in first-generation seed orchards, and up to 30% in a second generation (Bouffier et al. 2013). However, this response is obtained under strong artificial selection, having as a consequence a narrowing of the genetic basis in subsequent generations. In contrast, under natural conditions, we still do not know what the strength of natural selection is or what the main forces driving this process are. Also, it is not known if natural selection is stable over time and space for a given population. Therefore, it is not straightforward to make inferences about selection under natural conditions based on the knowledge produced by breeding programs and artificial selection (Clements et al. 2009; Hadfield et al. 2010; Morrissey et al. 2010).

Forest trees are long-lived organisms and thus the ontogenic stage at which selection takes place is also a topic of interest. Natural selection can be acting at early ages, but the individuals will only reproduce later in their lives. It is known that for some traits, such as height or wood density, there are trends in the variation of additive genetic and environmental variances over time (Kremer 1992; Danjon 1994; Costa and Durel 1996; Gaspar et al. 2008) that can affect their response to selection, or that these correlations can vary depending on the site conditions affecting the efficiency of early selection (Zas et al. 2004). As said above, for evolutionary change to take place, a heritable trait has to be correlated with fitness (or breeding objective in artificial selection). Variation of the genetic components of a trait with age can blur our understanding of these relationships.

5.3.5.2 Association Between Traits and Fitness

The association between traits and fitness can be estimated by the computation of phenotypic and genetic selection gradients. Phenotypic selection gradients (defined as the vector of partial regression coefficients of individual relative fitness on the traits; Lande and Arnold 1983) have been characterized for different plant species at the population level. In forest trees, while measuring adaptive phenotypes in the field is relatively straightforward, the estimation of fitness (or fitness components) is problematic. A common approach uses parentage analysis based on molecular markers and naturally regenerated offspring to estimate the contribution of each tree to the next generation, which can be considered a proxy of whole-life fitness in long-lived organisms such as forest trees.

In *Pinus pinaster*, significant positive female selection gradients for diameter and cone crop were reported (González-Martínez et al. 2006). In addition, in a related study in the same area (Alía et al. 2014), the higher female effective reproductive success was explained by differences in their production of offspring (due to seed quality) and, to a lesser extent, by seemingly better adapted seedlings. Also, it is important to mention that selection gradients and responses to selection for seedlings differed across experimental conditions. Therefore, despite the evidence of microevolutionary change in adaptive traits, directional or disruptive changes are difficult to predict due to variable selection at different life stages and environments. Distinct processes involved at different life stages (as can be mature trees or seedlings) together with environment-specific responses suggest that it would be difficult to predict evolutionary responses to environmental change in Mediterranean forest trees experimentally. Current ongoing work involving a larger number of populations evaluated in contrasted environments highlighted the complexity and case-specific nature of trait-fitness correlations in *Pinus pinaster* (Hurel 2020).

Currently, there is a focus on the study of trade-offs among traits and the extent of integration of the phenotypes (Santos-del-Blanco et al. 2015), which can also affect the response to selection under natural conditions. Indeed, trait correlation and phenotypic integration may both enhance and prevent evolutionary change, depending on the circumstances, with opposite patterns being possible in different environments.

5.3.5.3 Phenotypic Plasticity and Adaptation

Phenotypic plasticity, i.e., the ability of a genotype to generate a range of different phenotypes depending on the environment, is also an important evolutionary factor that determines the adaptability of a given genotype. It has been intensively studied in the last decade, because of its important implications in evolutionary biology. Indeed, there is a need to assess the extent and adaptive (or maladaptive) value of phenotypic plasticity, as its evolutionary impact is still far from clear (Sultan 2004; Miner et al. 2005). In *Pinus pinaster*, both significant effects of phenotypic plasticity (a recent study estimated the plastic component of height variation to exceed twice the genetic one (Archambeau et al. 2020), and genetic differences in plasticity among populations were reported for different traits (Alía et al. 1995; Chambel et al. 2004; Corcuera et al. 2010).

Theory predicts species adaptability will depend on phenotypic plasticity, relationship between fitness and traits, response to selection according to the additive and environmental variances, and the factors influencing them (Chevin et al. 2010; Walsh and Lynch 2018). This information is still lacking for *Pinus pinaster*, as well as for other Mediterranean pines, in order to effectively assess its evolutionary potential. Most studies have focused on specific populations, and the relative importance of phenotypic plasticity and adaptation is practically unknown for endangered populations, where low effective population sizes may affect adaptation patterns.

Based on the available information, we can expect that the existing standing genetic variation, and phenotypic plasticity would allow the populations to respond to different selection pressures. This expectation is illustrated by models forecasting adaptive potential of species that include additional aspects such as genetic structure, plasticity or local adaptation (Serra-Varela et al. 2015; Benito Garzón et al. 2019).

5.4 Perspectives

Although the current assembly of *Pinus pinaster* genome is still in a draft state, it is already a valuable resource for the conifer community. The assembly will benefit from improvements over the coming years with new datasets, resulting from existing or emerging technologies. An improved assembly will have considerable impact on many downstream analyses, ranging from gene-annotation to GWAS, thereby opening the full potential of the maritime pine genome to serve as a model system for conifer genomics. Additionally, advances in high throughput direct sequencing of full-length mRNA molecules are currently improving transcript identification, identifying alternative splicing events, simple sequence repeats and epitranscriptomic marks as well as long non-coding RNA prediction. Combination of spatial transcriptomics, based on, i.e., LCM, with methods for full-length single-molecule transcriptome analysis will provide accurate quantification of gene expression levels, which will enable researchers to study spatial organization of gene expression, as well as transcriptome dynamics associated with tree development under different environmental conditions. The integration of multi-omic information, combining genomics, transcriptomics, proteomics, metabolomics, and phenomics datasets, will require cross-talk among different omics communities (Aizat et al. 2018; Blokhina et al. 2019). This global-integrative analytical approach will lead to disentangle regulation of essential processes of paramount relevance in fundamental biology of maritime pine and other conifers, and thus unscramble the genetic basis of quantitative traits and the evolutionary forces that have shaped these traits.

In addition, maritime pine, as any living organism, does not live isolated in nature. Its growth, fitness, and health are also result of the complex and structured interaction with microbial communities they host in their phyllosphere, rhizosphere, and the endosphere, which comprise bacteria, fungi, protists, nematodes, and viruses. Therefore, to acquire a more holistic understanding of maritime pine productive and adaptive capacity, it is important to study their associated microbiomes and how this network of interactions is modulated by the environment.

Finally, studies under natural conditions are required to reliably estimate the short-term evolutionary changes expected under future climatic conditions, which help to implement in-situ conservation of maritime pine genetic resources as well as its sustainable management.

Acknowledgements The research leading to these results has received funding from the EU FP5 programme grant agreement QLK3-CT2002-01973 (TREESNIPS). The EU FP7 programme under REA grant agreements no 289841 (ProCoGen). The EU H2020 programme grant agreements no 676876 (GenTree); no 773383 (B4EST); no 676559 (ELIXIR-EXCELERATE); no 824110 (EASI-Genomics. PID 7825-ImPiONT); People Programme (Marie Curie Actions) of the EU FP7 is acknowledged under REA grant agreement PIEF-GA-2013-627761; Cost Action FP1406; Plant KBBE programme, Scientific and Technological Cooperation in Plant Genome Research PLE2009-0016 (SUSTAINPINE); ERA-NET Cofund ForestValue project MULTIFOREVER supported by ANR (France, ANR-19-SUM2-0002-01), FNR (Germany), MINCyT (Argentina), MINECO-AEI, RTA-2007-00084-00-00 and RTA2010-00120-C02 (Spain), MMM (Finland) and VINNOVA (Sweden) and EU H2020 programme grant agreements no 773324. Fundação para a Ciência e a Tecnologia (FCT) through grants BioISI (UIDB/04046/2020 and UIDP/04046/2020), the doctoral fellowships SFRH/BD/111687/2015 (I. Modesto), SFRH/BD/128827/2017 (A. Alves) and SFRH/BD/79779/2011 (A. Rodrigues), and project PTDC/BAA-MOL/28379/2017—LISBOA-01-0145-FEDER-028379 (FCT/MCTES and FEDER)); Conseil de la Région Nouvelle Aquitaine, France, through grant EMBRYOsoMATURE (17006494-0741); Spanish Ministries MEC, MICINN, MINECO and MICIU (BIO2007-29814-E; BIO2010-12302-E; AGL2014-54698R; BIO2015-69285-R; AGL2015-66048-C2; RTI2018-094041-B-I00; RTI2018-094691-B-C3; RTI2018-098015-B-I00), Junta de Andalucía grants (BIO-474 and BIO-114) and University of Alcalá grant (UAH-AE 2017-2), the doctoral contracts BES-2016-077347 (M. Callejas-Díaz), BES-2016-076833 (M. López-Hinojosa), PRE2019-090357 (L.F. Manjarrez) and 49-FPI-INIA-2014 (L. Hernández-Escribano) are also acknowledged for financial support.

References

Abad Viñas R, Caudullo G, Oliveira S, de Rigo D (2016) *Pinus pinaster* in Europe: distribution, habitat, usage and threats. In: San-Miguel-Ayanz J, de Rigo D, Caudullo G, Houston Durrant T, Mauri A (eds) European atlas of forest tree species. Publication Office of the European Union, Luxembourg, p e012d59+

Abarca D, Carneros E, Hernández H, Pizarro A, Trontin J-F, Díaz-Sala, C (2015) Phenotypic analysis of transgenic *Pinus pinaster* lines overexpressing *MYB5*. ProCoGen final open conference on promoting conifer genomic resources, Orléans, France, 30 November–2 December 2015, p 2

Abbott E, Hall D, Hamberger B, Bohlmann J (2010) Laser microdissection of conifer stem tissues: Isolation and analysis of high quality RNA, terpene synthase enzyme activity and terpenoid metabolites from resin ducts and cambial zone tissue of white spruce (*Picea glauca*). BMC Plant Biol 10(1):106. https://doi.org/10.1186/1471-2229-10-106

Ahuja MR (2017) Current status of forest tree biotechnology in a changing climate. In: Bonga JM, Park Y-S, Trontin J-F (eds) Proceedings of the 4th international conference of the IUFRO working party 2.09.02 on development and application of vegetative propagation technologies in plantation forestry to cope with a changing climate and environment, La Plata, Argentina, 19–23 September 2016. IUFRO, Vienna, Austria, pp 15–36

Ahuja MR, Neale DB (2005) Evolution of genome size in conifers. Silvae Genet 54(1–6):126–137. https://doi.org/10.1515/sg-2005-0020

Aizat W, Goh H-H, Baharum SN (2018) Omics applications for systems biology. Springer, Cham

Alberto FJ, Aitken SN, Alía R, González-Martínez SC, Hänninen H, Kremer A, Lefèvre F, Lenormand T, Yeaman S, Whetten R, Savolainen O (2013) Potential for evolutionary responses to climate change-evidence from tree populations. Glob Chang Biol 19(6):1645–1661. https://doi.org/10.1111/gcb.12181

Alía R, Chambel R, Notivol E, Climent J, González-Martínez SCSC (2014) Environment-dependent microevolution in a Mediterranean pine (*Pinus pinaster* Aiton). BMC Evol Biol 14(1):200. https://doi.org/10.1186/s12862-014-0200-5

Alía R, Gil LA, Pardos JA (1995) Performance of 43 *Pinus pinaster* Ait. provenances on 5 locations in Central Spain. Silvae Genet 44(2–3):75–81

Alía R, Moro J, Denis JB, Moro-Serrano J (1997) Performance of *Pinus pinaster* provenances in Spain: interpretation of the genotype by environment interaction. Can J for Res 27(10):1548–1559. https://doi.org/10.1139/cjfr-27-10-1548

Álvarez J, Cortizo M, Ordás R (2012) Cryopreservation of somatic embryogenic cultures of *Pinus pinaster*: effects on regrowth and embryo maturation. Cryo Letters 33(6):476–484

Alvarez JM, Bueno N, Cañas RA, Avila C, Cánovas FM, Ordás RJ (2018) Analysis of the WUSCHEL-RELATED HOMEOBOX gene family in *Pinus pinaster* : New insights into the gene family evolution. Plant Physiol Biochem 123:304–318. https://doi.org/10.1016/j.plaphy.2017.12.031

Álvarez JM, Majada J, Ordás RJ (2009) An improved micropropagation protocol for maritime pine (*Pinus pinaster* Ait.) isolated cotyledons. Forestry 82(2):175–184. https://doi.org/10.1093/forestry/cpn052

Alvarez JM, Ordás RJ (2013) Stable Agrobacterium-mediated transformation of maritime pine based on kanamycin selection. Sci World J 2013:1–9. https://doi.org/10.1155/2013/681792

Andivia E, Ruiz-Benito P, Díaz-Martínez P, Carro-Martínez N, Zavala MA, Madrigal-González J (2020) Inter-specific tolerance to recurrent droughts of pine species revealed in saplings rather than adult trees. For Ecol Manage 459:117848. https://doi.org/10.1016/j.foreco.2019.117848

Andivia E, Zuccarini P, Grau B, de Herralde F, Villar-Salvador P, Savé R (2019) Rooting big and deep rapidly: the ecological roots of pine species distribution in southern Europe. Trees 33(1):293–303. https://doi.org/10.1007/s00468-018-1777-x

Aranda I, Alía R, Ortega U, Dantas ÂK, Majada J (2010) Intra-specific variability in biomass partitioning and carbon isotopic discrimination under moderate drought stress in seedlings from four *Pinus pinaster* populations. Tree Genet Genomes 6(2):169–178. https://doi.org/10.1007/s11295-009-0238-5

Archambeau J, Benito Garzón M, Barraquand F, de Miguel Vega M, Plomion C, González-Martínez SC (2020) Combining climatic and genomic data improves range-wide tree height growth prediction in a forest tree. bioRxiv 2020.11.13.382515. https://doi.org/10.1101/2020.11.13.382515

Arrillaga I, Guevara MA, Muñoz-Bertomeu J, Lázaro-Gimeno D, Sáez-Laguna E, Díaz LM, Torralba L, Mendoza-Poudereux I, Segura J, Cervera MT (2014) Selection of haploid cell lines from megagametophyte cultures of maritime pine as a DNA source for massive sequencing of the species. Plant Cell, Tissue Organ Cult 118(1):147–155. https://doi.org/10.1007/s11240-014-0470-z

Arrillaga I, Morcillo M, Zanón I, Lario F, Segura J, Sales E (2019) New approaches to optimize somatic embryogenesis in maritime pine. Front Plant Sci 10:138. https://doi.org/10.3389/fpls.2019.00138

Assis TF, Fett-Neto AG, Alfenas AC (2004) Current techniques and prospects for the clonal propagation of hardwoods: emphasis on eucalyptus. In: Walter C, Carson M (eds) Plantation forestry for the 21st century, vol 1. Research SignPost. Trivandru, India, pp 303–333

Ávila C, Rueda-López M, Canales J, Cánovas FM, Michel R, Pillet-Emanuel H, Canlet F, Debille S, Trontin J-F (2016) Molecular characterization of transgenic maritime pine somatic plants overexpressing a cytosolic glutamine synthetase gene (*PsGS1a*) involved in nitrogen assimilation. IUFRO subdivision 2.4 conference on genomics and forest tree genetics, Arcachon, France, 30 May–3 June 2016, p 29 (S6–10)

Avila C, Suárez MF, Gómez-Maldonado J, Cánovas FM (2001) Spatial and temporal expression of two cytosolic glutamine synthetase genes in Scots pine: functional implications on nitrogen metabolism during early stages of conifer development. Plant J 25(1):93–102. https://doi.org/10.1046/j.1365-313x.2001.00938.x

Azeez A, Busov V (2021) CRISPR/Cas9-mediated single and biallelic knockout of poplar STERILE APETALA (PopSAP) leads to complete reproductive sterility. Plant Biotechnol J 19(1):23–25. https://doi.org/10.1111/pbi.13451

Ba M, Salin F, Fourcaud T, Stokes A (2010) Reorientation strategies in leaning stems of young maritime pine (*Pinus pinaster*) and loblolly pine (*Pinus taeda*). IAWA J 31(4):465–480. https://doi.org/10.1163/22941932-90000036

Baker EAG, Wegrzyn JL, Sezen UU, Falk T, Maloney PE, Vogler DR, Delfino-Mix A, Jensen C, Mitton J, Wright J, Knaus B, Rai H, Cronn R, Gonzalez-Ibeas D, Vasquez-Gross HA, Famula RA, Liu J-J, Kueppers LM, Neale DB (2018) Comparative transcriptomics among four white pine species. Genes Genomes Genet 8(5):1461–1474. https://doi.org/10.1534/g3.118.200257

Baradat P, Desprez-Loustau M (1997) Analyse diallèle et intégration de la sensibilité à la rouille courbeuse dans le programme d'amélioration du pin maritime (Diallel analysis and integration in the breeding program of maritime pine of sensitivity to twisting rust). Ann Des Sci for 54(1):83–106. https://doi.org/10.1051/forest:19970107

Bartholomé J, Bink MC, van Heerwaarden J, Chancerel E, Boury C, Lesur I, Isik F, Bouffier L, Plomion C (2016) Linkage and association mapping for two major traits used in the maritime pine breeding program: height growth and stem straightness. PLoS ONE 11(11):e0165323. https://doi.org/10.1371/journal.pone.0165323

Bartholomé J, Mabiala A, Burlett R, Bert D, Leplé J, Plomion C, Gion J (2020) The pulse of the tree is under genetic control: eucalyptus as a case study. Plant J 103(1):338–356. https://doi.org/10.1111/tpj.14734

Bartholomé J, Van Heerwaarden J, Isik F, Boury C, Vidal M, Plomion C, Bouffier L (2016b) Performance of genomic prediction within and across generations in maritime pine. BMC Genomics 17(1):604. https://doi.org/10.1186/s12864-016-2879-8

Batllori E, De Cáceres M, Brotons L, Ackerly DD, Moritz MA, Lloret F (2017) Cumulative effects of fire and drought in Mediterranean ecosystems. Ecosphere 8(8):e01906. https://doi.org/10.1002/ecs2.1906

Bautista R, Villalobos DP, Díaz-Moreno S, Cantón FR, Cánovas FM, Claros MG (2007) Toward a *Pinus pinaster* bacterial artificial chromosome library. Ann for Sci 64(8):855–864. https://doi.org/10.1051/forest:2007060

Beaulieu J, Doerksen TK, MacKay J, Rainville A, Bousquet J (2014) Genomic selection accuracies within and between environments and small breeding groups in white spruce. BMC Genomics 15(1):1048. https://doi.org/10.1186/1471-2164-15-1048

Beavis W (1998) QTL analyses: power, precision, and accuracy. In: Paterson AH (ed) Molecular dissection of complex traits, 1st edn. CRC Press, Boca Raton, Florida, USA, pp 145–162

Bedon F, Bomal C, Caron S, Levasseur C, Boyle B, Mansfield SD, Schmidt A, Gershenzon J, Grima-Pettenati J, Séguin A, MacKay J (2010) Subgroup 4 R2R3-MYBs in conifer trees: gene family expansion and contribution to the isoprenoid- and flavonoid-oriented responses. J Exp Bot 61(14):3847–3864. https://doi.org/10.1093/jxb/erq196

Behringer D, Zimmermann H, Ziegenhagen B, Liepelt S (2015) Differential gene expression reveals candidate genes for drought stress response in *Abies alba* (Pinaceae). PLoS ONE 10(4):e0124564. https://doi.org/10.1371/journal.pone.0124564

Benito Garzón M, Robson TM, Hampe A (2019) ΔTraitSDMs: species distribution models that account for local adaptation and phenotypic plasticity. New Phytol 222(4):1757–1765. https://doi.org/10.1111/nph.15716

Bentley DR, Balasubramanian S, Swerdlow HP, Smith GP, Milton J, Brown CG, Hall KP, Evers DJ, Barnes CL, Bignell HR, Boutell JM, Bryant J, Carter RJ, Keira Cheetham R, Cox AJ, Ellis DJ, Flatbush MR, Gormley NA, Humphray SJ, Irving LJ, Karbelashvili MS, Kirk SM, Li H, Liu X, Maisinger KS, Murray LJ, Obradovic B, Ost T, Parkinson ML, Pratt MR, Rasolonjatovo IMJ, Reed MT, Rigatti R, Rodighiero C, Ross MT, Sabot A, Sankar SV, Scally A, Schroth GP, Smith ME, Smith VP, Spiridou A, Torrance PE, Tzonev SS, Vermaas EH, Walter K, Wu X, Zhang L, Alam MD, Anastasi C, Aniebo IC, Bailey DMD, Bancarz IR, Banerjee S, Barbour SG, Baybayan PA, Benoit VA, Benson KF, Bevis C, Black PJ, Boodhun A, Brennan JS, Bridgham JA, Brown RC, Brown AA, Buermann DH, Bundu AA, Burrows JC, Carter NP, Castillo N, Chiara E. Catenazzi M, Chang S, Neil Cooley R, Crake NR, Dada OO, Diakoumakos KD, Dominguez-Fernandez B, Earnshaw DJ, Egbujor UC, Elmore DW, Etchin SS, Ewan MR, Fedurco M, Fraser LJ, Fuentes Fajardo KV, Scott Furey W, George D, Gietzen KJ, Goddard CP, Golda GS, Granieri PA, Green DE, Gustafson DL, Hansen NF, Harnish K, Haudenschild CD, Heyer NI, Hims MM, Ho JT, Horgan AM, Hoschler K, Hurwitz S, Ivanov DV, Johnson MQ, James T, Huw Jones TA, Kang G-D, Kerelska TH, Kersey AD, Khrebtukova I, Kindwall AP, Kingsbury Z, Kokko-Gonzales PI, Kumar A, Laurent MA, Lawley CT, Lee SE, Lee X, Liao AK, Loch JA, Lok M, Luo S, Mammen RM, Martin JW, McCauley PG, McNitt P, Mehta P, Moon KW, Mullens JW, Newington T, Ning Z, Ling Ng B, Novo SM, O'Neill MJ, Osborne MA, Osnowski A, Ostadan O, Paraschos LL, Pickering L, Pike AC, Pike AC, Chris Pinkard D, Pliskin DP, Podhasky J, Quijano VJ, Raczy C, Rae VH, Rawlings SR, Chiva Rodriguez A, Roe PM, Rogers J, Rogert Bacigalupo MC, Romanov N, Romieu A, Roth RK, Rourke NJ, Ruediger ST, Rusman E, Sanches-Kuiper RM, Schenker MR, Seoane JM, Shaw RJ, Shiver MK, Short SW, Sizto NL, Sluis JP, Smith MA, Ernest Sohna Sohna J, Spence EJ, Stevens K, Sutton N, Szajkowski L, Tregidgo CL, Turcatti G, VandeVondele S, Verhovsky Y, Virk SM, Wakelin S, Walcott GC, Wang J, Worsley GJ, Yan J, Yau L, Zuerlein M, Rogers J, Mullikin JC, Hurles ME, McCooke NJ, West JS, Oaks FL, Lundberg PL, Klenerman D, Durbin R, Smith AJ (2008) Accurate whole human genome sequencing using reversible terminator chemistry. Nature 456(7218):53–59. https://doi.org/10.1038/nature07517

Bernhardsson C, Vidalis A, Wang X, Scofield DG, Schiffthaler B, Baison J, Street NR, García-Gil MR, Ingvarsson PK (2019) An ultra-dense haploid genetic map for evaluating the highly fragmented genome assembly of Norway spruce (*Picea abies*). Genes Genomes Genet 9(5):1623–1632. https://doi.org/10.1534/g3.118.200840

Birol I, Raymond A, Jackman SD, Pleasance S, Coope R, Taylor GA, Saint YMM, Keeling CI, Brand D, Vandervalk BP, Kirk H, Pandoh P, Moore RA, Zhao Y, Mungall AJ, Jaquish B, Yanchuk A, Ritland C, Boyle B, Bousquet J, Ritland K, MacKay J, Bohlmann J, Jones SJM (2013) Assembling the 20 Gb white spruce (*Picea glauca*) genome from whole-genome shotgun sequencing data. Bioinformatics 29(12):1492–1497. https://doi.org/10.1093/bioinformatics/btt178

Blokhina O, Laitinen T, Hatakeyama Y, Delhomme N, Paasela T, Zhao L, Street NR, Wada H, Kärkönen A,

Fagerstedt K (2019) Ray parenchymal cells contribute to lignification of tracheids in developing xylem of norway spruce. Plant Physiol 181(4):1552–1572. https://doi.org/10.1104/pp.19.00743

Boller T, He SY (2009) Innate immunity in plants: an arms race between pattern recognition receptors in plants and effectors in microbial pathogens. Science 324(5928):742–744. https://doi.org/10.1126/science.1171647

Bömke C, Tudzynski B (2009) Diversity, regulation, and evolution of the gibberellin biosynthetic pathway in fungi compared to plants and bacteria. Phytochemistry 70(15–16):1876–1893. https://doi.org/10.1016/j.phytochem.2009.05.020

Bonga JM (2016) Conifer clonal propagation in tree improvement programs. In: Park Y-S, Bonga JM, Moon H-K (eds) Vegetative propagation of forest trees. National Institute of Forest Science (NIFoS), Seoul, Korea, pp 3–31

Bouffier L, Debille S, Harvengt L, Trontin J-F, Pastuszka P, Raffin A, Lelu-Walter M-A, Musch B (2017) Pollen contamination and mating structure in maritime pine clonal seed orchards. Proceedings of IUFRO seed orchard conference, Balsta, Sweden, 4–6 September 2017

Bouffier L, Klápště J, Suontama M, Dungey HS, Mullin TJ (2019) Evaluation of forest tree breeding strategies based on partial pedigree reconstruction through simulations: *Pinus pinaster* and *Eucalyptus nitens* as case studies. Can J for Res 49(12):1504–1515. https://doi.org/10.1139/cjfr-2019-0145

Bouffier L, Raffin A, Alia R (2013) Maritime Pine (*Pinus pinaster* Ait.). In: Mullin T, Lee S (eds) Best practice for tree breeding in Europe. Skogforsk, Uppsala, Sweden, pp 65–76

Brendel O, Pot D, Plomion C, Rozenberg P, Guehl J-M (2002) Genetic parameters and QTL analysis of δ 13 C and ring width in maritime pine. Plant Cell Environ 25 (8):945–953. https://doi.org/10.1046/j.1365-3040.2002.00872.x

Brodribb TJ, Pittermann J, Coomes DA (2012) Elegance versus speed: examining the competition between conifer and angiosperm trees. Int J Plant Sci 173 (6):673–694. https://doi.org/10.1086/666005

Bucci G, González-Martínez SC, Le Provost G, Plomion C, Ribeiro MM, Sebastiani F, Alía R, Vendramin GG (2007) Range-wide phylogeography and gene zones in *Pinus pinaster* Ait. revealed by chloroplast microsatellite markers. Mol Ecol 16 (10):2137–2153. https://doi.org/10.1111/j.1365-294X.2007.03275.x

Budde KB, Heuertz M, Hernández-Serrano A, Pausas JG, Vendramin GG, Verdú M, González-Martínez SC (2014) *In situ* genetic association for serotiny, a fire-related trait, in Mediterranean maritime pine (*Pinus pinaster*). New Phytol 201(1):230–241. https://doi.org/10.1111/nph.12483

Cabezas J, Morcillo M, Vélez M, Díaz L, Segura J, Cervera M, Arrillaga I (2016) Haploids in conifer species: characterization and chromosomal integrity of a maritime pine cell line. Forests 7(12):274. https://doi.org/10.3390/f7110274

Cabezas JA, González-Martínez SC, Collada C, Guevara MA, Boury C, de María N, Eveno E, Aranda I, Garnier-Géré PH, Brach J, Alía R, Plomion C, Cervera MT (2015) Nucleotide polymorphisms in a pine ortholog of the Arabidopsis degrading enzyme cellulase KORRIGAN are associated with early growth performance in *Pinus pinaster*. Tree Physiol 35(9):1000–1006. https://doi.org/10.1093/treephys/tpv050

Caminero L, Génova M, Camarero JJ, Sánchez-Salguero R (2018) Growth responses to climate and drought at the southernmost European limit of Mediterranean *Pinus pinaster* forests. Dendrochronologia 48:20–29. https://doi.org/10.1016/j.dendro.2018.01.006

Canales J, Bautista R, Label P, Gómez-Maldonado J, Lesur I, Fernández-Pozo N, Rueda-López M, Guerrero-Fernández D, Castro-Rodríguez V, Benzekri H, Cañas RA, Guevara M-A, Rodrigues A, Seoane P, Teyssier C, Morel A, Ehrenmann F, Le Provost G, Lalanne C, Noirot C, Klopp C, Reymond I, García-Gutiérrez A, Trontin J-F, Lelu-Walter M-A, Miguel C, Cervera MT, Cantón FR, Plomion C, Harvengt L, Avila C, Gonzalo Claros M, Cánovas FM (2014) *De novo* assembly of maritime pine transcriptome: implications for forest breeding and biotechnology. Plant Biotechnol J 12(3):286–299. https://doi.org/10.1111/pbi.12136

Cañas RA, Canales J, Gomez-Maldonado J, Avila C, Canovas FM (2014) Transcriptome analysis in maritime pine using laser capture microdissection and 454 pyrosequencing. Tree Physiol 34(11):1278–1288. https://doi.org/10.1093/treephys/tpt113

Cañas RA, Feito I, Fuente-Maqueda JF, Ávila C, Majada J, Cánovas FM (2015) Transcriptome-wide analysis supports environmental adaptations of two *Pinus pinaster* populations from contrasting habitats. BMC Genomics 16(1):909. https://doi.org/10.1186/s12864-015-2177-x

Cañas RA, Li Z, Pascual MB, Castro-Rodríguez V, Ávila C, Sterck L, Van de Peer Y, Cánovas FM (2017) The gene expression landscape of pine seedling tissues. Plant J 91(6):1064–1087. https://doi.org/10.1111/tpj.13617

Cañas RA, Pascual MB, de la Torre FN, Ávila C, Cánovas FM (2019) Resources for conifer functional genomics at the omics era. In: Cánovas FM (ed) Advances in botanical research. Academic Press, pp 39–76

Cardon LR, Bell JI (2001) Association study designs for complex diseases. Nat Rev Genet 2(2):91–99. https://doi.org/10.1038/35052543

Carneros E, Abarca D, del Amo A, Trontin J-F, Díaz-Sala C (2014) Genetic transformation of *Pinus pinaster* embryogenic lines: molecular characterization, optimization of embryo maturation-germination and plantlet micropropagation. In: Book of abstracts of the 3rd international conference of the IUFRO working party 2.09.02 on woody plant production integrating genetic and vegetative propagation technologies, Vitoria-Gasteiz, Spain, 8–12 September 2014, p 113

Carrasquinho I, Lisboa A, Inácio ML, Gonçalves E (2018) Genetic variation in susceptibility to pine wilt disease of maritime pine (*Pinus pinaster* Aiton) half-sib families. Ann For Sci 75(3):85. https://doi.org/10.1007/s13595-018-0759-x

Carrión JS, Navarro C, Navarro J, Munuera M (2000) The distribution of cluster pine (*Pinus pinaster*) in Spain as derived from palaeoecological data: relationships with phytosociological classification. The Holocene 10(2):243–252. https://doi.org/10.1191/095968300676937462

Castander-Olarieta A, Pereira C, Montalbán IA, Canhoto J, Moncaleán P (2020) Stress modulation in *Pinus spp*. Somatic embryogenesis as model for climate change mitigation: stress is not always a problem. In: Chong P, Newman D, Steinmacher D (eds) Agricultural, forestry and bioindustry biotechnology and biodiscovery. Springer, Cham, pp 117–130

Castro-Rodríguez V, García-Gutiérrez A, Cañas RA, Pascual M, Avila C, Cánovas FM (2015) Redundancy and metabolic function of the glutamine synthetase gene family in poplar. BMC Plant Biol 15(1):20. https://doi.org/10.1186/s12870-014-0365-5

Celedon JM, Yuen MMS, Chiang A, Henderson H, Reid KE, Bohlmann J (2017) Cell-type- and tissue-specific transcriptomes of the white spruce (*Picea glauca*) bark unmask fine-scale spatial patterns of constitutive and induced conifer defense. Plant J 92 (4):710–726. https://doi.org/10.1111/tpj.13673

Chagné D, Brown G, Lalanne C, Madur D, Pot D, Neale D, Plomion C (2003) Comparative genome and QTL mapping between maritime and loblolly pines. Mol Breed 12(3):185–195. https://doi.org/10.1023/A:1026318327911

Chambel MR, Climent J, Alía R (2004) Intra-specific variation of phenotypic plasticity for biomass allocation in Mediterranean pines. In: Arianoutsou M, Thanos C (eds) Proceedings 10th MEDECOS conference, Rhodes, Greece, April 2014. Millpress, Rhodas, Grecia

Chancerel E, Lamy J-B, Lesur I, Noirot C, Klopp C, Ehrenmann F, Boury C, Le PG, Label P, Lalanne C, Léger V, Salin F, Gion J-M, Plomion C (2013) High-density linkage mapping in a pine tree reveals a genomic region associated with inbreeding depression and provides clues to the extent and distribution of meiotic recombination. BMC Biol 11(1):50. https://doi.org/10.1186/1741-7007-11-50

Chaperon H, Hinschberger F, Haury P, Alazard P (1991) A comparative study of the development of maritime pine plants raised from cuttings or from seedlings. Ann Rech Sylvi. AFOCEL 1989–1990:115–133

Chávez Montes RA, de Rosas-Cárdenas FF, De Paoli E, Accerbi M, Rymarquis LA, Mahalingam G, Marsch-Martínez N, Meyers BC, Green PJ, de Folter S (2014) Sample sequencing of vascular plants demonstrates widespread conservation and divergence of microRNAs. Nat Commun 5(1):3722. https://doi.org/10.1038/ncomms4722

Chen X (2009) Small RNAs and their roles in plant development. Annu Rev Cell Dev Biol 25(1):21–44. https://doi.org/10.1146/annurev.cellbio.042308.113417

Chevin L-M, Collins S, Lefèvre F (2012) Phenotypic plasticity and evolutionary demographic responses to climate change: taking theory out to the field. Funct Ecol 27(4):967–979. https://doi.org/10.1111/j.1365-2435.2012.02043.x

Chevin L-M, Lande R, Mace GM (2010) Adaptation, plasticity, and extinction in a changing environment: towards a predictive theory. PLoS Biol 8(4):e1000357. https://doi.org/10.1371/journal.pbio.1000357

Clements MN, Morrissey MM, Wilson AJ, Re D, Postma E, Walling CA, Kruuk LEB, Nussey DH, Reale D, Clements MN, Morrissey MM, Postma E, Walling CA, Kruuk LEB, Nussey DH (2009) An ecologist's guide to the animal model. J Anim Ecol 79 (Pemberton 2008):13–26. https://doi.org/10.1111/j.1365-2656.2009.01639.x

Climent J, Prada MA, Calama R, Chambel MR, Sánchez de Ron D, Alía R, De Ron DS (2008) To grow or to seed: ecotypic variation in reproductive allocation and cone production by young female Aleppo pine (*Pinus halepensis*, Pinaceae). Am J Bot 95(7):833–842. https://doi.org/10.3732/ajb.2007354

Colina FJ, Carbó M, Álvarez A, Valledor L, Cañal MJ (2020) The analysis of *Pinus pinaster* SnRKs reveals clues of the evolution of this family and a new set of abiotic stress resistance biomarkers. Agronomy 10 (2):295. https://doi.org/10.3390/agronomy10020295

Consortium TG (2012) The tomato genome sequence provides insights into fleshy fruit evolution. Nature 485 (7400):635–641. https://doi.org/10.1038/nature11119

Corcuera L, Gil-Pelegrin E, Notivol E (2010) Phenotypic plasticity in *Pinus pinaster* δ13C: environment modulates genetic variation. Ann For Sci 67(8):812–812. https://doi.org/10.1051/forest/2010048

Corcuera L, Gil-Pelegrin E, Notivol E (2012) Differences in hydraulic architecture between mesic and xeric *Pinus pinaster* populations at the seedling stage. Tree Physiol 32(12):1442–1457. https://doi.org/10.1093/treephys/tps103

Costa P, Durel CE (1996) Time trends in genetic control over height and diameter in maritime pine. Can J for Res 26(7):1209–1217

Craven-Bartle B, Pascual MB, Cánovas FM, Ávila C (2013) A Myb transcription factor regulates genes of the phenylalanine pathway in maritime pine. Plant J 74(5):755–766. https://doi.org/10.1111/tpj.12158

Czemmel S, Höll J, Loyola R, Arce-Johnson P, Alcalde JA, Matus JT, Bogs J (2017) Transcriptome-wide identification of novel UV-B- and light modulated flavonol pathway genes controlled by VviMYBF1. Front Plant Sci 8:1084. https://doi.org/10.3389/fpls.2017.01084

Dai Y, Hu G, Dupas A, Medina L, Blandels N, San Clemente H, Ladouce N, Badawi M, Hernandez-Raquet G, Mounet F, Grima-Pettenati J, Cassan-Wang H (2020) Implementing the CRISPR/Cas9 technology in eucalyptus hairy roots using wood-related genes. Int J Mol Sci 21(10):3408. https://doi.org/10.3390/ijms21103408

Danjon F (1994) Heritabilities and genetic correlations for estimated growth curve parameters in maritime pine. Theor Appl Genet 89–89(7–8):911–921. https://doi.org/10.1007/BF00224517

De Bie T, Cristianini N, Demuth JP, Hahn MW (2006) CAFE: a computational tool for the study of gene family evolution. Bioinformatics 22(10):1269–1271. https://doi.org/10.1093/bioinformatics/btl097

De Diego N, Montalbán IA, Moncaleán P (2011) Improved micropropagation protocol for maritime pine using zygotic embryos. Scand J For Res 26 (3):202–211. https://doi.org/10.1080/02827581.2011.559174

De La Torre AR, Birol I, Bousquet J, Ingvarsson PK, Jansson S, Jones SJM, Keeling CI, MacKay J, Nilsson O, Ritland K, Street N, Yanchuk A, Zerbe P, Bohlmann J (2014) Insights into conifer giga-genomes. Plant Physiol 166(4):1724–1732. https://doi.org/10.1104/pp.114.248708

De La Torre AR, Li Z, Van de Peer Y, Ingvarsson PK (2017) Contrasting rates of molecular evolution and patterns of selection among gymnosperms and flowering plants. Mol Biol Evol 34(6):1363–1377. https://doi.org/10.1093/molbev/msx069

De La Torre AR, Wilhite B, Neale DB (2019) Environmental genome-wide association reveals climate adaptation is shaped by subtle to moderate allele frequency shifts in loblolly pine. Genome Biol Evol 11 (10):2976–2989. https://doi.org/10.1093/gbe/evz220

de María N, Guevara MÁ, Perdiguero P, Vélez MD, Cabezas JA, López-Hinojosa M, Li Z, Díaz LM, Pizarro A, Mancha JA, Sterck L, Sánchez-Gómez D, Miguel C, Collada C, Díaz-Sala MC, Cervera MT (2020) Molecular study of drought response in the Mediterranean conifer *Pinus pinaster* Ait.: differential transcriptomic profiling reveals constitutive water deficit-independent drought tolerance mechanisms. Ecol Evol 10(18):9788–9807. https://doi.org/10.1002/ece3.6613

de Miguel M, Bartholomé J, Ehrenmann F, Murat F, Moriguchi Y, Uchiyama K, Ueno S, Tsumura Y, Lagraulet H, de Maria N, Cabezas J-A, Cervera M-T, Gion JM, Salse J, Plomion C (2015) Evidence of intense chromosomal shuffling during conifer evolution. Genome Biol Evol 7(10):2799–2890. https://doi.org/10.1093/gbe/evv185

de Miguel M, Cabezas J-A, de María N, Sánchez-Gómez D, Guevara M-Á, Vélez M-D, Sáez-Laguna E, Díaz L-M, Mancha J-A, Barbero M-C, Collada C, Díaz-Sala C, Aranda I, Cervera M-T (2014) Genetic control of functional traits related to photosynthesis and water use efficiency in *Pinus pinaster* Ait. drought response: integration of genome annotation, allele association and QTL detection for candidate gene identification. BMC Genomics 15(1):464. https://doi.org/10.1186/1471-2164-15-464

de Miguel M, Guevara MÁ, Sánchez-Gómez D, de María N, Díaz LM, Mancha JA, Fernández de Simón B, Cadahía E, Desai N, Aranda I, Cervera M-T (2016) Organ-specific metabolic responses to drought in *Pinus pinaster* Ait. Plant Physiol Biochem 102:17–26. https://doi.org/10.1016/j.plaphy.2016.02.013

de Miguel M, Rodríguez-Quilón I, Heuertz M, Hurel A, Grivet D, Jaramillo-Correa J-P, Vendramin GG, Plomion C, Majada J, Alía R, Eckert AJ, González-Martínez SC (2020) Polygenic adaptation and negative selection across traits, years and environments in a long-lived plant species (*Pinus pinaster* Ait.). bioRxiv 2020.03.02.974113. https://doi.org/10.1101/2020.03.02.974113

de Miguel M, Sanchez-Gomez D, Cervera MT, Aranda I (2012) Functional and genetic characterization of gas exchange and intrinsic water use efficiency in a full-sib family of *Pinus pinaster* Ait. in response to drought. Tree Physiol 32(1):94–103. https://doi.org/10.1093/treephys/tpr122

de Vega-Bartol JJ, Simões M, Lorenz W, Rodrigues AS, Alba R, Dean JFD, Miguel CM (2013) Transcriptomic analysis highlights epigenetic and transcriptional regulation during zygotic embryo development of *Pinus pinaster*. BMC Plant Biol 13(1):123. https://doi.org/10.1186/1471-2229-13-123

Deamer D, Akeson M, Branton D (2016) Three decades of nanopore sequencing. Nat Biotechnol 34(5):518–524. https://doi.org/10.1038/nbt.3423

Desprez-Loustau M, Baradat P (1991) Variabilité interraciale de la sensibilité à la rouille courbeuse chez le pin maritime (Variation in susceptibility to twisting rust between maritime pine races). Ann Des Sci for 48 (5):497–511. https://doi.org/10.1051/forest:19910502

Di Matteo G, Voltas J (2016) Multienvironment evaluation of *Pinus pinaster* provenances: evidence of genetic trade-offs between sdaptation to optimal conditions and resistance to the maritime pine bast scale (*Matsucoccus feytaudi*). For Sci 62(5):553–563. https://doi.org/10.5849/forsci.15-109

Díaz-Sala C (2014) Direct reprogramming of adult somatic cells toward adventitious root formation in forest tree species: the effect of the juvenile-adult transition. Front Plant Sci 5:310. https://doi.org/10.3389/fpls.2014.00310

Díaz-Sala C (2019) Molecular dissection of the regenerative capacity of forest tree species: special focus on conifers. Front Plant Sci 9:1943. https://doi.org/10.3389/fpls.2018.01943

Díaz-Sala C (2020) A perspective on adventitious root formation in tree species. Plants 9(12):1789. https://doi.org/10.3390/plants9121789

Ding C, Park YS, Bonga J, Bartlett B, Li Y, Raley F (2019) A brief review of combining genomic selection and somatic embryogenesis for tree improvement. In: Bonga JM, Park YS, Trontin J-F (eds) Proceedings of the 5th international conference of the IUFRO working party 2.09.02 on clonal trees in the bioeconomy age: opportunities and challenges, Coimbra, Portugal,

10–15 September 2018. IUFRO, Vienna, Austria, pp 55–69

Dolgosheina EV, Morin RD, Aksay G, Sahinalp SC, Magrini V, Mardis ER, Mattsson J, Unrau PJ (2008) Conifers have a unique small RNA silencing signature. RNA 14(8):1508–1515. https://doi.org/10.1261/rna.1052008

Dubos C, Le Provost G, Pot D, Salin F, Lalane C, Madur D, Frigerio J-M, Plomion C (2003) Identification and characterization of water-stress-responsive genes in hydroponically grown maritime pine (*Pinus pinaster*) seedlings. Tree Physiol 23(3):169–179. https://doi.org/10.1093/treephys/23.3.169

Dubos C, Plomion C (2003) Identification of water-deficit responsive genes in maritime pine (*Pinus pinaster* Ait.) roots. Plant Mol Biol 51(2):249–262. https://doi.org/10.1023/A:1021168811590

Dumas E, Franclet A, Monteuuis O (1989) Microgreffage de méristèmes primaires caulinaires de pins maritimes (*Pinus pinaster* Ait.) âgés sur de jeunes semis cultivés *in vitro* (Apical meristem micrografting of mature maritime pines (*Pinus pinaster* Ait.) onto *in vitro* young seedlings.). C R Acad Sci (III), Paris, France 309(19):723–728

Dumas E, Monteuuis O (1995) *In vitro* rooting of micropropagated shoots from juvenile and mature *Pinus pinaster* explants: influence of activated charcoal. Plant Cell Tissue Organ Cult 40(3):231–235. https://doi.org/10.1007/BF00048128

Eid J, Fehr A, Gray J, Luong K, Lyle J, Otto G, Peluso P, Rank D, Baybayan P, Bettman B, Bibillo A, Bjornson K, Chaudhuri B, Christians F, Cicero R, Clark S, Dalal R, DeWinter A, Dixon J, Foquet M, Gaertner A, Hardenbol P, Heiner C, Hester K, Holden D, Kearns G, Kong X, Kuse R, Lacroix Y, Lin S, Lundquist P, Ma C, Marks P, Maxham M, Murphy D, Park I, Pham T, Phillips M, Roy J, Sebra R, Shen G, Sorenson J, Tomaney A, Travers K, Trulson M, Vieceli J, Wegener J, Wu D, Yang A, Zaccarin D, Zhao P, Zhong F, Korlach J, Turner S (2009) Real-time DNA sequencing from single polymerase molecules. Science 323(5910):133–138. https://doi.org/10.1126/science.1162986

El-Azaz J, de la Torre F, Ávila C, Cánovas FM (2016) Identification of a small protein domain present in all plant lineages that confers high prephenate dehydratase activity. Plant J 87(2):215–229. https://doi.org/10.1111/tpj.13195

El-Azaz J, de la Torre F, Pascual MB, Debille S, Canlet F, Harvengt L, Trontin J-F, Ávila C, Cánovas FM (2020) Transcriptional analysis of arogenate dehydratase genes identifies a link between phenylalanine biosynthesis and lignin biosynthesis. J Exp Bot 71(10):3080–3093. https://doi.org/10.1093/jxb/eraa099

Elvira-Recuenco M, Iturritxa E, Majada J, Alia R, Raposo R (2014) Adaptive potential of maritime pine (*Pinus pinaster*) populations to the emerging pitch canker pathogen *Fusarium circinatum*. Plos One 9(12): e114971. https://doi.org/10.1371/journal.pone.0114971

EPPO (2011) Plant quarantine data retrieval system (PQR version 5.3.5). Retrieved January 15, 2021 from http://www.eppo.int/DATABASES/pqr/pqr.htm

Estravis-Barcala M, Mattera MG, Soliani C, Bellora N, Opgenoorth L, Heer K, Arana MV (2020) Molecular bases of responses to abiotic stress in trees. J Exp Bot 71(13):3765–3779. https://doi.org/10.1093/jxb/erz532

Evans HF, McNamara DG, Braasch H, Chadoeuf J, Magnusson C (1996) Pest Risk Analysis (PRA) for the territories of the European Union (as PRA area) on *Bursaphelenchus xylophilus* and its vectors in the genus Monochamus. EPPO Bull 26(2):199–249. https://doi.org/10.1111/j.1365-2338.1996.tb00594.x

Eveno E, Collada C, Guevara MÁ, Léger V, Soto A, Díaz L, Leger P, González-Martínez SC, Cervera MT, Plomion C, Garnier-Gere PH (2008) Contrasting patterns of selection at *Pinus pinaster* Ait. drought stress candidate genes as revealed by genetic differentiation analyses. Mol Biol Evol 25(2):417–437. https://doi.org/10.1093/molbev/msm272

Fady B, Cottrell J, Ackzell L, Alía R, Muys B, Prada A, González-Martínez SC (2016) Forests and global change: what can genetics contribute to the major forest management and policy challenges of the twenty-first century? Reg Environ Chang 16(4):927–939. https://doi.org/10.1007/s10113-015-0843-9

Fan D, Liu T, Li C, Jiao B, Li S, Hou Y, Luo K (2015) Efficient CRISPR/Cas9-mediated targeted mutagenesis in Populus in the first generation. Sci Rep 5 (1):12217. https://doi.org/10.1038/srep12217

Fei Y, Xiao B, Yang M, Ding Q, Tang W (2016) MicroRNAs, polyamines, and the activities antioxidant enzymes are associated with *in vitro* rooting in white pine (*Pinus strobus* L.). Springerplus 5(1):416. https://doi.org/10.1186/s40064-016-2080-1

Fernando RL, Grossman M (1989) Marker assisted selection using best linear unbiased prediction. Genet Sel Evol 21(4):467. https://doi.org/10.1186/1297-9686-21-4-467

Feinard-Duranceau M, Berthier A, Vincent-Barbaroux C, Marin S, Lario F-J, Rozenberg P (2018) Plastic response of four maritime pine (*Pinus pinaster* Aiton) families to controlled soil water deficit. Ann For Sci 75(2):47. https://doi.org/10.1007/s13595-018-0719-5

Flachowsky H, Hanke M-V, Peil A, Strauss SH, Fladung M (2009) A review on transgenic approaches to accelerate breeding of woody plants. Plant Breed 128(3):217–226. https://doi.org/10.1111/j.1439-0523.2008.01591.x

Foissac S, Gouzy J, Rombauts S, Mathe C, Amselem J, Sterck L, de Peer Y, Rouze P, Schiex T (2008) Genome annotation in plants and fungi: EuGene as a model platform. Curr Bioinform 3(2):87–97. https://doi.org/10.2174/157489308784340702

Franclet A, Boulay M, Bekkaoui F, Fouret Y, Verschoore-Martouzet B, Walker N (1987) Rejuvenation. In: Bonga JM, Durzan DJ (eds) Cell and tissue culture in Forestry, Forestry Sciences, vol 24–26. Springer, Dordrecht, pp 232–248

Gaspar D, Trindade C, Usié A, Meireles B, Barbosa P, Fortes A, Pesquita C, Costa R, Ramos A (2017) Expression profiling in *Pinus pinaster* in response to infection with the pine wood nematode *Bursaphelenchus xylophilus*. Forests 8(8):279. https://doi.org/10.3390/f8080279

Gaspar D, Trindade C, Usié A, Meireles B, Fortes A, Guimarães J, Simões F, Costa R, Ramos A (2020) Comparative transcriptomic response of two *Pinus* species to infection with the pine wood nematode *Bursaphelenchus xylophilus*. Forests 11(2):204. https://doi.org/10.3390/f11020204

Gaspar MJ, Louzada JL, Silva ME, Aguiar A, Almeida MH (2008) Age trends in genetic parameters of wood density components in 46 half-sibling families of *Pinus pinaster*. Can J For Res 38 (6):1470–1477. https://doi.org/10.1139/X08-013

Gaspar MJ, Velasco T, Feito I, Alía R, Majada J (2013) Genetic variation of drought tolerance in *Pinus pinaster* at three hierarchical levels: a comparison of induced osmotic stress and field testing. PLoS ONE 8(11): e79094. https://doi.org/10.1371/journal.pone.0079094

Gómez A, Vendramin GG, González-Martínez SC, Alía R (2005) Genetic diversity and differentiation of two Mediterranean pines (*Pinus halepensis* Mill. and *Pinus pinaster* Ait.) along a latitudinal cline using chloroplast microsatellite markers. Divers Distrib 11(3):257–263. https://doi.org/10.1111/j.1366-9516.2005.00152.x

Gómez-Maldonado J, Avila C, Torre F, Cañas R, Cánovas FM, Campbell MM (2004) Functional interactions between a glutamine synthetase promoter and MYB proteins. Plant J 39(4):513–526. https://doi.org/10.1111/j.1365-313X.2004.02153.x

Gómez-Maldonado J, Crespillo R, ÉAvila C, Céanovas FM, (2001) Efficient preparation of maritime pine (*Pinus pinaster*) protoplasts suitable for transgene expression analysis. Plant Mol Biol Report 19(4):361–366. https://doi.org/10.1007/BF02772834

Gonçalves E, Figueiredo AC, Barroso JG, Henriques J, Sousa E, Bonifácio L (2020) Effect of *Monochamus galloprovincialis* feeding on *Pinus pinaster* and *Pinus pinea*, oleoresin and insect volatiles. Phytochemistry 169:112159. https://doi.org/10.1016/j.phytochem.2019.112159

González-Martínez SC, Alía R, Gil L (2002) Population genetic structure in a Mediterranean pine (*Pinus pinaster* Ait.): a comparison of allozyme markers and quantitative traits. Heredity 89(3):199–206. https://doi.org/10.1038/sj.hdy.6800114

González-Martínez SC, Gómez A, Carrión JS, Agúndez D, Alía R, Gil L (2007) Spatial genetic structure of an explicit glacial refugium of maritime pine (*Pinus pinaster* Aiton) in southeastern Spain. In: Weiss S, Ferrand N (eds) Phylogeography of southern European refugia. Springer, Dordrecht, pp 257–269

González-Martínez SC, Salvador L, Agúndez D, Alía R, Gil LA, Müller-Starck G, Schubert R et al (2001) Geographical variation of gene diversity of *Pinus pinaster* Ait. in the Iberian Peninsula. In: Muller-Starck H, Schubert R (eds) Genetic response of forest systems to changing environmental conditions. Kluwer Academic Publishers, Dordrecht, Boston, London, pp 161–171

González-Martínez SC, Burczyk J, Nathan R, Nanos N, Gil LA, Alía R (2006) Effective gene dispersal and female reproductive success in Mediterranean maritime pine (*Pinus pinaster* Aiton). Mol Ecol 15 (14):4577–4588. https://doi.org/10.1111/j.1365-294X.2006.03118.x

Goodwin S, McPherson JD, McCombie WR (2016) Coming of age: ten years of next-generation sequencing technologies. Nat Rev Genet 17:333–351. https://doi.org/10.1038/nrg.2016.49

Grattapaglia D (2014) Breeding forest trees by genomic selection: current progress and the way forward. In: Tuberosa R, Graner A, Frison E (eds) Genomics of plant genetic resources. Springer, Dordrecht, pp 651–682

Grattapaglia D, Silva-Junior OB, Resende RT, Cappa EP, Müller BSF, Tan B, Isik F, Ratcliffe B, El-Kassaby YA (2018) Quantitative genetics and genomics converge to accelerate forest tree breeding. Front Plant Sci 9(1693). https://doi.org/10.3389/fpls.2018.01693

Gravel-Grenier J, Lamhamedi MS, Beaulieu J, Carles S, Margolis HA, Rioux M, Stowe DC, Lapointe L (2011) Utilization of family genetic variability to improve the rooting ability of white spruce (*Picea glauca*) cuttings. Can J for Res 41(6):1308–1318. https://doi.org/10.1139/x11-044

Grivet D, Avia K, Vaattovaara A, Eckert AJ, Neale DB, Savolainen O, González-Martínez SC (2017) High rate of adaptive evolution in two widespread European pines. Mol Ecol 26(24):6857–6870. https://doi.org/10.1111/mec.14402

Grivet D, Sebastiani F, Alía R, Bataillon T, Torre S, Zabal-Aguirre M, Vendramin GG, González-Martínez SC (2011) Molecular footprints of local adaptation in two Mediterranean conifers. Mol Biol Evol 28 (1):101–116. https://doi.org/10.1093/molbev/msq190

Guan R, Zhao Y, Zhang H, Fan G, Liu X, Zhou W, Shi C, Wang J, Liu W, Liang X, Fu Y, Ma K, Zhao L, Zhang F, Lu Z, Lee SM-Y, Xu X, Wang J, Yang H, Fu C, Ge S, Chen W (2016) Draft genome of the living fossil *Ginkgo biloba*. Gigascience 5(1):49. https://doi.org/10.1186/s13742-016-0154-1

Guo L, Chen F (2014) A challenge for miRNA: multiple isomiRs in miRNAomics. Gene 544(1):1–7. https://doi.org/10.1016/j.gene.2014.04.039

Hadfield JD, Wilson AJ, Garant D, Sheldon BC, Kruuk LEB (2010) The misuse of BLUP in ecology and evolution. Am Nat 175(1):116–125. https://doi.org/10.1086/648604

Hamberger B, Ohnishi T, Hamberger B, Séguin A, Bohlmann J (2011) Evolution of diterpene metabolism: sitka spruce CYP720B4 catalyzes multiple oxidations in resin acid biosynthesis of conifer defense against insects. Plant Physiol 157(4):1677–1695. https://doi.org/10.1104/pp.111.185843

Harfouche A, Baradat P, Durel C, Pommery J (1995) Variabilité intraspécifique chez le pin maritime (*Pinus pinaster* Ait) dans le sud-est de la France. I. Variabilité des populations autochtones et des populations de l'ensemble de l'aire de l'espèce (Intraspecific variability in maritime pine (*Pinus pinaster* Ait.) in the south-east of France. Variability in autochthonous populations and in the whole range of the species.). Ann des Sci For 52(4):307–328. https://doi.org/10.1051/forest:19950402

Haslam E (1993) Shikimic acid: metabolism and metabolites. Wiley, Chichester, UK

Hassani SB, Benneckenstein T, Rupps A, Hensel G, Broeders S, Trontin J-F, Zoglauer K (2013a) Analysis of different promoters and reporter genes in somatic embryos of *Pinus pinaster* Ait. and *Larix decidua* Mill. In: Park YS, Bonga JM (eds) Proceeding of the 2nd international conference of the IUFRO working party 2.09.02 on integrating vegetative propagation, biotechnology and genetic improvement for tree production and sustainable forest management, Brno, Czech Republic, 25–28 June 2012). IUFRO, Vienna, Austria, p 192

Hassani SB, Rupps A, Arndt N, Cánovas FM, Trontin J-F, Zoglauer K (2013b) Overexpression of embryogenesis- and growth-related genes in transgenic *Pinus pinaster* embryos. Conference of German botanics (Botanikertagung), Tübingen, Germany, 30 September - 4 October 2013

Hassani, SB (2015). Gene transfer and expression analysis of genes related to growth and development of maritime pine (*Pinus pinaster*). PhD thesis, Humboldt University of Berlin, Germany

Hernández-Escribano L (2019) *Fusarium circinatum*—host interaction: ecological and molecular aspects of the pathogenic and endophytic association. Universidad Politécnica de Madrid

Hernández-Escribano L, Visser EA, Iturritxa E, Raposo R, Naidoo S (2020) The transcriptome of *Pinus pinaster* under *Fusarium circinatum* challenge. BMC Genomics 21(1):28. https://doi.org/10.1186/s12864-019-6444-0

Hughes-Jarlet E (1989). Recherches sur l'aptitude à l'embryogenèse somatique de matériel juvénile et de matériel issu d'arbres adultes de *Pinus pinaster* Sol (Research on the aptitude for somatic embryogenesis of juvenile and adult material from *Pinus pinaster* Sol.). PhD dissertation in plant biology and physiology, University Paris VI, France, 135 p

Hurel A (2020) Génomique écologique de l'adaptation locale chez le pin maritime *(Pinus pinaster*) (Ecological genomics of local adaptation in maritime pine (*Pinus pinaster* Aiton).) Agricultural Sciences. University of Bordeaux, France.

Hurel A, de Miguel M, Dutech C, Desprez-Loustau M-L, Plomion C, Rodríguez-Quilón I, Cyrille A, Guzman T, Alía R, González-Martínez SC, Budde KB (2021) Genetic basis of growth, spring phenology, and susceptibility to biotic stressors in maritime pine. Evol Appl. https://doi.org/10.1111/eva.13309

Ingvarsson PK, Street NR (2011) Association genetics of complex traits in plants. New Phytol 189(4):909–922. https://doi.org/10.1111/j.1469-8137.2010.03593.x

Isik F, Bartholomé J, Farjat A, Chancerel E, Raffin A, Sanchez L, Plomion C, Bouffier L (2016) Genomic selection in maritime pine. Plant Sci 242:108–119. https://doi.org/10.1016/j.plantsci.2015.08.006

Iturritxa E, Ganley RJ, Raposo R, García-Serna I, Mesanza N, Kirkpatrick SC, Gordon TR (2013) Resistance levels of Spanish conifers against *Fusarium circinatum* and *Diplodia pinea*. For Pathol 43 (6):488–495. https://doi.org/10.1111/efp.12061

Iturritxa E, Mesanza N, Elvira-Recuenco M, Serrano Y, Quintana E, Raposo R (2012) Evaluation of genetic resistance in *Pinus* to pitch canker in Spain. Australas Plant Pathol 41(6):601–607. https://doi.org/10.1007/s13313-012-0160-4

Jackman SD, Vandervalk BP, Mohamadi H, Chu J, Yeo S, Hammond SA, Jahesh G, Khan H, Coombe L, Warren RL, Birol I (2017) ABySS 2.0: resource-efficient assembly of large genomes using a Bloom filter. Genome Res 27(5):768–777. https://doi.org/10.1101/gr.214346.116

Jactel H, Kleinhentz M, Raffin A, Menassieu P (1999) Comparison of different selection methods for the resistance to *Dioryctria sylvestrella* Ratz. (Lepidoptera: Pyralidae) in *Pinus pinaster Ait.* In: Lieutier F, Mattson WJ, Wagner MR (eds) Proceedings of the physiology and genetics of tree-phytophage interactions international symposium, vol 90. INRA, Paris, France, pp 137–149

Jain M, Koren S, Miga KH, Quick J, Rand AC, Sasani TA, Tyson JR, Beggs AD, Dilthey AT, Fiddes IT, Malla S, Marriott H, Nieto T, O'Grady J, Olsen HE, Pedersen BS, Rhie A, Richardson H, Quinlan AR, Snutch TP, Tee L, Paten B, Phillippy AM, Simpson JT, Loman NJ, Loose M (2018) Nanopore sequencing and assembly of a human genome with ultra-long reads. Nat Biotechnol 36:338–345. https://doi.org/10.1038/nbt.4060

Jaramillo-correa JP, Prunier J, Vázquez-Lobo A, Keller S, Moreno-Letelier A (2015) Molecular signatures of adaptation and selection in forest trees. Adv Bot Res 74:265–306. https://doi.org/10.1016/bs.abr.2015.04.003

Jiao Y, Wickett NJ, Ayyampalayam S, Chanderbali AS, Landherr L, Ralph PE, Tomsho LP, Hu Y, Liang H, Soltis PS, Soltis DE, Clifton SW, Schlarbaum SE, Schuster SC, Ma H, Leebens-Mack J, DePamphilis CW (2011) Ancestral polyploidy in seed plants and angiosperms. Nature 473(7345):97–100. https://doi.org/10.1038/nature09916

Jones JDG, Dangl JL (2006) The plant immune system. Nature 444(7117):323–329. https://doi.org/10.1038/nature05286

Jones JT, Moens M, Mota M, Li H, Kikuchi T (2008) *Bursaphelenchus xylophilus*: opportunities in comparative genomics and molecular host-parasite interactions. Mol Plant Pathol 9(3):357–368. https://doi.org/10.1111/j.1364-3703.2007.00461.x

Joshi R, Wani SH, Singh B, Bohra A, Dar ZA, Lone AA, Pareek A, Singla-Pareek SL (2016) Transcription factors and plants response to drought stress: current understanding and future directions. Front Plant Sci 7:1029. https://doi.org/10.3389/fpls.2016.01029

Jyske TM, Suuronen J-P, Pranovich AV, Laakso T, Watanabe U, Kuroda K, Abe H (2015) Seasonal variation in formation, structure, and chemical properties of phloem in *Picea abies* as studied by novel microtechniques. Planta 242(3):613–629. https://doi.org/10.1007/s00425-015-2347-8

Kleinhentz M, Raffinz A, Jactel H (1998) Genetic parameters and gain expected from direct selection for resistance to *Dioryctria sylvestrella* Ratz. (Lepidoptera: Pyralidae) in *Pinus pinaster* Ait., using a full diallel mating design. For Genet 5(3):147–154

Klimaszewska K, Hargreaves C, Lelu-Walter M-A, Trontin J-F (2016) Advances in conifer somatic embryogenesis since year 2000. In: Germana M, Lambardi M (eds) *In vitro* embryogenesis in higher plants. Methods in molecular biology, vol 1359. Humana Press, New York, NY, pp 131–166

Klimaszewska K, Noceda C, Pelletier G, Label P, Rodriguez R, Lelu-Walter MA (2009) Biological characterization of young and aged embryogenic cultures of *Pinus pinaster* (Ait.). Vitr Cell Dev Biol Plant 45(1):20–33. https://doi.org/10.1007/s11627-008-9158-6

Kodan A, Kuroda H, Sakai F (2002) A stilbene synthase from Japanese red pine (*Pinus densiflora*): Implications for phytoalexin accumulation and down-regulation of flavonoid biosynthesis. Proc Natl Acad Sci 99(5):3335–3339. https://doi.org/10.1073/pnas.042698899

Kovach A, Wegrzyn JL, Parra G, Holt C, Bruening GE, Loopstra CA, Hartigan J, Yandell M, Langley CH, Korf I, Neale DB (2010) The *Pinus taeda* genome is characterized by diverse and highly diverged repetitive sequences. BMC Genomics 11(1):420. https://doi.org/10.1186/1471-2164-11-420

Kremer A (1992) Predictions of Age-Age correlations of total height based on serial correlations between height increments in maritime pine (*Pinus pinaster* Ait). Theor Appl Genet 85(2–3):152–158

Krivmane B, Šņepste I, Šķipars V, Yakovlev I, Fossdal CG, Vivian-Smith A, Ruņģis D (2020) Identification and *in silico* characterization of novel and conserved micrornas in methyl jasmonate-stimulated Scots pine (*Pinus sylvestris* L.) needles. Forests 11(4):384. https://doi.org/10.3390/f11040384

Krutovsky KV, Troggio M, Brown GR, Jermstad KD, Neale DB (2004) Comparative mapping in the Pinaceae. Genetics 168(1):447–461. https://doi.org/10.1534/genetics.104.028381

Kuromori T, Seo M, Shinozaki K (2018) ABA transport and plant water stress responses. Trends Plant Sci 23(6):513–522. https://doi.org/10.1016/j.tplants.2018.04.001

Kuzmin DA, Feranchuk SI, Sharov VV, Cybin AN, Makolov SV, Putintseva YA, Oreshkova NV, Krutovsky KV (2019) Stepwise large genome assembly approach: a case of Siberian larch (*Larix sibirica* Ledeb). BMC Bioinform 20(S1):37. https://doi.org/10.1186/s12859-018-2570-y

Laajanen K, Vuorinen I, Salo V, Juuti J, Raudaskoski M (2007) Cloning of *Pinus sylvestris* SCARECROW gene and its expression pattern in the pine root system, mycorrhiza and NPA-treated short roots. New Phytol 175(2):230–243. https://doi.org/10.1111/j.1469-8137.2007.02102.x

Lagraulet H (2015) Plasticité phenotypique et architecture genetique de la croissance et de la densite du bois du pin maritime (*Pinus pinaster* Ait.)(Phenotypic plasticity and genetic architecture of the growth and density of maritime pine wood (*Pinus pinaster* Ait.).). University of Bordeaux, France

Lambeth C, Lee B-C, O'Malley D, Wheeler N (2001) Polymix breeding with parental analysis of progeny: an alternative to full-sib breeding and testing. Theor Appl Genet 103(6–7):930–943. https://doi.org/10.1007/s001220100627

Lamy J-B, Bouffier L, Burlett R, Plomion C, Cochard H, Delzon S (2011) Uniform selection as a primary force reducing population genetic differentiation of cavitation resistance across a species range. PLoS ONE 6(8):e23476. https://doi.org/10.1371/journal.pone.0023476

Lande R, Arnold SJ (1983) The measurement of selection on correlated characters. Evolution 37(6):1210–1226. https://doi.org/10.1111/j.1558-5646.1983.tb00236.x

Landeras E, García P, Fernández Y, Braña M, Fernández-Alonso O, Méndez-Lodos S, Pérez-Sierra A, León M, Abad-Campos P, Berbegal M, Beltrán R, García-Jiménez J, Armengol J (2005) Outbreak of pitch canker caused by *Fusarium circinatum* on *Pinus spp.* in Northern Spain. Plant Dis 89(9):1015–1015. https://doi.org/10.1094/PD-89-1015A

Larkin PJ, Scowcroft WR (1981) Somaclonal variation—a novel source of variability from cell cultures for plant improvement. Theor Appl Genet 60(4):197–214. https://doi.org/10.1007/BF02342540

Le Provost G, Domergue F, Lalanne C, Ramos Campos P, Grosbois A, Bert D, Meredieu C, Danjon F, Plomion C, Gion J-M (2013) Soil water stress affects both cuticular wax content and cuticle-related gene expression in young saplings of maritime pine (*Pinus pinaster* Ait). BMC Plant Biol 13(1):95. https://doi.org/10.1186/1471-2229-13-95

Lee YG, Choi SC, Kang Y, Kim KM, Kang C-S, Kim C (2019) Constructing a reference genome in a single lab: the possibility to use Oxford nanopore technology. Plants 8(8):270. https://doi.org/10.3390/plants8080270

Lelu-Walter M-A, Bernier-Cardou M, Klimaszewska K (2006) Simplified and improved somatic embryogenesis for clonal propagation of *Pinus pinaster* (Ait.). Plant Cell Rep 25(8):767–776. https://doi.org/10.1007/s00299-006-0115-8

Lelu-Walter M-A, Klimaszewska K, Miguel C, Aronen T, Hargreaves C, Teyssier C, Trontin J-F (2016) Somatic embryogenesis for more effective breeding and deployment of improved varieties in *Pinus spp.*: bottlenecks and recent advances. In: Loyola-Vargas V, Ochoa-Alejo N (eds) Somatic embryogenesis:

fundamental aspects and applications. Springer, Cham, pp 319–365

Lepoittevin C, Harvengt L, Plomion C, Garnier-Géré P (2012) Association mapping for growth, straightness and wood chemistry traits in the *Pinus pinaster* Aquitaine breeding population. Tree Genet Genomes 8(1):113–126. https://doi.org/10.1007/s11295-011-0426-y

Lepoittevin C, Rousseau J-P, Guillemin A, Gauvrit C, Besson F, Hubert F, da Silva PD, Harvengt L, Plomion C (2011) Genetic parameters of growth, straightness and wood chemistry traits in *Pinus pinaster*. Ann For Sci 68(4):873–884. https://doi.org/10.1007/s13595-011-0084-0

Li Y, Wei H, Yang J, Du K, Li J, Zhang Y, Qiu T, Liu Z, Ren Y, Song L, Kang X (2020) High-quality *de novo* assembly of the *Eucommia ulmoides* haploid genome provides new insights into evolution and rubber biosynthesis. Hortic Res 7(1):183. https://doi.org/10.1038/s41438-020-00406-w

Li Z, Baniaga AE, Sessa EB, Scascitelli M, Graham SW, Rieseberg LH, Barker MS (2015) Early genome duplications in conifers and other seed plants. Sci Adv 1(10):e1501084. https://doi.org/10.1126/sciadv.1501084

Liu J, Schoettle AW, Sniezko RA, Yao F, Zamany A, Williams H, Rancourt B (2019) Limber pine (*Pinus flexilis* James) genetic map constructed by exome-seq provides insight into the evolution of disease resistance and a genomic resource for genomics-based breeding. Plant J 98(4):745–758. https://doi.org/10.1111/tpj.14270

Llebrés M-T, Pascual M-B, Debille S, Trontin J-F, Harvengt L, Avila C, Cánovas FM (2018) The role of arginine metabolic pathway during embryogenesis and germination in maritime pine (*Pinus pinaster* Ait.). Tree Physiol 38(3):471–484. https://doi.org/10.1093/treephys/tpx133

López-Goldar X, Villari C, Bonello P, Borg-Karlson AK, Grivet D, Zas R, Sampedro L (2018) Inducibility of plant secondary metabolites in the stem predicts genetic variation in resistance against a key insect herbivore in maritime pine. Front Plant Sci 9:1651. https://doi.org/10.3389/fpls.2018.01651

López-Goldar X, Villari C, Bonello P, Borg-Karlson AK, Grivet D, Sampedro L, Zas R (2019) Genetic variation in the constitutive defensive metabolome and its inducibility are geographically structured and largely determined by demographic processes in maritime pine. J Ecol 107(5):2464–2477. https://doi.org/10.1111/1365-2745.13159

López-Hinojosa M, de María N, Guevara MÁ, Vélez MD, Cabezas JA, Díaz LM, Mancha JA, Pizarro A, Manjarrez LF, Collada C, Díaz-Sala MC, Cervera MT (2021) Rootstock effects on scion gene expression in maritime pine. Sci Rep 11(1):11582. https://doi.org/10.1038/s41598-021-90672-y

Lv J, Yu K, Wei J, Gui H, Liu C, Liang D, Wang Y, Zhou H, Carlin R, Rich R, Lu T, Que Q, Wang WC, Zhang X, Kelliher T (2020) Generation of paternal haploids in wheat by genome editing of the centromeric histone CENH3. Nat Biotechnol 38 (12):1397–1401. https://doi.org/10.1038/s41587-020-0728-4

Mackay J, Dean JFD, Plomion C, Peterson DG, Cánovas FM, Pavy N, Ingvarsson PK, Savolainen O, Guevara MÁ, Fluch S, Vinceti B, Abarca D, Díaz-Sala C, Cervera M-T (2012) Towards decoding the conifer giga-genome. Plant Mol Biol 80(6):555–569. https://doi.org/10.1007/s11103-012-9961-7

Madrigal-González J, Herrero A, Ruiz-Benito P, Zavala MA (2017) Resilience to drought in a dry forest: insights from demographic rates. For Ecol Manage 389:167–175. https://doi.org/10.1016/j.foreco.2016.12.012

Majada J, Martínez-Alonso C, Feito I, Kidelman A, Aranda I, Alía R (2011) Mini-cuttings: an effective technique for the propagation of *Pinus pinaster* Ait. New For 41(3):399–412. https://doi.org/10.1007/s11056-010-9232-x

Marguerit E, Bouffier L, Chancerel E, Costa P, Lagane F, Guehl J-M, Plomion C, Brendel O (2014) The genetics of water-use efficiency and its relation to growth in maritime pine. J Exp Bot 65(17):4757–4768. https://doi.org/10.1093/jxb/eru226

Margulies M, Egholm M, Altman WE, Attiya S, Bader JS, Bemben LA, Berka J, Braverman MS, Chen Y-J, Chen Z, Dewell SB, Du L, Fierro JM, Gomes XV, Godwin BC, He W, Helgesen S, Ho CH, Irzyk GP, Jando SC, Alenquer MLI, Jarvie TP, Jirage KB, Kim J-B, Knight JR, Lanza JR, Leamon JH, Lefkowitz SM, Lei M, Li J, Lohman KL, Lu H, Makhijani VB, McDade KE, McKenna MP, Myers EW, Nickerson E, Nobile JR, Plant R, Puc BP, Ronan MT, Roth GT, Sarkis GJ, Simons JF, Simpson JW, Srinivasan M, Tartaro KR, Tomasz A, Vogt KA, Volkmer GA, Wang SH, Wang Y, Weiner MP, Yu P, Begley RF, Rothberg JM (2005) Genome sequencing in microfabricated high-density picolitre reactors. Nature 437:376–380. https://doi.org/10.1038/nature03959

Markussen T, Fladung M, Achere V, Favre JM, Faivre-Rampant P, Aragones A, Da Silva Perez D, Harvengt L, Espinel S, Ritter E (2003) Identification of QTLs controlling growth, chemical and physical wood property traits in *Pinus pinaster* (ait.). Silvae Genet 52 (1):8–15

Martínez-Alonso C, Kidelman A, Feito I, Velasco T, Alía R, Gaspar MJ, Majada J (2012) Optimization of seasonality and mother plant nutrition for vegetative propagation of *Pinus pinaster* Ait. New For 43(5–6):651–663. https://doi.org/10.1007/s11056-012-9333-9

Martinho C, Confraria A, Elias CA, Crozet P, Rubio-Somoza I, Weigel D, Baena-González E (2015) Dissection of miRNA pathways using arabidopsis mesophyll protoplasts. Mol Plant 8(2):261–275. https://doi.org/10.1016/j.molp.2014.10.003

Mauriat M, Le Provost G, Rozenberg P, Delzon S, Breda N, Clair B, Coutand C, Domec J-C, Fourcaud T,

Grima-Pettenati J, Herrera R, Leplé J-C, Richet N, Trontin J-F, Plomion C (2014) Wood formation in trees. In: Ramawat KG, Mérillon J-M, Ahuja MR (eds) Tree biotechnology. CRC Press, Boca Raton, Florida, USA, pp 56–111

Maury S, Le Gac A-L, Lafon-Placette C, Dia Sow M, Fichot R, Delaunay A, Le Jan I, Lelu-Walter M-A, Segura V, Rogier O, Trontin J-F, Plomion C, Le Provost G, Ehrenmann F, Salse J, Ambroise C, Gribkova S, Mirouze M, Grunau C, Chaparro C, Strauss SH, Conde D, Allona I, Tost J (2019a) Epigenetics in trees: a source of plasticity and adaptation in the context of climate change. In: Bonga JM, Park Y-S, Trontin J-F (eds) Proceedings of the 5th international conference of the IUFRO working party 2.09.02 on clonal trees in the bioeconomy age: opportunities and challenges, Coimbra, Portugal, 10–15 September 2018. IUFRO, Vienna, Austria, pp 110–115

Maury S, Sow MD, Le Gac A-L, Genitoni J, Lafon-Placette C, Iva Mozgova I (2019b) Phytohormone and chromatin crosstalk: the missing link for developmental plasticity? Front Plant Sci 10:395. https://doi.org/10.3389/fpls.2019.00395

Meijón M, Feito I, Oravec M, Delatorre C, Weckwerth W, Majada J, Valledor L (2016) Exploring natural variation of *Pinus pinaster* Aiton using metabolomics: Is it possible to identify the region of origin of a pine from its metabolites? Mol Ecol 25(4):959–976. https://doi.org/10.1111/mec.13525

Mendoza-Poudereux I, Cano M, Ávila C, Cánovas F, Lelu-Walter M-A, Trontin J-F, Segura J, Arrillaga I (2014) Generation of transgenic maritime pine somatic embryos with altered expression of genes involved in nitrogen metabolism and wood formation. In: Book of abstracts of the 3rd international conference of the IUFRO working party 2.09.02 on woody plant production integrating genetic and vegetative propagation technologies, Vitoria-Gasteiz, Spain, 8–12 September 2014, p 121

Menéndez-Gutiérrez M, Alonso M, Toval G, Díaz R (2017a) Variation in pinewood nematode susceptibility among *Pinus pinaster* Ait. provenances from the Iberian Peninsula and France. Ann For Sci 74(4):76. https://doi.org/10.1007/s13595-017-0677-3

Menéndez-Gutiérrez M, Alonso M, Toval G, Díaz R (2017b) Testing of selected *Pinus pinaster* half-sib families for tolerance to pinewood nematode (*Bursaphelenchus xylophilus*). For an Int J for Res 91 (1):38–48. https://doi.org/10.1093/forestry/cpx030

Miguel C, Marum L (2011) An epigenetic view of plant cells cultured *in vitro*: somaclonal variation and beyond. J Exp Bot 62(11):3713–3725. https://doi.org/10.1093/jxb/err155

Miner BG, Sultan SE, Morgan SG, Padilla DK, Relyea RA (2005) Ecological consequences of phenotypic plasticity. Trends Ecol Evol 20(12):685–692. https://doi.org/10.1016/j.tree.2005.08.002

Mitsuda N, Iwase A, Yamamoto H, Yoshida M, Seki M, Shinozaki K, Ohme-Takagi M (2007) NAC transcription factors, NST1 and NST3, are key regulators of the formation of secondary walls in woody tissues of Arabidopsis. Plant Cell 19(1):270–280. https://doi.org/10.1105/tpc.106.047043

Modesto I, Sterck L, Arbona V, Gómez-Cadenas A, Carrasquinho I, Van de Peer Y, Miguel CM (2021) Insights into the mechanisms implicated in *Pinus pinaster* resistance to pinewood nematode. Front Plant Sci. https://doi.org/10.3389/fpls.2021.690857

Monteuuis O, Dumas E (1992) Morphological features as indicators of maturity in acclimatized *Pinus pinaster* from different *in vitro* origins. Can J for Res 22 (9):1417–1421. https://doi.org/10.1139/x92-188

Moran E, Lauder J, Musser C, Stathos A, Shu M (2017) The genetics of drought tolerance in conifers. New Phytol 216(4):1034–1048. https://doi.org/10.1111/nph.14774

Morel A, Teyssier C, Trontin J-F, Eliášová K, Pešek B, Beaufour M, Morabito D, Boizot N, Le Metté C, Belal-Bessai L, Reymond I, Harvengt L, Cadene M, Corbineau F, Vágner M, Label P, Lelu-Walter M-A (2014a) Early molecular events involved in *Pinus pinaster* Ait. somatic embryo development under reduced water availability: transcriptomic and proteomic analyses. Physiol Plant 152(1):184–201. https://doi.org/10.1111/ppl.12158

Morel A, Trontin J-F, Corbineau F, Lomenech A-M, Beaufour M, Reymond I, Le Metté C, Ader K, Harvengt L, Cadene M, Label P, Teyssier C, Lelu-Walter M-A (2014b) Cotyledonary somatic embryos of *Pinus pinaster* Ait. most closely resemble fresh, maturing cotyledonary zygotic embryos: biological, carbohydrate and proteomic analyses. Planta 240 (5):1075–1095. https://doi.org/10.1007/s00425-014-2125-z

Moriguchi Y, Ujino-Ihara T, Uchiyama K, Futamura N, Saito M, Ueno S, Matsumoto A, Tani N, Taira H, Shinohara K, Tsumura Y (2012) The construction of a high-density linkage map for identifying SNP markers that are tightly linked to a nuclear-recessive major gene for male sterility in *Cryptomeria japonica* D. Don. BMC Genomics 13(1):95. https://doi.org/10.1186/1471-2164-13-95

Morozova O, Hirst M, Marra MA (2009) Applications of new sequencing technologies for transcriptome analysis. Annu Rev Genomics Hum Genet 10(1):135–151. https://doi.org/10.1146/annurev-genom-082908-145957

Morrissey MB, Kruuk LEB, Wilson AJ (2010) The danger of applying the breeder's equation in observational studies of natural populations. J Evol Biol 23 (11):2277–2288. https://doi.org/10.1111/j.1420-9101.2010.02084.x

Morse AM, Peterson DG, Islam-Faridi MN, Smith KE, Magbanua Z, Garcia SA, Kubisiak TL, Amerson HV, Carlson JE, Nelson CD, Davis JM (2009) Evolution of genome size and complexity in Pinus. PLoS ONE 4(2):1–11. https://doi.org/10.1371/journal.pone.0004332

Mosca E, Cruz F, Gómez-Garrido J, Bianco L, Rellstab C, Brodbeck S, Csilléry K, Fady B, Fladung M, Fussi B,

Gömöry D, González-Martínez SC, Grivet D, Gut M, Hansen OK, Heer K, Kaya Z, Krutovsky KV, Kersten B, Liepelt S, Opgenoorth L, Sperisen C, Ullrich KK, Vendramin GG, Westergren M, Ziegenhagen B, Alioto T, Gugerli F, Heinze B, Höhn M, Troggio M, Neale DB (2019) A reference genome sequence for the European silver fir (*Abies alba* Mill.): a community-generated genomic resource. Genes Genomes Genet 9(7):2039–2049. https://doi.org/10.1534/g3.119.400083

Mota MM, Braasch H, Bravo MA, Penas AC, Burgermeister W, Metge K, Sousa E (1999) First report of *Bursaphelenchus xylophilus* in Portugal and in Europe. Nematology 1(7):727–734. https://doi.org/10.1163/156854199508757

Muranty H, Jorge V, Bastien C, Lepoittevin C, Bouffier L, Sanchez L (2014) Potential for marker-assisted selection for forest tree breeding: lessons from 20 years of MAS in crops. Tree Genet Genomes 10(6):1491–1510. https://doi.org/10.1007/s11295-014-0790-5

Murat F, Armero A, Pont C, Klopp C, Salse J (2017) Reconstructing the genome of the most recent common ancestor of flowering plants. Nat Genet 49 (4):490–496. https://doi.org/10.1038/ng.3813

Naidoo S, Slippers B, Plett JM, Coles D, Oates CN (2019) The road to resistance in forest trees. Front Plant Sci 10:273. https://doi.org/10.3389/fpls.2019.00273

Nakamura M, Köhler C, Hennig L (2019) Tissue-specific transposon-associated small RNAs in the gymnosperm tree, Norway Spruce. BMC Genomics 20(1):997. https://doi.org/10.1186/s12864-019-6385-7

Nakashima K, Takasaki H, Mizoi J, Shinozaki K, Yamaguchi-Shinozaki K (2012) NAC transcription factors in plant abiotic stress responses. Biochim Biophys Acta Gene Regul Mech 2:97–103. https://doi.org/10.1016/j.bbagrm.2011.10.005

Nardini A, Lo Gullo MA, Trifilò P, Salleo S (2014) The challenge of the Mediterranean climate to plant hydraulics: responses and adaptations. Environ Exp Bot 103:68–79. https://doi.org/10.1016/j.envexpbot.2013.09.018

Neale DB, McGuire PE, Wheeler NC, Stevens KA, Crepeau MW, Cardeno C, Zimin AV, Puiu D, Pertea GM, Sezen UU, Casola C, Koralewski TE, Paul R, Gonzalez-Ibeas D, Zaman S, Cronn R, Yandell M, Holt C, Langley CH, Yorke JA, Salzberg SL, Wegrzyn JL (2017) The Douglas-fir genome sequence reveals specialization of the photosynthetic apparatus in Pinaceae. Genes Genomes Genet 7(9):3157–3167. https://doi.org/10.1534/g3.117.300078

Neale DB, Wegrzyn JL, Stevens KA, Zimin AV, Puiu D, Crepeau MW, Cardeno C, Koriabine M, Holtz-Morris AE, Liechty JD, Martínez-García PJ, Vasquez-Gross HA, Lin BY, Zieve JJ, Dougherty WM, Fuentes-Soriano S, Wu L-S, Gilbert D, Marçais G, Roberts M, Holt C, Yandell M, Davis JM, Smith KE, Dean JF, Lorenz W, Whetten RW, Sederoff R, Wheeler N, McGuire PE, Main D, Loopstra CA, Mockaitis K, DeJong PJ, Yorke JA, Salzberg SL, Langley CH (2014) Decoding the massive genome of loblolly pine using haploid DNA and novel assembly strategies. Genome Biol 15(3):R59. https://doi.org/10.1186/gb-2014-15-3-r59

Neilsen CT, Goodall GJ, Bracken CP (2012) IsomiRs—the overlooked repertoire in the dynamic microRNAome. Trends Genet 28(11):544–549. https://doi.org/10.1016/j.tig.2012.07.005

Nguyen NH, Jeong CY, Kang G, Yoo S, Hong S, Lee H (2015) MYBD employed by HY5 increases anthocyanin accumulation via repression of MYBL2 in Arabidopsis. Plant J 84(6):1192–1205. https://doi.org/10.1111/tpj.13077

Nguyen-Queyrens A, Bouchet-Lannat F (2003) Osmotic adjustment in three-year-old seedlings of five provenances of maritime pine (*Pinus pinaster*) in response to drought. Tree Physiol 23(6):397–404. https://doi.org/10.1093/treephys/23.6.397

Nguyen-Queyrens A, Costa P, Loustau D, Plomion C (2002) Osmotic adjustment in *Pinus pinaster* cuttings in response to a soil drying cycle. Ann For Sci 59 (7):795–799. https://doi.org/10.1051/forest:2002067

Niu S-H, Liu C, Yuan H-W, Li P, Li Y, Li W (2015) Identification and expression profiles of sRNAs and their biogenesis and action-related genes in male and female cones of *Pinus tabuliformis*. BMC Genomics 16(1):693. https://doi.org/10.1186/s12864-015-1885-6

Nystedt B, Street NR, Wetterbom A, Zuccolo A, Lin Y-C, Scofield DG, Vezzi F, Delhomme N, Giacomello S, Alexeyenko A, Vicedomini R, Sahlin K, Sherwood E, Elfstrand M, Gramzow L, Holmberg K, Hällman J, Keech O, Klasson L, Koriabine M, Kucukoglu M, Käller M, Luthman J, Lysholm F, Niittylä T, Olson Å, Rilakovic N, Ritland C, Rosselló JA, Sena J, Svensson T, Talavera-López C, Theißen G, Tuominen H, Vanneste K, Wu Z-Q, Zhang B, Zerbe P, Arvestad L, Bhalerao R, Bohlmann J, Bousquet J, Garcia Gil R, Hvidsten TR, de Jong P, MacKay J, Morgante M, Ritland K, Sundberg B, Lee Thompson S, Van de Peer Y, Andersson B, Nilsson O, Ingvarsson PK, Lundeberg J, Jansson S (2013) The Norway spruce genome sequence and conifer genome evolution. Nature 497(7451):579–584. https://doi.org/10.1038/nature12211

O'Connor LJ, Schoech AP, Hormozdiari F, Gazal S, Patterson N, Price AL (2019) Extreme polygenicity of complex traits is explained by negative selection. Am J Hum Genet 105(3):456–476. https://doi.org/10.1016/j.ajhg.2019.07.003

Ojeda DI, Mattila TM, Ruttink T, Kujala ST, Kärkkäinen K, Verta J-P, Pyhäjärvi T (2019) Utilization of tissue ploidy level variation in *de novo* transcriptome assembly of *Pinus sylvestris*. Genes Genomes Genet 9(10):3409–3421. https://doi.org/10.1534/g3.119.400357

Olsen JL, Rouzé P, Verhelst B, Lin Y-C, Bayer T, Collen J, Dattolo E, De Paoli E, Dittami S, Maumus F, Michel G, Kersting A, Lauritano C, Lohaus R, Töpel M, Tonon T, Vanneste K, Amirebrahimi M, Brakel J, Boström C, Chovatia M, Grimwood J, Jenkins JW,

Jueterbock A, Mraz A, Stam WT, Tice H, Bornberg-Bauer E, Green PJ, Pearson GA, Procaccini G, Duarte CM, Schmutz J, Reusch TBH, Van de Peer Y (2016) The genome of the seagrass *Zostera marina* reveals angiosperm adaptation to the sea. Nature 530(7590):331–335. https://doi.org/10.1038/nature16548

Ooka H (2003) Comprehensive analysis of NAC family genes in *Oryza sativa* and *Arabidopsis thaliana*. DNA Res 10(6):239–247. https://doi.org/10.1093/dnares/10.6.239

Park Y, Beaulieu J, Bousquet J (2016) Multi-varietal forestry integrating genomic selection and somatic embryogenesis. In: Park Y-S, Bonga J, Moon H (eds) Vegetative propagation of forest trees. National Institute of Forest Science (NIFoS), Seoul, Korea, pp 302–322

Parsons TJ, Sinkar VP, Stettler RF, Nester EW, Gordon MP (1986) Transformation of poplar by *Agrobacterium tumefaciens*. Nat Biotechnol 4(6):533–536. https://doi.org/10.1038/nbt0686-533

Pascual MB, Cánovas FM, Ávila C (2015) The NAC transcription factor family in maritime pine (*Pinus Pinaster*): molecular regulation of two genes involved in stress responses. BMC Plant Biol 15(1):254. https://doi.org/10.1186/s12870-015-0640-0

Pascual MB, Llebrés M-T, Craven-Bartle B, Cañas RA, Cánovas FM, Ávila C (2018) PpNAC1, a main regulator of phenylalanine biosynthesis and utilization in maritime pine. Plant Biotechnol J 16(5):1094–1104. https://doi.org/10.1111/pbi.12854

Pattison RJ, Csukasi F, Zheng Y, Fei Z, van der Knaap E, Catalá C (2015) Comprehensive tissue-specific transcriptome analysis reveals distinct regulatory programs during early tomato fruit development. Plant Physiol 168(4):1684–1701. https://doi.org/10.1104/pp.15.00287

Pavy N, Lamothe M, Pelgas B, Gagnon F, Birol I, Bohlmann J, Mackay J, Isabel N, Bousquet J (2017) A high-resolution reference genetic map positioning 8.8 K genes for the conifer white spruce: structural genomics implications and correspondence with physical distance. Plant J 90(1):189–203. https://doi.org/10.1111/tpj.13478

Pavy N, Pelgas B, Laroche J, Rigault P, Isabel N, Bousquet J (2012) A spruce gene map infers ancient plant genome reshuffling and subsequent slow evolution in the gymnosperm lineage leading to extant conifers. BMC Biol 10(1):84. https://doi.org/10.1186/1741-7007-10-84

Pelgas B, Beauseigle S, Acheré V, Jeandroz S, Bousquet J, Isabel N (2006) Comparative genome mapping among *Picea glauca, P. mariana* × *P. rubens* and *P. abies*, and correspondence with other Pinaceae. Theor Appl Genet 113(8):1371–1393. https://doi.org/10.1007/s00122-006-0354-7

Pelgas B, Bousquet J, Beauseigle S, Isabel N (2005) A composite linkage map from two crosses for the species complex *Picea mariana* × *Picea rubens* and analysis of synteny with other Pinaceae. Theor Appl Genet 111(8):1466–1488. https://doi.org/10.1007/s00122-005-0068-2

Perdiguero P, del Barbero M, C, Cervera MT, Collada C, Soto Á, (2013) Molecular response to water stress in two contrasting Mediterranean pines (*Pinus pinaster* and *Pinus pinea*). Plant Physiol Biochem 67:199–208. https://doi.org/10.1016/j.plaphy.2013.03.008

Perdiguero P, Barbero MC, Cervera MT, Soto Á, Collada C (2012a) Novel conserved segments are associated with differential expression patterns for Pinaceae dehydrins. Planta 236(6):1863–1874. https://doi.org/10.1007/s00425-012-1737-4

Perdiguero P, Collada C, del Barbero MC, García Casado G, Cervera MT, Soto Á (2012b) Identification of water stress genes in *Pinus pinaster* Ait. by controlled progressive stress and suppression-subtractive hybridization. Plant Physiol Biochem 50:44–53. https://doi.org/10.1016/j.plaphy.2011.09.022

Perdiguero P, Collada C, Soto Á (2014) Novel dehydrins lacking complete K-segments in Pinaceae. The exception rather than the rule. Front Plant Sci 5. https://doi.org/10.3389/fpls.2014.00682

Perdiguero P, Rodrigues AS, Chaves I, Costa B, Alves A, María N, Vélez MD, Díaz-Sala C, Cervera MT, Miguel CM (2020) Comprehensive analysis of the isomiRome in the vegetative organs of the conifer *Pinus pinaster* under contrasting water availability. Plant Cell Environ. https://doi.org/10.1111/pce.13976

Pérez-Oliver MA, Haro JG, Pavlovic I, Novák O, Segura J, Sales E, Arrillaga I (2021) Priming maritime pine megagametophytes during somatic embryogenesis improved plant adaptation to heat stress. Plants 10:446. https://doi.org/10.3390/plants10030446

Pérez-Rodríguez MJ, Suárez MF, Heredia R, Ávila C, Breton D, Trontin J-F, Filonova L, Bozhkov P, von Arnold S, Harvengt L, Cánovas FM (2006) Expression patterns of two glutamine synthetase genes in zygotic and somatic pine embryos support specific roles in nitrogen metabolism during embryogenesis. New Phytol 169:35–44. https://doi.org/10.1111/j.1469-8137.2005.01551.x

Peterson DG, Tomkins JP, Frisch DA, Wing RA, Paterson AH (2000) Construction of plant bacterial artificial chromosome (BAC) libraries: an illustrated guide. J Agric Genomics 5:1–100

Pizarro A, Díaz-Sala C (2019) Cellular dynamics during maturation-related decline of adventitious root formation in forest tree species. Physiol Plant 165(1):73–80. https://doi.org/10.1111/ppl.12768

Plomion C, Bartholomé J, Lesur I, Boury C, Rodríguez-Quilón I, Lagraulet H, Ehrenmann F, Bouffier L, Gion JM, Grivet D, de Miguel M, de María N, Cervera MT, Bagnoli F, Isik F, Vendramin GG, González-Martínez SC (2016a) High-density SNP assay development for genetic analysis in maritime pine (*Pinus pinaster*). Mol Ecol Resour 16(2):574–587. https://doi.org/10.1111/1755-0998.12464

Plomion C, Bastien C, Bogeat-Triboulot M-B, Bouffier L, Déjardin A, Duplessis S, Fady B, Heuertz M, Le Gac A-L, Le Provost G, Legué V, Lelu-Walter M-A, Leplé J-C, Maury S, Morel A, Oddou-Muratorio S, Pilate G, Sanchez L, Scotti I, Scotti-Saintagne C, Segura V, Trontin J-F, Vacher C (2016b) Forest tree genomics: 10 achievements from the past 10 years and future prospects. Ann For Sci 73(1):77–103. https://doi.org/10.1007/s13595-015-0488-3

Plomion C, Chancerel E, Endelman J, Lamy J-B, Mandrou E, Lesur I, Ehrenmann F, Isik F, Bink MC, van heerwaarden J, Bouffier L, (2014) Genome-wide distribution of genetic diversity and linkage disequilibrium in a mass-selected population of maritime pine. BMC Genomics 15(1):171. https://doi.org/10.1186/1471-2164-15-171

Plomion C, Durel C-E, O'Malley DM (1996a) Genetic dissection of height in maritime pine seedlings raised under accelerated growth conditions. Theor Appl Genet 93(5–6):849–858. https://doi.org/10.1007/BF00224085

Plomion C, Yani A, Marpeau A (1996b) Genetic determinism of δ3-carene in maritime pine using RAPD markers. Genome 39(6):1123–1127. https://doi.org/10.1139/g96-141

Pot D, Rodrigues J-C, Rozenberg P, Chantre G, Tibbits J, Cahalan C, Pichavant F, Plomion C (2006) QTLs and candidate genes for wood properties in maritime pine (*Pinus pinaster* Ait.). Tree Genet Genomes 2(1):10–24. https://doi.org/10.1007/s11295-005-0026-9

Pritchard JK, Pickrell JK, Coop G (2010) The genetics of human adaptation: hard sweeps, soft sweeps, and polygenic adaptation. Curr Biol 20(4):R208–R215. https://doi.org/10.1016/j.cub.2009.11.055

Ralph SG, Chun H, Kolosova N, Cooper D, Oddy C, Ritland CE, Kirkpatrick R, Moore R, Barber S, Holt RA, Jones SJ, Marra MA, Douglas CJ, Ritland K, Bohlmann J (2008) A conifer genomics resource of 200,000 spruce (*Picea spp.*) ESTs and 6,464 high-quality, sequence-finished full-length cDNAs for Sitka spruce (*Picea sitchensis*). BMC Genomics 9(1):484. https://doi.org/10.1186/1471-2164-9-484

Ren R, Wang H, Guo C, Zhang N, Zeng L, Chen Y, Ma H, Qi J (2018) Widespread whole genome duplications contribute to genome complexity and species diversity in angiosperms. Mol Plant 11(3):414–428. https://doi.org/10.1016/j.molp.2018.01.002

Rigault P, Boyle B, Lepage P, Cooke JEK, Bousquet J, MacKay JJ (2011) A white spruce gene catalog for conifer genome analyses. Plant Physiol 157(1):14–28. https://doi.org/10.1104/pp.111.179663

Rincent R, Charpentier J-P, Faivre-Rampant P, Paux E, Le Gouis J, Bastien C, Segura V (2018) Phenomic selection is a low-cost and high-throughput method based on indirect predictions: proof of concept on wheat and poplar. Genes Genomes Genet 8(12):3961–3972. https://doi.org/10.1534/g3.118.200760

Riov J, Fox H, Attias R, Shklar G, Farkash-Haim L, Sitbon R, Moshe Y, Abu-Abied M, Sadot E, David-Schwartz R (2020) Improved method for vegetative propagation of mature *Pinus halepensis* and its hybrids by cuttings. Isr J Plant Sci 67(1–2):5–15. https://doi.org/10.1163/22238980-20191118

Rodrigues AM, Mendes MD, Lima AS, Barbosa PM, Ascensão L, Barroso JG, Pedro LG, Mota MM, Figueiredo AC (2017) *Pinus halepensis, Pinus pinaster, Pinus pinea* and *Pinus sylvestris* essential oils chemotypes and monoterpene hydrocarbon enantiomers, before and after inoculation with the pinewood nematode *Bursaphelenchus xylophilus*. Chem Biodivers 14(1):e1600153. https://doi.org/10.1002/cbdv.201600153

Rodrigues AS, Chaves I, Costa BV, Lin Y-C, Lopes S, Milhinhos A, Van de Peer Y, Miguel CM (2019) Small RNA profiling in *Pinus pinaster* reveals the transcriptome of developing seeds and highlights differences between zygotic and somatic embryos. Sci Rep 9(1):11327. https://doi.org/10.1038/s41598-019-47789-y

Rodrigues AS, De Vega JJ, Miguel CM (2018) Comprehensive assembly and analysis of the transcriptome of maritime pine developing embryos. BMC Plant Biol 18(1):379. https://doi.org/10.1186/s12870-018-1564-2

Rodríguez-Quilón I, Santos-del-Blanco L, Serra-Varela MJ, Koskela J, González-Martínez SC, Alía R (2016) Capturing neutral and adaptive genetic diversity for conservation in a highly structured tree species. Ecol Appl 26(7):2254–2266. https://doi.org/10.1002/eap.1361

Romero IG, Ruvinsky I, Gilad Y (2012) Comparative studies of gene expression and the evolution of gene regulation. Nat Rev Genet 13(7):505–516. https://doi.org/10.1038/nrg3229

Roodt D, Lohaus R, Sterck L, Swanepoel RL, Van de Peer Y, Mizrachi E (2017) Evidence for an ancient whole genome duplication in the cycad lineage. PLoS ONE 12(9):e0184454. https://doi.org/10.1371/journal.pone.0184454

Rosvall O, Bradshaw RHW, Egertsdotter U, Ingvarsson PK, Wu H (2019) Using Norway spruce clones in Swedish forestry: introduction. Scand J for Res 34(5):333–335. https://doi.org/10.1080/02827581.2018.1562565

Rowe DB, Blazich FA, Raper CD (2002) Nitrogen nutrition of hedged stock plants of loblolly pine. I. Tissue nitrogen concentrations and carbohydrate status. New For 24(1):39–51. https://doi.org/10.1023/A:1020551029894

Rozas V, Zas R, García-González I (2011) Contrasting effects of water availability on *Pinus pinaster* radial growth near the transition between the Atlantic and Mediterranean biogeographical regions in NW Spain. Eur J For Res 130(6):959–970. https://doi.org/10.1007/s10342-011-0494-4

Rubiales JM, García-Amorena I, García Álvarez S, Morla C (2009) Anthracological evidence suggests naturalness of *Pinus pinaster* in inland southwestern Iberia. Plant Ecol 200(2):155–160. https://doi.org/10.1007/s11258-008-9439-5

Ruffault J, Curt T, Moron V, Trigo RM, Mouillot F, Koutsias N, Pimont F, Martin-StPaul N, Barbero R, Dupuy J-L, Russo A, Belhadj-Khedher C (2020) Increased likelihood of heat-induced large wildfires in the Mediterranean Basin. Sci Rep 10(1):13790. https://doi.org/10.1038/s41598-020-70069-z

Rupps A, Raschke J, Rümmler M, Linke B, Zoglauer K (2016) Identification of putative homologs of *Larix decidua* to *BABYBOOM (BBM), Leafy Cotyledon1 (LEC1), Wuschel-related HOMEOBOX2 (WOX2)* and *Somatic Embryogenesis Receptor-like Kinase (SERK)* during somatic embryogenesis. Planta 243:473–488. https://doi.org/10.1007/s00425-015-2409-y

Sampedro L, Moreira X, Llusia J, Peñuelas J, Zas R (2010) Genetics, phosphorus availability, and herbivore-derived induction as sources of phenotypic variation of leaf volatile terpenes in a pine species. J Exp Bot 61(15):4437–4447. https://doi.org/10.1093/jxb/erq246

Sampedro L, Moreira X, Martíns P, Zas R (2009) Growth and nutritional response of *Pinus pinaster* after a large pine weevil (*Hylobius abietis*) attack. Trees 23(6):1189–1197. https://doi.org/10.1007/s00468-009-0358-4

Sampedro L, Moreira X, Zas R (2011) Costs of constitutive and herbivore-induced chemical defences in pine trees emerge only under low nutrient availability. J Ecol 99(3):818–827. https://doi.org/10.1111/j.1365-2745.2011.01814.x

Sánchez-Gómez D, Majada J, Alía R, Feito I, Aranda I (2010) Intraspecific variation in growth and allocation patterns in seedlings of *Pinus pinaster* Ait. submitted to contrasting watering regimes: can water availability explain regional variation? Ann For Sci 67(5):505. https://doi.org/10.1051/forest/2010007

Sánchez-Salguero R, Camarero JJ, Rozas V, Génova M, Olano JM, Arzac A, Gazol A, Caminero L, Tejedor E, de Luis M, Linares JC (2018) Resist, recover or both? Growth plasticity in response to drought is geographically structured and linked to intraspecific variability in *Pinus pinaster*. J Biogeogr 45(5):1126–1139. https://doi.org/10.1111/jbi.13202

Santos-del-Blanco L, Alía R, González-Martínez SCSC, Sampedro L, Lario F, Climent J (2015) Correlated genetic effects on reproduction define a domestication syndrome in a forest tree. Evol Appl 8(4):403–410. https://doi.org/10.1111/eva.12252

Schveste D, Ughetto F (1986) Différences de sensibilité à *Matsucoccus feytaudi* Duc (Homoptera: Margarodidae) selon les provenances de pin maritime (*Pinus pinaster* AIT)(Differences in susceptibility to *Matsucoccus feytaudi* Duc (Homoptera: Margarodidae) in maritime pine (*Pinus pinaster* AIT) according to provenance.). Ann des Sci For 43(4):459–474. https://doi.org/10.1051/forest:19860403

Scott AD, Zimin AV, Puiu D, Workman R, Britton M, Zaman S, Caballero M, Read AC, Bogdanove AJ, Burns E, Wegrzyn J, Timp W, Salzberg SL, Neale DB (2020) A reference genome sequence for giant sequoia. Genes Genomes Genet 10(11):3907–3919. https://doi.org/10.1534/g3.120.401612

Seidl R, Thom D, Kautz M, Martin-Benito D, Peltoniemi M, Vacchiano G, Wild J, Ascoli D, Petr M, Honkaniemi J, Lexer MJ, Trotsiuk V, Mairota P, Svoboda M, Fabrika M, Nagel TA, Reyer CPO (2017) Forest disturbances under climate change. Nat Clim Chang 7(6):395–402. https://doi.org/10.1038/nclimate3303

Seoane-Zonjic P, Cañas RA, Bautista R, Gómez-Maldonado J, Arrillaga I, Fernández-Pozo N, Claros MG, Cánovas FM, Ávila C (2016) Establishing gene models from the *Pinus pinaster* genome using gene capture and BAC sequencing. BMC Genomics 17:148. https://doi.org/10.1186/s12864-016-2490-z

Serra-Varela MJ, Grivet D, Vincenot L, Broennimann O, Gonzalo-Jiménez J, Zimmermann NE (2015) Does phylogeographical structure relate to climatic niche divergence? A test using maritime pine (*Pinus pinaster* Ait.). Glob Ecol Biogeogr 24(11):1302–1313. https://doi.org/10.1111/geb.12369

Shao C, Ma X, Xu X, Wang H, Meng Y (2012) Genome-wide identification of reverse complementary microRNA genes in plants. PLoS ONE 7(10):e46991. https://doi.org/10.1371/journal.pone.0046991

Sharma SK, Verma SK (2011) Seasonal influences on the rooting response of Chir pine (*Pinus roxburghii* Sarg.). Ann For Res 54(2):241–247. https://doi.org/10.15287/afr.2011.93

Shepherd M, Mellick R, Toon P, Dale G, Dieters M (2005) Genetic control of adventitious rooting on stem cuttings in two *Pinus elliottii* × *P. caribaea* hybrid families. Ann For Sci 62(5):403–412. https://doi.org/10.1051/forest:2005036

Shimizu T, Tanizawa Y, Mochizuki T, Nagasaki H, Yoshioka T, Toyoda A, Fujiyama A, Kaminuma E, Nakamura Y (2017) Draft sequencing of the heterozygous diploid genome of satsuma (*Citrus unshiu* Marc.) using a hybrid assembly approach. Front Genet 8:180. https://doi.org/10.3389/fgene.2017.00180

Shin H, Lee H, Woo K-S, Noh E-W, Koo Y-B, Lee K-J (2009) Identification of genes upregulated by pinewood nematode inoculation in Japanese red pine. Tree Physiol 29(3):411–421. https://doi.org/10.1093/treephys/tpn034

Simpson JT, Wong K, Jackman SD, Schein JE, Jones SJM, Birol I (2009) ABySS: a parallel assembler for short read sequence data. Genome Res 19(6):1117–1123. https://doi.org/10.1101/gr.089532.108

Singh KB (1998) Transcriptional regulation in plants: the importance of combinatorial control. Plant Physiol 118(4):1111–1120. https://doi.org/10.1104/pp.118.4.1111

Sniezko RA, Koch J (2017) Breeding trees resistant to insects and diseases: putting theory into application. Biol Invasions 19(11):3377–3400. https://doi.org/10.1007/s10530-017-1482-5

Sohn J, Nam J-W (2018) The present and future of *de novo* whole-genome assembly. Brief Bioinform 19(1):23–40. https://doi.org/10.1093/bib/bbw096

Solla A, Aguín O, Cubera E, Sampedro L, Mansilla JP, Zas R (2011) Survival time analysis of *Pinus pinaster* inoculated with *Armillaria ostoyae*: genetic variation and relevance of seed and root traits. Eur J Plant Pathol 130(4):477–488. https://doi.org/10.1007/s10658-011-9767-5

Stein L (2001) Genome annotation: from sequence to biology. Nat Rev Genet 2(7):493–503. https://doi.org/10.1038/35080529

Stevens KA, Wegrzyn JL, Zimin A, Puiu D, Crepeau M, Cardeno C, Paul R, Gonzalez-Ibeas D, Koriabine M, Holtz-Morris AE, Martínez-García PJ, Sezen UU, Marçais G, Jermstad K, McGuire PE, Loopstra CA, Davis JM, Eckert A, de Jong P, Yorke JA, Salzberg SL, Neale DB, Langley CH (2016) Sequence of the sugar pine megagenome. Genetics 204 (4):1613–1626. https://doi.org/10.1534/genetics.116.193227

Street NR (2019) Genomics of forest trees. In: Cánovas FM (ed) Advances in botanical research, vol 89. Academic Press, pp 1–37

Suárez-Vidal E, López-Goldar X, Sampedro L, Zas R (2017) Effect of light availability on the interaction between maritime pine and the pine weevil: light drives insect feeding behavior but also the defensive capabilities of the host. Front Plant Sci 8:1452. https://doi.org/10.3389/fpls.2017.01452

Sultan SE (2004) Promising directions in plant phenotypic plasticity. Perspect Plant Ecol Evol Syst 6(4):227–233. https://doi.org/10.1078/1433-8319-00082

Takahashi F, Kuromori T, Sato H, Shinozaki K (2018) Regulatory gene networks in drought stress responses and resistance in plants. In: Iwaya-Inoue M, Sakurai M, Uemura M (eds) Survival strategies in extreme cold and desiccation. Advances in experimental medicine and biology, vol 1081. Springer, Singapore, pp 189–214

Tereso S, Gonçalves S, Marum L, Oliveira M, Maroco J, Miguel C (2006a) Improved axillary and adventitious bud regeneration from Portuguese genotypes of *Pinus pinaster* AIT. Propag Ornam Plants 6(1):24–33

Tereso S, Miguel C, Maroco J, Oliveira MM (2006b) Susceptibility of embryogenic and organogenic tissues of maritime pine (*Pinus pinaster*) to antibiotics used in Agrobacterium-mediated genetic transformation. Plant Cell Tiss Org Cult 87:33–40

Tereso S, Miguel C, Zoglauer K, Valle-Piquera C, Oliveira MM (2006c) Stable Agrobacterium-mediated transformation of embryogenic tissues from *Pinus pinaster* Portuguese genotypes. J Plant Growth Reg 50:57–68

Tereso S, Zoglauer K, Miguel C, Oliveira MM (2003) Establishing a genetic transformation system in *Pinus pinaster*. In: Espinel S, Barreto Y, Ritter E (eds) Sustainable forestry, wood products and biotechnology. DFA-AFA Press, Vitoria-Gasteiz, Spain, pp 195–204

Toda T, Kurinobu S (2002) Realized genetic gains observed in progeny tolerance of selected red pine (*Pinus densiflora*) and black pine (*P. thunbergii*) to pine wilt disease. Silvae Genet 51(1):42–44

Trontin J-F, Alazard P, Dumas E, Quniou S, Canlet F, Chantre G, Harvengt L (2004) Prospects for clonal propagation of selected maritime pine (*Pinus pinaster* Ait.) using micropropagation techniques. 9th international conference on biotechnology in the pulp and paper industry, Durban, South Africa, 10–14 October 2004, p 9.5

Trontin J-F, Aronen T, Hargreaves C, Montalbán I, Moncaleán P, Reeves C, Quoniou S, Lelu-Walter, M-A, Klimaszewska K (2016a) International effort to induce somatic embryogenesis in adult pine trees. In: Park Y-S, Bonga J, Moon H-K (eds) Vegetative propagation of forest trees. National Institute of Forest Science (NIFoS), Seoul, Korea, pp 211–260

Trontin J-F, Ávila C, Debille S, De La Torre F, El-Azaz J, Pascual B, Canlet F, Teyssier C, Boizot N, Le Metté C, Lesage-Descauses M-C, Da Silva Perez D, Cañas R, Le Provost G, Plomion C, Harvengt L, Label P, Lelu-Walter M-A, Cánovas F (2015a) Towards functional genomics of transcription factor genes associated to growth and wood formation in maritime pine. ProCoGen final open conference, on promoting conifer genomic resources, Orléans, France, 30 November–2 December 2015, p 11

Trontin J-F, Ávila C, Debille S, Teyssier C, Canlet F, Rueda-López M, Canales J, De la Torre F, El-Azaz J, Pascual B, Cañas R, Boizot N, Le Metté C, Lesage-Descauses M-C, Abarca D, Carneros E, Rupps A, Hassani SB, Zoglauer K, Arrillaga I, Mendoza-Poudereux I, Cano M, Segura J, Miguel C, De Vega-Bartol J, Tonelli M, Rodrigues A, Label P, Le Provost G, Plomion C, da Silva Perez D, Harvengt L, Díaz-Sala C, Cánovas FM, Lelu-Walter M-A (2017) Somatic embryogenesis as an enabling technology for reverse genetics: achievements and prospects for breeding maritime pine (*Pinus pinaster* Ait.). In: Bonga JM, Park YS, Trontin J-F (eds) Proceedings of the 4th international conference of the IUFRO working party 2.09.02 on development and application of vegetative propagation technologies in plantation forestry to cope with a changing climate and environment, La Plata, Argentina, 19–23 September 2016. IUFRO, Vienna, Austria, p 338

Trontin J-F, Debille S, Canlet F, Harvengt L, Lelu-Walter M-A, Label P, Teyssier C, Lesage-Descause MC, Le Metté C, Miguel C, De Vega-Bartol J, Tonelli M, Santos R, Rupps A, Hassani SB, Zoglauer K, Carneros E, Diaz-Sala C, Abarca D, Arrillaga I, Mendoza-Poudereux I, Segura J, Ávila C, Rueda M, Canales J, Cánovas FM (2013) Somatic embryogenesis as an effective regeneration support for reverse genetics in maritime pine: the Sustainpine collaborative project as a case study. In: Park YS, Bonga JM (eds) Proceeding of the 2nd international conference of the IUFRO working party 2.09.02 on integrating vegetative propagation, biotechnology and genetic improvement for tree production and sustainable forest management, Brno, Czech Republic, 25–28 June 2012. IUFRO, Vienna, Austria, pp 184–187

Trontin J-F, Debille S, Vallance M, Quoniou S, Lesage-Descause M-C, Label P, Harvengt L, Lelu-Walter M-A (2009) Basal medium formulation strongly affects transformation efficiency in maritime pine. IUFRO tree biotechnology conference, Whistler, British Columbia, Canada, 28 June–2 July 2009, p 62 (P 122)

Trontin J-F, Harvengt L, Garin E, Lopez-Vernaza M, Arancio L, Hoebeke J, Canlet F, Pâques M (2002) Towards genetic engineering of maritime pine (*Pinus pinaster* Ait.). Ann For Sci 59(5–6):687–697. https://doi.org/10.1051/forest:2002057

Trontin J-F, Klimaszewska K, Morel A, Hargreaves C, Lelu-Walter M-A (2016b) Molecular aspects of conifer zygotic and somatic embryo development: a review of genome-wide approaches and recent insights. In: Germana M, Lambardi M (eds) *In vitro* embryogenesis in higher plants. Methods in molecular biology, vol 1359. Humana Press, New York, NY, pp 167–207

Trontin J-F, Raschke J, Rupps A (2020) Tree "memory": new insights on temperature-induced priming effects during early embryogenesis. Tree Physiol 41:906–911. https://doi.org/10.1093/treephys/tpaa150

Trontin J-F, Reymond I, Canlet F, Sow MD, Delaunay A, Maury S, Le Metté C, Teyssier C, Lelu-Walter M-A (2019a) Temperature affects somatic embryo development in maritime pine. In: Bonga JM, Park Y-S, Trontin J-F (eds) Proceedings of the 5th international conference of the IUFRO working party 2.09.02 on clonal trees in the bioeconomy age: opportunities and challenges, Coimbra, Portugal, 10–15 September 2018. IUFRO, Vienna, Austria, p 274

Trontin JF, Rupps A, Raschke J, Lelu-Walter MA, Teyssier C, Gauchat ME, Vera Bravo C, Aronen T, Varis S, Tikkinen M, Moncaleán P, Montalbán I, Egertsdotter U, Dobrowolska I, Street N (2019b) MULTIFOREVER: towards intensification of conifer production through multi-varietal forestry based on somatic embryogenesis. ForestValue kick-off seminar, 23–24 May 2019, Helsinki, Finland, https://forestvalue.org/

Trontin J-F, Teyssier C, Avila C, Debille S, Le Metté C, Lesage-Descauses M-C, Boizot N, Canlet F, Le Provost G, Harvengt L, Plomion C, Label P, Cánovas F, Lelu-Walter M-A (2015b) Molecular phenotyping of maritime pine somatic plants transformed with an RNAi construct targeting cinnamyl alcohol dehydrogenase (CAD). In: Park YS, Bonga JM (eds) Proceedings of the 3rd IUFRO unit 2.09.02 international conference on woody plant production integrating genetic and vegetative propagation technologies, Vitoria-Gasteiz, Spain, 8–12 September 2014, p 131

Trontin J-F, Teyssier C, Morel A, Harvengt L, Lelu-Walter M-A (2016c) Prospects for new variety deployment through somatic embryogenesis in maritime pine. In: Park Y-S, Bonga JM, Moon H-K (eds) Vegetative propagation of forest trees. National Institute of Forest Science (NIFoS), Seoul, Korea, pp 572–606

Trontin J-F, Walter C, Klimaszewska K, Park Y-S, Lelu-Walter M-A (2007) Recent progress on genetic transformation of four *Pinus* spp. Transgenic Plant J 1:314–329

Tsavkelova E, Oeser B, Oren-Young L, Israeli M, Sasson Y, Tudzynski B, Sharon A (2012) Identification and functional characterization of indole-3-acetamide-mediated IAA biosynthesis in plant-associated Fusarium species. Fungal Genet Biol 49 (1):48–57. https://doi.org/10.1016/j.fgb.2011.10.005

Tuskan GA, DiFazio S, Jansson S, Bohlmann J, Grigoriev I, Hellsten U, Putnam N, Ralph S, Rombauts S, Salamov A, Schein J, Sterck L, Aerts A, Bhalerao RR, Bhalerao RP, Blaudez D, Boerjan W, Brun A, Brunner A, Busov V, Campbell M, Carlson J, Chalot M, Chapman J, Chen G-L, Cooper D, Coutinho PM, Couturier J, Covert S, Cronk Q, Cunningham R, Davis J, Degroeve S, Dejardin A, DePamphilis C, Detter J, Dirks B, Dubchak I, Duplessis S, Ehlting J, Ellis B, Gendler K, Goodstein D, Gribskov M, Grimwood J, Groover A, Gunter L, Hamberger B, Heinze B, Helariutta Y, Henrissat B, Holligan D, Holt R, Huang W, Islam-Faridi N, Jones S, Jones-Rhoades M, Jorgensen R, Joshi C, Kangasjarvi J, Karlsson J, Kelleher C, Kirkpatrick R, Kirst M, Kohler A, Kalluri U, Larimer F, Leebens-Mack J, Leple J-C, Locascio P, Lou Y, Lucas S, Martin F, Montanini B, Napoli C, Nelson DR, Nelson C, Nieminen K, Nilsson O, Pereda V, Peter G, Philippe R, Pilate G, Poliakov A, Razumovskaya J, Richardson P, Rinaldi C, Ritland K, Rouze P, Ryaboy D, Schmutz J, Schrader J, Segerman B, Shin H, Siddiqui A, Sterky F, Terry A, Tsai C-J, Uberbacher E, Unneberg P, Vahala J, Wall K, Wessler S, Yang G, Yin T, Douglas C, Marra M, Sandberg G, Van de Peer Y, Rokhsar D (2006) The genome of black cottonwood, *Populus trichocarpa* (Torr. & Gray). Science 313(5793):1596–1604. https://doi.org/10.1126/science.1128691

Ullah A, Manghwar H, Shaban M, Khan AH, Akbar A, Ali U, Ali E, Fahad S (2018) Phytohormones enhanced drought tolerance in plants: a coping strategy. Environ Sci Pollut Res 25(33):33103–33118. https://doi.org/10.1007/s11356-018-3364-5

Umezawa T, Nakashima K, Miyakawa T, Kuromori T, Tanokura M, Shinozaki K, Yamaguchi-Shinozaki K (2010) Molecular basis of the core regulatory network in ABA responses: sensing, signaling and transport. Plant Cell Physiol 51(11):1821–1839. https://doi.org/10.1093/pcp/pcq156

Varis S, Klimaszewska K, Aronen T (2018) Somatic embryogenesis and plant regeneration from primordial shoot explants of *Picea abies* (L.) H. Karst. Somatic Trees. Front Plant Sci 9:1551. https://doi.org/10.3389/fpls.2018.01551

Vázquez-González C, López-Goldar X, Zas R, Sampedro L (2019) Neutral and climate-driven adaptive processes contribute to explain population variation in resin duct traits in a Mediterranean pine species. Front Plant Sci 10:1613. https://doi.org/10.3389/fpls.2019.01613

Vázquez-González C, Sampedro L, López-Goldar X, Zas R (2021) Genetic and ecological basis of resistance to herbivorous insects in Mediterranean pines.

In: Ne'eman G, Osem Y (eds) Pines and their mixed forest ecosystems in the Mediterranean basin. Springer Nature Switzerland AG, p in press

Velasco-Conde T, Yakovlev I, Majada JP, Aranda I, Johnsen Ø (2012) Dehydrins in maritime pine (*Pinus pinaster*) and their expression related to drought stress response. Tree Genet Genomes 8(5):957–973. https://doi.org/10.1007/s11295-012-0476-9

Vicente C, Espada M, Vieira P, Mota M (2012) Pine wilt disease: a threat to European forestry. Eur J Plant Pathol 133(1):89–99. https://doi.org/10.1007/s10658-011-9924-x

Vidal M, Plomion C, Harvengt L, Raffin A, Boury C, Bouffier L (2015) Paternity recovery in two maritime pine polycross mating designs and consequences for breeding. Tree Genet Genomes 11(5):105. https://doi.org/10.1007/s11295-015-0932-4

Vidal M, Plomion C, Raffin A, Harvengt L, Bouffier L (2017) Forward selection in a maritime pine polycross progeny trial using pedigree reconstruction. Ann For Sci 74(1):21. https://doi.org/10.1007/s13595-016-0596-8

Vilasboa J, Da Costa CT, Fett-Neto AG (2019) Rooting of eucalypt cuttings as a problem-solving oriented model in plant biology. Prog Biophys Mol Biol 146:85–97. https://doi.org/10.1016/j.pbiomolbio.2018.12.007

Vivas M, Zas R, Solla A (2012) Screening of Maritime pine (*Pinus pinaster*) for resistance to *Fusarium circinatum*, the causal agent of pitch canker disease. Forestry 85(2):185–192. https://doi.org/10.1093/forestry/cpr055

Walsh B, Lynch M (eds) (2018) Evolution and selection of quantitative traits, vol 2. Oxford University Press, Cary, N.C

Wan L-C, Wang F, Guo X, Lu S, Qiu Z, Zhao Y, Zhang H, Lin J (2012) Identification and characterization of small non-coding RNAs from Chinese fir by high throughput sequencing. BMC Plant Biol 12 (1):146. https://doi.org/10.1186/1471-2229-12-146

Wan T, Liu Z-M, Li L-F, Leitch AR, Leitch IJ, Lohaus R, Liu Z-J, Xin H-P, Gong Y-B, Liu Y, Wang W-C, Chen L-Y, Yang Y, Kelly LJ, Yang J, Huang J-L, Li Z, Liu P, Zhang L, Liu H-M, Wang H, Deng S-H, Liu M, Li J, Ma L, Liu Y, Lei Y, Xu W, Wu L-Q, Liu F, Ma Q, Yu X-R, Jiang Z, Zhang G-Q, Li S-H, Li R-Q, Zhang S-Z, Wang Q-F, Van de Peer Y, Zhang J-B, Wang X-M (2018) A genome for gnetophytes and early evolution of seed plants. Nat Plants 4(2):82–89. https://doi.org/10.1038/s41477-017-0097-2

Wang X-Q, Ran J-H (2014) Evolution and biogeography of gymnosperms. Mol Phylogenet Evol 75(1):24–40. https://doi.org/10.1016/j.ympev.2014.02.005

Warren RL, Keeling CI, Saint YMM, Raymond A, Taylor GA, Vandervalk BP, Mohamadi H, Paulino D, Chiu R, Jackman SD, Robertson G, Yang C, Boyle B, Hoffmann M, Weigel D, Nelson DR, Ritland C, Isabel N, Jaquish B, Yanchuk A, Bousquet J, Jones SJM, MacKay J, Birol I, Bohlmann J (2015) Improved white spruce (*Picea glauca*) genome assemblies and annotation of large gene families of conifer terpenoid and phenolic defense metabolism. Plant J 83 (2):189–212. https://doi.org/10.1111/tpj.12886

Wenger AM, Peluso P, Rowell WJ, Chang P-C, Hall RJ, Concepcion GT, Ebler J, Fungtammasan A, Kolesnikov A, Olson ND, Töpfer A, Alonge M, Mahmoud M, Qian Y, Chin C-S, Phillippy AM, Schatz MC, Myers G, DePristo MA, Ruan J, Marschall T, Sedlazeck FJ, Zook JM, Li H, Koren S, Carroll A, Rank DR, Hunkapiller MW (2019) Accurate circular consensus long-read sequencing improves variant detection and assembly of a human genome. Nat Biotechnol 37:1155–1162. https://doi.org/10.1038/s41587-019-0217-9

Wingfield MJ, Brockerhoff EG, Wingfield BD, Slippers B (2015) Planted forest health: the need for a global strategy. Science 349(6250):832–836. https://doi.org/10.1126/science.aac6674

Wingfield MJ, Hammerbacher A, Ganley RJ, Steenkamp ET, Gordon TR, Wingfield BD, Coutinho TA (2008) Pitch canker caused by *Fusarium circinatum*—a growing threat to pine plantations and forests worldwide. Australas Plant Pathol 37(4):319. https://doi.org/10.1071/AP08036

Wisser RJ, Fang Z, Holland JB, Teixeira JEC, Dougherty J, Weldekidan T, de Leon N, Flint-Garcia S, Lauter N, Murray SC, Xu W, Hallauer A (2019) The genomic basis for short-term evolution of environmental adaptation in maize. Genetics 213(4):1479–1494. https://doi.org/10.1534/genetics.119.302780

Woodcock P, Cottrell JE, Buggs RJA, Quine CP (2018) Mitigating pest and pathogen impacts using resistant trees: a framework and overview to inform development and deployment in Europe and North America. For an Int J For Res 91(1):1–16. https://doi.org/10.1093/forestry/cpx031

Workman R, Fedak R, Kilburn D, Hao S, Liu K, Timp W (2018) High molecular weight DNA extraction from recalcitrant plant species for third generation sequencing. Protoc Exch Version 1:1–15. https://doi.org/10.1038/protex.2018.059

Xu L, Huang H (2014) Genetic and epigenetic controls of plant regeneration. Curr Top Dev Biol 108:1–33

Xu L, Zhang J, Gao J, Chen X, Jiang C, Hao Y (2012) Study on the disease resistance of candidate clones in *Pinus massoniana* to *Bursaphelenchus xylophilus*. China For Sci Technol 26:27–30

Xu S (2003) Theoretical basis of the beavis effect. Genetics 165(4):2259–2268

Yakovlev IA, Fossdal CG, Johnsen Ø (2010) MicroRNAs, the epigenetic memory and climatic adaptation in Norway spruce. New Phytol 187(4):1154–1169. https://doi.org/10.1111/j.1469-8137.2010.03341.x

Yamaguchi M, Demura T (2010) Transcriptional regulation of secondary wall formation controlled by NAC domain proteins. Plant Biotechnol 27(3):237–242. https://doi.org/10.5511/plantbiotechnology.27.237

Yandell M, Ence D (2012) A beginner's guide to eukaryotic genome annotation. Nat Rev Genet 13 (5):329–342. https://doi.org/10.1038/nrg3174

Yi L, Liang Z-T, Peng Y, Yao X, Chen H-B, Zhao Z-Z (2012) Tissue-specific metabolite profiling of alkaloids in *Sinomenii caulis* using laser microdissection and liquid chromatography-quadrupole/time of flight-mass spectrometry. J Chromatogr A 1248:93–103. https://doi.org/10.1016/j.chroma.2012.05.058

Zapata-Valenzuela J, Isik F, Maltecca C, Wegrzyn J, Neale D, McKeand S, Whetten R (2012) SNP markers trace familial linkages in a cloned population of *Pinus taeda*-prospects for genomic selection. Tree Genet Genomes 8(6):1307–1318. https://doi.org/10.1007/s11295-012-0516-5

Zas R, Merlo E, Fernández-López J (2004) Juvenile—mature genetic correlations in *Pinus pinaster* ait. under different nutrient x water regimes. Silvae Genet 53(1–6):124–129. https://doi.org/10.1515/sg-2004-0022

Zas R, Moreira X, Ramos M, Lima MRM, Nunes da Silva M, Solla A, Vasconcelos MW, Sampedro L (2015) Intraspecific variation of anatomical and chemical defensive traits in maritime pine (*Pinus pinaster*) as factors in susceptibility to the pinewood nematode (*Bursaphelenchus xylophilus*). Trees 29(3):663–673. https://doi.org/10.1007/s00468-014-1143-6

Zas R, Sampedro L, Prada E, Fernández-López J (2005) Genetic variation of *Pinus pinaster* Ait. seedlings in susceptibility to the pine weevil *Hylobius abietis* L. Ann For Sci 62(7):681–688. https://doi.org/10.1051/forest:2005064

Zas R, Solla A, Sampedro L (2007) Variography and kriging allow screening *Pinus pinaster* resistant to *Armillaria ostoyae* in field conditions. Forestry 80 (2):201–209. https://doi.org/10.1093/forestry/cpl050

Zeng J, de Vlaming R, Wu Y, Robinson MR, Lloyd-Jones LR, Yengo L, Yap CX, Xue A, Sidorenko J, McRae AF, Powell JE, Montgomery GW, Metspalu A, Esko T, Gibson G, Wray NR, Visscher PM, Yang J (2018) Signatures of negative selection in the genetic architecture of human complex traits. Nat Genet 50(5):746–753. https://doi.org/10.1038/s41588-018-0101-4

Zhang J, Wu T, Li L, Han S, Li X, Zhang S, Qi L (2013) Dynamic expression of small RNA populations in larch (*Larix leptolepis*). Planta 237(1):89–101. https://doi.org/10.1007/s00425-012-1753-4

Zhang M, Zhang Y, Scheuring CF, Wu C-C, Dong JJ, Zhang H-B (2012) Preparation of megabase-sized DNA from a variety of organisms using the nuclei method for advanced genomics research. Nat Protoc 7 (3):467–478. https://doi.org/10.1038/nprot.2011.455

Zhang W, Gao S, Zhou X, Xia J, Chellappan P, Zhou X, Zhang X, Jin H (2010) Multiple distinct small RNAs originate from the same microRNA precursors. Genome Biol 11(8):R81. https://doi.org/10.1186/gb-2010-11-8-r81

Zhu Y, Li H, Bhatti S, Zhou S, Yang Y, Fish T, Thannhauser TW (2016) Development of a laser capture microscope-based single-cell-type proteomics tool for studying proteomes of individual cell layers of plant roots. Hortic Res 3(1):16026. https://doi.org/10.1038/hortres.2016.26

Zimin A, Stevens KA, Crepeau MW, Holtz-Morris A, Koriabine M, Marçais G, Puiu D, Roberts M, Wegrzyn JL, de Jong PJ, Neale DB, Salzberg SL, Yorke JA, Langley CH (2014) Sequencing and assembly of the 22-Gb loblolly pine genome. Genetics 196(3):875–890. https://doi.org/10.1534/genetics.113.159715

Zimmer A, Lang D, Richardt S, Frank W, Reski R, Rensing SA (2007) Dating the early evolution of plants: detection and molecular clock analyses of orthologs. Mol Genet Genomics 278(4):393–402. https://doi.org/10.1007/s00438-007-0257-6

Zonneveld BJM (2012) Conifer genome sizes of 172 species, covering 64 of 67 genera, range from 8 to 72 picogram. Nord J Bot 30(4):490–502. https://doi.org/10.1111/j.1756-1051.2012.01516.x

Zwaenepoel A, Van de Peer Y (2019) Inference of ancient whole-genome duplications and the evolution of gene duplication and loss rates. Mol Biol Evol 36(7):1384–1404. https://doi.org/10.1093/molbev/msz088

Understanding the Genetic Architecture of Complex Traits in Loblolly Pine

6

Mengmeng Lu and Carol A. Loopstra

Abstract

Understanding the genetic architecture of loblolly pine complex traits is key to identify the causal genes that underlie various fitness and phenotypic traits and to develop tools for breeding practices. Toward this end, a number of genetics studies and breeding programs have been deployed over the past 70 years. For the last three decades, a series of quantitative trait locus (QTL) mapping and genetic association studies have been performed to link genetic with phenotypic variation, thanks to the rapidly developing genotyping technologies and statistical methods. Researchers have developed an array of genomic resources such as high-density linkage maps and candidate loci associated with a variety of traits, which are useful tools to practice genomic selection and marker-assisted breeding. Though we have made some progress in understanding the composition of complex traits, the causal genes or alleles still remain largely unknown.

M. Lu (✉)
Department of Biological Sciences, University of Calgary, Calgary, Canada

C. A. Loopstra
Department of Ecology and Conservation Biology, Texas A&M University, College Station, TX, USA

6.1 Introduction

Loblolly pine (*Pinus taeda*) is a diploid (2n = 24), long-lived, and largely outcrossing conifer species with a natural range extending from eastern Texas throughout the southeastern United States and north into Delaware and thus is distributed across diverse environmental gradients (Prasad et al. 2007-ongoing). It is the leading timber species for wood and pulp products and provides ecological services including preservation of biodiversity, carbon sequestration, and preservation of water quality; thus, a number of genetics studies and breeding programs have been deployed over the past 70 years (Baker and Langdon 1990; Johnsen et al. 2004).

A major goal of genetics studies on loblolly pine is to identify the causal genes that underlie various fitness and phenotypic traits, which can be used as molecular markers to breed trees for desirable traits. Loblolly pine populations harbor a remarkable diversity of phenotypic variation for morphology, physiology, and disease susceptibility (Quesada et al. 2010; Lu et al. 2017), which is due to underlying genetic complexity in different environmental conditions. Understanding the genotype–phenotype relationship can yield insights that are important for accelerating selective breeding and predicting adaptive evolution.

The common path toward this goal is to link genetic with phenotypic variation, either through

A. R. De La Torre (ed.), *The Pine Genomes*, Compendium of Plant Genomes,
https://doi.org/10.1007/978-3-030-93390-6_6

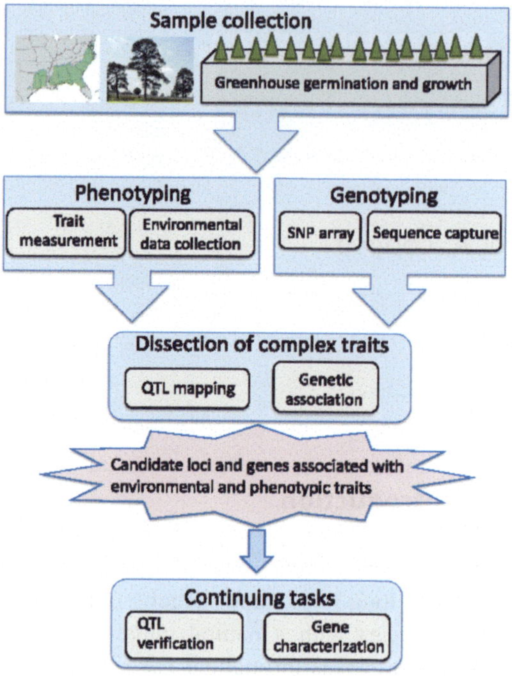

Fig. 6.1. A typical workflow in identifying the genetic architecture of complex traits in loblolly pine. Options per step are horizontally aligned in each box

quantitative trait locus (QTL) mapping or through genetic association by scanning a genome-wide set of genetic variants in different individuals (Fig. 6.1). As with other conifer species, loblolly pine is a difficult organism for complex trait dissection and gene identification due to its long generation time, large genome, limited funding, and lack of well-defined mutants for reverse genetic experimentation. Hence genetics research on loblolly pine has lagged behind that of a few model organisms. However, owing to cost-effective genotyping methods and the development of statistical methods, we have begun to decipher the genomic basis of complex traits that are related to economic and adaptive values.

In this chapter, we begin by introducing genotyping methods and development of linkage maps and then review progress in research leading to an understanding of the genetic architecture of different complex traits. We summarize what we have learned from the past genetic dissection of complex traits. Finally, we end the chapter by highlighting potential future tasks in the coming decade.

6.2 Genotyping Tools

Genotyping enables identification of variations in DNA sequence in a sample. Currently, genotyping tools can analyze a substantial number of markers, followed by statistical analyses to figure out genomic diversity specifically prevalent in certain groups of samples. The information helps to understand genome organization and identify candidate genes. In the last four decades, we have witnessed the rapid development of genotyping tools in loblolly pine, from low-throughput gel-based approaches to high-throughput fixed array systems, then to genotyping by sequencing (GBS) methods, generating a variety of markers.

The advances in marker development allow the construction of high-resolution genetic linkage maps, which are important tools in breeding efforts. Genetic linkage maps containing the information of markers' physical arrangement have been used in biology for over 100 years (Sturtevant 1913). They are based on the frequencies of recombination between markers during crossover of homologous chromosomes, with a high recombination rate between two genetic markers indicating they are further apart on the map. The genetic linkage map for loblolly pine is one of the most developed for coniferous species thanks to the great amount of genetic information obtained over many years of intensive efforts.

The advent of loblolly pine genome assembly and annotation (Neale et al. 2014; Wegrzyn et al. 2014), paired with high-throughput sequencing, enable a vast number of genetic markers, which can be used to generate high-density linkage maps. Simultaneously, the linkage maps can help to anchor, orient, and order the fragmented de novo assembled sequences into chromosome-scale sequences.

In this section, we will review the development of genetic markers and linkage maps of loblolly pine. A combination of various types of markers resulted in different linkage maps.

6.2.1 Low-Throughput Genotyping and Genetic Linkage Maps

Prior to the development of high-throughput sequencing, genotyping mainly relied on low-throughput gel-based approaches. Genetic linkage map construction of loblolly pine began in the early 1980s using 12 allozyme loci (Adams and Joly 1980a, b), resulting in four pairs of linked loci. It was not until 1994 that formal loblolly pine linkage maps became available. One was constructed using 73 restriction fragment length polymorphism (RFLP) markers and two isozyme loci based on the pedigree *base1* including 95 progeny (Devey et al. 1994). Two maps were constructed based on the pedigree *qtl1* including 177 progeny, representing maternal and paternal parent gamete segregations as inferred from diploid progeny RFLP genotypes, with maternal map containing 87 loci and paternal map containing 75 loci (Groover et al. 1994). The symbols "*base*" and "*qtl*" represent two three-generation outbred pedigrees, which were obtained from the North Carolina State University Cooperative Tree Improvement Program. Two sets of progeny for each pedigree (*base1*, *base2*, *qtl1*, *qtl2*) were generated by crossing the parents of each pedigree at different times (Martínez-García et al. 2013). Wilcox et al. (1996) constructed a map for fusiform rust disease resistance study in loblolly pine based on the 10–5 pedigree, which was selected at age 32 from a naturally regenerated forest in 1958 by the North Carolina State University Industry Cooperative Tree Improvement Program. This map had a map length of 1,727 cM, with four unlinked random amplified polymorphic DNA (RAPD) markers.

In 1999, Remington et al. (1999) reported a genetic map including 12 linkage groups, which is equal to the loblolly pine haploid chromosome number. This map was generated using 508 amplified fragment length polymorphism (AFLP) markers segregating in haploid megagametophytes (the haploid, maternally derived nutritive tissue of conifer seeds) from a single parent. Sewell et al. (1999) constructed a consensus map based on integration of linkage data from two pedigrees, *base1* (95 progeny) and *qtl1* (172 progeny), using 357 genetic markers, including RFLP, RAPD, and isozyme markers.

An alternative to genetic markers like RFLPs, RAPDs, and AFLPs, is the PCR-based markers from expressed sequence tags (ESTs), which were based on coding sequences (cDNAs) and were later developed as EST polymorphisms (ESTPs) in loblolly pine (Harry et al. 1998). Temesgen et al. (2001) constructed a consensus map using 56 ESTP markers along with RFLP markers based on two pedigrees *base1* (95 progeny) and *qtl1* (172 progeny). With the same pedigrees, Brown et al. (2001) constructed a consensus map using 235 ESTP, RFLP, and isozyme markers. The orthologous ESTP and RFLP markers enabled the comparative mapping across different genera in conifers. Krutovsky et al. (2004) identified ten homologous linkage groups in loblolly pine and Douglas-fir [*Pseudotsuga menziesii* (Mirb.) Franco] using orthologous ESTP and RFLP markers, revealing extensive synteny and collinearity between these two *Pinaceae* species. The loblolly pine marker pool was expanded by microsatellite markers, also known as simple sequence repeats (SSR). Zhou et al. (2003) constructed a loblolly pine linkage map using 51 SSR markers without observing clustering or uneven distribution across the genome. Echt et al. (2011) constructed a consensus map using 429 SSR markers based on two pedigrees, *base1* (98 progeny) and *qtl1* (172 progeny).

Molecular markers evolved rapidly. When single nucleotide polymorphisms (SNPs), a type of allelic co-dominant marker, came into use in the early 2000s, they soon gained popularity among molecular biologists because of their high abundance, relatively low mutation rate, and amenability to automation. The early SNPs of loblolly pine were discovered using the candidate-gene-based approach. For example, 196 SNPs were discovered using 18 candidate genes for drought-stress response (González-Martínez et al. 2006); 58 SNPs were discovered using 20 wood- and drought-related candidate genes (González-Martínez et al. 2007); and 46 SNPs were discovered using 41 disease and abiotic stress-inducible genes (González-Martínez et al. 2008).

6.2.2 High-Throughput Genotyping and High-Density Maps

Low-throughput genotyping cannot meet the requirements for fine mapping in loblolly pine. Since 2009, multiplexed high-throughput SNP genotyping assays have been widely used, which led to high-density linkage maps.

Eckert et al. (2009) reported the first multiplexed high-throughput SNP genotyping array of loblolly pine implemented by Illumina's GoldenGate Assay (San Diego, CA, United States). The SNPs used in this array were generated through resequencing of previously identified candidate genes related to wood quality, drought tolerance, and disease resistance. This genotyping array allowed 384 SNPs to be assayed and 27 candidate genes subsequently mapped onto the existing loblolly pine consensus map. Loblolly pine genotyping was soon enhanced by a new array implemented by Illumina's Infinium platform (Eckert et al. 2010a). The SNPs used in this array were generated through resequencing of 7,535 unique EST contigs in 18 loblolly pine haploid megagametophytes. This genotyping array allowed 7,216 SNPs to be assayed and increased the genomic resolution of the linkage map with a total of 1635 SNPs mapped on 12 linkage groups based on the *qtl1* pedigree (172 progeny). Using the same Infinium platform, SNPs were genotyped in an outbred full-sibling family (217 progeny) provided by ArborGen Inc. (Ridgeville, SC, United States), resulting in 409 SNP markers mapped on 12 linkage groups (Xiong et al. 2016).

Combining SNPs and classical markers further increased the genomic resolution of linkage maps (Table 6.1). Martínez-García et al. (2013) constructed a consensus map using integration of SNP, RFLP, RAPD, isozymes, ESTP, and SSR markers and incorporated 2,466 markers across 12 linkage groups based on pedigrees *base* (202 progeny) and *qtl* (487 progeny). De La Torre et al. (2019a) developed an Affymetrix Axiom SNP genotyping array comprised of 635k SNP markers derived from whole-genome resequencing data from 10 loblolly pine individuals. The great SNP density resulted in the buildup of the most complete and dense linkage map for loblolly pine, which includes 44,722 anchored SNPs. The average number of SNPs per linkage group was 3,726 based on pedigrees *base* (100 progeny) and *qtl* (92 progeny) (De La Torre et al. 2019b).

6.2.3 Genotyping by Exome Sequencing

Given the large size and complexity of the loblolly pine genome (Neale et al. 2014; Wegrzyn et al. 2014), the genome complexity reduction method, mainly in the form of sequence capture, has been used in resequencing loblolly pine. Neves et al. (2013) designed 54,773 oligonucleotides probes (6.57 Mbp), which avoided exon–intron boundaries, using 14,729 EST-derived genes. Each probe had a length of 120 bp. This probe set was used to prepare a SureSelect Custom assay by Agilent Technologies (Santa Clara, CA, United States). The sequence capture was proven to be efficient in different types of samples including complementary DNA (cDNA) and genomic DNA (gDNA) from both haploid and diploid tissue. It was successfully applied in a mapping population including 72 haploid samples for discovering sequence variants, including SNPs, short insertions and deletions (indels), multiple-nucleotide polymorphisms (MNPs), and presence/absence variations (PAVs) (Neves et al. 2014). The segregated markers were used to generate a gene-based linkage map composed of 2,841 genes distributed among 12 linkage groups. Westbrook et al. (2015a) merged this

Table 6.1 Summary and description of the genetic linkage maps obtained for loblolly pine (*Pinus taeda*) since 2013*

Authors	Type of markers†	No. of markers	Population (no. of progeny used)‡	Type of map	LG	cM	Density (cM/Marker)
Martínez-García et al. (2013)	RFLP, RAPD, ESTP, SSR, isozyme, SNP	2,466	*qtl* (487), *base* (202)	Consensus	12	1,476	0.62
Neves et al. (2014)	PAV, SNP, short indels, MNP	2,841	10–5 (72)	Gene-based	12	1,637	0.58
Westbrook et al. (2015a)	RFLP, ESTP, SSR, PAV, SNP	3,856	*qtl-base1* (267), *qtl-base2* (689), BC1 (490), 10–5 (72)	Consensus	12	2,305	0.60
Xiong et al. (2016)	SNP	409	ArborGen (217)	Consensus	12	1,622	3.90
De La Torre et al. (2019b)	SNP	26,021	*qtl* (92), *base* (100)	Consensus	12	2,270	0.09

* Summary and description of the genetic linkage maps obtained for loblolly pine (*Pinus taeda*) prior to 2013 can be found in Martínez-García et al. (2013)

†RFLP—restriction fragment length polymorphism; RAPD—random amplified polymorphic DNA; ESTP—expressed sequence tags polymorphism; SSR—simple sequence repeat; SNP—single nucleotide polymorphism; PAV—presence/absence variation; indels—insertions and deletions; MNP—multiple-nucleotide polymorphism

‡*base* and *qtl* represent two mapping populations belonging to two three-generation outbred pedigrees. They were obtained from the North Carolina State University Cooperative Tree Improvement Program. For each pedigree (*base* and *qtl*), both parents were crossed at different times to obtain two sets of progeny for each pedigree (*base1*, *base2*, *qtl1*, *qtl2*); 10–5 represents a mapping population selected at age 32 from a naturally regenerated forest in 1958 by the North Carolina State University Industry Cooperative Tree Improvement Program; BC1 represents (*P. taeda* × *P. elliottii*) × *P. elliottii* pseudo-backcross of 345 full-sib progeny

gene-based map and two other published loblolly pine maps (Echt et al. 2011; Martínez-García et al. 2013) with a linkage map from a pseudo-backcross between loblolly pine and slash pine (*Pinus elliottii*) and constructed a consensus map, which positioned 3,856 markers including SNPs, PAVs, SSRs, ESTPs, and RFLPs on 12 linkage groups based on four pedigrees *qtl-base1* (267 progeny), *qtl-base2* (689 progeny), BC1 (490 progeny), and 10–5 (72 progeny).

With the loblolly pine reference genome and annotation (Neale et al. 2014; Wegrzyn et al. 2014), which can be found at https://treegenesdb.org/FTP/Genomes/Pita/, as well as the advanced sequence capture technique, Lu et al. (2016) designed approximately 2.1 million single strand oligonucleotide probes using 199,723 loblolly pine exons (~49 Mbp) in the loblolly pine reference genome v1.01. Each probe had a length between 55 and 105 bp. The final probe set is available from Roche NimbleGen as custom SeqCap EZ design "140422_Ptaeda_Exome_ML_EZ_HX3". When this probe set was applied to genotype a population of 375 loblolly pine trees, a total of 972,720 high quality SNPs were identified and more than 70% of the captured target bases had at least 10X sequencing depth.

6.3 Dissection of Complex Traits

Complex traits, also known as quantitative traits, show a continuous, as opposed to discrete, distribution of phenotypic values. They do not follow the Mendelian inheritance pattern, namely the genetic segregation of a single gene. Variation in these complex traits are influenced by multiple genes of relatively small effects or by a few genes with large effects and other genes with small effects (Flint and Mackay 2009). Dissecting the genetic architecture of complex traits can help us understand the forces maintaining the variation in phenotypic traits.

Loblolly pine is one of the most studied conifer species in terms of complex traits. Variation in most traits show a polygenic basis with a large number of genes of small effect sizes coupled with environmental and epistatic interactions (Neale and Savolainen 2004), but it is also possible that genes with major effects remain undetected. Dissection of complex traits in loblolly pine began in the 1990s for wood property traits using QTL mapping (Groover et al. 1994). Since the 2000s, studies on loblolly pine and other forest trees have moved to an alternative approach for complex trait dissection —genetic association. Newly generated genomic technologies and resources help us gain insights into the genetic basis of a large number of complex traits through genetic association and QTL mapping. There are significant differences in the number, effect size, genomic location, and frequency of alleles underlying complex trait variation.

For dissecting complex traits in loblolly pine, both QTL mapping and genetic association show strengths and weaknesses. Both approaches require measuring individuals' marker genotype and phenotype. The premise of QTL mapping is that the most predictive markers and causal loci reside in proximity and therefore tend to segregate together after recombination. Hence, QTL mapping can identify the number, position, and effect size of chromosomal regions that affect phenotypic variation in a segregating population. However, the creation of such segregating populations (e.g. F2 and backcross) for loblolly pine requires many years. Moreover, QTL mapping studies are primarily relevant within the pedigrees being evaluated, severely limiting their utility to make broad evolutionary inferences. Genetic association takes advantage of historical recombination to identify trait-marker relationships on the basis of linkage disequilibrium. It is regarded as a more appropriate tool for dissecting complex traits in loblolly pine, which has an outcrossing mating system, ample genetic diversity, and minimal impacts from domestication (Neale and Kremer 2011). Linkage disequilibrium decays rapidly in loblolly pine, thus markers showing genetic association are likely to be located in close proximity to the causative polymorphisms. The identified genetic variants can manifest themselves by affecting protein structure and hence protein function through modification of coding and non-coding genes. However, the identified markers often explain only a small proportion of the genetic variation in complex traits (Brachi et al. 2011; Zuk et al. 2012; Pallares 2019). Therefore, a large sample size is needed to capture a large proportion of the genetic variation (Yengo et al. 2018; López-Cortegano and Caballero 2019). The early genetic association studies of loblolly pine focused on candidate genes or specific genomic regions. The advent of the reference genome and annotation brings an opportunity for genome wide association studies (GWAS) and high-density linkage map construction for loblolly pine.

Next, we will summarize the state of knowledge of the genetic architecture of complex traits in loblolly pine, based on research literature in the last three decades (Table 6.2).

6.3.1 Wood Property Traits

Wood property traits play fundamental roles when it comes to plant growth, environmental stress resistance, and lumber and pulp quality (Peter and Neale 2004). The physical properties of wood mainly affect the quality and end use of sawed lumber, which are evaluated by wood-specific gravity and microfibril angle. The chemical properties of wood mainly affect the pulping process, which are evaluated by the relative amounts of cellulose, hemicellulose, and lignin.

Since the 1990s, a series of QTL mapping studies have been conducted to inform breeding strategy. Groover et al. (1994) identified five QTLs for wood specific gravity. Sewell et al. (2000) detected nine unique QTLs for wood specific gravity, five for volume percentage of

Table 6.2 Summary and description of the QTL mapping and genetic association studies for complex traits of loblolly pine (*Pinus taeda*)

Traits	Authors	Number of individuals of the mapping population	Type of markers*	Method	Main outcomes
Wood property					
	Groover et al. (1994)	177	RFLP	QTL mapping	Five QTLs for wood specific gravity
	Sewell et al. (2000)	172	RFLP	QTL mapping	Nine QTLs for wood specific gravity, five QTLs for volume percentage of late wood, five QTLs for microfibril angle
	Sewell et al. (2002)	172	RFLP	QTL mapping	Eight QTLs for cell wall chemistry
	Brown et al. (2003)	172 + 457 + 445	RFLP, ESTP	QTL mapping	Eight QTLs for wood specific gravity, five QTLs for percentage of late wood, one QTL for cell wall chemistry
	González-Martínez et al. (2007)	435	SNP	Genetic association	Two SNPs for earlywood specific gravity, one SNP for percentage of latewood, and one SNP for earlywood microfibril angle
	Westbrook et al. (2015b)	520	SNP	Genetic association	16 well-supported candidate regulators of resin canal number
Growth					
	González-Martínez et al. (2008)	961	SNP	Genetic association	Four significant associations between carbon isotope discrimination and SNPs on disease and abiotic stress-inducible genes
	Cumbie et al. (2011)	380	SNP	Genetic association	Seven SNPs for carbon isotope discrimination, one SNP for height, and six SNPs for foliar nitrogen concentration
	Lu et al. (2017)	375	SNP	Genetic association	Four SNPs for carbon isotope discrimination, nine SNPs for height, two SNPs for foliar nitrogen concentration, five SNPs for specific leaf area, two SNPs for branch angle, three SNPs for crown width, four SNPs for stem diameter
	De La Torre et al. (2019a)	377	SNP	Genetic association	Two SNPs for carbon isotope discrimination
Disease					
	Wilcox et al. (1996)	386	RAPD	QTL mapping	A major gene for resistance to fusiform rust disease
	Quesada et al. (2010)	498	SNP	Genetic association	Ten SNPs associated with pitch canker resistance
	Lu et al. (2017)	375	SNP	Genetic association	Seven SNPs associated with pitch canker resistance
	De La Torre et al. (2019a)	377	SNP	Genetic association	26 SNPs associated with pitch canker resistance

(continued)

Table 6.2 (continued)

Traits	Authors	Number of individuals of the mapping population	Type of markers*	Method	Main outcomes
Molecular					
	Eckert et al. (2012)	384	SNP	Genetic association	28 single locus associations between 24 and 20 SNPs and metabolites; 2998 multilocus associations between 1617 SNPs and 255 metabolites
	Palle et al. (2013)	400	SNP	Genetic association	88 significant associations between 80 SNPs and 33 xylem development genes
	Lu et al. (2018)	278 for gene expression, 212 for metabolites	SNP	Genetic association	1841 SNPs associated with 191 gene expression mRNA phenotypes, and 524 SNPs associated with 53 metabolite level phenotypes
	De La Torre et al. (2019a)	377	SNP	Genetic association	53 SNPs associated with 12 expression of xylem development genes, and 1513 SNPs associated with 97 metabolite level phenotypes
Environment					
	Eckert et al. (2010a)	907	SNP	Genetic association	Five loci associated with aridity
	Eckert et al. (2010b)	682	SNP	Genetic association	481 correlations between 394 SNPs and five multivariate measures of climate
	Lu et al. (2019)	328	SNP	Genetic association	611 SNPs associated with 56 climate and geographic variables
	De La Torre et al. (2019b)	377	SNP	Genetic association	821 SNPs showed significant association with climate variables

*RFLP—fragment length polymorphism; ESTP—expressed sequence tags polymorphism; RAPD—random amplified polymorphic DNA; SNP—single nucleotide polymorphism

latewood, and five for microfibril angle. Sewell et al. (2002) detected eight unique QTLs for chemical wood properties, namely cell wall chemistry, related to lignin and cellulose content. Loblolly pine trees experience a variety of environmental conditions and developmental stages during their long lifespan, so the expression of some QTLs may be influenced by genotype-by-environment interactions (G × E) and/or the state of development. Brown et al. (2003) identified eight QTLs for wood specific gravity, five for percentage of latewood, and one for cell wall chemistry by repeated detection. The QTLs that were detectable over multiple genetic backgrounds, environments, and growing seasons may be most valuable in a wide range of breeding programs.

With the advancement in genotyping technologies, researchers began to dissect the genetic architecture of wood property traits in loblolly pine. They found wood properties are determined by the activity of the genes involved in xylogenesis (Whetten et al. 2001), and these genes are potential targets for the modification of wood properties (Yang and Loopstra 2005). Using the candidate-gene-based association mapping method, González-Martínez et al. (2007) found four SNPs, which reside in lignification and other wood- and drought-related genes. These genes were associated with wood property traits, such as wood specific gravity, percentage of latewood, and microfibril angle. In the wood of conifer stems, resin canals can help to defend against insects and pathogens by synthesizing and storing oleoresin, a viscous mixture of terpenes (Franceschi et al. 2005). Westbrook et al. (2015b) found that the constitutive development of axial resin canals in loblolly pine stems was regulated by many genes of small effects and these genes were influenced by environmental conditions. They found 16 well-supported candidate genes that were detected across sites and ages, which can map to resin canal number QTLs in an independent population. The identified candidate genes can be used to accelerate genomic selection across environments.

6.3.2 Growth and Biomass Traits

Growth traits, like height and diameter at breast height (DBH), have long been used as fitness proxies in loblolly pine breeding practices. In addition, carbon isotopic discrimination ($\Delta^{13}C$) has been used to estimate water use efficiency (WUE) in forest trees. Plants with higher WUE discriminate less against $\Delta^{13}C$ when they are exposed to the same fluctuations in environmental conditions (Farquhar et al. 1989; Aitken et al. 1995; Baltunis et al. 2008; Cumbie et al. 2011). However, since growth traits are mostly influenced by many genes with each having a small effect, dissecting these traits are difficult tasks, and little progress has been made.

González-Martínez et al. (2008) employed a candidate-gene-based association mapping method and found four SNPs, which reside in disease and abiotic stress-inducible genes, were associated with $\Delta^{13}C$. To increase the SNP numbers, Cumbie et al. (2011) employed a high-throughput SNP array containing 3938 EST-derived SNPs and identified seven SNPs associated with $\Delta^{13}C$, one with height, and six with foliar nitrogen concentration. These SNPs were found in genes such as AP2 transcription factor gene, which was related to $\Delta^{13}C$, and glutamate decarboxylase gene, which was related to foliar nitrogen concentration. Using 2.8 million exome-based SNPs, Lu et al. (2017) found four SNPs associated with $\Delta^{13}C$, two with foliar nitrogen concentration, nine with height, five with specific leaf area, two with branch angle, three with crown width, and four with stem diameter. However, only 5 to 27% of the clonal variance and 2–6% of the phenotypic variance could be explained by the associated SNPs. They also found that non-additive effects imposed by dominance and epistasis account for a larger proportion of the genetic variance for the complex traits than additive effects. De La Torre et al. (2019a) used genome-wide SNPs derived from whole-genome resequencing to run the genetic association and identified two SNPs associated with $\Delta^{13}C$. Since so few loci have been identified, we have little knowledge about the genetic architecture of growth traits.

6.3.3 Resistance to Disease

Dissecting the genetic architecture of disease resistance can help us identify candidate genes for resistance as well as quantify their magnitude of effect. Investigations of disease resistance in loblolly pine have involved fusiform rust and pitch canker. Different genes appear to be involved in resistance to these diseases. The fusiform rust disease is caused by the endemic rust fungus *Cronartium quercuum* f. sp. *fusiforme*. Symptoms of infected trees include branch and stem galls, which subsequently result in reduced wood quality and weakened trees. A major gene was detected for resistance to fusiform rust disease in loblolly pine by genomic mapping (Wilcox et al. 1996), suggesting the resistance to fusiform rust disease is controlled by a few major genes and follows a gene-for-gene model (Kubisiak et al. 2005).

Pitch canker disease is caused by the fungal pathogen *Fusarium circinatum* and is manifested as resinous lesions in stems and branches. Pitch canker disease resistance in loblolly pine appears to be quantitative as a continuous distribution of resistance phenotypes has been observed within a large family-based population (Kayihan et al. 2005). Quantitative resistance is generally considered to be more durable, but the genes and mechanisms of quantitative disease resistance are difficult to study, in part due to the small effects of individual genes underlying the resistance phenotype. Quesada et al. (2010) inoculated 498 largely unrelated, clonally propagated genotypes with the pathogen *Fusarium circinatum* and measured lesion length four, eight, and 12 weeks after inoculation. The high-throughput SNP array containing 3938 EST-derived SNPs was used to test for association with disease resistance phenotype. They identified ten SNPs, which had small effects and putative roles in basal resistance, direct defense, and signal transduction. Using 2.8 million exome-based SNPs, Lu et al. (2017) found seven SNPs associated with pitch canker resistance. De La Torre et al. (2019a) used genome-wide SNPs derived from whole-genome resequencing to run the genetic association and identified 26 SNPs associated with lesion length at eight or 12 weeks post inoculation. The genetic basis of pitch canker disease resistance still remains elusive. Do major genes exist but remain undetected, or is resistance conditioned entirely by genes that exert minor effects? These questions need to be explored further.

6.3.4 Molecular Traits

Some genetic variation does not directly affect whole-plant traits, but instead acts through molecular phenotypes, such as gene expression and metabolite levels, which in turn cause changes in higher-order traits (Zhou et al. 2019; Zhang et al. 2020). Gene expression and metabolite levels vary genetically in populations and are themselves complex traits; thus, studies associating gene expression and metabolite phenotypes with genetic variations may enhance our understanding to the biological processes that underlie variations of whole-plant phenotypes.

An array of genetic association studies has been performed to explore the genetic architecture of molecular phenotypes in loblolly pine. Palle et al. (2013) identified 80 SNPs associated with the expression of 33 xylem development genes, including those involved in cell wall formation, lignin biosynthesis, and transcription factors that are preferentially expressed in loblolly pine xylem tissue. Using the same set of genotyped SNPs and mapping population, Eckert et al. (2012) found metabolite levels were heritable and relatively few SNPs (n = 1–23) explained a large fraction of the heritability of the metabolites. Using more SNPs from exome and whole-genome genotyping, Lu et al. (2018) and De La Torre et al. (2019a) further verified that the molecular phenotypes in loblolly pine have a polygenic basis and there are prevalent pleiotropic effects. Eckert et al. (2013) found that molecular phenotypes appeared to be different with regard to their genetic architecture and patterns of selection compared with whole plant phenotypes. For example, patterns in associated genetic variants for molecular phenotypes displayed segregating deleterious variation, whereas this was not likely the case for whole plant phenotypes. Unfortunately, gene expression and metabolite levels were measured in clones of trees grown in different environments in the aforementioned studies. Future experiments should consider measuring gene expression, metabolite levels, and whole-plant phenotypic traits of the same samples collected at the same time and incorporating their correlations for the biological network inferences.

6.3.5 Environmental Traits

Loblolly pine is distributed across diverse precipitation and temperature gradients from coast to mountains, showing high adaptability to local conditions. The outlier test, a method based on identifying genomic regions of high genetic differentiation among populations, has frequently been used to detect local adaptation patterns in populations. Yet in the case of polygenic adaptation, the outlier test has difficulty detecting those alleles with subtle changes. An alternative method is to treat environment variation as a phenotype and to apply genetic association approaches to link environmental factors and genetic variation. This method is referred to as environmental association, which identifies genetic variants associated with particular environmental factors and has the potential to uncover specific selection pressures that underlie adaptation (Rellstab et al. 2015). Nonetheless, the use of environmental association is often hindered by the high rates of false positives in significant associations because the genetic structure, which is mainly caused by geographic and demographic processes, can lead to patterns that mimic those observed as a consequence of selection. A solution to this issue is to incorporate the genetic structure as a covariate in the environmental association analysis.

The first environmental association study of loblolly pine was conducted by Eckert et al. (2010a) using aridity index, the ratio of precipitation to potential evapotranspiration. A total of 7,216 EST-derived SNPs were genotyped in a mapping population including 907 largely unrelated trees sampled across the natural range of loblolly pine. They studied genetic structure in the studied populations using SSR markers and identified three clusters along the geographical trends: clusters of trees in the Florida-Atlantic region, the Gulf Coastal Plain region, and in the region west of the Mississippi River. After accounting for the genetic structure, they identified five SNPs associated with aridity. The SNPs resided in genes encoding proteins primarily involved with biotic and abiotic stress responses. In another study, Eckert et al. (2010b) associated a genome-wide dataset of SNPs, which were genotyped from 1730 loci in 682 loblolly pine trees sampled from the full range, with five multivariate measures of climate, and identified several well-supported SNPs associated with principal components corresponding to geography, temperature, growing degree-days, precipitation, and aridity. The identified SNPs resided in a diverse set of abiotic stress response genes including those encoding for transmembrane proteins and proteins involved in sugar metabolism. Using 2.8 million whole exome-based SNPs, Lu et al. (2019) identified 611 SNPs associated with 56 climate and geographical variables. Functions of candidate genes related to terpenoid synthesis, pathogen defense, transcription factors, etc. The identified SNPs may also contribute to the genetic architecture of phenotypic traits such as height, diameter, metabolite, and gene expression levels. De La Torre et al. (2019b) tested associations between 87k SNPs, which were obtained from whole-genome resequencing of loblolly pine individuals, and 270 environmental variables and combinations of them. They found that water availability was the main climatic variable shaping local adaptation of the species and identified 821 SNPs associated with climatic variables.

6.4 Lessons We Learned

6.4.1 Many Loci with Small Effects and Pleiotropy

With the exception of fusiform rust resistance, all complex traits studied to date in loblolly pine show a polygenic basis, with differences in number, effect size, and genomic location of alleles contributing to phenotypic variation in different types of traits. This is consistent with the vast majority of traits in other forest trees (Neale and Kremer 2011). It is also possible that genes with major effects remain undetected, but with the deep sequencing targeted on the exome (Lu et al. 2017) and whole genome (De La Torre et al. 2019a), it seems unlikely that we would have missed the genes with major effects for every trait if they exist.

Alleles composing genetic architecture of complex traits in loblolly pine generally have a small effect size. For example, the effects of five identified QTLs explained 23% of the total phenotypic variance of wood specific gravity (Groover et al. 1994). Most QTLs explained < 5% of the phenotypic variance of wood property traits (Brown et al. 2003). Using genetic association, each identified SNP associated with wood property trait explained <5% of the phenotypic variation (González-Martínez et al. 2007). A genetic association study showed that constitutive development of axial resin canals in loblolly pine stems is regulated by many genes with small effects that are influenced by environmental conditions (Westbrook et al. 2015b). For the growth traits, only 5 to 27% of the clonal variance and 2 to 6% of the phenotypic variance could be explained by the robustly associated SNPs (Lu et al. 2017). Such genetic control by many loci with relatively small individual effects for the phenotypic variance of growth traits were also seen in other studies (González-Martínez et al. 2008; Cumbie et al. 2011; De La Torre et al. 2019a). Based on genetic association studies, pitch canker disease resistance in loblolly pine exhibited a quantitative basis with single SNPs explaining <0.1% up to 6% of the phenotypic variance (Quesada et al. 2010; Lu et al. 2017; De La Torre et al. 2019a). Environmental associations suggest that, with the exception of a small number of SNPs with large effect size (related to moisture deficit, temperature, and precipitation), a large number of SNPs with small to moderate effect sizes are associated with climate variables or combinations of them (De La Torre et al. 2019b).

Alleles composing genetic architecture of molecular traits are different from that of other traits in effect size and genomic location. Many more SNPs were found to be associated with molecular traits than with other traits. Up to 448 SNPs were identified to be associated with metabolite phenotypes (De La Torre et al. 2019a) and up to 181 SNPs were identified to be associated with gene expression phenotypes (Lu et al. 2018). Single SNPs explained 9 to 57% of the phenotypic variance of metabolite phenotypes (Eckert et al. 2012; Lu et al. 2018; De La Torre et al. 2019a), and 5 to 85% of the phenotypic variance of gene expression phenotypes (Palle et al. 2013; Lu et al. 2018; De La Torre et al. 2019a). Eckert et al. (2012) reported 93 SNPs associated with five or more metabolites, which suggested the existence of hotspots for metabolites regulation. De La Torre et al. (2019a) verified that widely dispersed metabolite-associated SNPs were found in several distinctive physical clusters (metabolic hotspots). They also found clusters in alleles associated with the expression of xylem development genes.

It seems some genes underlying phenotypic traits are likely involved in adaptability to environment. Some environment-associated SNPs also composed the genetic architecture of phenotypic traits including height, diameter, metabolite, and gene expression phenotypes (Lu et al. 2019). Two genomic clusters, which are comprised of climate-associated alleles of small effect, were also considered as "metabolic hotspots" (De La Torre et al. 2019b).

Pleiotropy refers to the effect of a gene on more than one phenotype and also to the effect of a particular allele on more than one phenotype (Mackay et al. 2009). In genetic association studies of loblolly pine, many SNPs were found to be associated with several gene expression and metabolite phenotypes, suggesting the pleiotropic nature of these SNPs (Lu et al. 2018; De La Torre et al. 2019a). The "hub" SNPs that are associated with multiple gene expression and metabolite phenotypes can be highlighted through gene networks (Lu et al. 2018). One thing worth noting is that for QTL mapping, we should take caution to disentangle close linkage from pleiotropy as the intervals to which QTLs map may contain multiple genes (Mackay et al. 2009). For example, several cell wall chemistry QTLs (Sewell et al. 2002) co-mapped with QTLs for physical wood property traits (Sewell et al. 2000), but there was little phenotypic correlation and little congruence among the QTL effects for these traits, which indicated different QTLs are represented in these QTL clusters.

The current reference genome sequences of loblolly pine have not been assembled into chromosomes, and the linkage map is not complete. We expect a higher density linkage map, and a chromosome-scale reference genome will help to locate more candidate loci and advance our understanding to the genetic architecture of complex traits.

6.4.2 Genotype-by-Environment Interactions, Dominance, and Epistasis

Alleles that comprise complex traits may differ in magnitude or direction under different genetic backgrounds or different environments (Mackay et al. 2009), including genotype-by-environment interactions (G × E), dominance, and genotype-by-genotype interactions (epistasis). These context dependent effects are important as epistasis can help to identify genetic networks that impact complex traits, while G × E can imply the potential mechanisms that maintain the genetic variation of complex traits in natural populations. The context dependent effects might complicate validation of alleles for complex traits as estimates of allelic effects are relevant to the environment and genetic background in which the phenotypes were assessed.

As long-lived trees, loblolly pine trees experience different developmental stages and diverse environmental conditions over their lifespans, which impacts the physiology and growth status and may result in inconsistent QTLs detection. Experimental designs including large numbers of samples across multiple test sites and different growing seasons may help to identify repeatedly detected QTLs, which are most valuable in a wide range of breeding programs. Sewell et al. (2000) detected QTLs for wood property traits in both earlywood and latewood. They found different subsets of QTLs were specific to either earlywood or latewood, and two QTLs were detected for both earlywood and latewood. Sewell et al. (2002) detected QTL × environment interactions for multiple cell wall chemistry components. In a verification test, Brown et al. (2003) found half of QTLs influencing wood specific gravity and percentage of latewood were consistently detected over multiple growing seasons. Using a genetic association approach, Westbrook et al. (2015b) found the majority of resin canal number associations were specific to individual sites and ages. González-Martínez et al. (2008) identified different functional candidate gene types associated with carbon isotope discrimination at two different test sites. Environment may modify the effects of alleles controlling traits by influencing gene expression, epigenomic states, or post-translational modification of proteins (Bollati and Baccarelli 2010; Henikoff and Greally 2016), thus increasing the complexity of complex trait dissection, but we lack evidence of epigenetic regulation in loblolly pine.

Epistasis and dominance are often referred to as non-additive effects, which also contribute to the genetic architecture of complex traits. Epistasis refers to the masking of genotype effects at one locus by the genotype of another locus and also to any statistical interaction between the genotypes at two or more loci (Mackay et al. 2009). Dominance refers to the masking of allele effect at one locus by another allele of the same locus. In loblolly pine, although the additive effects are the main focus for complex trait dissection, studies indicated that non-additive effects (dominance and epistasis) also contributed to the clonal and phenotypic variance (Cumbie et al. 2011; Lu et al. 2017), suggesting that non-additive effects might play important roles in capturing desirable traits. Epistatic SNP-SNP interactions were identified to be associated with gene expression and metabolite phenotypes, indicating clusters of candidate genes in biosynthetic pathways contributing to complex traits (Lu et al. 2018). In QTL mapping studies, Sewell et al. (2002) found some degree of non-additive (i.e. dominant or epistatic) expression for alleles at five cell wall chemistry QTLs as evidenced by the strong non-zero interaction effect at these QTLs.

6.4.3 Population Structure and Selective Forces

Loblolly pine displays clear patterns of local adaptation (Schmidtling 2003), which likely drive environmental association and genotype–phenotype association. However, the selective forces that shape current association patterns remain largely unknown.

Population structure of loblolly pine across its natural range is mainly caused by a dual-Pleistocene refugia that separated populations located east and west of the Mississippi River (Schmidtling 2003; Soltis et al. 2006; Eckert et al. 2010a; Lu et al. 2016). With adaptive and selectively neutral processes, east and west populations became adapted to their distinct environments, resulting in considerable diversity in phenotypic performance across geographically distant populations. Populations east of the Mississippi river grow faster and taller and are less drought tolerant than populations on the west side (Schmidtling 2001; Lu et al. 2017). The population structure can give rise to false positives of genetic association models. Experiments incorporating efficient designs to remove the influence from population structure are used in genetic association studies.

One line of evidence shows that different patterns of selective forces shaped the genetic architecture of molecular phenotypes and whole plant phenotypes. Both segregation of slightly deleterious variants and positive selection can give rise to the pattern in alleles associated with phenotypic traits. Eckert et al. (2013) estimated the proportion of adaptive nonsynonymous substitutions (α), a long-term selection measure, in amplicons (described from ESTs) associated with phenotypic traits. They found α values of whole plant phenotypes (including the environmental associations) were almost twofold greater than the genome-wide average, whereas that of the molecular phenotypes were almost twofold less than the genome-wide average. This result implied selective constraint for amplicons associated with molecular phenotypes, whereas this was not likely the case for whole plant phenotypes.

Another line of evidence shows that mutation selection balance may shape the genetic variation in metabolite phenotypes, while balancing selection may shape the genetic variation in gene expression phenotypes (De La Torre et al. 2019a). Rare alleles (maintained by mutation) and common alleles (maintained by selection) that underlie phenotypic variation can help to imply the evolutionary forces maintaining genetic variation in complex traits (Gibson 2012). De La Torre et al. (2019a) found effect sizes and minor allele frequencies were negatively associated for the majority of metabolite phenotypes, which probably was a result of purifying selection across metabolite networks. This result is consistent with mutation-selection balance theory. On the other hand, effect sizes and minor allele frequencies tended to be positively associated for the SNPs associated with expression of xylem development genes, which is consistent with balancing selection theory.

6.5 Tasks We Need to Consider

6.5.1 Verification

Though loblolly pine can be easily propagated to create large detection and verification populations; the long generation time, size, and cost of experiments have hampered the verification of the results from QTL mapping and genetic association studies.

QTL verification experiments deploying large amounts of clonal material from multiple families over multiple test sites are essential to confirm the prospect of marker-assisted selection (MAS), to provide precise estimates of the magnitude of QTL effects, and to predict whether the QTL will be expressed at a given developmental stage or in a particular environment. In a QTL verification study for wood property traits, Brown et al. (2003) defined QTL verification as repeated detection of QTLs at a similar position on the genetic map under multiple experimental conditions. They found half of the QTLs could be detected in multiple seasons, but fewer QTL were common to different populations. This verification can help to assess how well the genetic architecture of wood property traits have been described; nonetheless, these QTLs still need to be verified in mature wood with a large mapping population.

Nearly all the genetic association studies performed on loblolly pine so far were based solely on a single common garden experiment. The allele frequency distributions are likely to change from population to population. Future experiments should incorporate efficient designs with replicated sampling across multiple sites, in order to validate the genotype–phenotype association and find out the causal variants. For example, the loci associated with carbon isotope discrimination need be verified in stressed environments; loblolly pine genotypes should be inoculated with diverse isolates of the pitch canker fungus to test whether the resistance loci are isolate-specific. For environmental association, we have identified a large number of environment associated alleles and we need to perform fitness measurements to tell if loblolly pine individuals carrying these alleles can better adapt to one environment or not.

Studies based on other conifer species may provide new ideas on verification of genotype–phenotype associations. Cullingham et al. (2020) combined results from population genetics, phenotype association, and gene expression analyses with independent natural and experimental populations as multiple lines of evidence to validate genetic variation related to host susceptibility to mountain pine beetle in lodgepole and jack pine. Liu et al. (2019) proposed a quantitative phenotype model based on core and peripheral genes. They suggested most heritability in complex traits is driven by trans-acting variants, whose effects are mediated through peripheral genes to impact the expression of core genes. A systems genetics approach integrating multiple levels of biological organization can help to understand the genetic pathways that affect the variation of complex traits in loblolly pine.

6.5.2 Gene Characterization

Nearly all complex trait dissection on loblolly pine stopped at presenting a list of interesting loci and lacked follow-ups to characterize functions of candidate genes that underlie complex traits. We can infer the functions of candidate genes if they have high similarity with known genes in other organisms, yet functions of a large number of genes remain unknown because they do not share similarity to known genes. In both cases, the functions of genes merit further investigation.

Molecular experiments have been performed to understand the genes underlying complex traits in a few plant species, but such experiments in loblolly pine have not been performed. For model organisms, mutagenesis screens have been employed to identify the genes required for producing the wild-type phenotype. Alternatively, positional cloning of QTLs was used to reveal the specific genes underlying complex traits. However, for loblolly pine, a gymnosperm species without well-defined mutant systems but with a long generation time for crosses, methods like mutagenesis screens and positional cloning of QTLs are challenging. Cutting-edge technologies revolutionized the bioengineering field, hence providing new options for characterizing genes in non-model organisms. RNAi has been employed in sweetgum to reduce the expression of floral organ identity genes, which led to phenotypes of impaired fertility and modified reproductive development (Klocko et al. 2020). CRISPR/Cas9 has been employed in poplar to generate loss-of-function mutations targeting flowering related genes, which showed promising use for generating reliably infertile trees (Elorriaga et al. 2018). The identification of gene function in loblolly pine and other conifer species using molecular experiments is still an unsolved question.

6.5.3 Exploring Other Genotyping and Phenotyping Methods

With the advancement of new sequencing technologies, we might opt for genotyping other types of genetic polymorphisms. Most recent studies on loblolly pine and other conifer species have focused on SNPs, whereas other forms of polymorphisms, like insertions/deletions (indels), rearrangements, copy number variations (CNVs), have long been overlooked. However, analyzing complex traits using indels and CNVs may bring novel insight to the genetic architecture of complex traits, especially as the improved method has been developed for linking a wide range of genetic variants with phenotypes for plants (Voichek and Weigel 2020). Prunier et al. (2017b) identified CNVs in spruce species and suggested CNVs play a role in climate adaptation (Prunier et al. 2017a). As individual genotyping for large populations is still expensive, pooling individuals based on phenotypes for genotyping has shown to be an effective and efficient method for discovering loci underlying phenotypes (Schlötterer et al. 2014). Combining pool-seq and exome capture methods, Rellstab et al. (2019) estimated allele frequencies of population pools in Swiss stone pine (*Pinus cembra*) at a reasonable cost. The output can be used to identify candidate loci for future studies based upon allele frequency differences.

While we step into the high-throughput genotyping era, the phenotyping of loblolly pine and other forest tree species has become one of the major bottlenecks in the dissection of complex traits. Phenotyping young seedlings of loblolly pine in laboratory environments has limited value. Foresters work in outdoor plantations to measure trees, including tree height, diameter, presence or absence of disease, and other traits. However, it is not an easy task to measure mature and large forest trees in the field. The development of high-throughput phenotyping in crop species may bring solutions to phenotyping loblolly pine trees (Yang et al. 2020). For example, remote sensing technology using drones or unmanned aerial vehicles equipped with multiple sensors can be used for rapid measurement for crop growth, yield, stress response (Maes and Steppe 2019). Multidisciplinary cooperation among geneticists, physiologists, ecologists, and engineers is urgently needed to develop precise, standardized, and feasible high-throughput phenotyping technologies for forest tree phenotyping.

6.6 Concluding Remarks

Loblolly pine is one of the most studied conifer species due to its great economic and ecological value. Over the past three decades, the intensive efforts from a series of research projects, and dedicated researchers have resulted in a wealth of genomic resources, including the reference genome sequences, transcriptomes, gene annotation, high-density linkage maps, and candidate loci associated with a variety of traits. These are useful tools to practice genomic selection and marker-assisted breeding in loblolly pine. With the ever-advancing "omics" and bioengineering technologies, we expect to discover and validate more genetic variations associated with complex traits, as well as to characterize the functions of some key genes. In doing so, we can accelerate the breeding practices that satisfy major economic needs and develop diagnostic tools for the management of natural and breeding populations.

References

Adams WT, Joly RJ (1980a) Linkage relationships among twelve allozyme loci in loblolly pine. J Hered 71 (3):199–202

Adams WT, Joly RJ (1980b) Genetics of allozyme variants in loblolly pine. J Hered 71(1):33–40

Aitken SN, Kavanagh KL, Yoder BJ (1995) Genetic variation in seedling water-use efficiency as estimated by carbon isotope ratios and its relationship to sapling growth in Douglas-fir. For Genet 2(4):199–206

Baker JB, Langdon OG (1990) *Pinus taeda* L., loblolly pine. In: Burns RM, Honkala BH (eds) Silvics of North America: 1. Conifers. Agriculture Handbook 654. U.S. Department of Agriculture, Forest Service, Washington, D.C., pp 497–512

Baltunis BS, Martin TA, Huber DA, Davis JM (2008) Inheritance of foliar stable carbon isotope discrimination and third-year height in *Pinus taeda* clones on contrasting sites in Florida and Georgia. Tree Genet Genomes 4(4):797–807

Bollati V, Baccarelli A (2010) Environmental epigenetics. Heredity 105(1):105–112

Brachi B, Morris GP, Borevitz JO (2011) Genome-wide association studies in plants: the missing heritability is in the field. Genome Biol 12(10):232

Brown GR, Kadel EE, 3rd, Bassoni DL, Kiehne KL, Temesgen B, Van Buijtenen JP, Sewell MM, Marshall KA, Neale DB (2001) Anchored reference loci in loblolly pine (Pinus taeda L.) for integrating pine genomics. Genetics 159(2): 799–809

Brown GR, Bassoni DL, Gill GP, Fontana JR, Wheeler NC, Megraw RA, Davis MF, Sewell MM, Tuskan GA, Neale DB (2003) Identification of quantitative trait loci influencing wood property traits in loblolly pine (*Pinus taeda* L.). III. QTL verification and candidate gene mapping. Genetics 164(4): 1537–1546

Cullingham CI, Peery RM, Fortier CE, Mahon EL, Cooke JEK, Coltman DW (2020) Linking genotype to phenotype to identify genetic variation relating to host susceptibility in the mountain pine beetle system. Evol Appl 13(1):48–61

Cumbie WP, Eckert A, Wegrzyn J, Whetten R, Neale D, Goldfarb B (2011) Association genetics of carbon isotope discrimination, height and foliar nitrogen in a natural population of *Pinus taeda* L. Heredity 107 (2):105–114

De La Torre AR, Puiu D, Crepeau MW, Stevens K, Salzberg SL, Langley CH, Neale DB (2019) Genomic architecture of complex traits in loblolly pine. New Phytol 221(4):1789–1801

De La Torre AR, Wilhite B, Neale DB (2019) Environmental genome-wide association reveals climate Adaptation is shaped by subtle to moderate allele frequency shifts in loblolly pine. Genome Biol Evol 11(10):2976–2989

Devey ME, Fiddler TA, Liu BH, Knapp SJ, Neale DB (1994) An RFLP linkage map for loblolly pine based on a three-generation outbred pedigree. Theor Appl Genet 88(3–4):273–278

Echt CS, Saha S, Krutovsky KV, Wimalanathan K, Erpelding JE, Liang C, Nelson CD (2011) An annotated genetic map of loblolly pine based on microsatellite and cDNA markers. BMC Genet 12:17

Eckert AJ, Pande B, Ersoz ES, Wright MH, Rashbrook VK, Nicolet CM, Neale DB (2009) High-throughput genotyping and mapping of single nucleotide polymorphisms in loblolly pine (*Pinus taeda* L.). Tree Genet Genomes 5(1):225–234

Eckert AJ, van Heerwaarden J, Wegrzyn JL, Nelson CD, Ross-Ibarra J, González-Martínez SC, Neale DB (2010a) Patterns of population structure and environmental associations to aridity across the range of loblolly pine (*Pinus taeda* L., Pinaceae). Genetics 185 (3):969–982

Eckert AJ, Bower AD, González-Martínez SC, Wegrzyn JL, Coop G, Neale DB (2010) Back to nature: ecological genomics of loblolly pine (*Pinus taeda*, Pinaceae). Mol Ecol 19(17):3789–3805

Eckert AJ, Wegrzyn JL, Cumbie WP, Goldfarb B, Huber DA, Tolstikov V, Fiehn O, Neale DB (2012) Association genetics of the loblolly pine (*Pinus taeda*, Pinaceae) metabolome. New Phytol 193(4):890–902

Eckert AJ, Wegrzyn JL, Liechty JD, Lee JM, Cumbie WP, Davis JM, Goldfarb B, Loopstra CA, Palle SR, Quesada T, Langley CH, Neale DB (2013) The evolutionary genetics of the genes underlying phenotypic associations for loblolly pine (*Pinus taeda*, Pinaceae). Genetics 195(4):1353–1372

Elorriaga E, Klocko AL, Ma C, Strauss SH (2018) Variation in mutation spectra among CRISPR/Cas9 mutagenized poplars. Front Plant Sci 9:594

Farquhar GD, Ehleringer JR, Hubick KT (1989) Carbon isotope discrimination and photosynthesis. Annu Rev Plant Physiol Plant Mol Biol 40(1):503–537

Flint J, Mackay TF (2009) Genetic architecture of quantitative traits in mice, flies, and humans. Genome Res 19(5):723–733

Franceschi VR, Krokene P, Christiansen E, Krekling T (2005) Anatomical and chemical defenses of conifer bark against bark beetles and other pests. New Phytol 167(2):353–376

Gibson G (2012) Rare and common variants: twenty arguments. Nat Rev Genet 13(2):135–145

González-Martínez SC, Ersoz E, Brown GR, Wheeler NC, Neale DB (2006) DNA sequence variation and selection of tag single-nucleotide polymorphisms at candidate genes for drought-stress response in *Pinus taeda* L. Genetics 172(3):1915–1926

González-Martínez SC, Wheeler NC, Ersoz E, Nelson CD, Neale DB (2007) Association genetics in *Pinus taeda* L. I. Wood Property Traits. Genetics 175(1):399–409

González-Martínez SC, Huber D, Ersoz E, Davis JM, Neale DB (2008) Association genetics in *Pinus taeda* L. II. Carbon Isotope Discrimination. Heredity 101 (1):19–26

Groover A, Devey M, Fiddler T, Lee J, Megraw R, Mitchel-Olds T, Sherman B, Vujcic S, Williams C, Neale D (1994) Identification of quantitative trait loci influencing wood specific gravity in an outbred pedigree of loblolly pine. Genetics 138(4):1293–1300

Harry DE, Temesgen B, Neale DB (1998) Codominant PCR-based markers for *Pinus taeda* developed from mapped cDNA clones. Theor Appl Genet 97(3):327–336

Henikoff S, Greally JM (2016) Epigenetics, cellular memory and gene regulation. Curr Biol 26(14): R644–R648

Johnsen K, Teskey B, Samuelson L, Butnor J, Sampson D, Sanchez F, Maier C, Mckeand S (2004) Carbon sequestration in loblolly pine plantations: Methods, limitations, and research needs for estimating storage pools. In: Gen. Tech. Rep. SRS–75. U.S. Department of Agriculture, Forest Service, Southern Research Station, Asheville, NC, pp 373–381

Kayihan GC, Huber DA, Morse AM, White TL, Davis JM (2005) Genetic dissection of fusiform rust and pitch canker disease traits in loblolly pine. Theor Appl Genet 110(5):948–958

Klocko AL, Brunner AM, Ma C, Etherington E, Rosenstiel K, Magnuson A, Taylor BJ, Cappellazzi J, Lockwood T, Covarrubias N, Bao M, Morrell JJ, Strauss SH, Bao M, Covarrubias N (2020) RNAi of AGAMOUS genes in sweetgum alters reproductive organ identity and decreases fruit persistence. Plant Direct 4:1–16

Krutovsky KV, Troggio M, Brown GR, Jermstad KD, Neale DB (2004) Comparative mapping in the Pinaceae. Genetics 168(1):447–461

Kubisiak TL, Amerson HV, Nelson CD (2005) Genetic interaction of the fusiform rust fungus with resistance gene *Fr*1 in loblolly pine. Phytopathology 95(4):376–380

Liu X, Li YI, Pritchard JK (2019) Trans effects on gene expression can drive omnigenic inheritance. Cell 177 (4):1022-1034.e6

López-Cortegano E, Caballero A (2019) Inferring the nature of missing heritability in human traits using data from the GWAS catalog. Genetics 212(3):891–904

Lu M, Krutovsky KV, Nelson CD, Koralewski TE, Byram TD, Loopstra CA (2016) Exome genotyping, linkage disequilibrium and population structure in loblolly pine (*Pinus taeda* L.). BMC Genomics 17 (1):730

Lu M, Krutovsky KV, Nelson CD, West JB, Reilly NA, Loopstra CA (2017) Association genetics of growth and adaptive traits in loblolly pine (*Pinus taeda* L.) using whole-exome-discovered polymorphisms. Tree Genet Genomes 13:57

Lu M, Seeve CM, Loopstra CA, Krutovsky KV (2018) Exploring the genetic basis of gene transcript abundance and metabolite levels in loblolly pine (*Pinus taeda* L.) using association mapping and network construction. BMC Genetics 19(1):100

Lu M, Loopstra CA, Krutovsky KV (2019) Detecting the genetic basis of local adaptation in loblolly pine (*Pinus taeda* L.) using whole exome-wide genotyping and an integrative landscape genomics analysis approach. Ecol Evol 9(12):6798–6809

Mackay TFC, Stone EA, Ayroles JF (2009) The genetics of quantitative traits: challenges and prospects. Nat Rev Genet 10(8):565–577

Maes WH, Steppe K (2019) Perspectives for remote sensing with unmanned aerial vehicles in precision agriculture. Trends Plant Sci 24(2):152–164

Martínez-García PJ, Stevens KA, Wegrzyn JL, Liechty J, Crepeau M, Langley CH, Neale DB (2013) Combination of multipoint maximum likelihood (MML) and regression mapping algorithms to construct a high-density genetic linkage map for loblolly pine (*Pinus taeda* L.). Tree Genet Genomes 9:1529–1535

Neale DB, Savolainen O (2004) Association genetics of complex traits in conifers. Trends Plant Sci 9(7):325–330

Neale DB, Kremer A (2011) Forest tree genomics: growing resources and applications. Nat Rev Genet 12(2):111–122

Neale DB, Wegrzyn JL, Stevens KA, Zimin AV, Puiu D, Crepeau MW, Cardeno C, Koriabine M, Holtz-Morris AE, Liechty JD, Martínez-García PJ, Vasquez-Gross HA, Lin BY, Zieve JJ, Dougherty WM, Fuentes-Soriano S, Wu LS, Gilbert D, Marçais G, Roberts M, Holt C, Yandell M, Davis JM, Smith KE, Dean JF, Lorenz WW, Whetten RW, Sederoff R, Wheeler N, Mcguire PE, Main D, Loopstra CA, Mockaitis K, Dejong PJ, Yorke JA, Salzberg SL, Langley CH (2014) Decoding the massive genome of loblolly pine using haploid DNA and novel assembly strategies. Genome Biol 15(3):R59

Neves LG, Davis JM, Barbazuk WB, Kirst M (2013) Whole-exome targeted sequencing of the uncharacterized pine genome. Plant J 75(1):146–156

Neves LG, Davis JM, Barbazuk WB, Kirst M (2014) A high-density gene map of loblolly pine (*Pinus taeda* L.) based on exome sequence capture genotyping. G3 (Bethesda) 4(1):29–37

Pallares LF (2019) Searching for solutions to the missing heritability problem. eLife 8:e53018

Palle SR, Seeve CM, Eckert AJ, Wegrzyn JL, Neale DB, Loopstra CA (2013) Association of loblolly pine xylem development gene expression with single-nucleotide polymorphisms. Tree Physiol 33(7):763–774

Peter G, Neale D (2004) Molecular basis for the evolution of xylem lignification. Curr Opin Plant Biol 7(6):737–742

Prasad AM, Iverson LR, Matthews S, Peters M (2007-ongoing) A climate change atlas for 134 forest tree species of the eastern United States [database]. https://www.nrs.fs.fed.us/atlas/tree. Northern Research Station, USDA Forest Service, Delaware, Ohio

Prunier J, Caron S, Lamothe M, Blais S, Bousquet J, Isabel N, Mackay J (2017) Gene copy number variations in adaptive evolution: The genomic distribution of gene copy number variations revealed by genetic mapping and their adaptive role in an undomesticated species, white spruce (*Picea glauca*). Mol Ecol 26(21):5989–6001

Prunier J, Caron S, Mackay J (2017b) CNVs into the wild: screening the genomes of conifer trees (*Picea spp.*) reveals fewer gene copy number variations in hybrids and links to adaptation. BMC Genomics 18 (1):97

Quesada T, Gopal V, Cumbie WP, Eckert AJ, Wegrzyn JL, Neale DB, Goldfarb B, Huber DA, Casella G, Davis JM (2010) Association mapping of quantitative disease resistance in a natural population of loblolly pine (*Pinus taeda* L.). Genetics 186 (2):677–686

Rellstab C, Gugerli F, Eckert AJ, Hancock AM, Holderegger R (2015) A practical guide to environmental association analysis in landscape genomics. Mol Ecol 24(17):4348–4370

Rellstab C, Dauphin B, Zoller S, Brodbeck S, Gugerli F (2019) Using transcriptome sequencing and pooled exome capture to study local adaptation in the giga-genome of *Pinus cembra*. Mol Ecol Resour 19 (2):536–551

Remington DL, Whetten RW, Liu BH, O'Malley DM (1999) Construction of an AFLP genetic map with nearly complete genome coverage in *Pinus taeda*. Theor Appl Genet 98:1279–1292

Schlötterer C, Tobler R, Kofler R, Nolte V (2014) Sequencing pools of individuals—mining genome-wide polymorphism data without big funding. Nat Rev Genet 15(11):749–763

Schmidtling RC (2001) Southern pine seed sources. General technical report SRS-44. USDA Forest Service, Southern Research Station, Asheville, NC, pp 6–9

Schmidtling RC (2003) Determining seed transfer guidelines for southern pines. In: Riley LE, Dumroese RK, Landis TD (eds) National proceedings: forest and conservation nursery associations-2002. Proceedings RMRS-P-28. USDA Forest Service, Rocky Mountain Research Station, Ogden, UT, pp 8–11

Sewell MM, Sherman BK, Neale DB (1999) A consensus map for loblolly pine (*Pinus taeda* L.). I. Construction and integration of individual linkage maps from two outbred three-generation pedigrees. Genetics 151 (1):321–330

Sewell MM, Bassoni DL, Megraw RA, Wheeler NC, Neale DB (2000) Identification of QTLs influencing wood property traits in loblolly pine (*Pinus taeda* L.). I. Physical wood properties. Theor Appl Genet 101 (8):1273–1281

Sewell MM, Davis MF, Tuskan GA, Wheeler NC, Elam CC, Bassoni DL, Neale DB (2002) Identification of QTLs influencing wood property traits in loblolly pine (*Pinus taeda* L.). II. Chemical wood properties. Theor Appl Genet 104(2):214–222

Soltis DE, Morris AB, McLachlan JS, Manos PS, Soltis PS (2006) Comparative phylogeography of unglaciated eastern North America. Mol Ecol 15(14):4261–4293

Sturtevant AH (1913) The linear arrangement of six sex-linked factors in Drosophila, as shown by their mode of association. J Exp Zool 14:43–59

Temesgen B, Brown GR, Harry DE, Kinlaw CS, Sewell MM, Neale DB (2001) Genetic mapping of expressed sequence tag polymorphism (ESTP) markers in loblolly pine (*Pinus taeda* L.). Theor Appl Genet 102(5):664–675

Voichek Y, Weigel D (2020) Identifying genetic variants underlying phenotypic variation in plants without complete genomes. Nat Genet 52(5):534–540

Wegrzyn JL, Liechty JD, Stevens KA, Wu LS, Loopstra CA, Vasquez-Gross HA, Dougherty WM, Lin BY, Zieve JJ, Martínez-García PJ, Holt C, Yandell M, Zimin AV, Yorke JA, Crepeau MW, Puiu D, Salzberg SL, de Jong PJ, Mockaitis K, Main D, Langley CH, Neale DB (2014) Unique features of the loblolly pine (*Pinus taeda* L.) megagenome revealed through sequence annotation. Genetics 196(3):891–909

Westbrook JW, Chhatre VE, Wu L-S, Chamala S, Neves LG, Muñoz P, Martínez-García PJ, Neale DB, Kirst M, Mockaitis K, Nelson CD, Peter GF, Davis JM, Echt CS (2015a) A consensus genetic map for *Pinus taeda* and *Pinus elliottii* and extent of linkage disequilibrium in two genotype-phenotype discovery populations of *Pinus taeda*. G3 (Bethesda) 5 (8):1685–1694

Westbrook JW, Walker AR, Neves LG, Munoz P, Resende MFR Jr, Neale DB, Wegrzyn JL, Huber DA, Kirst M, Davis JM, Peter GF (2015) Discovering candidate genes that regulate resin canal number in *Pinus taeda* stems by integrating genetic analysis across environments, ages, and populations. New Phytol 205(2):627–641

Whetten R, Sun Y-H, Zhang Y, Sederoff R (2001) Functional genomics and cell wall biosynthesis in loblolly pine. Plant Mol Biol 47(1–2):275–291

Wilcox PL, Amerson HV, Kuhlman EG, Liu BH, O'Malley DM, Sederoff RR (1996) Detection of a major gene for resistance to fusiform rust disease in loblolly pine by genomic mapping. Proc Natl Acad Sci U S A 93(9):3859–3864

Xiong JS, Mckeand SE, Isik F, Wegrzyn J, Neale DB, Zeng Z-B, da Costa e Silva L, Whetten RW, (2016) Quantitative trait loci influencing forking defects in an outbred pedigree of loblolly pine. BMC Genet 17 (1):138

Yang S-H, Loopstra CA (2005) Seasonal variation in gene expression for loblolly pines (*Pinus taeda*) from different geographical regions. Tree Physiol 25 (8):1063–1073

Yang W, Feng H, Zhang X, Zhang J, Doonan JH, Batchelor WD, Xiong L, Yan J (2020) Crop phenomics and high-throughput phenotyping: Past decades, current challenges, and future perspectives. Mol Plant 13(2):187–214

Yengo L, Sidorenko J, Kemper KE, Zheng Z, Wood AR, Weedon MN, Frayling TM, Hirschhorn J, Yang J, Visscher PM, GIANT Consortium (2018) Meta-analysis of genome-wide association studies for height and body mass index in ~700000 individuals of European ancestry. Hum Mol Genet 27(20):3641–3649

Zhang L, Yu Y, Shi T, Kou M, Sun J, Xu T, Li Q, Wu S, Cao Q, Hou W, Li Z (2020) Genome-wide analysis of expression quantitative trait loci (eQTLs) reveals the regulatory architecture of gene expression variation in the storage roots of sweet potato. Hortic Res 7:90

Zhou S, Kremling KA, Bandillo N, Richter A, Zhang YK, Ahern KR, Artyukhin AB, Hui JX, Younkin GC, Schroeder FC, Buckler ES, Jander G (2019) Metabolome-scale genome-wide association studies reveal chemical diversity and genetic control of maize specialized metabolites. Plant Cell 31(5):937–955

Zhou Y, Gwaze DP, Reyes-Valdés MH, Bui T, Williams CG (2003) No clustering for linkage map based on low-copy and undermethylated microsatellites. Genome 46(5):809–816

Zuk O, Hechter E, Sunyaev SR, Lander ES (2012) The mystery of missing heritability: Genetic interactions create phantom heritability. Proc Natl Acad Sci U S A 109(4):1193–1198

Genomics of Disease Resistance in Loblolly Pine

7

Daniel Ence, Tania Quesada, Jeremy T. Brawner, Gary F. Peter, C. Dana Nelson,and John M. Davis

Abstract

Decades of genetic research and breeding in southern pines, principally focused on loblolly pine (*Pinus taeda* L.) and slash pine (*Pinus elliottii* var. *elliottii*), have resulted in germplasm with improved resistance to the fungal pathogens that cause fusiform rust and pitch canker disease. Fusiform rust resistance is regulated by gene-for-gene, or major gene, resistance. Multiple clusters of major fusiform rust resistance genes (*Fr* genes) have been genetically mapped in loblolly pine. One candidate *Fr* gene is predicted to encode a TIR-NBS-LRR protein. This suggests a similar model of NLR evolution to that found in model plants where NLR gene clusters evolve through homologous recombination mediated by transposable elements. In contrast, studies of pitch canker resistance have identified no major resistance genes, supporting a model of quantitative resistance. For loblolly pine, candidate genes involved in resistance to fusiform rust and pitch canker are summarized and functional annotation data are provided along with relevant studies from other *Pinus* spp. Avenues of future research are discussed including advanced genomic technologies for simultaneously characterizing resistance genes in the host and avirulence genes in the pathogen and the development of prediction models of resistance and virulence from heterogeneous data sources.

D. Ence · T. Quesada · G. F. Peter · J. M. Davis (✉)
School of Forest, Fisheries, and Geomatics Sciences, University of Florida, Gainesville, FL, USA
e-mail: jmdavis@ufl.edu

D. Ence
e-mail: d.ence@ufl.edu

T. Quesada
e-mail: tquesada@ufl.edu

G. F. Peter
e-mail: gfpeter@ufl.edu

J. T. Brawner
Department of Plant Pathology, University of Florida, Gainesville, FL, USA
e-mail: jeremybrawner@ufl.edu

C. D. Nelson
USDA Forest Service, Southern Research Station, Southern Institute of Forest Genetics, Saucier, MS, USA
e-mail: charles.d.nelson@usda.gov

C. D. Nelson
Forest Health Research and Education Center, Lexington, KY, USA

7.1 Introduction

Resistance to fusiform rust and pitch canker is of great economic and ecological importance in southern pines, including loblolly pine (*Pinus taeda* L.) and slash pine (*Pinus elliottii* var *elliottii*). The architecture of resistance to these two fungal diseases is being characterized, building on decades of breeding and genetics research. In

A. R. De La Torre (ed.), *The Pine Genomes*, Compendium of Plant Genomes,
https://doi.org/10.1007/978-3-030-93390-6_7

contrast to the evidence for gene-for-gene resistance to the fusiform rust pathogen, current evidence supports a model of quantitative genetic resistance to the pitch canker pathogen. The assembled loblolly pine genome has already improved our understanding of DNA marker association for genes involved in resistance to both diseases leading to the identification of candidate resistance genes.

This chapter describes the current understanding of the genomics of resistance to fusiform rust and pitch canker diseases, principally in loblolly pine but also with reference to other pine species. We describe the infrastructure and resources that support large-scale analysis of disease resistance in greenhouse screening and field trials. We then describe the genomic evidence in loblolly pine for qualitative or major gene resistance for fusiform rust, and the evidence supporting a quantitative or polygenic model of resistance to pitch canker. We detail the candidate genes suspected to be involved in resistance to both diseases and the genomic evidence supporting them. Finally, we conclude with a discussion of potential applications made possible with advancements of sequencing and other genomics technologies that may be used to improve our understanding and management of resistance in these pathosystems.

7.2 Genetics of Disease Resistance in Loblolly Pine

7.2.1 Introduction

In the United States, forestry has a very large economic and ecological impact. Forestry supports rural economies at scale with annual values on par with that of major row crops such as corn and soybeans (McKeever and Howard 1996; Howard and Liang 2019). Forest products represent approximately 1.5% of the overall U.S. economy and contribute approximately 5% of the country's manufacturing outputs (Alvarez 2018). In the southern region, loblolly pine is the most widely planted conifer, accounting for three-quarters of the annual seedling production in the U.S., with greater than 95% of those seedlings being genetically improved (McKeand et al. 2003). In addition to plantations, loblolly pine is ecologically significant in the region, as it is the predominant forest tree species on over ten million hectares of native forest (Baker and Langdon 1990) providing important ecosystem services that impact the welfare of nearly 400 species of vertebrates (Schultz 1999). Loblolly pine is also host to two fungal pathogens endemic to the southeastern U.S. that are economically and ecologically important—*Cronartium quercuum* Berk. Miyable ex Shirai f. sp. *fusiforme* (referred to as Cqf; (Burdsall and Snow 1977), and *Fusarium circinatum* Nirenberg et O'Donnell (referred to as Fc; (Nirenberg and O'Donnell 1998). These pathogens demonstrate contrasting life histories and both cause severe damage in pine stands. Cqf is an obligate biotrophic pathogen that incites fusiform rust, a disease with estimated losses of $134 million each year (Cubbage et al. 2000). In contrast, Fc is a facultative necrotrophic pathogen that incites pitch canker, a disease that is highly episodic with most serious damage often associated with conditions of elevated nitrogen levels (Barnard and Blakeslee 2006).

7.2.2 Phenotyping for Disease Resistance

An important feature that underlies all studies of genetic resistance to disease in loblolly pine, is the importance of comprehensive and accurate phenotyping. Disease phenotypes in the field are variable across sites often occurring more in some years than others. Artificial inoculation under controlled conditions is a more consistent and repeatable means to evaluate disease resistance. Another benefit is that uninfected "escapes" can be minimized to provide more accurate estimates of genetic parameters and breeding values for resistance than under field test conditions. Despite more variability in the field, southern pine breeding programs use fusiform rust gall phenotypes from progeny tests to select for more resistant parents and progeny and rankings are similar between field and artificial inoculations (Isik et al. 2008; Walker and McKeand 2017). A successful public–private partnership is the Resistance Screening Center (RSC) near Asheville, NC which plays a pivotal role in ensuring high-quality phenotyping of pine seedlings for resistance to fusiform rust or pitch canker. The RSC was established by the USDA Forest Service nearly 50 years ago (Cowling and Young 2013), and screens seed lots for university and industrial scientists using spores systematically collected from across the entire region. In the case of fusiform rust resistance screening, the inoculations are complicated by the need to first inoculate Cqf aeciospores onto the uredinial (i.e., "repeating") oak host, where Cqf undergoes meiosis and produces the basidiospores that can infect pine seedlings. This requires careful attention to seedling quality and timing to ensure the phenology of the oak and pine seedlings are optimal for infection of both hosts, and at the appropriate time. In the case of pitch canker screening, the inoculations take the form of a wound inoculation method on hedged seedlings to promote Fc infection (see Fig. 7.1a). In both cases, the RSC has standard phenotyping protocols that allow the comparison of test materials with both positive (susceptible) and negative (resistant) control "check lots" to ensure that the screen was successful. The RSC is highly effective at supporting all breeding efforts, because it has created valuable services that can be carried out at low cost for collaborators.

Fig. 7.1 a Wounded seedlings with pitch canker disease, **b** Pitch canker symptoms on mature tree, **c** Seedling with fusiform rust gall producing pycniospores, and **d** Fusiform rust gall producing aeciospores on mature tree. Photos a, b, and c taken by T. Quesada. Photo d taken by J.M. Davis

7.2.3 Status of Breeding for Disease Resistance

Genetic resistance to fusiform rust and to pitch canker have been the goals of breeding efforts in the region for several decades. Southern pine breeding is one example of how public–private partnerships can strengthen fundamental forest science, and promote its immediate applications to stakeholder needs. Most forest products companies with land holdings in the region belong to one or more cooperatives hosted by a university in the region. Themes of forestry-related cooperatives vary widely, but there are three in the region focused on genetic improvement of loblolly pine, hosted at North Carolina State University, Texas A&M Forest Service, and the University of Florida. Both companies and universities have seed orchards and parental archives that can be leveraged by university researchers to carry out experiments to identify trees with desirable alleles, and to select for the valuable traits they express in the field. Much of the land for these studies is owned and managed by cooperative members or other industrial partners. These cooperatives and partners enable experiments to be conducted at a scale that would be difficult if not impossible to carry out by a single organization. One example is the study known by the acronym CCLONES (Comparing Clonal Lines On Experimental Sites; (Baltunis et al. 2005), which leveraged a complex but highly efficient mating design that has proven to be powerful in dissecting the genetic architecture of resistance to fusiform rust and pitch canker (Kayihan et al. 2005), to train and test the accuracy of genomic selection methods in loblolly pine (Resende Junior et al. 2012), and to identify a candidate gene for pathotype-specific resistance to fusiform rust (Neale et al. 2014). Another example is an association population of unrelated loblolly pine genotypes known by the acronym ADEPT2 (Allele Discovery for Economic Pine Traits), which was a powerful clonal population that enabled the identification of candidate genes for resistance to pitch canker (Quesada et al. 2010).

The success of breeding to improve fusiform rust resistance has resulted from data collected over the past decades by university/industry cooperatives, where the data have guided parental selection, mate allocation, and seedling deployment (Schmidt 2003; Walker and McKeand 2017). The durability of disease resistance is an unanswered question in this pathosystem. In contrast to most agricultural crops where rust pathogens cause major economic losses due to strong selection for virulence on the uredinial (repeating stage) host in a single season, loblolly pine is the aecial (not the repeating stage) host of fusiform rust, which may increase durability of resistance. Consistent with this view, fusiform rust resistance ratings appear to be generally consistent across locations within the range of loblolly pine, suggesting that genetic resistance has not been defeated to any noticeable extent (Spitzer et al. 2017). On the other hand, we have only begun to unravel the genetic architecture of fusiform rust resistance, where it appears there are resistance gene clusters that may segregate, assort, and evolve in complex ways (Amerson et al. 2015). Cqf is highly variable and appears to evolve quickly (Anderson et al. 2010; Pendleton et al. 2014). Current research objectives are to track the transmission of specific resistance genes through existing pedigrees, in the fourth breeding cycle, to assess whether and how the resistance genes mapped to date have been introgressed and/or "stacked" in elite materials, and to continue to monitor field sites for potential breakdown in predicted levels of fusiform rust resistance. In conjunction with phenotyping at the RSC, genetic markers will be used to track the presence and efficacy of pathotype-specific resistance and avirulence genes to guide seedling deployment to the landscape with appropriate spatial and temporal resolution.

The potential of breeding efforts to improve pitch canker resistance in loblolly pine has been demonstrated by estimating its heritability and family rankings or breeding values and identifying candidate genes (Kayihan et al. 2005; Quesada et al. 2010). These results suggest that the current breeding programs have useful levels of genetic resistance and families can be selected for deployment into areas where the pathogen is problematic. Many open questions remain, including whether Fc has evolved virulence to defeat what appears to be quantitative resistance in the host. Its life history as a facultative pathogen suggests there would be little selection pressure for virulence, however, we are only now beginning to characterize the Fc genomic variability present in the population (Fulton et al. 2020). More sampling of Fc genotypes in the field, coupled with screening at the RSC, is needed to assess risk of genetic resistance being defeated (Quesada et al. 2019). Interestingly, most of the research in this regard has been focused on pine species that are highly susceptible to pitch canker disease, notably Monterey or radiata pine (*Pinus radiata* D. Don; (Zamora-Ballesteros et al. 2019). A breeding approach that involves the entire suite of genes available through interspecific hybridization may be a path forward in highly susceptible species (Hodge and Dvorak 2000; Martín-García et al. 2019).

7.3 Genomics of Resistance to Fusiform Rust

7.3.1 Introduction

Fusiform rust is a disease of loblolly and slash pines caused by the endemic fungus, Cqf. Cqf has a complex life-cycle with five spore types that alternate between southern pines (*Pinus spp.*) and southern oaks (*Quercus* spp.). In the repeating stage of the life-cycle on the oak host, Cqf grows telia on leaves which produce basidiospores that infect the pine host. In the pine host, Cqf forms galls that produce pycniospores and aeciospores (see Fig. 7.1c, d). Aeciospores infect the oak host. Because both the oak and pine hosts are widespread throughout the southeastern U.S., genetic resistance has been accepted as the most feasible method of control of infection.

Resistance for fusiform rust has been a major focus of pine breeding programs resulting in improved pine families with higher rust resistance (Cubbage et al. 2000; McKeand et al. 2003; Isik et al. 2008; Walker and McKeand 2017). Starting in the 1990s, evidence emerged to support a gene-for-gene model for resistance to fusiform rust in loblolly pine (Wilcox et al. 1996; Amerson et al. 1997; Stelzer et al. 1999) following early work suggesting the same in slash pine (Kinloch and Walkinshaw 1991; Amerson et al. 2015). Several studies have identified markers linked to several fusiform rust resistance genes (*Fr*), which appear to be organized in at least two clusters in the loblolly pine genome (Nelson et al. 2010; Isik et al. 2012; Neves et al. 2014; Amerson et al. 2015). With the recent publication of the assembled and annotated genome of the loblolly pine, those markers were placed in a genomic context, allowing for the identification of a candidate *Fr* gene model which encodes an NLR protein (Wegrzyn et al. 2014; Neale et al. 2014; Zimin et al. 2017). The organization of NLR resistance genes in genomic clusters mirrors results from the model plant system, arabidopsis (Van de Weyer et al. 2019; Lee and Chae 2020). As future studies take advantage of advancements in sequencing technology, more candidate gene models will continue to be identified. This will allow for not only the development of improved families with multiple "pyramided" *Fr* resistance alleles, but also the querying of extant families to inventory their specific *Fr* gene repertoires.

7.3.2 Major Gene Resistance in Fusiform Rust Resistance

Early studies of fusiform rust resistance measured resistance as the percentage of diseased progeny after inoculation with mixes of spores from multiple Cqf galls (Zobel and Talbert 1984; Wilcox et al. 1996; Nelson et al. 2010). This approach treated resistance as a quantitative trait, controlled by multiple genes. Protocols to examine the interaction between specific fungal isolates and specific pine families used inoculation of seedlings with spores from single galls, single aeciospore isolates, or single uredinial pustules (Zobel and Talbert 1984; Nelson et al. 1993, 2010; Wilcox et al. 1996). In this way, researchers could identify specific host alleles that controlled the expression of resistance (nongalled) or susceptible (galled) phenotypes in response to inoculation with a single, genotypically homogenous fungal inoculum (Zobel and Talbert 1984; Wilcox et al. 1996; Nelson et al. 2010). These interactions describe a gene-for-gene model of resistance, where resistance is controlled by dominant genes of major effect (Flor 1956, 1971; Kubisiak et al. 2005; Jones and Dangl 2006). Under the gene-for-gene model, Cqf galls do not form when the host is heterozygous at the resistance gene locus or when the pathogen is homozygous for avirulent alleles at a single locus that interacts with the resistant gene.

Wilcox et al. (1996) identified a RAPD marker, J7-485A, that segregated with resistance in greenhouse-grown seedlings descended from tree 10–5 when they were inoculated with basidiospores from a single-aeciospores rust isolate, SC20-21. In the same study, the authors presented evidence from field trials in South Carolina that marker J7-485A was associated with absence of fusiform rust galls, that the marker segregated close to 1:1, and that both the marker and resistance to SC20-21 were rare in the pollen pool and inherited from tree 10–5 (Wilcox et al. 1996). This was the first major gene in loblolly pine associated with resistance to fusiform rust and was denoted as *Fr1*.

In another application of markers, Amerson et al. (2015) examined host–pathogen interactions between seven loblolly pine families and five single-spore fusiform rust fungal isolates, including family 10–5 and isolate SC20-21. This work identified markers linked to 9 different *Fr* genes (including the previously identified *Fr1*), organized in clusters on linkage groups 2, 3, and 10 of the loblolly pine reference genetic map (Amerson et al. 2015). Markers linked to *Fr1* were placed on linkage group 2. Quesada et al. (2014) later mapped SNPs linked to fusiform rust resistance to linkage groups 2, 10, and 11. In that same study, Quesada et al. (2014) identified two SNPs in trees descended from 10–5 that mapped near *Fr1*. The SNP with greatest LOD score for resistance was later mapped to a gene model in the loblolly pine v1.0 genome assembly (Neale et al. 2014). That gene model encoded a full-length NLR protein, while the EST lacked the TIR and NB domains. This was the first candidate gene model for a fusiform rust resistance gene in loblolly pine (see Table 7.1).

7.3.3 Avirulence Genes in the Fusiform Rust Fungal Pathogen

In the gene-for-gene model of host–pathogen interaction, major resistance genes in the host implies the existence of corresponding avirulence genes in the pathogen. After the identification of markers linked to *Fr* genes, marker studies began on the pathogen. By using open-pollinated pine seedlings from resistant mother trees inoculated with spores from single uredinial pustule (SUP) isolates of Cqf, markers for genes controlling avirulence in *Fr* genes (*Avr* genes) were developed and a genome-wide map enriched for markers linked to *Avr1*, the avirulence gene controlling gall formation in the presence of *Fr1* (Nelson et al. 2010; Kubisiak et al. 2011). Subsequent analysis of the assembled Cqf genome identified families of putative effector genes, likely candidates for avirulence genes, that were expanded specifically in Cqf relative to other rust fungi genomes (Pendleton et al. 2014), although markers for *Avr1* still need to be linked to specific gene models in the Cqf genome.

7.3.4 Quantitative Resistance to Fusiform Rust

It is now well-established that the development measured by lesion length of Cqf galls on loblolly pine (gene-for-gene resistance) provides an example of genetic control over Cqf gall development (Quesada et al. 2014) (see Table 7.1). Instead of single genes of large dominant effect, multiple genes of small additive effect appear to control gall size (i.e., gall length, gall width, and gall volume).

Table 7.1 List of candidate SNPs associated with resistance to fusiform rust in loblolly pine and annotations with respect to the loblolly pine genome

Marker ID	Source	Genomic Location (v2.01)	Linkage Group	Predicted Gene
0-17,383-01-149	Cumbie et al. (2020)	scaffold140683	1	DExH18
2_5345	Neale et al. (2014)	scaffold55875	2	TIR NB-LRR
PtIFG_2538_B	Amerson et al. (2015)	C3828663	2	No domain
estPTflFG_9022_a	Amerson et al. (2015)	scaffold56001	2	TIF
0-13,348-01-203	Cumbie et al. (2020)	C4826185	2	None
UMN_4596_01-83	Cumbie et al. (2020)	scaffold166216	3	HSP70
CL573Contig1-03-68	Cumbie et al. (2020)	scaffold101243	7	NADP-dependent oxidoreductase domain
0-17,971-01-642	Cumbie et al. (2020)	super3642	10	SRRA
CL1208Contig1-03-75	Cumbie et al. (2020)	scaffold27023	11	F-Box
UMN_CL789Contig1	Quesada et al. (2014)	C31300012	No LG	None
CL1196Contig1-03-56	Cumbie et al. (2020)	C4915959	No LG	DUF4228_plant
0_16664_01-01-386	Cumbie et al. (2020)	C3876229	No LG	Polyprenyl synthetase
0_12076_01-311	Cumbie et al. (2020)	C5097911	No LG	Formate dehydrogenase1
0_3078_01-53	Cumbie et al. (2020)	C4485385	No LG	Proton-dependent oligopeptide transporter family
0-13,762-01-190	Cumbie et al. (2020)	super3018	No LG	FES1

It is suspected that quantitative resistance would be more durable in the field and identifying markers and genes that confer partial resistance to pathogens are of great interest (Poland et al. 2009; Kou and Wang 2010; Poland 2010).

Another recent study identified multiple dominant and additive effect SNPs that control rust resistance in clonal lines from full-sib trees at field sites across a wide range of sites in the southern United States (Cumbie et al. 2020). This study also evaluated the resistance of two full-sib families where seedlings were inoculated with mixed-spore sources. Although these SNPs evidently control resistance in the field, without single-spore inoculation studies, it is unclear what specific *Fr* genes are present in those families (see Table 7.1).

7.3.5 Implications of Resistance Gene Clusters

Aside from the *Fr1* candidate, it is unknown how many of the other *Fr* genes clustered throughout the loblolly pine genome also encode NLR proteins. Evidence from other conifer genomes supports the hypothesis that NLR genes in the loblolly pine genome are organized into clusters that may correspond to clusters of *Fr* genes (Liu et al. 2019; Scott et al. 2020). Studies in Arabidopsis support a model of genome evolution where clusters of NLR genes evolve through intragenic recombination and gene conversion (Choi et al. 2016, 2018; Jiao and Schneeberger 2020). This process can be mediated by transposable elements (Leister 2004; Seidl and Thomma 2017; Krasileva 2019). Although it is likely that a similar model occurs in the loblolly pine genome, the difference in size between the Arabidopsis genome and the loblolly pine genome, and the fragmentation of loblolly pine genome assembly make it difficult to conclude whether the mechanisms are conserved between the two species.

7.4 Genomics of Quantitative Resistance to Pitch Canker

7.4.1 Introduction

Pitch canker disease, caused by *Fusarium circinatum*, is one of the most widespread diseases in pine with outbreaks occurring episodically in the native range and causing extensive damage in pine plantations established overseas. It is caused by the necrotrophic fungus Fc (Blakeslee et al. 1980), and it is characterized by the production of resinous lesions or cankers in stems and branches, as well as in reproductive structures (Barnard and Blakeslee 2006). This disease affects 57 species of pines (Dwinell and Barrows-Broaddus 1985) as well as Douglas-fir (Gordon et al. 2006), and can cause large economic losses in the forest industry due to high seedling mortality in nurseries or deformation, girdling and loss of wood quality in more mature trees (see Fig. 7.1b).

Overcoming this disease is a much desired goal for the forest industry, given the magnitude of loss at different stages of the life cycles of commercially planted pine species such as loblolly, slash, and Monterey pines. Fungicides are effective in preventing or delaying the presence of pitch canker disease in nurseries; however, once in the field, it seems that the most cost-effective approach is breeding for resistance. Previous studies show that pitch canker resistance is a quantitative and heritable trait (Kayihan et al. 2005; Quesada et al. 2010), as families show a continuous distribution of lesion lengths based on susceptibility, where highly susceptible genotypes have greater lesion lengths in greenhouse experiments (Gordon et al. 1998a, b; Kayihan et al. 2005; Quesada et al. 2010).

Most of the studies on quantitative resistance of pitch canker have been conducted in loblolly and Monterey pines, Monterey pine is the main commercially planted species in South America, Spain and New Zealand (Roux et al. 2007), and loblolly pine is the main species planted in the southeastern United States.

7.4.2 Quantitative Resistance in Monterey Pine

While there is some evidence that pitch canker originated in Mexico (Desjardins et al. 2000), pitch canker was first described in the southeastern United States in the 1940s (Hepting and Roth 1946) and has rapidly spread worldwide in the past decades, likely via the movement of seed material (Wingfield et al. 2008), to impact forest plantations in different countries around the world. In the 1980s, the disease was also observed in California (Correll 1991), causing a massive outbreak in the highly susceptible Monterey pine (Gordon et al. 2001). Although initial studies on the pathogen from this outbreak determined that the isolates from California were not vegetatively compatible with those from Florida, they did not completely exclude those of southeastern US origin (Correll et al. 1992). Since then, studies have focused on recording the incidence and severity of pitch canker, quantifying disease resistance among surviving trees, and evaluating the host range of the pathogen (Storer et al. 1999; Bonello et al. 2001; Carter and Gordon 2020).

Initial field and greenhouse studies in Monterey pine in California showed a differential response to infection among trees, suggesting a genetic component of resistance (Correll 1991; Correll et al. 1992). Further studies on Monterey pine seedlings from Australia, Chile, and New Zealand have supported a quantitative type of resistance to pitch canker disease, with heritabilities between 0.34 and 0.78 (Matheson et al. 2012). More recent research on the Chilean population of Monterey pine has resulted in the development of a genetic linkage map where quantitative trait loci (QTL) related to pitch canker resistance were detected (Moraga-Suazo et al. 2014). In this work, Moraga-Suazo and colleagues screened 86 full-sib individuals of a single Monterey pine family. These individuals were clonally replicated to obtain five replicated blocks in a randomized complete block design and inoculated with Fc conidia. Genetic linkage maps were constructed for each parent, obtaining the expected 12 linkage groups (Moraga-Suazo et al. 2014). The results showed that only linkage groups LG-1 and LG-9 for one parent and LG-9 for the other had strong statistical significance for at least one QTL related to pitch canker resistance.

7.4.3 Quantitative Resistance in Loblolly Pine

Loblolly pine is relatively resistant to pitch canker, in comparison to the more susceptible Monterey and slash pine species. Greenhouse studies using rooted cuttings inoculated with Fc spores have shown that resistance of loblolly pine to pitch canker disease is quantitative and heritable (Kayihan et al. 2005; Quesada et al. 2010). These observations were consistent in both a population with known pedigree, consisting of full-sib and half-sib families (Kayihan et al. 2005) as well as in an "association" population where individuals were assumed unrelated (Quesada et al. 2010). The polygenic architecture of resistance to pitch canker was confirmed in a recent study which conducted GWAS between the phenotypic data from Quesada et al. 2010 with a novel genotyping dataset in ADEPT2 material and found 18 SNPs on five different linkage groups that were significantly associated with pitch canker lesion length at 4, 8, or 12 weeks after inoculation (De La Torre et al. 2018).

7.4.4 Candidate Genes for Pitch Canker Resistance in Pinus Spp.

As discussed earlier, pitch canker disease is a quantitative trait and as such, it is likely that many genes with small effects control this trait. In addition, the broad-sense heritability of this trait is around 0.4 (Kayihan et al. 2005; Quesada et al. 2010), which suggests that approximately 40% of the phenotypic variance is due to genetic effects, whereas the remaining 60% of the variance is due to the environment. This has not discouraged the search for causative genes, and for this purpose, several studies have used molecular markers and candidate genes from metabolic pathways related to disease resistance. One of the first assays that used a targeted genomics approach was that of Morse et al. (2004). This study used a cDNA array to examine differential expression of candidate genes in slash pine at different points after inoculation with Fc. Of 311 candidate genes tested, 29 were significantly upregulated after infection during the disease state. Of these, seven had similar sequences to genes involved in defense or stress response pathways, while others had unknown function (Morse et al. 2004).

The use of molecular markers has helped clarify the quantitative nature of pitch canker resistance, supporting the hypothesis that resistance of loblolly pine is due to the action of multiple genes with small effects (Lu et al. 2017; Quesada et al. 2010; De La Torre et al. 2018). As genotyping became more efficient and affordable, SNP markers were developed to genotype over 7,000 expressed sequence tags (ESTs) in a population of over 600 unrelated clones that were phenotyped for multiple traits, including lesion length following inoculation with Fc. After screening for polymorphic markers among 498 clones inoculated with Fc, a final number of 3,938 SNP markers were tested (Quesada et al.

2010). Of the 3,938 SNP markers tested, ten showed significant associations with reduced lesion length; however, the sum of all ten SNPs explained only about 14.6% of the clonal variance observed, which translates to roughly 3.7% of the overall phenotypic variance. Four of these SNPs showed no changes to the resulting protein sequence (synonymous substitutions), whereas another four resulted in nonsynonymous substitutions and none of these were directly annotated to known disease or stress response pathways. The other two significant SNPs were in non-coding regions, presumably in the 3'-UTR (Quesada et al. 2010). The same phenotypic data from the study by Quesada et al. (2010) was also used in 2017 to test for associations with over 2.8 million markers (Lu et al. 2017). Seven SNPs showed significant associations with pitch canker lesion length, which explained between 7.14% and 14.84% of clonal variance. The percentage of phenotypic variance explained by these seven SNPs corresponded to about 17%, (Lu et al. 2017). In addition, significant SNP-SNP interactions were observed between three pairs of SNPs, which explained around 14% of the phenotypic variance. Together, the significant associations and epistatic effects explained about 24% of the phenotypic variance (Lu et al. 2017). Four of these SNPs mapped to coding sequences, of which two resulted in synonymous substitutions in genes annotated to be involved with disease or stress response. The other two resulted in nonsynonymous substitutions, a predicted nitrate transporter and a gene of unknown function. Of the remaining significant SNPs, one was within an intron, one was within a putative 3' regulatory sequence and a third one was unclassified (Lu et al. 2017; see Table 7.2). A recent GWAS study genotyped material from the ADEPT2 study dataset with a custom Affymetrix genotyping array, and identified 13 SNPs associated with pitch canker resistance with small to moderate effect sizes (De La Torre et al. 2018) (see Table 7.2).

In Monterey pine, research has focused on the analysis of relative expression of candidate genes related to pathogen response. The authors inoculated ten genotypes of Monterey pine and analyzed the response through a time course using quantitative PCR (Donoso et al. 2015). Significant differences between resistant and susceptible genotypes were found for a subset of these candidate genes, which were overexpressed in the resistant genotype. However, while these were found responsive to infection, these may be only a subset of all genes involved in pitch canker resistance (Donoso et al. 2015). More recently, Carrasco et al (2017) analyzed the expression profiles of monterey pine candidate genes between susceptible and resistant genotypes after inoculation with Fc using RNA-seq. After correcting for genes related to wounding, they found a total of 294 putative candidate genes that were upregulated in response to infection sampling at 2, 6, and 12 days after inoculation (Carrasco et al. 2017). Of these, 32 upregulated genes showed sequence similarity to defense response genes. Validation of a subset of these genes using quantitative PCR showed that resistant genotypes had increased expression of these genes at six days after inoculation, suggesting that resistance to pitch canker involves an early response to the pathogen.

7.5 Applications of Technology Advancement to Improve Forest Health

While the forest genetics community has been working on the genomic characterization of disease resistance in southern pines and their pathogens for decades, major advances followed the accumulation of genomic and transcriptomic resources in a collaboration on wood formation in the pine genome that was managed by Prof. Ron Sederoff (Allona et al. 1998; Retzel et al. 2003; Egertsdotter et al. 2004). Subsequent NSF projects on Allele Discovery of Economic Pine Traits (ADEPT and ADEPT2) built upon these resources to examine resistance to rust and pitch canker in greater detail (Quesada et al. 2010, 2014; Eckert et al. 2012). More recently, a SNP chip has been developed to provide a tool that may be used by the community to consistently target the same set of markers from loblolly pine

Table 7.2 List of candidate SNPs associated with resistance to pitch canker in loblolly pine and annotations with respect to the loblolly pine genome version 2.01

SNP	Location	Change	Gene/Scaffold	Annotation	References*
0_15227_01_159	Coding region	Synonymous	scaffold146257	ATP binding protein, lectin-like protein kinase	Quesada et al. (2010)
0_15382_01_99	Coding region	V to A	scaffold102118	Geranylgeranyl transferase type I beta subunit	Quesada et al. (2010)
0_2234_01_128	Coding region	D to Y	scaffold94451	Putative long-chain acyl-CoA synthetase	Quesada et al. (2010)
0_6323_01_240	Coding region	Synonymous	scaffold719208*	DELLA protein	Quesada et al. (2010)
0_9288_01_370	Coding region	Synonymous	C3563749	No hits found	Quesada et al. (2010)
1_3327_01_113	Coding region	C to Y	scaffold89078.1*	No hits found	Quesada et al. (2010)
2_4484_02_622	Coding region	C to Y	C4249085	Plastid hexose transporter	Quesada et al. (2010)
2_6181_02_400	Putative 3' regulatory sequence	N/A	scaffold170722	Hexokinase	Quesada et al. (2010)
2_8946_02_435	Putative 3' regulatory sequence	N/A	scaffold125728	Cucumber peeling Cupredoxin	Quesada et al. (2010)
CL4336Contig1_01_180	Coding region	Synonymous	scaffold94596	Unknown	Quesada et al. (2010)
C32571366_88829_G_A	Coding region	Synonymous	PITA_000019796	Predicted, Leucine-rich repeat	Lu et al. (2017)
scaffold394991_238149_C_A	Unclassified	N/A	C4826907	Probable quinone oxidoreductase	Lu et al. (2017)
scaffold490830_154983_T_A	Coding region	I to K	PITA_000030834	Predicted, probable peptide/nitrate transporter At3g53960-like	Lu et al. (2017)
scaffold771268_7109_G_A	Putative 3' regulatory sequence	N/A	PITA_000095986	Predicted, probable aminotransferase TAT2-like	Lu et al. (2017)
tscaffold758_520984_C_T	Coding region	A to V	PITA_000009116	Unknown	Lu et al. (2017)
tscaffold1272_36798_C_T	Intron	N/A	PITAhm_002502	Unknown	Lu et al. (2017)
tscaffold5506_305141_G_A	Coding region	Synonymous	PITA_000020926	Predicted, dehydration-responsive protein RD22-like isoform 2	Lu et al. (2017)

(continued)

Table 7.2 (continued)

SNP	Location	Change	Gene/Scaffold	Annotation	References*
AX-173214925	Unclassified	N/A	C4313391	N/A	De La Torre et al. (2019)
AX-173022044	Unclassified	N/A	C4925709	N/A	De La Torre et al. (2019)
AX-173375787	Unclassified	N/A	scaffold105617	N/A	De La Torre et al. (2019)
AX-172799963	Unclassified	N/A	scaffold109981	N/A	De La Torre et al. (2019)
AX-172799964	Unclassified	N/A	scaffold109981	N/A	De La Torre et al. (2019)
AX-173169704	Unclassified	N/A	scaffold109981	N/A	De La Torre et al. (2019)
AX-173179448	Unclassified	N/A	scaffold109981	N/A	De La Torre et al. (2019)
AX-173199529	Unclassified	N/A	scaffold147406	N/A	De La Torre et al. (2019)
AX-173078399	Intron	N/A	PITA_17759	RFT1 homolog, membrane sugar-transporter	De La Torre et al. (2)019
AX-173172841	Unclassified	N/A	scaffold21579	N/A	De La Torre et al. (2019)
AX-172916974	Unclassified	N/A	scaffold22951	N/A	De La Torre et al. (2019)
AX-173190165	Unclassified	N/A	scaffold53216	N/A	De La Torre et al. (2019)
AX-173208588	Coding Region	S to P	PITA_06805	LRR receptor kinase	De La Torre et al. (2019)

*Sequences were aligned against the loblolly pine genome v.1.01 after publication

(Isik et al. 2018). A similar chip was subsequently designed for many different tropical pines including *P. patula* Schiede ex Schltdl., *P. caribaea* Morelet, and *P. tecunumanii* Eguiluz and Perry (Jackson 2019). While these community wide standards provide an excellent platform for making comparisons among populations of the target species, the high set-up costs complicate modifications or improvements to the design. A more flexible approach using exome capture is based on oligonucleotide hybridization probes enriching target loci prior to sequencing (Lu et al. 2016). Several new probe designs have focused and expanded these panels to sequence a diversity of loci of interest. The flexibility of probe based capture and sequencing provides an excellent tool to focus genotyping efforts on specific gene families and may be paired with algorithms to bioinformatically identify gene families of interest in genomes to focus sequencing on loci of interest. For example, given the important role that NLR genes likely play in fusiform rust resistance and the large number of those genes in conifer genomes, a "pan NLRome" like that developed for Arabidopsis would be a valuable resource for developing resistance markers in loblolly pine (Van de Weyer et al. 2019; Kourelis et al. 2020). Such a resource could be developed by sequencing RNA or DNA from pine families with improved resistance.

A greater understanding of the host and pathogen interaction will require further development of genomic resources for both host and pathogen. Endemic pathogens in natural systems are likely to be highly genetically diverse relative to introduced pathogens. This diversity complicates the development of disease resistance. Improved understanding of gene-for-gene interactions and their impact on disease requires the evaluation of large host and pathogen populations. For pathogens, deeper sequencing is needed and long-read technologies may offer advantages over the cost-effectiveness of pooling multiple libraries for short-read sequencing (Jiao and Schneeberger 2017; Diao et al. 2019; Dumschott et al. 2020). Prediction models may be used to identify genes incorporating variants from host and pathogens to reflect both resistance and virulence, respectively. The genomic connectivity that is provided in relationship matrices may also be utilized to connect progeny trials established across the range of pathogen populations to help identify pedigrees and markers within pedigrees that are more susceptible to one population or another. This provides the southern pine breeding community with an opportunity to leverage information from many different populations that have experienced different pathogen populations. The goal of targeting different resistance mechanisms to the pathogen populations where they are most effective requires the ability to characterize pathogen populations and track changes over time. The decades of university and industry cooperative tree improvement that delivered disease resistant pine seed for reforestation may be accelerated using selection methods based on pathosystem understanding.

7.6 Conclusions

We summarized current knowledge of fungal disease resistance in loblolly pine, with a focus on the genetics and genomics of resistance to fusiform rust and pitch canker. Within living memory, the field has transitioned from a quantitative model of resistance to fusiform rust disease to a model that includes aspects of both quantitative, or partial resistance, and major gene resistance (Wilcox et al. 1996; Isik et al. 2008; Nelson et al. 2010; Quesada et al. 2014; Cumbie et al. 2020). On the other hand, resistance to pitch canker disease appears to be entirely quantitative (Quesada et al. 2010; De La Torre et al. 2018). Public–private partnerships in the form of pine cooperative breeding programs and facilities like the Rust Screening Center have been and will continue to be essential to advance research into disease resistance in loblolly pine (Schmidt 2003; Cowling and Young 2013).

A recurring shortcoming of genomic results is the difficulty of rapidly placing markers and candidate genes in linkage maps (Cumbie et al. 2020; Neves et al. 2014, 2013; Westbrook et al. 2015; De La Torre et al. 2019). In QTL studies, the high-scoring intervals may include so many scaffolds and genes that clear causative candidates are difficult to identify (Beavis 1998; Xu 2003; Slate 2013). Alternatively, GWAS studies may identify genes with SNPs associated with resistance, but may be hindered by the fragmented genome assembly from identifying interesting candidate genes located near the high-scoring SNP (Marjoram et al. 2014; King and Long 2017). As improvements to the loblolly pine genome assembly continue to be implemented (Zimin et al. 2017), the identification of candidate genes for both fusiform rust and pitch canker resistance will likely continue as the effort benefits from both the integration of new sequencing technology (Moll et al. 2017; Scott et al. 2020) and deeper exploration of the genomic diversity of improved pine accessions as has been done in model plants and crops (McKeand et al. 1989; Quesada et al. 2010; Van de Weyer et al. 2019).

References

Allona I, Quinn M, Shoop E, Swope K, St. Cyr S, Carlis J, Riedl J, Retzel E, Campbell MM, Sederoff R, Whetten RW (1998) Analysis of xylem formation in pine by cDNA sequencing. Proc Natl Acad Sci USA 95(16):9693–9698

Alvarez M (2018) The State of America's Forests. U.S. Endowment for Forestry and Communities, Inc., Greenville, SC www.usaforests.org

Amerson H, Nelson C, Kubisiak T, Kuhlman E, Garcia S (2015) Identification of Nine Pathotype-Specific Genes Conferring Resistance to Fusiform Rust in

Loblolly Pine (*Pinus taeda* L.). Forests 6(8):2739–2761

Amerson HV, Jordan AP, Kuhlman EG, O'Malley DM, Sederoff RR (1997) Genetic basis of fusiform rust disease resistance in loblolly pine. In: Proceedings of the 24th Southern Forest Tree Improvement Conference. University of Florida, Springfield, VA, p 403

Anderson CL, Kubisiak TL, Nelson CD, Smith JA, Davis JM (2010) Genome size variation in the pine fusiform rust pathogen *Cronartium quercuum* f. sp. *fusiforme* as determined by flow cytometry. Mycologia 102(6):1295–1302

Baker JB, Langdon OG (1990) Pinus taeda L. Loblolly Pine. Silvics of North America 1:497–512

Baltunis BS, Huber DA, White TL (2005) Genetic effects of rooting loblolly pine stem cuttings from a partial diallel mating design. Can J for Res 35(5):1098–1108

Barnard EL, Blakeslee GM (2006) Pitch canker of southern pines. Plant Pathology Circular

Beavis WD (1998) QTL analyses: power, precision, and accuracy. Molecular Dissection of Complex Traits 1998:145–162

Blakeslee GM, Dwinell LD, Anderson RL (1980) Pitch canker of southern pines: identification and management considerations. US Department of Agriculture Forest Service

Bonello P, Gordon TR, Storer AJ (2001) Systemic induced resistance in Monterey pine. Forest Pathol 32(2):99–106

Burdsall HH, Snow GA (1977) Taxonomy of *Cronartium quercuum* and *C. fusiforme*. Mycologia 69(3):503–508

Carrasco A, Wegrzyn JL, Durán R, Fernández M, Donoso A, Rodriguez V, Neale D, Valenzuela S (2017) Expression profiling in *Pinus radiata* infected with *Fusarium circinatum*. Tree Genet Genomes 13(2):46

Carter JW, Gordon TR (2020) Infection of the Native California Grass, *Bromus carinatus*, by *Fusarium circinatum*, the Cause of Pitch Canker in Pines. Plant Dis 104(1):194–197

Choi K, Reinhard C, Serra H, Ziolkowski PA, Underwood CJ, Zhao X, Hardcastle TJ, Yelina NE, Griffin C, Jackson M, Mézard C, McVean G, Copenhaver GP, Henderson IR (2016) Recombination Rate Heterogeneity within Arabidopsis Disease Resistance Genes. PLoS Genetics 12(7):e1006179

Choi K, Zhao X, Tock AJ, Lambing C, Underwood CJ, Hardcastle TJ, Serra H, Kim J, Cho HS, Kim J, Ziolkowski PA, Yelina NE, Hwang I, Martienssen RA, Henderson IR (2018) Nucleosomes and DNA methylation shape meiotic DSB frequency in *Arabidopsis thaliana* transposons and gene regulatory regions. Genome Res 28(4):532–546

Correll JC (1991) The relationship between formae speciales, races, and vegetative compatibility groups in *Fusarium oxysporum*. Phytopathology 81(9):1061–1064

Correll JC, Gordon TR, McCain AH (1992) Genetic diversity in California and Florida populations of the pitch canker fungus *Fusarium subglutinans* f. sp. *pini*. Phytopathology 82(4):415–420

Cowling E, Young C (2013) Narrative History of the resistance screening center: it's origins, leadership and partial list of public benefits and scientific contributions. Forests 4(3):666–692

Cubbage FW, Pye JM, Holmes TP, Wagner JE (2000) An Economic Evaluation of Fusiform Rust Protection Research. South J Appl for 24(2):77–85

Cumbie WP, Huber DA, Steel VC, Rottmann W, Cannistra C, Pearson L, Cunningham M (2020) Marker associations for fusiform rust resistance in a clonal population of loblolly pine (*Pinus taeda*, L.). Tree Genet Genomes 16(6):86

De La Torre AR, Puiu D, Crepeau MW, Stevens K, Salzberg SL, Langley CH, Neale DB (2018) Genomic architecture of complex traits in loblolly pine. New Phytol 221(4):1789–1801

De La Torre AR, Wilhite B, Neale DB (2019) Environmental genome-wide association reveals climate adaptation is shaped by subtle to moderate allele frequency shifts in loblolly Pine. Genome Biol Evol 11(10):2976–2989

Desjardins AE, Plattner RD, Gordon TR (2000) *Gibberella fujikuroi* mating population A and *Fusarium subglutinans* from teosinte species and maize from Mexico and Central America. Mycol Res 104(7):865–872

Diao S, Ding X, Luan Q, Jiang J (2019) A complete transcriptional landscape analysis of *Pinus elliottii* Engelm. Using third-generation sequencing and comparative analysis in the *Pinus* phylogeny. Forests 10(11):942

Donoso A, Rodriguez V, Carrasco A, Ahumada R, Sanfuentes E, Valenzuela S (2015) Relative expression of seven candidate genes for pathogen resistance on *Pinus radiata* infected with *Fusarium circinatum*. Physiol Mol Plant Pathol 92:42–50

Dumschott K, Schmidt MH-W, Chawla HS, Snowdon R, Usadel B (2020) Oxford Nanopore Sequencing: new opportunities for plant genomics? J Exp Bot 71(18):5313–5322

Dwinell LD, Barrows-Broaddus J (1985) Infection of fusiform rust galls on slash and loblolly pines by pitch canker fungus, *Fusarium moniliforme* var subglutinans. See Ref 3:239–248

Eckert AJ, Wegrzyn JL, Cumbie WP, Goldfarb B, Huber DA, Tolstikov V, Fiehn O, Neale DB (2012) Association genetics of the loblolly pine (*Pinus taeda*, Pinaceae) metabolome. New Phytol 193(4):890–902

Egertsdotter U, van Zyl LM, MacKay J, Peter G, Kirst M, Clark C, Whetten R, Sederoff R (2004) Gene expression during formation of earlywood and latewood in loblolly pine: expression profiles of 350 genes. Plant Biol 6(6):654–663

Flor HH (1956) The complementary genic systems in flax and flax rust. Adv Genet 8:29–54

Flor HH (1971) Current status of the gene-for-gene concept. Annu Rev Phytopathol 9(1):275–296

Fulton JC, Huguet-Tapia JC, Adams SM, Dufault NS, Quesada T, Brawner JT (2020) Draft genome sequences of three *Fusarium circinatum* isolates used to inoculate a pedigreed population of *Pinus elliottii* seedlings. Microbiol Resour Announc 9(30):e00631-e720

Gordon TR, Kirkpatrick SC, Aegerter BJ, Wood DL, Storer AJ (2006) Susceptibility of Douglas fir (*Pseudotsuga menziesii*) to pitch canker, caused by *Gibberella circinata* (anamorph = *Fusarium circinatum*). Plant Pathol 55(2):231–237

Gordon TR, Okamoto D, Storer AJ, Wood DL (1998a) Susceptibility of five landscape pines to pitch canker disease, caused by *Fusarium subglutinans* f. sp. *pini*. HortScience 33(5):868–871

Gordon TR, Storer AJ, Wood DL (2001) The Pitch Canker Epidemic in California. Plant Dis 85 (11):1128–1139

Gordon TR, Wikler KR, Clark SL, Okamoto D, Storer AJ, Bonello P (1998b) Resistance to pitch canker disease, caused by *Fusarium subglutinans* f. sp. *pini*, in Monterey pine (*Pinus radiata*). Plant Pathol 47 (6):706–711

Hepting GH, Roth ER (1946) Pitch canker, a new disease of some southern pines. J for 44(10):742–744

Hodge GR, Dvorak WS (2000) Differential responses of Central American and Mexican pine species and *Pinus radiata* to infection by the pitch canker fungus. New Forest 19:241–258

Howard JL, Liang S (2019) U.S. timber production, trade, consumption, and price statistics, 1965–2017. US Department of Agriculture—Forest Service

Isik F, Acosta JJ, Eckert AJ, Sniezko R, Wegrzyn J (2018) Pine SNP Chip Consortium: Progress on Pine SNP Discovery and Array Design in Loblolly Pine. In: Plant and Animal Genome XXVI Conference (January 13–17, 2018). PAG

Isik F, Amerson HV, Whetten RW, Garcia SA, Li B, McKeand S (2008) Resistance assessments of elite loblolly pine families to fusiform rust inocula in greenhouse testing. Can J for Res 38:2687–2696

Isik F, Amerson HV, Whetten RW, Garcia SA, McKeand SE (2012) Interactions of *Fr* genes and mixed-pathogen inocula in the loblolly pine-fusiform rust pathosystem. Tree Genet Genomes 8(1):15–25

Jackson CA (2019) Single Nucleotide Polymorphism Discovery and Probe Design Using RNA-Seq and Targeted Capture in Tropical and Subtropical Pine Species. North Carolina State University, MSc

Jiao W-B, Schneeberger K (2020) Chromosome-level assemblies of multiple Arabidopsis genomes reveal hotspots of rearrangements with altered evolutionary dynamics. Nat Commun 11(1):1–10

Jiao W-B, Schneeberger K (2017) The impact of third generation genomic technologies on plant genome assembly. Curr Opin Plant Biol 36:64–70

Jones JDG, Dangl JL (2006) The plant immune system. Nature 444(7117):323–329

Kayihan GC, Huber DA, Morse AM, White TL, Davis JM (2005) Genetic dissection of fusiform rust and pitch canker disease traits in loblolly pine. Theor Appl Genet 110(5):948–958

King EG, Long AD (2017) The Beavis Effect in Next-Generation Mapping Panels in Drosophila melanogaster. Genes Genomes Genet 7(6):1643–1652

Kinloch BB, Walkinshaw CH (1991) Resistance to fusiform rust in southern pines: how is it inherited. In: Hiratsuka Y, Laishley BL (eds) Rusts of Pine, pp 219–228

Kourelis J, Sakai T, Adachi H, Kamoun S (2020) RefPlantNLR: a comprehensive collection of experimentally validated plant NLRs. bioRxiv 2020.07.08.193961

Kou Y, Wang S (2010) Broad-spectrum and durability: understanding of quantitative disease resistance. Curr Opin Plant Biol 13(2):181–185

Krasileva KV (2019) The role of transposable elements and DNA damage repair mechanisms in gene duplications and gene fusions in plant genomes. Curr Opin Plant Biol 48:18–25

Kubisiak TL, Amerson HV, Nelson CD (2005) Genetic interaction of the fusiform rust fungus with resistance gene *Fr1* in Loblolly Pine. Phytopathology 95(4):376–380

Kubisiak TL, Anderson CL, Amerson HV, Smith J a., Davis JM, Nelson CD (2011) A genomic map enriched for markers linked to *Avr1* in *Cronartium quercuum* f.sp. fusiforme. Fungal Genet Biol 48 (3):266–274

Lee RRQ, Chae E (2020) Variation Patterns of NLR Clusters in *Arabidopsis thaliana* Genomes. Plant Communications 1(4). https://doi.org/10.1016/j.xplc.2020.100089

Leister D (2004) Tandem and segmental gene duplication and recombination in the evolution of plant disease resistance genes. Trends Genet 20(3):116–122

Liu J-J, Schoettle AW, Sniezko RA, Yao F, Zamany A, Williams H, Rancourt B (2019) Limber pine (*Pinus flexilis* James) genetic map constructed by exome-seq provides insight into the evolution of disease resistance and a genomic resource for genomics-based breeding. Plant J 98(4):745–758

Lu M, Krutovsky KV, Nelson CD, Koralewski TE, Byram TD, Loopstra CA (2016) Exome genotyping , linkage disequilibrium and population structure in loblolly pine (*Pinus taeda* L.). BMC Genomics 17:730

Lu M, Krutovsky KV, Nelson CD, West JB, Reilly NA, Loopstra CA (2017) Association genetics of growth and adaptive traits in loblolly pine (*Pinus taeda* L.) using whole-exome-discovered polymorphisms. Tree Genet Genomes 13(3):57

Marjoram P, Zubair A, Nuzhdin SV (2014) Post-GWAS: where next? More samples, more SNPs or more biology? Heredity 112(1):79–88

Martín-García J, Zas R, Solla A, Woodward S, Hantula J, Vainio EJ, Mullett M, Morales-Rodríguez C, Vannini A, Martínez-Álvarez P, Pinto G, Alves A, Amaral J, Wingfield MJ, Fourie G, Steenkamp ET, Ahumada R, Šerá B, Sanz-Ros AV, Raposo R, Elvira-Recuenco M, Iturritxa E, Gordon TR, Diez JJ (2019) Environmentally friendly methods for controlling pine pitch canker. Plant Pathol 68(5):843–860

Matheson AC, Mark WR, Stovold G, Balocchi C, Smith N, Brassey C (2012) Genetic variation in resistance to pine pitch canker and western gall rust in Monterey pine (Pinus radiata D. Don): Results from a three-country collaborative field trial. In: Sniezko RA, Yanchuk AD, Kliejunas JT, Palmieri KM, Alexander JM, Frankel SJ (eds) Proceedings of the 4th international workshop on genetics of host-parasite interactions in forestry. U.S. Department of Agriculture Forest Service, pp 253–255

McKeand SE, Weir RJ, Hatcher AV (1989) Performance of diverse provenances of loblolly pine throughout the Southeastern United States. South J Appl for 13 (1):46–51

McKeand S, Mullin T, Byram T, White T (2003) Deployment of genetically improved loblolly and slash pines in the south. J for 101(3):32–37

McKeever DB, Howard JL (1996) Value of timber and agricultural products in the United States. For Prod J 46(10):45–50

Moll KM, Zhou P, Ramaraj T, Fajardo D, Devitt NP, Sadowsky MJ, Stupar RM, Tiffin P, Miller JR, Young ND, Silverstein KAT, Mudge J (2017) Strategies for optimizing BioNano and Dovetail explored through a second reference quality assembly for the legume model. Medicago Truncatula. BMC Genomics 18(1):578

Moraga-Suazo P, Orellana L, Quiroga P, Balocchi C, Sanfuentes E, Whetten RW, Hasbún R, Valenzuela S (2014) Development of a genetic linkage map for *Pinus radiata* and detection of pitch canker disease resistance associated QTLs. Trees 28(6):1823–1835

Morse AM, Nelson CD, Covert SF, Holliday AG, Smith KE, Davis JM (2004) Pine genes regulated by the necrotrophic pathogen *Fusarium circinatum*. Theor Appl Genet 109(5):922–932

Neale DB, Wegrzyn JL, Stevens KA, Zimin AV, Puiu D, Crepeau MW, Cardeno C, Koriabine M, Holtz-Morris AE, Liechty JD, Martínez-García PJ, Vasquez-Gross HA, Lin BY, Zieve JJ, Dougherty WM, Fuentes-Soriano S, Wu L-S, Gilbert D, Marçais G, Roberts M, Holt C, Yandell M, Davis JM, Smith KE, Dean JFD, Lorenz WW, Whetten RW, Sederoff R, Wheeler N, McGuire PE, Main D, Loopstra CA, Mockaitis K, deJong PJ, Yorke JA, Salzberg SL, Langley CH (2014) Decoding the massive genome of loblolly pine using haploid DNA and novel assembly strategies. Genome Biol 15(3):R59

Nelson CD, Doudrick RL, Nance WL, Hamaker JM, Capoz B (1993) Specificity of host: pathogen genetic interaction for fusiform rust disease on slash pine. In: Lantz CW, Moorhead DJ (eds) Proceedings of the 22nd Southern forest tree improvement conference. southern forest tree improvement committee, Springfield, VA 22161, pp 403–410

Nelson CD, Kubisiak TL, Amerson HV (2010) Unravelling and managing fusiform rust disease: a model approach for coevolved forest tree pathosystems. For Pathol 40(1):64–72

Neves LG, Davis JM, Barbazuk WB, Kirst M (2013) Whole-exome targeted sequencing of the uncharacterized pine genome. Plant J 75(1):146–156

Neves LG, Davis JM, Barbazuk WB, Kirst M (2014) A high-density gene map of loblolly pine (*Pinus taeda* L.) based on exome sequence capture genotyping. Genes Genomes Genet 4(1):29–37

Nirenberg HI, O'Donnell K (1998) New Fusarium species and combinations within the *Gibberella fujikuroi* species complex. Mycologia 90(3):434–458

Pendleton AL, Smith KE, Feau N, Martin FM, Grigoriev IV, Hamelin R, Nelson CD, Burleigh JG, Davis JM (2014) Duplications and losses in gene families of rust pathogens highlight putative effectors. Front Plant Sci 5:299

Poland J (2010) The genetic architecture of quantitative disease resistance In maize. Cornell University, Doctor of Philosophy

Poland JA, Balint-Kurti PJ, Wisser RJ, Pratt RC, Nelson RJ (2009) Shades of gray: the world of quantitative disease resistance. Trends Plant Sci 14(1):21–29

Quesada T, Gopal V, Cumbie WP, Eckert AJ, Wegrzyn JL, Neale DB, Goldfarb B, Huber DA, Casella G, Davis JM (2010) Association mapping of quantitative disease resistance in a natural population of loblolly pine (*Pinus taeda* L.). Genetics 186 (2):677–686

Quesada T, Lucas S, Smith K, Smith J (2019) Response to temperature and virulence assessment of *Fusarium circinatum* Isolates in the context of climate change. Forests 10(1):40

Quesada T, Resende M Jr, Muñoz P, Wegrzyn J, Neale D, Kirst M, Peter G, Gezan S, Nelson C, Davis J (2014) Mapping fusiform rust resistance genes within a complex mating design of loblolly pine. Forests 5 (2):347–362

Resende Junior MFR, Muñoz P, Resende MDV, Garrick DJ, Fernando RL, Davis JM, Jokela EJ, Martin TA, Peter GF, Kirst M (2012) Accuracy of genomic selection methods in a standard dataset of loblolly pine (*Pinus taeda* L.). Genetics 190:1503–1510

Retzel E, Whetten R, Sederoff R (2003) Apparent homology of expressed genes from wood-forming tissues of loblolly pine (*Pinus taeda* L.) with *Arabidopsis thaliana*. Proc Natl Acad Sci USA 100 (12):7383–7388

Roux J, Eisenberg B, Kanzler A, Nel A, Coetzee V, Kietzka E, Wingfield MJ (2007) Testing of selected South African Pinus hybrids and families for tolerance to the pitch canker pathogen. *Fusarium Circinatum*. New Forests 33(2):109–123

Schmidt RA (2003) Fusiform rust of southern pines: a major success for forest disease management. Phytopathology 93(8):1048–1051

Schultz RP (1999) Loblolly—the pine for the twenty-first century. New Forest 17(1):71–88

Scott AD, Zimin AV, Puiu D, Workman R, Britton M, Zaman S, Caballero M, Read AC, Bogdanove AJ, Burns E, Wegrzyn J, Timp W, Salzberg SL, Neale DB

(2020) A reference genome sequence for giant Sequoia. Genes Genomes Genet 10(11):3907–3919

Seidl MF, Thomma BPHJ (2017) Transposable Elements Direct The coevolution between plants and microbes. Trends Genet 33(11):842–851

Slate J (2013) From beavis to beak color: a simulation study to examine how much QTL mapping can reveal about the genetic architecture of quantitative traits. Evolution 67(5):1251–1262

Spitzer J, Isik F, Whetten RW, Farjat AE, McKeand SE (2017) Correspondence of loblolly pine response for fusiform rust disease from local and wide-ranging tests in the Southern United States. For Sci 63(5):496–503

Stelzer HE, Doudrick RL, Kubisiak TL, Nelson CD (1999) Prescreening slash pine and Cronartium pedigrees for evaluation of complementary gene action in fusiform rust disease. Plant Dis 83(4):385–389

Storer AJ, Bonello P, Gordon TR, Wood DL (1999) Evidence of resistance to the pitch canker pathogen (*Fusarium circinatum*) in native stands of Monterey pine (*Pinus radiata*). For Sci 45(4):500–505

Van de Weyer A-L, Monteiro F, Furzer OJ, Nishimura MT, Cevik V, Witek K, Jones JDG, Dangl JL, Weigel D, Bemm F (2019) A Species-wide inventory of NLR genes and alleles in *Arabidopsis thaliana*. Cell 178(5):1260-1272.e14

Walker TD, McKeand SE (2017) Fusiform rust hazard mapping for loblolly pine in the southeastern United States using progeny test data. J for 116(2):117–122

Wegrzyn JL, Liechty JD, Stevens KA, Wu LS, Loopstra CA, Vasquez-Gross HA, Dougherty WM, Lin BY, Zieve JJ, Martínez-García PJ, Holt C, Yandell M, Zimin AV, Yorke JA, Crepeau MW, Puiu D, Salzberg SL, de Jong PJ, Mockaitis K, Main D, Langley CH, Neale DB (2014) Unique features of the loblolly pine (*Pinus taeda* L.) megagenome revealed through sequence annotation. Genetics 196(3):891–909

Westbrook JW, Chhatre VE, Wu L-S, Chamala S, Neves LG, Muñoz P, Martínez-García PJ, Neale DB, Kirst M, Mockaitis K, Nelson CD, Peter GF, Davis JM, Echt CS (2015) A consensus genetic map for *Pinus taeda* and *Pinus elliottii* and extent of linkage disequilibrium in two genotype-phenotype discovery populations of *Pinus taeda*. Genes Genomes Genet 5(8):1685–1694

Wilcox PL, Amerson HV, Kuhlmant EG, Liu B-H, Malley DMO (1996) Detection of a major gene for resistance to fusiform rust disease in loblolly pine by genomic mapping. Proc Natl Acad Sci USA 93 (9):3859–3864

Wingfield MJ, Hammerbacher A, Ganley RJ, Steenkamp ET, Gordon TR, Wingfield BD, Coutinho TA (2008) Pitch canker caused by *Fusarium circinatum*—a growing threat to pine plantations and forests worldwide. Australas Plant Pathol 37(4):319–334

Xu S (2003) Theoretical basis of the Beavis effect. Genetics 165(4):2259–2268

Zamora-Ballesteros C, Diez JJ, Martín-García J, Witzell J, Solla A, Ahumada R, Capretti P, Cleary M, Drenkhan R, Dvořák M, Elvira-Recuenco M, Fernández-Fernández M, Ghelardini L, Gonthier P, Hernández-Escribano L, Ioos R, Markovskaja S, Martínez-Álvarez P, Muñoz-Adalia EJ, Nowakowska JA, Oszako T, Raposo R, Santini A, Hantula J (2019) Pine Pitch Canker (PPC): Pathways of pathogen spread and preventive measures. Forests 10(12):1158

Zimin AV, Stevens KA, Crepeau MW, Puiu D, Wegrzyn JL, Yorke JA, Langley CH, Neale DB, Salzberg SL (2017) An improved assembly of the loblolly pine mega-genome using long-read single-molecule sequencing. Gigascience 6(1):1–4

Zobel B, Talbert J (1984) Applied forest tree improvement. Wiley, New York, USA

8 Genomic Advances in Research on Genetic Resistance to White Pine Blister Rust in North American White Pines

Jun-Jun Liu, Jeremy S. Johnson, and Richard A. Sniezko

Abstract

Cronartium ribicola, the exotic pathogen responsible for the devastating white pine blister rust (WPBR) disease, has been killing native white pines in North America since the early 1900s. Large-scale seedling inoculation trials have been underway for more than 50 years, and have revealed major gene resistance (MGR) and quantitative disease resistance (QDR) in several species. Integration of resistance (R) genes and QDR-related quantitative trait loci (QTLs) is needed for long-term WPBR control. The genomic resources and knowledge created in recent years will help improve tree health through novel breeding approaches and selection tools for better utilization of various resistance resources. We review recent advances in genomic research on tree resistance with a focus on putative host R genes and defense response genes. Outputs from omics studies and plant resistance theory are integrated into conceptual models to decipher molecular interactions of white pines with *C. ribicola*. The R candidate genes identified by fine genetic mapping and association studies are highlighted for their applications in developing breeding tools for marker-assisted selection (MAS). In addition, we discuss the challenges and future directions to realize the full potential of omics-based breeding for broad-spectrum and durable resistance against the various *C. ribicola* pathotypes.

J.-J. Liu (✉)
Canadian Forest Service, Natural Resources Canada, 506 West Burnside Road, Victoria, BC V8Z 1M5, Canada
e-mail: jun-jun.liu@NRCan-RNCan.gc.ca

J. S. Johnson
Department of Environmental Studies, Prescott College, 220 Grove Ave, Prescott, Arizona 86301, USA

R. A. Sniezko (✉)
USDA Forest Service, Dorena Genetic Resource Center, 34963 Shoreview Road, Cottage Grove, Oregon 97424, USA
e-mail: richard.sniezko@usda.gov

8.1 Introduction

Five-needle pines, also known as white pines, (in the genus *Pinus* and the subgenus *Strobus*) are important conifer trees in many forest environments, providing an array of ecosystem services. Several of the white pine species are also notable for their economic significance in the forest industry, as a source of timber, fiber, wood biomass, seeds, resin, and other related products. White pine blister rust (WPBR), caused by the fungus *Cronartium ribicola* (J.C. Fisch.), is a devastating disease of white pines. This native Eurasian pathogen was inadvertently introduced to North America in the early 1900s. All North American white pine species are highly

A. R. De La Torre (ed.), *The Pine Genomes*, Compendium of Plant Genomes,
https://doi.org/10.1007/978-3-030-93390-6_8

susceptible to WPBR. Since its introduction, WPBR has become one of the most damaging forest diseases in North America (Kinloch 2003; Maloy 1997, 2001), affecting natural stands of eight of the nine white pine species in the U.S. and Canada, including western white pine (*Pinus monticola* Dougl. Ex D. Don), eastern white pine (*P. strobus L.*), whitebark pine (*P. albicaulis* Engelm.), limber pine (*P. flexilis* James), sugar pine (*P. lambertiana* Dougl.), southwestern white pine (*P. strobiformis* Engelm.), foxtail pine (*P. balfouriana Grev. and Balf.*) and Rocky Mountain bristlecone pine (*P. aristata* Engelm.) (Sniezko et al. 2004; Tomback and Achuff 2010). Several of these species, including western white pine, eastern white pine, and sugar pine are economically important, while others are keystone species in mountainous ecosystems. WPBR continues to spread and cause tree mortality, killing both large trees and trees in regenerated stands, and can result in tree mortality of 95% or more in the areas of highest impact. Due to WPBR and other biotic agents such as mountain pine beetle (MPB, *Dendroctonus ponderosae* Hopkins) and abiotic factors such as fire suppression and climate change, whitebark pine and limber pine are listed as endangered species in Canada, and whitebark pine is classified as endangered by the International Union for Conservation of Nature (IUCN) (Mahalovich and Stritch 2013), and also proposed for listing as 'threatened' in the U.S. (Mahalovich and Stritch 2013; COSEWIC 2014; Government of Alberta 2014; Federal Register 2020). Developing effective strategies to control WPBR will help restore the native white pine ecosystems, as well as encourage the continued use of the species for economic interests in reforestation. The development of populations of white pines with genetic resistance to WPBR will be the key step to future restoration and reforestation efforts.

Over the past 80 years, large, concerted research and tree improvement efforts have been undertaken to breed genetic resistance in the white pine species of highest economic importance. More recently, efforts to develop populations with resistance have expanded to other white pine species. Good progress has been made in screening and breeding for genetic resistance to WPBR (Bingham 1983; Sniezko and Koch 2017), but further work is needed to increase the level of resistance and ensure its durability. In the future, with the availability of expanded genetic knowledge, resistant parents of various *Strobus* species, and new selection approaches facilitated by new tools, there is great potential to shorten breeding cycles and increase WPBR resistance further. In the several decades since the first R gene (*Cr1*) was mapped in sugar pine, using random amplified polymorphic DNA (RAPD) markers (Devey et al. 1995), WPBR pathosystems have become an active area of research on the molecular basis of tree-microbe interactions. Previous reviews (Liu et al. 2004; Richardson et al. 2010; Sniezko et al. 2014) summarized earlier research on the genetic resistance and molecular mechanisms underlying tree defense responses in WPBR pathosystems.

Until relatively recently, the very large genome size has precluded the sequencing of conifer genomes and the development of genomic resources in these species. However, recent and anticipated advances in technologies and falling costs of next generation sequencing (NGS) have helped overcome this barrier, providing the opportunity to more fully understand the underlying nature of resistance to WPBR and how to most efficiently advance disease resistance programs. With rapid advances of omics (genomics, transcriptomics, proteomics, metabolomics, metagenomics, metatranscriptomics, effectromics, and phenomics)-related technologies in recent years, a significant body of white pine research has been undertaken, covering a wide range of scientific disciplines. Very recently, NGS technologies were applied to whole genome sequencing (WGS) of sugar pine (Gonzalez-Ibeas et al. 2016; Stevens et al. 2016), RNA-seq-based transcriptome profiling in sugar pine, western white pine, whitebark pine, and limber pine (Liu et al. 2013b, 2016b, c; Baker et al. 2018), and capture sequencing of the exomes in whitebark pine (Syring et al. 2016) and limber pine (Liu et al. 2019a). Efforts to sequence the genomes of whitebark pine have recently been initiated (David Neale, personal communication). NGS technologies have revealed a large amount of DNA variations throughout genomes of white pines, including single nucleotide polymorphisms (SNPs), for use in high throughput genotyping using various SNP arrays. Genotypic data have been generated for hundreds of thousands of SNP loci using large populations. Fine genetic maps were constructed with high-density DNA markers in sugar pine (Jermstad et al. 2011; Weiss et al. 2020), western white pine (Liu et al. 2017a), limber pine (Liu et al. 2016b, 2019a), and foxtail pine (Friedline et al. 2015). Most recently, genome-wide association (GWAS) studies and fine genetic mapping of QTLs were reported in sugar pine, thus finely dissecting quantitative disease resistance (QDR) against WPBR (Weiss et al. 2020). Omics-related studies have also been performed on different pathotypes of *C. ribicola* to understand their genetic variations and associations with fungal pathogenicity and virulence (Brar et al. 2015; Liu et al. 2015, 2021b).

Genomic progress in both the pine hosts and *C. ribicola* has provided significant insight into genetic architecture across their whole genomes, which will permit a more comprehensive understanding of molecular defense mechanisms underlying the resistance of white pine species to WPBR. This rapid increase in omics knowledge and resources is sharpening strategies that will aid breeding white pines with a broad-spectrum and durable resistance using cost-efficient practices through development and deployment of marker-assisted selection (MAS) and genomic selection tools in genomics-based breeding. This review chapter focuses on the recent understanding of the molecular interactions of white pines with *C. ribicola* at the genome level and applications of omics technologies in breeding programs to improve pine resistance against WPBR in North America.

8.2 Germplasm Resources and Diversity of White Pine Resistance to WPBR

The pathogen *C. ribicola* requires two hosts to complete its life cycle (Fig. 8.1). White pine species are the host species most impacted by this pathogen. The main alternate host plants in North America are species of the genus *Ribes*, commonly known as currants and gooseberries. Five spore stages are required by *C. ribicola* to complete its complex life cycle, with spermatia and aeciospores occurring on the pines, and development of urediniospores, teliospores, and basidiospores on an alternate host (Fig. 8.1). Basidiospores develop on the underside of infected Ribes leaves in late summer or early fall when the weather is cool and wet. The basidiospores are wind-dispersed onto the pine needles, initiating infection by spore germination and penetration of the germ tubes into needle tissues through stomatal pores. The majority of needle spots (or lesions) are developed by the next spring. Fungal hyphae grow and spread from needle infection sites to stems through vascular tissues over the winter, and WPBR symptoms of stem cankers appear in the summer or fall, ultimately resulting in mortality of susceptible trees from stem girdling. Hundreds of infections and subsequent cankers can affect a single tree.

Fortunately, genetic resistance to WPBR is present in North American white pine species, albeit at low frequency. Both major gene resistance (MGR) and quantitative disease resistance (QDR) have now been documented in several white pine species and QDR in others. Host resistance functions at various stages of the rust life cycle throughout the infection process from initial fungal penetration to late stages of disease development. MGR is a type of race-specific resistance occurring at low frequency in natural populations in some portions of the range of some of the white pine species (Kinloch et al. 2003). Like other plants, white pine MGR usually exhibits hypersensitive response (HR)-like reaction, a phenomenon of the rapid and

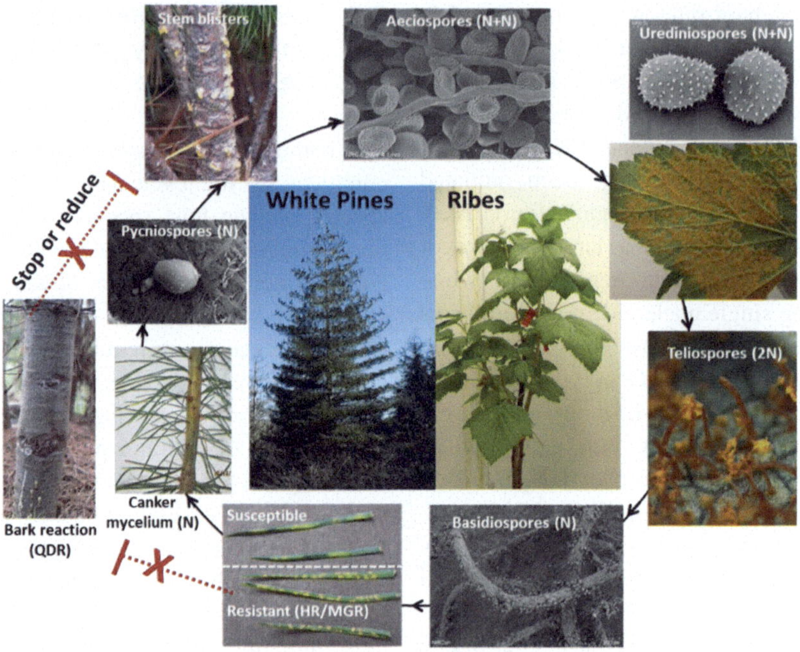

Fig. 8.1 Life cycle of *Cronartium ribicola*. This heteroecious pathogen requires two plant hosts to complete its life cycle. Throughout its five-spore developmental stages, it causes diseases on white pine trees, as well as on alternate hosts (mainly species of *Ribes* in North America). Resistance to white pine blister rust interrupts this life cycle, limiting or preventing its completion. Pine infection by *C. ribicola* occurs by basidiospores, which develop on the alternate host and are transferred by the wind onto pine needles, where their germination produces germ tubes, penetrating into needles through stomata in late summer or early fall. Major gene resistance (MGR) usually functions in the needles and kills the fungus at the local infection site through a hypersensitive-like response (HR), and thus no stem disease symptoms develop in seedlings with MGR genotypes. Susceptible trees show yellow diseased spots that grow in size as the fungal hyphae grow in spring. The fungal hyphae invade the stem bark through vascular tissue, usually leading to swollen stem cankers in the summer or fall. Pycniospores develop at the canker margin and their insect-facilitated mating generates dikaryotic mycelia. Subsequently, aecial blisters develop on these cankers, and release aeciospores, in the early spring at low elevation (and later at high elevation) causing infection in *Ribes*. Quantitative disease resistance (QDR) functions in both needle and stem tissues, slowing or stopping the progression of the fungus, and impairing the development of pycniospores and aeciospores. Urediniospores develop on the lower side of *Ribes* leaves which were infected by aeciospores, and the uredia continue to multiply on infected plants and spread to adjacent currants and gooseberries (*Ribes*) throughout the growing season. In the late summer or early fall, when the weather is cool and wet, telia develop on the infected *Ribes* and generate teliospores, with karyogamy occurring soon after teliospore formation. During germination, teliospores undergo meiosis and produce basidiospores, which then disseminate to white pines, completing the life cycle of the fungus

programmed cell death occurring at the pathogen infection site that results in some suppression of disease progress (Goodman and Novacky 1994). Upon *C. ribicola* infection, HR-like reactions on resistant needles in sugar pine are often expressed as "fleck" spots, characterized by a pale, yellow margin with a necrotic fleck, in the center, or yellow spot with necrotic border, while susceptible diseased needles develop typical yellow and/or red spots. The onset of HR-like spots in resistant white pines differs from that in classical HR observed in model plants, and it may not be observed for several weeks or even months after initial entry of *C. ribicola* (Kinloch and Littlefield 1977). Prior to the onset of cell necrosis, *C. ribicola* hyphae extensively proliferate and spread into needle vascular tissue of the resistant seedlings (Sweeney et al. 2012).

Compared to susceptible spots, HR-like spots of secondary needles showed relatively few dense tannin deposits. HR reactions that were caused by *C. ribicola* infection have been reported in four Strobus species: sugar pine, western white pine, southwestern white pine, and limber pine (Kinloch 1970; Kinloch et al. 1999; Kinloch and Littlefield 1977; Kinloch and Dupper 2002). Host cell death in the HR or HR-like spots commonly stops the progression of the pathogen (Balint-Kurti 2019). The presence of MGR generally results in HR-like reactions, leading to canker-free seedlings. Within seedling families from a MGR parent tree, the Mendelian segregation ratios observed for cankered versus canker-free seedlings indicate one dominant genetic locus controls MGR in these four species. These MGR loci are designated as *Cr1* in sugar pine (Kinloch 1970), *Cr2* in western white pine (Kinloch et al. 1999); *Cr3* in southwestern white pine (Kinloch and Dupper 2002), and *Cr4* in limber pine (Schoettle et al. 2014; Sniezko et al. 2016), respectively. MGR in sugar pine and western white pine show typical HR-like spots on the needles infected by avirulent (Avr) pathotypes, while the needle phenotypes in limber pine and southwestern white pine are often less diagnostic. Generally, MGR blocks fungal invasion into stems from needles, and expresses stem canker-free phenotypes, preventing the completion of the *C. ribicola* life cycle (Fig. 8.1). However, in sugar pine and western white pine, separate virulent pathotypes of *C. ribicola* have arisen in some environments, rendering MGR ineffective at those sites (Kinloch et al. 2004).

By contrast, current data suggests that QDR provides resistance against all known *C. ribicola* races and generally confers a reduction, or in some cases absence of WPBR disease. Strong QDR to WPBR has been well documented in sugar pine, western white pine, whitebark pine, and southwestern white pine breeding programs in the USA and Canada (Kinloch et al. 2012; Sniezko et al. 2008, 2014, 2020), and the other white pine species are currently under investigation. The level of QDR varies by species, with resistant families of western white pine generally showing higher survival than those of sugar pine (Sniezko et al. 2020). Expression of QDR-related phenotypic traits is a complex process, usually occurring in stems one to several years following initial needle infection. In the most susceptible seedlings, the stem cankers are perennial and usually produce aecia in the 2nd spring, often for several subsequent years before eventually killing the trees. QDR-related phenotypic traits are expressed in a continuous distribution of disease scores across a segregating population, but with the levels varying by species. QDR interrupts the *C. ribicola* life cycle in a range of stem responses with a continuum, from completely stopping to reducing or delaying spore outputs at varying degrees. For example, QDR trees may have cankers, but reduced quantities of spores. Pycniospores and aeciospores may develop in some QDR trees with partial bark reaction, in some cases with just a small amount of aecia produced. In other cases, bark reaction may seem to have inactivated the canker, but it becomes active several years later, with the fungus growing vigorously, indicating that this type of QDR only inhibits fungal growth temporarily, probably depending on environmental conditions. In many other cases, the complete bark reaction persists, eventually fading as the tree ages and grows. Normal canker development, partial bark reaction, or complete bark reactions are sometimes displayed in the branches and stems of the same tree. These observed variations of phenotypic traits indicate that many rust races with high genotypic diversity have the potential to infect the same tree during the same infection season, making the assessment of QDR-related phenotypes more difficult due to host's genotype x pathogen's genotype x environment interactions. QDR is influenced by the genetic contribution of many genes with small to moderate effects, thus it is considered as a potential source of durable resistance in white pine breeding.

In seedling inoculation trials of open-pollinated seed collections from natural stands of white pines, most seedling families are very susceptible to WPBR, usually with 95–100% mortality (Kegley and Sniezko 2004; Sniezko et al. 2008). However, a subset of seedlings in a small number of families shows resistance that manifests in various ways, the most notable being (a) fewer needle spots, (b) fewer stem symptoms, (c) no stem symptoms, (d) delayed appearance (one or two years later) of stem symptoms, complete or partial bark reactions, and (e) survival of cankered trees (Hoff et al. 1980; Sniezko et al. 2014). Several of these observed traits of resistant seedlings suggest that the progress of the fungus in both the needle and stem is being slowed relative to that in the highly susceptible seedlings. Many of these resistance traits occur within the same seedling families, leading to higher survival of these families (Sniezko et al. 2014). The inheritance of most of these traits is still uncertain, but potentially quantitative in nature. The strong QDR families have at least some canker-free seedlings or trees —whether this is due to 'needle shed' or 'short-shoot fungicidal' reaction or some other mechanism(s) is unknown.

The level of QDR varies by species. For example, in Pacific Northwest U.S. populations, progenies of thousands of western white pine and sugar pine parent trees have been screened in seedling inoculation trials, as well as in over 1300 progenies of whitebark pine. From these seedling inoculation trials, it was established that there is a higher frequency and level of resistance in whitebark pine than in western white pine, and a higher level in western white pine than in sugar pine. In field trials, WWP also shows higher resistance than sugar pine (Sniezko et al. 2020; Kinloch et al. 2008), and field trials of WBP are just beginning. All of these observations imply a complex genetic basis underlying white pine QDR.

8.3 Major Gene Resistance

8.3.1 Genetic Mapping

Of the four MGR loci (*Cr1-Cr4*), three (*Cr1*, *Cr2*, and *Cr4*) have been mapped in different populations of the corresponding species using several types of DNA markers, including RAPD, AFLP, and SNPs (Devey et al. 1995; Jermstad et al. 2011; Liu et al. 2006; Liu and Ekramoddoullah 2007, 2008). In recent years, NGS and high throughput SNP genotyping approaches have been used to generate a large number of SNPs for fine genetic mapping in conifers (De Miguel et al. 2015; Westbrook et al. 2015; Pavy et al. 2017; Liu et al. 2017a, 2019a), allowing comparative genetic mappings through orthologous genes bridging between species (Fig. 8.2). Comparisons of fine genetic maps between sugar pine, western white pine, limber pine, and loblolly pine (*P. taeda*) anchored *Cr1*, *Cr2*, and *Cr4* on the *Pinus* consensus LG-2, LG-1, and LG-8, respectively (Jermstad et al. 2011; Liu et al. 2016b, 2017a, 2021c; Stevens et al. 2016). The loblolly pine R locus (*Fr1*) against the fusiform rust pathogen *C. quercuum* sp. fusiforme was also mapped to *Pinus* consensus LG-2 (Wilcox et al. 1996; Quesada et al. 2014), but in a position different from *Cr1* when compared using syntenic markers (Neale et al. 2014). Other *Fr* genes (*Fr2* ~ *Fr10*) have since been discovered and mapped to single loci in the loblolly pine genome, with *Fr8* and *Fr10* on LG-10, and LG-11, respectively (Quesada et al. 2014).

Because the majority of functionally well characterized R genes belong to the plant nucleotide-binding site leucine-rich repeat (NLR) family (Kourelis and van der Hoorn 2018), members of this gene family were selected as the focus for identification of R candidates in the syntenic genomic regions of the R locus in sugar pine (*Cr1*), western white pine (*Cr2*), and limber pine (*Cr4*). In sugar pine, physical linkage through WGS revealed the top *Cr1* candidate was a NLR gene (Pila_lg017786) in contig super 6135 of the genome sequence (Pila-v1.0 scafSeq), and its transcript (TR43508|c1_g1_i2|m.82078, later ID updated as Pilahq_017371) was well expressed (Stevens et al. 2016). An association study and fine linkage mapping of the *Cr2* region localized multiple NLR genes in the western white pine, two of them (contig_41490 and contig_58688) were the most promising *Cr2* candidates based on a large survey of seed families originating in western North America (Liu et al. 2014b, 2017a). Several SNPs of the limber pine NLR gene M287456 co-segregated with *Cr4*-conferred MGR-related phenotypes in a fine

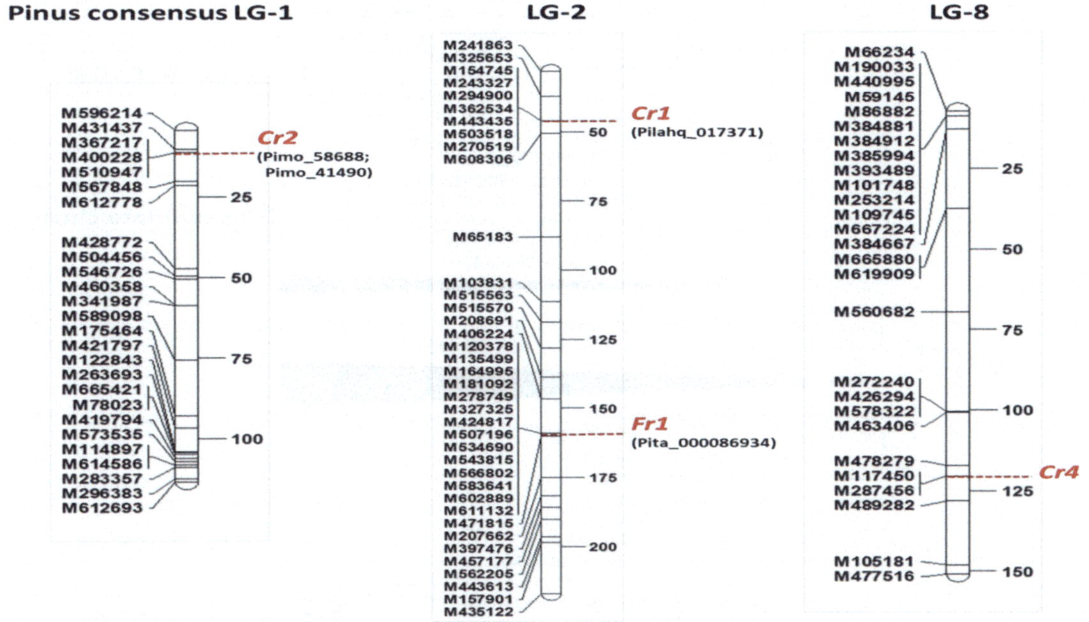

Fig. 8.2 Comparative mapping of *Pinus R* loci. Genetic maps of *Pinus* consensus linkage groups (LGs) are adapted from limber pine genetic maps constructed previously using exome-seq (Liu et al. 2019a). Limber pine putative nucleotide-binding site leucine-rich repeat (NLR) genes were labeled on the left side of the genetic maps. Orthologous positions of *Pinus R* loci were labeled on the right side to show their syntenic relationships. The NLR genes of R candidates are shown in parentheses for the corresponding species: sugar pine (Pila), western white pine (Pimo), and loblolly pine (Pita)

genetic mapping population (Liu et al. 2019a). A loblolly pine NLR gene, with transcript ID: Evg1_1A_all_VO_L_3760_240252 (later ID updated as Pita_000086934), was also considered as the *Fr1* candidate based on the annotation of the draft genome sequence with the RNA-seq-based transcriptome assembly (Neale et al. 2014).

Following the identification of orthologs of these *Pinus* R candidates in related pine species, the conserved nucleotide-binding site (NBS) domain of these NLR genes was used to reveal the phylogenetic relationship among them (Fig. 8.3). Pila_lg017786 and its loblolly pine ortholog (Pita_000073654-RA) encoded NLR-like proteins with an N-terminal Toll/interleukin-1 receptor (TIR) domain (TNL), but both lacked a typical NBS domain. In contrast, its limber pine ortholog (Pifl-DN4702) contains a typical NBS domain, and it was included in the phylogenetic analysis with other limber pine putative TNL genes genetically mapped in a previous study. As shown in the phylogenetic tree (Fig. 8.3), NLR candidates of *Cr1*, *Cr2*, *Cr4*, and *Fr1* grouped with their orthologs and formed separate clusters. Long tree branches among these R candidates suggest they arose from distinct ancestors during the origin of the subgenus *Strobus*, or the genus *Pinus*.

Both phylogenetic analysis of the NLR candidates and comparative genetic mapping indicated that *Pinus* R loci are not orthologous, but paralogous, suggesting that they arose through convergent evolution rather than the conservation of an ancient specificity in the *Pinus* genus (Fig. 8.3). Since *C. ribicola* was only introduced to North America a little more than 100 years ago, the specific R-alleles must have evolved through exaptation long ago, before these white pine species were exposed to this exotic pathogen in North America. The R genes against specific pathogenic races are likely involved in independent evolution among different white pine species, and presumably without a co-

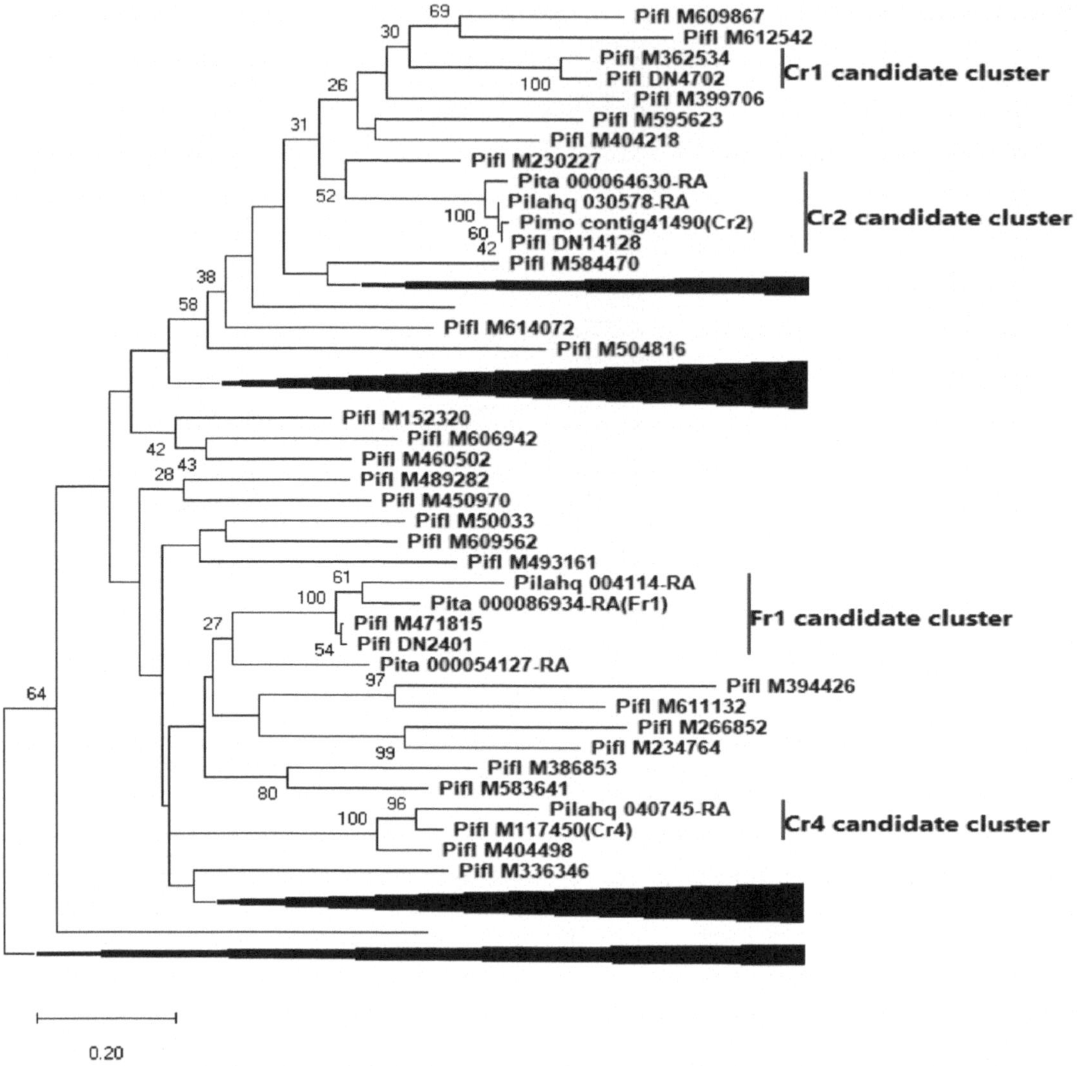

Fig. 8.3 Phylogenetic and evolutionary relationships of putative nucleotide-binding site leucine-rich repeat (NLR) proteins of *Pinus R* candidates. A phylogenetic tree was constructed by maximum likelihood (ML) method (Guindon and Gascuel 2003) with JTT model in 100 bootstrap replicates using MEGA X (Kumar et al. 2018). Only amino acid sequences of the nucleotide-binding site (NBS) domain were included in the alignment analysis using Clustal Omega (Sievers et al. 2011). *Cr1, Cr2, Cr4,* and *Fr1* candidates and their orthologs from sugar pine (Pila), limber pine (Pifl), western white pine (Pimo), and loblolly pine (Pita) were included in the analysis except for the *Cr1* candidate and its loblolly pine ortholog because both lacked a typical NBS domain. Other genes were selected from putative TIR-NB-LRR (TNL) proteins of limber pine because their sequences were genetically mapped in a previous study (Liu et al. 2019a). Scale bar is substitutions per site. The bootstrap values are shown under the tree branches, and branch lengths are proportional to amino acid variation rates. Four large clusters with only limber pine TNL genes were suppressed

evolutionary relationship with *C. ribicola*. This could be explained by the balancing model (Stahl et al. 1999), which suggests that variations of genetic disease resistance are maintained in populations, even if there is a fitness cost for the preservation of temporarily non-functional R-alleles. This hypothesis has some support by other lines of evidence. In limber pine, MGR was documented in seed families originating both in the southern Rocky Mountain region and Alberta, Canada, with presumably little or no gene flow due to geographical isolation since the last glaciation event. The MGR in limber pine in both these areas is controlled by the same R locus (*Cr4*) or genetic loci very close to *Cr4*, as shown by a combination of genetic linkage mapping and an association study using seed families from both areas (Liu et al. 2020a). This observation implies that the specificity of *Cr4* has been conserved since a time point before diversification of the isolated populations. A survey of *Cr4* distribution also found its frequency in Hardy–Weinberg equilibrium across populations in the absence of WPBR, indicating little or no significant pressure on natural *Cr4* selection (Schoettle et al. 2014). These R-alleles may have been co-opted with pleiotropic effects in North American white pines, with unknown functions (such as response to abiotic stress or diseases other than WPBR) before exposure to WPBR. At least two Eurasian white pines, *P. armandii* and *P. morrisonicola*, also appear to possess MGR (Hoff and McDonald 1975; Sniezko et al. 2008), where Eurasian white pines and *C. ribicola* are believed to have co-evolved for a longer period of time.

8.3.2 MGR-Mediated Defense Mechanisms

R-seedlings develop MGR-related phenotypes with HR-like reactions in response to infection by corresponding fungal *Avcr* races (Kinloch et al. 2004). Based on the gene to gene model (Flor 1971), recognition of avirulent effectors by corresponding R proteins (such as *Cr1* vs. *Avcr1*, and *Cr2* vs. *Avcr2*) activates effector-triggered immunity (ETI), resulting in programmed cell death (PCD) around the pathogens' local infection sites through a complex network of signaling cascades (Jones and Dangl 2006). Incompatible interactions of white pines with *C. ribicola* have been investigated in several reports, mainly in western white pine and sugar pine. Both proteomic and transcriptomic approaches were used to characterize host defense responses during incompatible *Cr2*-*Avcr2* interaction at the early stages (0∼96 h) post *C. ribicola* infection (Zamany et al. 2012; Liu et al. 2013b).

Proteomic profiling detected dynamic changes of the needle proteomes in the *Cr2*-resistant seedlings post infection by *Avcr2* races (Zamany et al. 2012). This study revealed that 78 and 10 proteins of *Cr2*-seedlings were significantly regulated by rust infection compared to either the controls at time 0-hour or susceptible genotypes (*cr2/cr2*), respectively (Zamany et al. 2012). Based on protein annotation, these *Avcr2*-regulated host proteins covered well characterized components functioning throughout the plant ETI system, from recognition of pathogenic effectors to expression of downstream defense-responsive genes (Fig. 8.4). The *Avcr2*-regulated host proteins were annotated with putative functions in defense signaling cascades (such as receptor-like kinase-RLK, GTP-binding proteins-GBPs, F-box/FBD/LRR protein, chloroplast ribonucleoprotein, and 14-3-3-like protein), enzymes for reactive oxygen species (ROS) generation and scavenging mechanisms (chloroplastic Cu–Zn superoxide dismutase-SOD, peroxisomal S-2 hydroxy-acid oxidase-HAO, peroxiredoxin-PRX, serine hydroxymethyltransferase-SHMT, pyridoxine biosynthesis protein 1-PDX1), heat shock proteins (HSPs)/chaperonins of various types (HSP90, HSP82, HSP81, HSP70, and mitochondrial CPN60), regulation of gene expression at various levels from transcription (DNA-directed RNA polymerase-like protein, histones), translation (ribosomal proteins), to protein destination (20S proteasome α subunit-like protein, plastid lipid-associated protein). *Cr2*-mediated downstream defense also included altered metabolisms, such as photosynthesis, as well related energy metabolism (RuBisCO large chain-like proteins, fructose-bisphosphate aldolase, ferredoxin-NADP reductase, apocytochrome f, malate dehydrogenase, photosynthesis II 22 kDa protein, and chlorophyll a/b-binding protein), amino acid metabolism (mitochondrial aminomethyltransferase, glutamate dehydrogenase, glutamine synthetase, aspartate aminotransferase), and cell wall reinforcement (nucleoside-diphosphate-sugar epimerase).

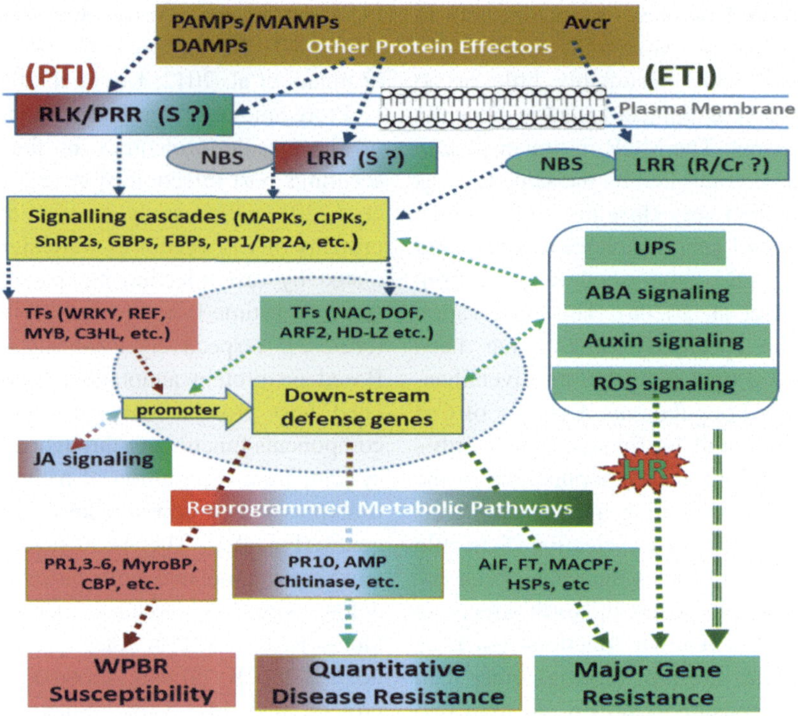

Fig. 8.4 Schematic representation of plant innate immunity. Effector-triggered immunity (ETI) and PAMP-triggered immunity (PTI) are modeled with the molecular signaling mechanisms underlying white pine major gene resistance (MGR), quantitative disease resistance (QDR), and susceptibility to *Cronartium ribicola*. The ETI or PTI model includes three phases as **a** host perception of pathogenic signals, **b** signaling cascades for transduction networks, and **c** downstream defense response. PTI and ETI may share some components of their signaling cascades and defense-responsive genes. **a** Host R proteins of putative NLRs (NBS-LRR) or RLKs perceive avirulent effectors to activate the ETI system during an incompatible interaction (such *Cr1* vs. *Avcr1* or *Cr2* vs. *Avcr2*). In contrast, host pattern recognition receptors (PRRs) or susceptible (S) proteins perceive PAMPs/MAMPs/DAMPs, or other effectors to activate the PTI system during compatible interactions. Some characterized plant PRRs or S proteins are members of the RLK or NLR families. **b** Signaling cascades, including mitogen-activated protein kinases (MAPKs) and other intermediate signaling components, and a series of transcription factors (TFs). These signaling intermediate factors jointly regulate the expression of downstream defense-responsive genes. **c** Expression of downstream defense-responsive gene results in specific defense responses, including induction of various pathogenesis-related (PR) proteins, dynamic changes of plant hormones and metabolic pathways, and accumulation of secondary metabolites

Compared to the proteomic approach, the RNA-seq-based transcriptome profiling provided a more comprehensive insight into incompatible *Cr2-Avcr2* interactions. A total of 789 differentially expressed genes (DEGs) were identified in the *Cr2*-seedlings post infection by *Avcr2* races (Liu et al. 2013b). These DEGs provided details of the *Avcr2*-triggered immunity in western white pine (Fig. 8.4). A group of DEGs encoded NLR, RLK, calcineurin B-like (CBL)-interacting protein kinases (CIPK), F-box family proteins (FBP), and membrane attack complex component/perforin (MACPF) domain proteins. They may be the direct binding targets of *C. ribicola* effectors/avirulent proteins or may play roles as signaling molecules of ETI-activated signaling cascades, given that their homologs have been well characterized as factors of signal pathway networks during plant defense responses (Thordal-Christensen 2020). In addition, several gene families of plant transcriptional factors (TF), including NAC, DOF, PLATZ,

ARF2, zinc-finger HD, nuclear transcription factor Y, and others, were activated during Avcr2-triggered immunity. In turn, these *Avcr2*-activated TFs either downregulated or upregulated the expression of a large set of defense-response genes in western white pine, including apoptosis-inducing factor (AIF), flowering locus T-like protein (FT), subtilisin-like protease, abscisic acid (ABA) receptor, pathogenesis-related (PR) proteins of different families, and a wide diversity of enzymes that putatively participate in several distinct metabolic pathways (such as biosynthesis of cellulose, flavonoid, flavone, and flavonol).

Evidence from both proteomic and transcriptomic studies indicated that the western white pine antioxidant system exhibited a two-phase oxidative burst, similar to the ROS signaling outcome as reported in other well-known plant pathosystems. Plant R-Avr interactions activate intracellular mitogen-activated protein kinase (MAPK) cascades for the rapid generation of ROS (Kawasaki et al. 2017). Dynamic expression of ROS production and scavenging enzymes contributed to this oxidative burst pattern observed in western white pine. ROS are well known as signaling molecules for activation of both local and systemic resistance (Waszczak et al. 2018), and likely contributed to HR-like programmed cell death, as well as cell wall reinforcement processes through callose deposition in the WPBR system. Thus, signaling mechanisms are included in the ETI model and are undoubtedly involved in phenotypic HR expression in *Cr2*-genotypes in response to infection by *Avcr2* races (Fig. 8.4). Other *Avcr2*-activated genes were involved in the ubiquitin-proteasome system (UPS), auxin signaling, and ABA-signaling, suggesting that various signaling pathways might contribute to *Cr2*-mediated MGR in western white pine. For example, of the 22 DEGs in the 26S Ubiquitin-proteasome pathway, 14 were upregulated only in *Cr2*-resistant seedlings.

Interestingly, a large-scale GWAS identified 17 polymorphic genes with significant association with MGR status in sugar pine parental trees (Weiss et al. 2020). The effect sizes of their SNP variations ranged from 8.3 to 21.4%, implying that these genes quantitatively contribute to *Cr1*-*Avcr1* interactions, although they are not candidates of *Cr1*. The main functional categories of these significantly associated genes included NLR, RLK, cytochrome P450, glutathione peroxidase, MORC family CW-type zinc finger, WRKY-TF, and E3 ubiquitin-protein ligase, GTP-binding protein, and stress-regulated genes; putative functions of these genes were consistent with those of the DEGs detected by RNA-seq analysis in the *Cr2*-*Avcr2* interaction.

Despite the significant progress in understanding molecular details of the incompatible *Cr*-*Avcr* interactions underlying white pine MGR, no *R* gene has been functionally characterized yet in the WPBR pathosystems. Fine genetic mapping and identification of NLR genes at the *Pinus* R loci provided candidate genes for further positional cloning and functional verification. Future manipulations of R candidate genes via cutting-edge omics technologies will determine if any of the NLR candidates confer MGR against *C. ribicola* in a white pine species.

8.3.3 Molecular Markers for MGR Selection

In western white pine, two *Cr2* candidates of NLR genes (contig41490 and contig58688) were selected for the development of MAS tools because they were mapped with genetic distances of 1.66 cM and 0.59 cM from *Cr2*, respectively (Liu et al. 2017a). The SNP sites of contig41490-1778M and contig58688-82Y were used to design TaqMan and Kompetitive Allele Specific PCR (KASP) genotyping assays (Liu et al. 2017a, 2020b). The contig41490-1778M-based TaqMan array was first validated using 749 individual samples, a core set of breeding germplasm with wide geographical distribution in western North America (British Columbia, Washington, and Oregon). SNP genotyping confirmed these results with a genotype x phenotype match rate at 95% for those samples with known phenotypic data, including parental trees and progenies of composite *Cr2*-resistant families and 25 QDR-related open-pollinated seed families, which were selected from the current breeding program at the USDA Forest Service's Dorena Genetic Resource Center. This high accuracy for MGR genotype prediction indicated that the contig41490-1778M-based TaqMan array would be a powerful MAS tool for *Cr2*-selection in the germplasm collected from Oregon and Washington (Liu et al. 2017a).

TaqMan and KASP arrays of both contig41490-1778M and contig58688-82Y were further tested in 40 putative Cr2-parental trees and 46 open-pollinated bulk seed-lots randomly selected from the breeding program in British Columbia (Liu et al. 2020b). This study proved both SNP genotyping technologies were consistent with each other, and further confirmed the effectiveness for the SNP marker of contig41490-1778M to predict *Cr2* with 85 and 91% accuracy for seed-lots originating from BC's coastal and interior regions, respectively. The SNP marker of contig58688-82Y provided a highly reliable MAS tool, having a *Cr2*-prediction accuracy of up to 100% across the whole BC landscape, which included both the coastal and interior regions of BC. The current known natural distribution of *Cr2* in western white pine is limited to only portions of the species geographic range. *Cr2* frequencies range from 0.005 to 0.008 in the Sierra Nevada, and 0.001 in the central Cascades from southern Oregon to southern Washington (Kinloch et al. 2003), but no incidence of *Cr2* was detected in other US regions or BC (Kinloch et al. 2003; Hunt et al. 2018). BC parent trees with *Cr2* originate from crosses with *Cr2* parents from Oregon and Washington. These TaqMan tools allow screening of wild parental trees to search for additional MGR parents. However, their utility awaits further validation in western white pine populations originating in California and potentially elsewhere.

Similar investigations were used to select two *Cr4* tightly linked genes (M304083 and M287456) for the development of MAS tools using TaqMan technology for breeding of limber pine MGR (Liu et al. 2020a). The M304083-485Y locus-based TaqMan array was able to detect with high genotype-phenotype matching rates across four MGR segregating seed families from British Columbia: family #7027 at 100%, family PB #2 at 97%, family #6665 at 97%, and family #6470 at 96%. The M287456-700R locus-based TaqMan array was valid with a genotype-phenotype matching rate of 95.4% in family PB #2, but this locus did not match well with genotypic segregation in the other three MGR seed families. Gene M287456 is an NLR candidate of *Cr4*, and thus provides a gene target to search for more SNPs throughout this gene using targeted NGS technologies, such as RenSeq (Jupe et al. 2013) and targeted amplicon-seq (Onda et al. 2018; Liu et al. 2021c).

Integration of R-alleles or R-allelic combinations into elite seed orchards is a long-term process in the conventional breeding of white pines. Because MGR is controlled by single major genetic loci, molecular selection of genotypes with a focus on a few targeted genes allowed accurate prediction of phenotypes in natural stands and offspring of cross-pollinated seed families. Based on *Cr1*-linked genomic DNA sequences (Stevens et al. 2016), a SNP array for *Cr1* selection is currently under validation in sugar pine populations (Arnaldo R. Ferreira, personal communication). DNA markers, genotyping technologies, and related omics resources are transferable from one species to another for molecular breeding of these closely related white pine species. Genomics-based breeding can help shorten conventional breeding cycles. An immediate advantage is that needle tissues can be collected from wild parent trees and genotyped using MAS tools. Knowing which new selections are the strongest MGR candidates in parental trees would reduce the rust inoculation phenotyping, and also confirm the *Cr2* status of parental trees that are used in breeding. In addition, many more trees could be assayed simply by needle collections rather than having to wait for cone production and subsequent seedling inoculation trials. MGR selection by MAS tools would also support the pyramiding of multiple R genes when more R loci in a specific white pine species are found, as well as pyramiding of MGR and QDR. Within-family QDR selections are made in tests that use virulent pathotypes, genotyping of MGR x QDR offspring by MGR-related MAS tools will allow selection of individuals with both QDR and MGR, where their phenotypes may be masked during phenotyping due to infection by virulent races.

8.4 Quantitative Disease Resistance

8.4.1 QTL Mapping and Association Studies

Currently, planting MGR seedlings is an effective and environmentally friendly way to control WPBR in some white pine species when a virulent race is not present. MGR stops *C. ribicola* infection at the needle stage and generally precludes the development of stem symptoms. However, successful breeding for plant disease resistance has often been countered by the pathogens' ability to evolve greater virulence in their population. Plantations with MGR genotypes may enhance selection pressure on *C. ribicola*, potentially resulting in the rapid spread of virulent races (Kinloch et al. 2008). *Cronartium ribicola vcr1* and *vcr2* pathotypes overcome MGR and kill trees of sugar pine with *Cr1/-* genotypes and western white pine with *Cr2/-* genotypes, respectively (Kinloch et al. 2003, 2004). The *vcr2* pathotype is now widespread in western Oregon, negating the use of *Cr2* in many areas (Kinloch et al. 2004; Sniezko et al. 2020). The R gene (*Cr*) of *Ribes nigrum* was also broken down by virulent races, posing a threat to the white pine resource and to cultivated *Ribes* production (Frederick et al. 2011; Munck et al. 2015). Although forecasting the effects of climate change on individual pathogens is a challenging task, especially in natural forest settings with long-lived hosts and short generation times of pathogens (Sturrock et al. 2011), fungal pathogens are able to rapidly adapt to environmental change and host resistance due to their fast generation times and high levels of genetic recombination. Fortunately, the resistance programs for the white pine species have been simultaneously searching for both MGR and QDR. Understanding the genetic architecture of QDR will support white pine breeders in their efforts to more fully utilize QDR-related traits.

The genetic architecture underlying white pine QDR is presumed to contain many genes with small to moderate phenotypic effects (Fig. 8.4). So far, genetic mapping of QTLs for QDR to WPBR has been reported only in sugar pine (Weiss et al. 2020). This work estimated the heritability of sugar pine QDR-related traits at 0.247 and identified four QTLs, (two on LG-5 and a single QTL each on LG-7 and LG-12) using a second-generation full-sib cross, and one additional QTL on LG-4 using a second pseudo-back cross. Gene ontology analysis revealed that genomic regions of these five QTLs were involved in several biological processes, including signal transductions, defense responses to oxidative stress, metabolic processes, and developmental processes. Four of the five QTL regions contained NLR genes, including one on LG-7 (59.81 cM), two on LG-12 (66.08 and 68.89 cM), two on LG-4 (89.18 and 94.01 cM), and three on LG-5 (40.60 ~ 43.75 cM). Oxidative stress-related defense-responsive genes included zf-C3HC4, guanine nucleotide-binding protein, MAPK, fatty acid desaturase 2, thioredoxin, zf_PARP protein, and SMART (COIL) protein.

Weiss et al. (2020) also used GWAS and identified 30 SNPs with significant association to QDR, with effect sizes ranging from 6.2 to 14.5%. These QDR-associated SNPs were annotated with functional categories of stress and defense response (valine-glutamine motif protein, no apical meristem, microsomal glutathione S-transferase gene, and ubiquitin PPAR signaling), carbohydrate metabolism (glucan endo-1-3-beta-glucosidase, UDP-glycosyltransferase, and phosphoglycerate gene), and secondary metabolism (protochlorophyllide reductase and serine carboxypeptidase-like, and pectin biosynthesis (ARAD1 exostosin).

An earlier association study applied a candidate gene-based approach with a focus on genotyping to investigate sugar pine QDR to WPBR, using 384 SNPs selected from genes with putative functions in disease resistance and defense responses to biotic stress and general stress (Vázquez-Lobo et al. 2017). This work found 20 significant alleles of 12 genes significantly associated with sugar pine QDR-related traits such as normal active cankers and blights. Following correction of association significance for multiple testing, three genes with four SNPs were still significantly associated with normal active cankers and blights. The three genes were annotated as ATPase AAA domain protein, protein of unknown function DUF789, and Erythronate-4-phosphate dehydrogenase like protein, respectively. Both QTL mapping and association study suggest that defense-related genes contributed to sugar pine QDR to WPBR. Of all the sugar pine SNPs identified by association studies, their positions on a genetic map have not yet been reported, except the valine-glutamine motif gene (PILA_28470) that localized on LG-3 (79.12 cM).

A candidate gene-based approach was also used to explore the potential association of defense- and resistance-related genes with QDR to WPBR in western white pine and whitebark pine. PR genes of various families are well documented for their roles in plant defense and resistance (van Loon et al. 2006; Liu et al. 2010a, 2014a). Sanger resequencing analysis of PR gene families revealed that several western white pine defense-response genes significantly contributed to QDR, which encoded PR10 proteins (PmPR10-2 and PmPR10-3), antimicrobial protein (PmAMP1), and chitinase (PmCh4B); and DNA variations of these PR genes explained phenotypic variations of QDR-related traits in a range of 5–22 % (Liu et al. 2011, 2021a, 2013a, c).

Unlike sugar pine and western white pine in which both MGR and QDR have been documented, only QDR has been documented at this point in eastern white pine and whitebark pine. In whitebark pine, an association study was performed by genotyping 216 nsSNPs within candidate genes annotated with putative functions in disease resistance and plant defense (Liu et al. 2016c). Nine genetic groups (GG1 to GG9) were identified among 371 whitebark pine seedlings of seedling families, collected from parental trees in British Columbia, Oregon, and Washington. Of the nine genetic groups, GG1 and GG9 had QDR levels significantly higher than GG2, GG4, and GG8. A significant association was also detected between genetic components and geographic variables. Evidence from all these association studies demonstrated that candidate genes of putative resistance-related functions had small to moderate phenotypic effects, and they might act as important genetic factors contributing to the polygenic inheritance of white pine QDR to WPBR.

8.4.2 QDR-Mediated Defense Mechanisms

It is well acknowledged that plant QDR is conferred by the plant immune system (PTI) triggered by pathogen/microbe- or damage-associated molecular patterns (PAMPs/MAMPs, or DAMPs). The PTI signaling cascades regulate the dynamic expression of host defense-responsive genes, which leads to the development of wide-ranging phenotypic traits from disease susceptibility to durable disease resistance, depending on the type of plant-microbe interactions (Corwin and Kliebenstein 2017). Putative pattern recognition receptors (PRRs), either cell surface-localized receptors (such as RLKs), or intracellular receptors of NLRs, are proposed to perceive PAMPs, MAMPs, and DAMPs in the modeled QDR mechanisms (Fig. 8.4). PAMPs/MAMPs include effector proteins secreted by pathogens (Thomma et al. 2011). The QTL-mapping and association studies summarized above identified resistance- and defense-related genes, including NLRs, RLKs, MAPK, and TFs which are putatively functional in PTI-triggered signaling cascades. Although each of the polygenic QTLs or individual genes within a QTL locus may contribute only a small part of the total phenotypic effect, a network of all involved QTLs and genes would function together in a concerted way, potentially explaining a large part of phenotypic variations of the QDR-related traits (Roux et al. 2014).

Because the evolutionary selective pressures are key factors shaping plant disease resistance, genome-scale detection of positive selection adds to our understanding of white pine's adaptation of molecular mechanisms against WPBR infection. A comparative transcriptomic analysis revealed positive selection on a set of resistance-related genes among four white pine species: western white pine, limber pine, whitebark pine, and sugar pine (Barker et al. 2018). This work found 2,025 conserved gene families, and signatures of evolutionary selection were examined. Among 39 conserved gene families under positive selection, the majority (33) were functionally annotated as TFs, chloroplast proteins, and various protein kinases. Of these positively selected white pine genes, NLR, Kelch motif bearing F-box protein (F-box-SKP6), WRKY-TF, E3 ubiquitin ligase, and family 18 glycoside hydrolase are known to be involved in disease resistance in angiosperm plants.

An early proteomic comparison between QDR and susceptible families of eastern white pine also revealed 19 polypeptides specific to resistant seedlings, including six homologs of the leucine-rich repeat (LRR) superfamily and one homolog of HSP90 (Smith et al. 2006a). A recent RNA-seq analysis was performed for a comparative transcriptomic study of whitebark pine using tree groups of uninoculated susceptible, uninoculated resistant, inoculated susceptible, and inoculated resistant needles (Bair 2017). This work identified a variety of differentially expressed genes with annotations related to the plant defense response to fungal diseases, including several encoding PR proteins, enzymes involved in the biosynthesis of secondary metabolic compounds, and a suite of other gene products putatively related to WPBR resistance. In another study, compared to dead trees killed by multiple disturbances including MPB and WPBR, live whitebark pine trees had significantly greater resin duct size, duct area, and relative duct area; suggesting that survivors of recent disturbance invest more in defensive structures than growth (Kichas et al. 2021).

Jasmonic acid (JA) and methyl jasmonates (MeJAs) are well known to have multifunctional roles in plant resistance to biotic and abiotic stressors, including triggering induced systemic resistance (ISR) in plants (Ahmad et al. 2016). Protein phosphatases 1 and 2A (PP1 and PP2A) are the most ubiquitous and abundant serine/threonine phosphatases and play fundamental signaling roles for regulating various biological processes in eukaryotic cells (Takai et al. 2018). Following treatments of either MeJA or PP1/PP2A inhibitor okadaic acid on western white pine seedlings, chitinase proteins of class IV (PmCh4A) and PR10 proteins were significantly induced (Liu and Ekramoddoullah 2006), and both signaling chemicals induced one isoform that was exclusively expressed in seedlings with a QDR trait of slow canker growth (SCG) following WPBR infection (Liu et al. 2005). These results suggest that hyper-phosphorylation of target proteins is a step in the signaling pathway for western white pine QDR against WPBR.

A transcriptomic profiling study provided more comprehensive insight into the roles of the JA/MeJA-triggered ISR against WPBR in whitebark pine (Liu et al. 2017b). RNA-seq-based transcriptomic profiling demonstrated that the transcriptomic reprogramming substantially overlapped between MeJA treatments and infection by WPBR or insect infestation. Annotations of differentially expressed genes (DEGs) revealed several biological processes, including defense signaling (RLKs, NLRs, and phytohormone signaling-related proteins), transcription regulation (TFs of different families), biosynthesis of secondary metabolites, and others that were regulated by MeJA treatment. In particular, four biosynthetic pathways (biosynthesis of antibiotics; flavonoid biosynthesis; phenylalanine; tyrosine and tryptophan biosynthesis; and phenylpropanoid biosynthesis) were coordinately upregulated by treatments of MeJA and attacks from both the WPBR fungus and pests, while biosynthetic pathways of isoquinoline alkaloid, ubiquinone, and other terpenoid quinones were notable among the defense responses limited to both WPBR infection and MeJA treatment. This is not surprising, because several types of secondary metabolites (such as lignin, tannins, and some phenolics) in foliar tissue were positively associated with increased resistance to foliar pathogens in lodgepole pine (Wallis and Reich 2010). Similarly, treatment of Chinese white pine (*P. armandi* Fr) by MeJA increased the total monoterpene in the phloem and xylem at levels significantly higher than control or untreated saplings (Pham et al. 2015). Whitebark pine resistant seedlings contained higher concentrations of multiple terpenes, showing a negative correlation with disease severity, but a positive correlation with the number of needle spots (Bullington et al. 2018). Evaluation of needle epicuticular wax between resistant and susceptible eastern white pine implicated the role of epicuticular waxes as a resistance mechanism (Smith et al. 2006b). Comparisons of volatile organic compounds (VOCs) consistently distinguished resistant species from those susceptible to WPBR or MPB, suggesting VOCs play a central role in the environmental adaptation of white pines and in their defense against *C. ribicola* and other biotic stresses (Runyon et al. 2020). These results imply that a set of secondary metabolites may contribute to white pine QDR against WPBR (Liu et al. 2017b). Relationships between induced accumulations of secondary metabolites and QDR can be better understood by employing multiple omics approaches, such as metabolomics-related high-throughput analyses.

Evidence from all these studies provides support that the predominant group of QDR-involved genes within white pine PTI were those potentially involved in downstream defense responses. Several defensive downstream genes (such as PR proteins of different families and enzymes of different metabolisms) were significantly up-regulated by WPBR infection and wounding treatment at transcriptional or/and translational levels (Liu and Ekramoddoullah 2006; Liu et al. 2010a, b, Liu et al. 2013b; Zamany et al. 2012). However, the biological complexity underlying QDR and the technical difficulty for functional analysis in the WPBR pathosystems have hindered molecular verification of the causal genes. Transient gene expression by Agroinfiltration provided an experimental approach in western white pine for functional genomics studies (Ma et al. 2020). Up to now, only a few western white pine genes have been functionally confirmed for their contribution to QDR. These include PmAMP1 and PmPR10-3, which showed inhibitory activity to *C. ribicola* as well as a wide range of other fungal pathogens (such as *Alternaria brassicae*, *Leptosphaeria maculans*, *Coniferiporia sulphurascens*, *Phoma exigua*, *Phoma argillacea*, *Ophiostoma clavigerum*, *Ophiostoma montium*, and *Sclerotinia sclerotiorum*) by in vitro antifungal arrays using purified proteins (Zamany et al. 2011; Liu et al. 2020a). Our laboratory also expressed PmCh4A and thaumatin-like protein (PmTLP) in *E. coli*, and both recombinant proteins were shown to have in vitro antifungal activities (Liu et al. unpublished data). Heterogeneous gene over-expression of PmAMP1 enhanced disease resistance against several fungal pathogens in transgenic *Brassica napus* plants (Verma et al. 2012).

PTI and ETI shared some components of the signaling cascades (such as NLRs, RLKs, and MAPKs), and many defense-responsive genes (such as PR proteins) during downstream defense responses (Fig. 8.4). Nevertheless, PTI activates downstream components of defense responses in a slower and less robust manner compared to ETI (Naveed et al. 2020). Although a large array of candidate genes was detected by QTL mapping, GWAS, and other omics approaches, further analyses are required for extensive functional validation of more components that potentially contribute to white pine QDR, especially for factors up-stream of the PTI system.

8.4.3 Molecular Markers for QDR Selection

As discussed above, QTL mapping and association studies identified a number of protein and DNA markers, genes, and QTLs associated with white pine QDR in sugar pine, western white pine, and whitebark pine. These QDR candidates and genomic regions provide a potential for the development of molecular tools for MAS in breeding and conservation programs. In several cases, pyramiding resistance QTLs has been applied to increase resistance levels in agricultural crops (Pilet-Nayel et al. 2017). Application of molecular markers with the help of MAS is a promising avenue toward breeding integration of multiple R genes and QTLs into elite seed-lots with durable resistance.

Although conventional white pine breeding programs are making good progress with the identification of seed families with QDR in several species, only a few effective MAS tools have been developed from QTLs and QDR-associated genes to use in breeding. In BC's western white pine breeding program, *Cr2* was incorporated using MGR parent trees from Oregon. The Oregon MGR parents were intermated with local parent trees to establish a *Cr2* × SCG seed orchard. Hybrid offspring carrying *Cr2* were identified using SNP markers of Cr2 candidates of the NLR genes as mentioned above. QDR phenotypes (SCG traits expressed in the stems) would have been masked by the *Cr2*-mediated MGR, because no *vcr2* has yet been found in BC (Hunt et al. 2018). The SCG genotypes expressed two isoforms, while susceptible genotypes expressed only one isoform of a class IV chitinase, PmCh4A (Liu et al. 2005). Therefore, a 26-kDa PmCh4A isozyme, detected exclusively in the SCG genotypes, provides a typical protein marker for MAS in BC's western white pine breeding program.

Compared to protein markers, DNA markers, especially large numbers of SNPs, are more convenient in white pine QDR selection. Currently, MAS tools are becoming accessible to breeders by using QDR-tightly linked SNPs, simple sequence repeats (SSRs), and related DNA markers. All DNA markers available from the previously mentioned QTL mapping and association studies can be used to design genotyping arrays to detect and exploit naturally occurring DNA polymorphisms in different white pine species (Weiss et al. 2020). For example, an indel marker of *PmCh4B*, associated with QDR levels in selected seed families originating in various regions of Oregon and Washington, was adapted into a gene-specific PCR marker with 74-bp length difference. Of three expected genotypes: L/L (318 bp), L/S (318 and 244 bp), and S/S (244 bp); Tukey's HSD test indicated that the heterozygous genotype (L/S) had QDR levels significantly higher than genotype L/L (Liu et al. 2011).

SSR markers are co-dominant DNA markers and are widely used in MAS in agricultural breeding programs because of their ease of availability, abundance, and ubiquitous distribution throughout the genome. SSRs are also a simpler and more cost-effective genotyping technique than other molecular techniques and possess high allelic diversity. A large number of SRR markers are available in white pine species (Liu and Hammett 2014), and single SSR markers may detect genotypes much more readily than a single SNP marker. The discovery of a rare octanucleotide (AATTATTT)n SSR marker within the PR10 gene intron allowed detection of six alleles (H1 ~ H6) of the *PmPR10-2* gene (Liu et al. 2013a). Of 15 potential PmPR10-2-related diploid genotypes, 11 were actually detected among the seed families of the USDA Forest Service's breeding program, and the genotype H3:H3 had the highest mean levels of QDR against WPBR. Although SNPs are replacing SSRs as the preferred DNA markers (Vieira et al. 2016), integrating high-level PCR multiplexing, such as Amplicon-seq and the Fluidigm Access Array system (Fluidigm Corporation 2015; Onda et al. 2018), with NGS technologies would increase SSR screening efficiency. Some NGS technologies, such as next generation-targeted amplicon sequencing (NG-TAS) and targeted amplicon-seq can be adapted to genotype thousands of SSR loci using one run of NGS for large populations. SSR markers will serve as practical molecular tools for germplasm characterization, gene tagging, and molecular breeding studies in white pine species (Liu and Hammett 2014).

The results presented above clearly highlight that application of these kinds of molecular tools would shorten the selection of QDR genotypes from at least six or seven years, the time period for inoculation trials, to a few months at an earlier stage of development for within-family or across-family selection. However, it is worthy to point out that MAS tools targeting a few genes may be too narrow for QDR selection, because QDR involves many genes for the development of complex traits with environmental interactions. MAS tools such as SNP arrays containing hundreds of targeted genes, or genomic selection would be more powerful for QDR selection, especially if underlying QTLs have only small phenotypic effects and strong environmental effects. So far QTLs for QDR have been mapped only in sugar pine. Identification of robust QTLs enables the design of reliable marker systems to select for these QTLs. Through marker-assisted forward selective breeding, these QTLs can be integrated into elite seed orchards with complex genomic backgrounds for their deployment in breeding without associated genetic drag.

As the genomic basis is characterized for QDR in additional white pine species, especially for those species where no MGR has been documented (such as whitebark pine and eastern white pine), further comparative omics studies may discover more durable resistance mechanisms conserved across the *Strobus* subgenus. For example, a few characterized plant QDR loci are canonical immune receptors such as NLRs and PRRs (Zhang and Coaker 2017). Due to the large genome size of conifers and complicated QDR inheritance, further work will be needed in the development of multiple QTL-mapping populations in individual white pine species. Previous studies provided new insight into the regulation of gene expression, genetic variations, and evolutionary mechanisms of the NLR families in white pines (Liu and Ekramoddoullah 2003, 2007, 2011; Liu and Xiang 2019; Liu et al. 2019a). A more detailed and comparative dissection of white pine QDR will elucidate molecular interactions of PPRs-PAMPs in the WPBR pathosystem. Once such NLRs and PPRs are identified in white pines, stacking multiple PRRs and/or NLRs will provide the potential for broad-spectrum QDR to diverse *C. ribicola* races, helping to provide a significant promise of durable resistance in field plantations and forest restoration.

8.5 Novel Approaches and Future Directions

8.5.1 Phenome-Wide Association Study (PheWAS) to Identify New Resistance Resources

An integration of omics-based approaches in different disciplines with machine learning is proposed for advanced breeding (Fig. 8.5). Phenotyping has been recognized as a bottleneck of selecting and breeding for plant disease resistance. QDR-related traits are complex, manifesting themselves in a range of phenotypes over several years, and may be strongly affected by environmental factors. Including a large number of different phenotypes is necessary for a comprehensive dissection of the genomic basis underlying white pine QDR through the application of phenome-wide association study (PheWAS). GWAS usually considers a few predefined phenotypes to test if any DNA markers have phenotypic associations by exploring genome-wide DNA variations, while PheWAS utilizes all the available phenotypic data to explore both marker-trait associations and associations among multiple phenotypic networks (Roden 2017). Currently, several phenomic technologies are showing promise of large-scale application for disease resistance screening, including non-invasive imaging followed by artificial intelligence (AI)-assisted image analysis, spectroscopy, robotics, and high-performance computing management of phenomics databases (Mir et al. 2019). These high throughput phenotyping approaches with systematic collection of a large number of different phenotypes will help to alleviate the phenomics bottlenecks in the breeding of white pines. Open-source genomics and phenomics databases, such as TreeGene (Falk et al. 2019), provide central hubs of repositories to communicate and integrate genetic, phenotypic, and environmental data from different tree species, coordinating necessary information for comparative studies of forest trees

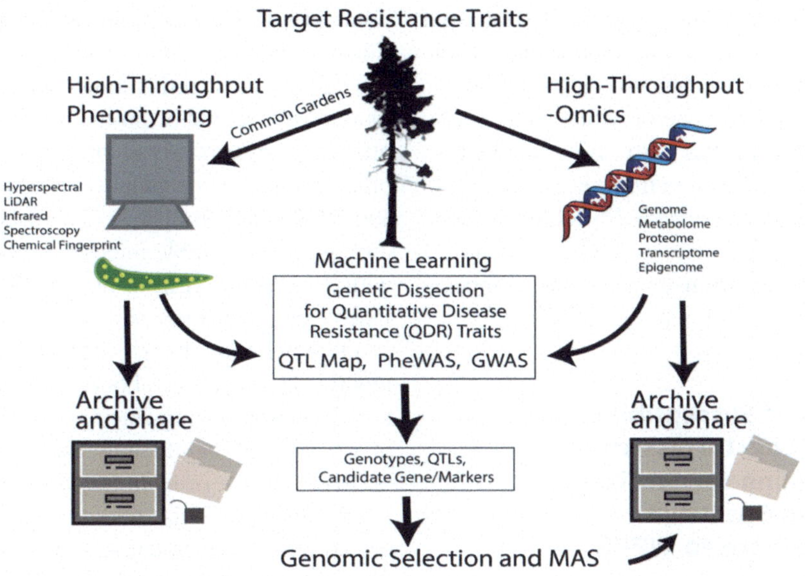

Fig. 8.5 A general workflow for combining omics-based approaches coupled with machine learning for advanced breeding with the application of genomic selection and marker associated selection (MAS). In order to speed up the process of identifying and breeding genetically resistant five-needle pines to WPBR, both genomics and phenomics must be combined. Other omics related to metabolome, proteome, transcriptome, and epigenome use will depend on the target resistance trait. Genomic SNPs can be identified and correlated with phenotypes identified with hyperspectral remote sensing and LiDAR in a genome-wide association study (GWAS), or phenome-wide association study (PheWAS). The identified candidate genes can then be used for marker-aided selection (MAS). Genomic and phenomic analysis may proceed independently; findings should be archived and shared via an open and accessible database that includes raw data and metadata. Likewise, the combined analysis and genomic selection/MAS should also be archived and shared in an appropriate database

(including white pines). More importantly, current MAS tools are more effectively applied to the breeding selection of the large-effect QTLs. Small-effect QTLs are easily ignored during MAS, but they may contribute to the trait either in a synergistic or additive way. Availability of small-effect QTLs is essential for the breeding program to maintain the full gamut of resistance factors in the germplasm, which may be vital in ensuring the durability of resistance in this long-lived tree species. Before the convenient application of genomic selection becomes feasible, phenotypic selection through cutting-edge phenomic technologies provides potential methods toward developing desired phenotypes that are contributed by small-effect QTLs.

A chemical fingerprinting technique, Fourier-transform infrared (FT-IR) spectroscopy, is under development for high throughput phenotyping of chemical traits post disease infection. Preliminary results indicate that FT-IR can show distinct profiles between different phenotypes, potentially predicting resistance-associated traits in the progenies of sampled trees in a few forest pathosystems, including WPBR-infected whitebark pine (Villari et al. 2017). Hyperspectral imagery is another approach to detect differences in response of white pine trees following infection by WPBR. Infection by WPBR usually turns diseased needles yellow and then rusty red, which can be captured from the airborne HyMap sensor. Hyperspectral remotely sensed images were used to determine the locations of stress and mortality in whitebark pine crowns (Hatala et al. 2010). Hyperspectral imaging was also used to track the progression and severity of WPBR infection in southwestern white pine seedlings and was 87% accurate in detecting infection over the entire duration of the experiment (Haagsma et al. 2020). Despite these preliminary efforts with high throughput phenotyping and availability of a large number of white pine SNPs, currently, PheWAS has only been applied in humans. The application in trees, including white pines, still awaits the availability of phenotypic data from plant phenomics studies conducted over a range of environmental factors (temperature, precipitation, soil nutrients, etc.).

8.5.2 Loss of Susceptibility and Effector-Assisted Breeding for Strobus Broad-Spectrum Resistance to WPBR

Loss of susceptibility breeding (Pavan et al. 2010) and effector-assisted breeding (Vleeshouwers and Oliver 2014) have great potential for application in white pine breeding to facilitate the development of broad-spectrum resistance to WPBR. Susceptible (S) genes enable compatible plant–microbe interaction and facilitate pathogen penetration, establishment, and proliferation, leading to successful pathogenic disease development on susceptible plants (van Schie and Takken 2014). They encode PRRs as targeted by PAMPs or effector proteins and have a dominant inheritance. Loss of an S gene function results in recessively inherited resistance against a broad-spectrum of pathogenic races. There are cases where this has happened naturally, as well as cases where genetic manipulation has induced it through gene mutagenesis in several agricultural crops (Gorash et al. 2021). Although naturally occurring recessive resistance to WPBR was postulated for 'needle shed' and 'short shoot' resistance (McDonald and Hoff 1970; Hoff and McDonald 1971), the hypothesis of recessive WPBR resistance is still in question and needs to be confirmed (Kinloch et al. 2008). New resistance screening approaches (such as high throughput phenotyping mentioned above) may speed up the discovery of S genes in the WPBR pathosystem. Effector-assisted breeding may provide another strategy for the identification of S genes, because transcriptomes and/or proteomes have been well profiled in susceptible genotypes of several white pine species. Some putative PRRs of the RLK and NLR gene families were induced specifically during the compatible interactions of white pines with *C. ribicola* (Liu et al. 2013b), which provides candidate S genes.

Of potentially higher interest are the recent reports documenting about 1,000 secreted proteins and other pathogenicity-related genes, some of which were significantly associated with *vcr2* or *Avcr2* in *C. ribicola* populations (Liu et al. 2015, 2021b). This finding opens up many opportunities for functional studies to validate the roles of effectors. One of the *C. ribicola* secreted proteins was functionally demonstrated to be an effector, and its transient expression in tobacco and Arabidopsis affected host resistance to *Pseudomonas syringae* through regulating a set of defense-responsive genes (Ma et al. 2019). This case study provided an experimental system for the successful identification of *C. ribicola* effector genes. Such effectors with validated virulent or avirulent functions can be used to identify and then deploy white pine S genes. Omics knowledge accumulated so far for both host defense response and pathogen infection mechanisms should allow the application of candidate gene approaches to explore S genes and PRRs in a white pine species. Traditional resistance selection depends on the frequency of existing genetic variation. Knockout of candidate S genes will expand genetic resources for enhanced resistance to WPBR in the near future. A multi-locus allele replacement can be realized through genome editing technologies, such as the CRISPR/Cas-9 platform (Cong et al. 2013). Because some S genes (such as *Mlo*) are conserved between monocots and dicots (Kusch and Panstruga 2017), the conserved S genes identified in one white pine species may have orthologous genes among a wide range of conifer species for broad-spectrum resistance.

8.5.3 Epigenetic Resistance and Climate Change

The heritability of QDR-related traits is limited in white pines. Phenotypic variations of white pine QDR-related traits, explained by QTLs and associated genes identified so far, were at relatively low to moderate levels. Therefore, the effects of epigenetic changes need to be considered for WPBR management during tree-pathogen-environment interactions. Epigenetics investigates changes in gene expression resulting from (de)methylation of DNA and modifications of histones/chromatin (López Sánchez et al. 2016; Moler et al. 2018; Agarwal et al. 2020). Epigenetically modulated gene expression is probably the major factor contributing to the phenotypic variation that does not result from DNA sequence variation. Spontaneous changes of DNA (de)methylation can lead to a phenotypic change in plants (Johannes and Schmitz 2019). During evolution, plants have adapted their immune system for a quick response to environmental changes through adjusting DNA methylation patterns (Chinnusamy and Zhu 2009). This adaptation to environmental changes is especially important for tree species, as their life spans hundreds to thousands of years. Climate change is expected to affect both host plants and their pathogens. For example, temperature fluctuation is a key determinant for microbial invasion and host evasion (Cheng et al. 2013). Evidence from currently available data points toward a conserved role of epigenetics in responding to temperature change (MaCaw et al. 2020).

Comparison of DNA methylation (an important epigenetic mark) between leaves from different branches of a 330 year old poplar tree found that epimutations accumulate continuously as a function of the tree's age, and some of the identified epimutations are associated with variations in gene expression (Hofmeister et al. 2020). Interestingly, grafts from 10 to 80 year old eastern white pine trees were infected by WPBR at progressively lower rates as the age of the graft increased (Patton 1961). This research led us to ask if epigenetic variation arising during the lifetime of long-lived trees makes any contribution to the age-related resistance in white pines. No epigenetic nor epigenomic study has been reported on a white pine species yet. Future works, such as comparative epigenomic profiling between white pine QDR genotypes and case studies of susceptible seed families by treatment of 5-azadeoxycytidine (an effective inhibitor of methyltransferase), can be performed as a starting point to understand the environmental effects on the development of QDR-related traits and tree fitness in WPBR pathosystems. Elucidation of epigenetic effects and underlying molecular mechanisms is an essential part of understanding how white pines might adapt to climate change and may highlight novel strategies for long-term WPBR management.

8.5.4 Microbiota-Mediated Resistance Against WPBR

Plants live together with a diverse microbial community, including bacteria, archaea, protists, fungi, and viruses, collectively referred to as the microbiome (Dastogeer et al. 2020). Of the thousands of microbial species that exist along with a single plant species, very few are pathogenic to this host plant, but other organisms of the microbiome can have a positive (mutualistic), or neutral (commensalistic) influence on plant fitness and health (Vorholt et al. 2017; Hassani et al. 2018). Beneficial plant-associated microbes play important roles to enhance host plant growth, nutrient uptake, stress tolerance, and resistance to pathogens. Variation in endophyte community structure in whitebark pine correlated with host locations, tree sizes, and the presence of alternate host plants (*Ribes* spp. or *Castilleja* spp.) near sampled pines (Moler and Aho 2018). Whitebark pine seedlings colonized by host-specific fungi grew faster with higher biomass and higher foliar nitrogen content (Jenkins et al. 2018). Characterization of fungal endophyte communities revealed that whitebark pine genotypes filtered the locally available fungal propagules and structured fungal endophyte communities in a common garden. Furthermore, significant correlations between multiple fungal endophytes and WPBR disease characteristics and terpene concentrations suggested that endophytes have the potential to be developed as biocontrol agents to protect whitebark pine from *C. ribicola* infection (Bullington et al. 2018). Metatranscriptomics studies of western white pine cankered tissues showed that over 8% of the total transcripts were assigned to organisms (including viruses, bacteria, protists, and others) other than western white pine and *C. ribicola*, indicating that a complex microbiome associates with the development of WPBR disease (Liu et al. 2016a, 2021b). Interestingly, these metatranscriptomics studies also found that mycoviruses and RNA-dependent RNA polymerases were associated with *C. ribicola* virulence (Liu et al. 2019b, 2021b). Several viruses have been developed as biological control tools for the effective management of pathogenic fungi and parasites (Nuss 2005; Ghabrial et al. 2015).

Application of related high-throughput technologies (metagenomics, metatranscriptomics, metametabolomics, 16S rRNA gene sequencing, 18S rDNA/ITS gene sequencing, etc.) has the potential to identify thousands of microbial taxa and to compare profiles of microbiomes across a large number of samples, increasing omics knowledge of the complex WPBR systems. With the integration of multi-omics approaches, this type of metagenome-wide association study (MWAS) allows exploration of associations between microbial taxa or microbial functions and plant phenotypic traits (Gilbert et al. 2016; Beilsmith et al. 2019). Following the identification of microorganisms with beneficial or harmful effects on host plants, synthetic microbial communities (SynComs) can be designed with more flexibility to improve microbiota-mediated resistance (Vorholt et al. 2017; Pascale et al. 2020; Vannier et al. 2019). Thus, successful implementation of microbiota-mediated resistance for improved resistance to WPBR will depend on our further investigations to understand white pine-probiotics-*C. ribicola*-environment interactions in natural ecosystems. Availability of more omics resources about the baseline of microbiome diversity in natural white pine stands, as well as their roles and interactions in alleviating abiotic and biotic stressors, should provide guidance on how to help ensure seedlings are best prepared for the environments where they will be planted for reforestation and restoration.

8.6 Conclusions

Because white pines provide important ecological and economic roles where they occur, we must continue to fight to protect them from the spread and evolution of *C. ribicola*. We illustrate the role of genomics in identifying patterns of natural genetic resistance in white pines. The use of omics, including characterizations of genomic SNPs, transcripts, and proteins, has allowed the identification of candidate genes associated with MGR and QDR. By associating genomic regions, proteins, and metabolites with phenotypic traits, patterns of defense mechanisms, including defense signaling and regulatory genes, have been used to identify functional components of host defense to WPBR. The information synthesized in this review expands the knowledge about target genes for the identification of white pine resistance to WPBR and supports innovations for disease control in North American white pines. Despite the progress to date, future studies must speed up the process of QDR candidate gene identification along with the functional role of the genes. On balance, rapid, automated, high spatial and temporal phenotyping must be used to further characterize the traits associated with QDR. Novel research frontiers in epigenetics and microbiota associations could further shed light on white pine resistance to WPBR. The merger of omics and phenotyping will allow a more nuanced understanding of QDR to emerge and will further speed up the identification and breeding of elite trees that can be put on the landscape for restoration and reforestation in the face of an evolving pathogen (Fig. 8.5).

Utilizing these new tools for maximum benefit will take continued support, as well as further focused collaborations between those working directly with omics and those in tree breeding. With the wealth of materials and background accumulated, and with further developments, the WPBR system has the potential to serve as a model for developing disease resistance in forest trees.

Acknowledgments Projects in the J-JL laboratory are funded mainly by the Pest Risk Management Project of the Canadian Forest Service. JSJ is supported by a USDA STDP grant #STDP-R6-2019-1. The authors would like to acknowledge the collaboration facilitated by USDA-FS, and helpful comments by Holly Williams, Arezoo Zamany, and Angelia Kegley greatly improved this review.

References

Agarwal G, Kudapa H, Ramalingam A, Choudhary D, Sinha P, Garg V, Singh V, Patil GB, Pandey MK, Nguyen HT, Guo B, Sunkar R, Niederhuth CE, Varshney RK (2020) Epigenetics and epigenomics: underlying mechanisms, relevance, and implications in crop improvement. Funct Integr Genomics 20:739–761

Ahmad P, Rasool S, Gul A, Sheikh SA, Akram NA, Ashraf M, Kazi AM, Gucel S (2016) Jasmonates: multifunctional roles in stress tolerance. Front Plant Sci 7:813

Bair ZJ (2017) Candidate genes associated with blister rust resistance in whitebark pine (*Pinus albicaulis*) revealed through de novo transcriptome assembly and differential gene expression analysis. https://ir.library.oregonstate.edu/concern/graduate_thesis_or_dissertations/dj52w9689

Baker EAG, Wegrzyn JL, Sezen UU, Falk T, Maloney PE, Vogler DR, Delfino-Mix A, Jensen C, Mitton J, Wright J, Knaus B, Rai H, Cronn R, Gonzalez-Ibeas D, Vasquez Gross HA, Famula RA, Liu J-J, Kueppers LM, Neale DB (2018) Comparative transcriptomics among four white pine species. G3 (Bethesda) 8(5):1461–1474

Balint-Kurti P (2019) The plant hypersensitive response: concepts, control and consequences. Mol Plant Pathol 20(8):1163–1178

Beilsmith K, Thoen MPM, Brachi B, Gloss AD, Khan MH, Bergelson J (2019) Genome-wide association studies on the phyllosphere microbiome: embracing complexity in host-microbe interactions. Plant J 97:164–181

Bingham RT (1983) Blister rust resistant western white pine for the Inland Empire: The story of the first 25 years of research and development program. In: USDA For Serv Gen Tech Rep INT-146. p 41

Brar S, Tsui CK, Dhillon B, Bergeron MJ, Joly DL, Zambino PJ, El-Kassaby YA, Hamelin RC (2015) Colonization, history, host distribution, anthropogenic influence and landscape features shape populations of white pine blister rust, an invasive alien tree pathogen. PLoS One 10:e0127916

Bullington LS, Lekberg Y, Sniezko R, Larkin B (2018) The influence of genetics, defensive chemistry and the fungal microbiome on disease outcome in whitebark pine trees. Mol Plant Pathol 19(8):1847–1858

Cheng C, Gao X, Feng B, Sheen J, Shan L, He P (2013) Plant immune response to pathogens differs with changing temperatures. Nat Commun 4:2530

Chinnusamy V, Zhu J (2009) Epigenetic regulation of stress responses in plants. Curr Opin Plant Biol 12 (2):133–139

Cong L, Ran FA, Cox D, Lin S, Barretto R, Habib N, Zhang F, Zhang F (2013) Multiplex genome engineering using CRISPR/Cas systems. Science 339:819–823

Corwin JA, Kliebenstein DJ (2017) Quantitative resistance: more than just perception of a pathogen. Plant Cell 29:655–665

COSEWIC (2014) COSEWIC assessment and status report on the limber pine *Pinus flexilis* in Canada. Committee on the Status of Endangered Wildlife in Canada, Ottawa, pp 9–49

Dastogeer KM, Tumpa FH, Sultana A, Akter MA, Chakraborty A (2020) Plant microbiome–an account of the factors that shape community composition and diversity. Curr Plant Biol 23:100161

De Miguel M, Bartholome J, Ehrenmann F, Murat F, Moriguchi Y, Uchiyama K, Ueno S, Tsumura Y, Lagraulet H, de Maria N, Cabezas JA, Cervera MT, Gion JM, Salse J, Plomion C (2015) Evidence of intense chromosomal shuffling during conifer evolution. Genome Bio Evol 7:2799–2809

Devey ME, Mix AD, Kinloch BBJr, Neale DB, (1995) Random amplified polymorphic DNA markers tightly linked to a gene for resistance to white pine blister rust in sugar pine. Proc Natl Acad Sci USA 92:2066–2070

Falk T, Herndon N, Grau E, Buehler S, Richter P, Zaman S, Baker EM, Ramnath R, Ficklin S, Staton M, Feltus FA, Jung S, Main D, Wegrzyn JL (2019) Growing and cultivating the forest genomics database, TreeGenes. Database (Oxford) 2019:baz043

Federal Register (2020) Endangered and threatened wildlife and plants; Threatened species status for *Pinus albicaulis* (Whitebark Pine) with section 4(d) Rule. A Proposed Rule by the Fish and Wildlife Service on 12/02/2020

Flor HH (1971) Current status of the gene-for-gene concept. Annu Rev Phytopathol 9:275–296

Fluidigm Corporation (2015) Access array system for illumina sequencing systems user guide. PN 100–3770 H1. Fluidigm Corporation, South San Francisco, CA

Frederick ZA, Villani S, Cox KD (2011) First report of white pine blister rust caused by *Cronartium ribicola* on immune black currant *Ribes nigrum* Cv. Titania in Preston, Connecticut. Plant Dis 95(12):1589

Friedline CJ, Lind BM, Hobson EM, Harwood DE, Mix AD, Maloney PE, Eckert AJ (2015) The genetic architecture of local adaptation I: the genomic landscape of foxtail pine (*Pinus balfouriana* Grev. & Balf.) as revealed from a high-density linkage map. Tree Genet. Genomes 11:49

Ghabrial SA, Castón JR, Jiang D, Nibert ML, Suzuki N (2015) 50-plus years of fungal viruses. Virol 479–480:356–368

Gilbert JA, Quinn RA, Debelius J, Xu ZZ, Morton J, Garg N, Jansson JK, Dorrestein PC, Knight R (2016) Microbiome-wide association studies link dynamic microbial consortia to disease. Nature 535:94–103

Gonzalez-Ibeas D, Martinez-Garcia PJ, Famula RA, Delfino-Mix A, Stevens KA, Loopstra CA, Langley CH, Neale DB, Wegrzyn JL (2016) Assessing the gene content of the megagenome: sugar pine (*Pinus lambertiana*). G3: Genes Genomes Genet 6:3787-3802

Goodman RN, Novacky AJ (1994) The hypersensitive reaction in plants to pathogens: a resistance phenomenon. American Phytopathological Society, Chicago

Gorash A, Armonienė R, Kazan K (2021) Can effectoromics and loss-of-susceptibility be exploited for improving Fusarium head blight resistance in wheat? Crop J 9:1–16

Government of Alberta (2014) Species assessed by Alberta's endangered species conservation committee: short list. Edmonton, AB, Canada

Guindon S, Gascuel O (2003) A simple, fast, and accurate algorithm to estimate large phylogenies by maximum likelihood. Syst Biol 52:696–704

Haagsma M, Page GFM, Johnson JS, Still C, Waring KM, Sniezko RA, Selker JS (2020) Using hyperspectral imagery to detect an invasive fungal pathogen and symptom severity in *Pinus strobiformis* seedlings of different genotypes. Remote Sens 12:4041

Hassani MA, Durán P, Hacquard S (2018) Microbial interactions within the plant holobiont. Microbiome 6 (1):58

Hatala JA, Crabtree RL, Halligan KQ, Moorcroft PR (2010) Landscape-scale patterns of forest pest and pathogen damage in the Greater Yellowstone Ecosystem. Remote Sens Environ 114(2):375–384

Hoff RJ, McDonald GI (1971) Resistance to *Cronartium ribicola* in *Pinus monticola*: short shoot fungicidal reaction. Can J Bot 49:1235–1239

Hoff RJ, McDonald GI (1975) Hypersensitive reaction in *Pinus armandii* caused by *Cronartium ribicola*. Can J Forest Res 5(1):146–148

Hoff R, Bingham RT, McDonald GI (1980) Relative blister rust resistance of white pines. Eur J For Path 10:307–316

Hofmeister BT, Denkena J, Colomé-Tatché M, Shahryary Y, Hazarika R, Grimwood J, Mamidi S, Jenkins J, Grabowski PP, Sreedasyam A, Shu S, Barry K, Lail K, Adam C, Lipzen A, Sorek R, Kudrna D, Talag J, Wing R, Hall DW, Jacobsen D, Tuskan GA, Schmutz J, Johannes F, Schmitz RJ (2020) A genome assembly and the somatic genetic and epigenetic mutation rate in a wild long-lived perennial *Populus trichocarpa*. Genome Biol 21:259

Hunt RS, Murray M, Reich R, Rusch D, Woods A, Zeglen S (2018) Persistence of major gene resistance in western white pine (*Pinus monticola*) in British Columbia. In: Schoettle AW, Sniezko RA, Kliejunas JT (eds) Proceedings of the IUFRO joint conference: genetics of five-needle pines, rusts of forest trees, and Strobusphere; 2014 Jun 15–20; Fort Collins, CO. Proc. RMRS-P-76; Fort Collins (CO): U.S. Department of Agriculture, Forest Service, Rocky Mountain Research Station. pp. 149–150

Jenkins ML, Cripps CL, Gains-Germain L (2018) Scorched Earth: Suillus colonization of *Pinus albicaulis* seedlings planted in wildfire-impacted soil affects seedling biomass, foliar nutrient content, and isotope signatures. Plant Soil 425:113–131

Jermstad KD, Eckert AJ, Wegrzyn JL, Delfino-Mix A, Davis DA, Burton DC, Neale DB (2011) Comparative mapping in *Pinus*: sugar pine (*Pinus lambertiana* Dougl.) and loblolly pine (*Pinus taeda* L.). Tree Genet Genomes 7:457–468

Johannes F, Schmitz RJ (2019) Spontaneous epimutations in plants. New Phytol 221(3):1253–1259

Jones JD, Dangl JL (2006) The plant immune system. Nature 444:323–329

Jupe F, Witek K, Verweij W, Sliwka J, Pritchard L, Etherington GJ, Maclean D, Cock PJ, Leggett RM, Bryan GJ, Cardle L, Hein I, Jones JD (2013) Resistance gene enrichment sequencing (RenSeq) enables reannotation of the NB-LRR gene family from sequenced plant genomes and rapid mapping of resistance loci in segregating populations. Plant J 76(3):530–44

Kawasaki T, Yamada K, Yoshimura S, Yamaguchi K (2017) Chitin receptor-mediated activation of MAP kinases and ROS production in rice and Arabidopsis. Plant Signal Behav 12(9):e1361076

Kegley AJ, Sniezko RA (2004) Variation in blister rust resistance among 226 *Pinus monticola* and 217 *P. lambertiana* seedling families in the Pacific Northwest. In: Sniezko RA, Samman S, Schlarbaum SE, Kriebel HB (eds) Breeding and genetic resources of five-needle pines: genetics, breeding, and adaptability. Proceedings of the IUFRO 2.02.15 Working Party Conference. Proceedings RMRS-P-32.USDA Forest Service Rocky Mountain Research Station, Ft. Collins, CO, pp 209-226

Kichas NE, Hood SM, Pederson GT, Everett RG, McWethy DB (2021) Whitebark pine (*Pinus albicaulis*) growth and defense in response to mountain pine beetle outbreaks. For Ecol Manag 457:117736

Kinloch BB Jr (1970) White pine blister rust: simply inherited resistance in sugar pine. Science 167:193–195

Kinloch BB Jr (2003) White pine blister rust in North America: Past and prognosis. Phytopathology 93:1044–1047

Kinloch BB Jr, Dupper GE (2002) Genetic specificity in the white pine-blister rust pathosystem. Phytopathology 92(3):278–80

Kinloch BB Jr, Littlefield JL (1977) White pine blister rust: hypersensitive resistance in sugar pine. Can J Bot 55(9):1148–1155

Kinloch BB Jr, Sniezko RA, Barnes GD, Greathouse TE (1999) A major gene for resistance to white pine blister rust in western white pine from the western cascade range. Phytopathology 889:861–867

Kinloch BB Jr, Sniezko RA, Dupper GE (2003) Origin and distribution of *Cr2*, a gene for resistance to white pine blister rust in natural populations of western white pine. Phytopathology 93:691–694

Kinloch BB Jr, Sniezko RA, Dupper GE (2004) Virulence gene distribution and dynamics of the white pine blister rust pathogen in western North America. Phytopathology 94:751–758

Kinloch BB Jr, Davis DA, Burton D (2008) Resistance and virulence interactions between two white pine species and blister rust in a 30-year field trial. Tree Genet Genomes 4:65–74

Kinloch BB Jr, Burton D, Davis DA, Westfall R, Dunlap J, Vogler DR (2012) Strong partial resistance to white pine blister rust in sugar pine. In: Sniezko RA, Yanchuk AD, Kliejunas JT, Palmieri KM, Alexander JM, Frankel SJ (eds) Proceedings of the fourth international workshop on the genetics of host-parasite interactions in forestry: Disease and insect resistance in forest trees, Gen Tech Rep. PSW-GTR-240. US Department of Agriculture, Forest Service, Pacific Southwest Research Station, Albany, CA, pp 80–91

Kourelis J, van der Hoorn RAL (2018) Defended to the nines: 25 years of resistance gene cloning identifies nine mechanisms for R protein function. Plant Cell 30 (2):285–299

Kumar S, Stecher G, Li M, Knyaz C, Tamura K (2018) MEGA X: molecular evolutionary genetics analysis across computing platforms. Mol Biol Evol 35:1547–1549

Kusch S, Panstruga R (2017) mlo-based resistance: an apparently universal 'weapon' to defeat powdery mildew disease. Mol Plant Microbe Interact 30:179–189

Liu J-J, Ekramoddoullah A (2003) Isolation, genetic variation and expression of TIR-NBS-LRR resistance gene analogs from western white pine (*Pinus monticola* Dougl. ex. D. Don.). Mol Genet Genomics 270:432–441

Liu J-J, Ekramoddoullah A (2006) The family 10 of plant pathogenesis-related proteins: their structure, regulation, and function in response to biotic and abiotic stresses. Physiol Mol Plant Pathol 68:3–13

Liu J-J, Ekramoddoullah A (2007) The CC-NBS-LRR subfamily in western white pine (*Pinus monticola* D. Don.): targeted identification, gene expression and genetic linkage with disease resistance against *Cronartium ribicola*. Phytopathology 97:728–736

Liu J-J, Ekramoddoullah A (2008) Development of LRR polymorphism, AFLP, and SCAR markers to the *Cronartium ribicola* resistance gene *Cr2* in western white pine (*Pinus monticola*). Tree Genet Genomes 4:601–610

Liu J-J, Ekramoddoullah A (2011) Genomic organization, induced expression and promoter activity of a resistance gene analog (*PmTNL1*) in western white pine (*Pinus monticola*). Planta 233:1041–1053

Liu J-J, Hammett C (2014) Development of novel polymorphic microsatellite markers by technology of next generation sequencing in western white pine. Conserv Genet Resourc 6(3):647–648

Liu J-J, Xiang Y (2019) Characterization of the western white pine TIR-NBS-LRR (*PmTNL2*) gene by transcript profiling and promoter analyses. Genome 62 (7):477–488

Liu J-J, Hunt R, Ekramoddoullah A (2004) Recent insights into western white pine genetic resistance to white blister rust. Rec Res Dev Biotechnol Bioeng 6:65–76

Liu J-J, Ekramoddoullah A, Hunt R, Zamani A (2006) Identification and Characterization of RAPD markers linked to a major gene (*Cr2*) for resistance to *Cronartium ribicola* (Fish.) in *Pinus monticola* (D. Don). Phytopathology 96:395–399

Liu J-J, Sturrock R, Ekramoddoullah A (2010a) The superfamily of thaumatin-like proteins: its origin, evolution, and expression towards biological function. Plant Cell Rep 29(5):419–36

Liu J-J, Zamani A, Ekramoddoullah A (2010b) Expression profiling of a complex thaumatin-like protein family in western white pine. Planta 231:637–651

Liu J-J, Sniezko RA, Ekramoddoullah AK (2011) Association of a novel *Pinus monticola* chitinase gene (*PmCh4B*) with quantitative resistance to *Cronartium ribicola*. Phytopathol 101:904–911

Liu J-J, Hammett C, Sniezko RA (2013a) *Pinus monticola* pathogenesis-related gene *PmPR10-2* alleles as defence candidates for stem quantitative disease resistance against white pine blister rust (*Cronartium ribicola*). Tree Genet Genomes 9(2):397–408

Liu J-J, Sturrock RN, Benton R (2013b) Transcriptome analysis of *Pinus monticola* primary needles by RNA-seq provides novel insight into host resistance to *Cronartium ribicola*. BMC Genomics 14:884

Liu J-J, Zamany A, Sniezko RA (2013c) Anti-microbial peptide (AMP): nucleotide variation, expression, and association with resistance in the white pine-blister rust pathosystem. Planta 237:43–54

Liu J-J, Chan D, Sturrock RN, Sniezko RA (2014a) Genetic variation and population differentiation of the endochitinase gene family in *Pinus monticola*. Plant Syst Evol 300(6):1313–1322

Liu J-J, Sniezko RA, Sturrock RN, Chen H (2014b) Western white pine SNP discovery and high-throughput genotyping for breeding and conservation applications. BMC Plant Biol 14:380

Liu J-J, Sturrock RN, Sniezko RA, Williams H, Benton R, Zamany A (2015) De novo sequencing of *Cronartium ribicola* transcriptome for understanding molecular white pine blister rust interactions. BMC Genomics 16:678

Liu J-J, Schoettle AW, Sniezko RA, Sturrock RN, Zamany A, Williams H, Ha A, Chan D, Danchok D, Savin DP, Kegley A (2016b) Genetic mapping of *Pinus flexilis* major gene (*Cr4*) for resistance to white pine blister rust using transcriptome-based SNP genotyping. BMC Genomics 17:753

Liu J-J, Sniezko RA, Zamany A, Williams H, Wang N, Kegley A, Savin DP, Chen H, Sturrock RN (2017a) Saturated genic SNP mapping identified functional candidates and selection tools for the *Pinus monticola Cr2* locus controlling resistance to white pine blister rust. Plant Biotechnol J 15:1149–1162

Liu J-J, Williams H, Li XR, Schoettle AW, Sniezko RA, Murray M, Zamany A, Roke G, Chen H (2017b) Profiling methyl jasmonate-responsive transcriptome for understanding induced systemic resistance in whitebark pine (*Pinus albicaulis*). Plant Mol Biol 95:359–374

Liu J-J, Schoettle AW, Sniezko RA, Yao F, Zamany A, Williams H, Rancourt B (2019a) Limber pine (*Pinus flexilis* James) genetic map constructed by exome-seq provides insight into evolution of disease resistance and a genomic resource for genomics-based breeding. Plant J 98(4):745–758

Liu J-J, Xiang Y, Sniezko RA, Schoettle AW, Williams H, Arezoo Z (2019b) Characterization of *Cronartium ribicola* dsRNAs reveals novel taxa of the family Totiviridae and viral association with fungal virulence. BMC Virol J 16:118

Liu J-J, Ekramoddoullah A, Zamani A (2005) A class IV chitinase is up-regulated upon fungal infection and abiotic stresses and associated with slow-canker-growth resistance to *Cronartium ribicola* in western white pine (*Pinus monticola,* Dougl. Ex D. Don). Phytopathology 95(3):284–291

Liu J-J, Chan D, Xiang Y, Williams H, Li XR, Sniezko RA, Sturrock RN (2016a) Characterization of five novel mitoviruses in the white pine blister rust fungus *Cronartium ribicola*, PLoS One 11(5): e0154267

Liu J-J, Sniezko RA, Murray M, Wang N, Chen H, Zamany A, Sturrock RN, Savin D, Kegley A (2016c) Genetic diversity and population structure of whitebark pine (*Pinus albicaulis* Engelm.) in western North America. PLoS One 11(12):e0167986

Liu J-J, Sniezko RA, Sissons R, Krakowski J, Alger G, Schoettle AW, Williams H, Zamany A, Zitomer R, Kegley A (2020a) Association mapping and development of marker-assisted selection tools for the resistance to white pine blister rust in the Alberta limber pine populations. Front Plant Sci 11:557672

Liu J-J, Williams H, Zamany A, Li X-R, Gellner S, Sniezko RA (2020b) Development and application of marker-assisted selection (MAS) tools for breeding of western white pine (*Pinus monticola* Douglas ex D. Don) resistance to blister rust (*Cronartium ribicola* J. C. Fisch.) in British Columbia. Can J Plant Pathol 42 (2):250–259

Liu J-J, Fernandes H, Zamany A, Sikorski M, Jaskolski M, Sniezko RA (2021a) In-vitro anti-fungal assay and association analysis reveal a role for the *Pinus monticola* PR10 gene (*PmPR10-3.1*) in quantitative disease resistance to white pine blister rust. Genome https://doi.org/10.1139/gen-2020-0080

Liu J-J, Sniezko RA, Zamany A, Williams H, Omendja K, Kegley A, Savin DP (2021b) Comparative transcriptomics and RNA-seq-based bulked segregant analysis (BSR-Seq) reveals genomic basis underlying *Cronartium ribicola vcr2* virulence. Front Microbiol 12:602812

Liu J-J, Schoettle AW, Sniezko RA, Williams H, Arezoo Zamany A, Rancourt B (2021c) Fine dissection of limber pine resistance to *Cronartium ribicola* using targeted sequencing of the NLR family. BMC Genomics

López Sánchez A, Stassen JH, Furci L, Smith LM, Ton J (2016) The role of DNA (de)methylation in immune responsiveness of *Arabidopsis*. Plant J 88(3):361–374

Ma Z, Liu J-J, Zamany A (2019) Identification and functional analysis of an effector secreted by *Cronartium ribicola*. Phytopatholoy 109(6):942–951

Ma Z, Liu J-J, Zamany A, Williams H (2020) Transient gene expression in western white pine by Agroinfiltration. J Forest Res 31(5):1823–1832

Mahalovich M, Stritch L (2013) *Pinus albicaulis*. IUCN Red List of Threatened Species 2013:e. T39049A2885918

Maloy OC (1997) White pine blister rust control in North America: a case history. Annu Rev Phytopathol 35:87–109

Maloy OC (2001) White pine blister rust. Online. Plant Health Prog https://doi.org/10.1094/PHP-2001-0924-01-HM

McCaw BA, Stevenson TJ, Lancaster LT (2020) Epigenetic responses to temperature and climate. Integ Comp Biol 60(6):1469–1480

McDonald GI, Hoff RJ (1970) Resistance to *Cronartium ribicola* in *Pinus monticola*: early shedding of infected needles. USDA Forest Service Research Note INT-124. Intermountain Forest and Range Experiment Station, Ogden, Utah, p 8

Mir RR, Reynolds M, Pinto F, Khan MA, Bhat MA (2019) High-throughput phenotyping for crop improvement in the genomics era. Plant Sci 282: 60–72

Moler ERV, Aho K (2018) Whitebark pine foliar fungal endophyte communities in the southern Cascade Range, USA: host mycobiomes and white pine blister rust. Fungal Ecol 33:104–114
Moler EV, Abakir A, Eleftheriou M, Johnson JS, Krutovsky KV, Lewis LC, Ruzov A, Whipple AV, Rajora OP (2018) Population epigenomics: Advancing understanding of phenotypic plasticity, acclimation, adaptation, and diseases. In: Rajora OP, (ed.) Population Genomics: Concepts, Approaches and Applications. Springer
Munck IA, Tanguay P, Weimer J, Villani SM, Cox KD (2015) Impact of white pine blister rust on resistant cultivated *Ribes* and neighboring eastern white pine in New Hampshire. Plant Dis 99:1374–1382
Naveed ZA, Wei X, Chen J, Mubeen H, Ali GS (2020) The PTI to ETI continuum in Phytophthora-plant interactions. Front Plant Sci 11:593905
Neale DB, Wegrzyn JL, Stevens KA, Zimin AV, Puiu D, Crepeau MW, Cardeno C, Koriabine M, Holtz-Morris AE, Liechty JD, Martínez-García PJ, Vasquez-Gross HA, Lin BY, Zieve JJ, Dougherty WM, Fuentes-Soriano S, Wu LS, Gilbert D, Marçais G, Roberts M, Holt C, Yandell M, Davis JM, Smith KE, Dean JF, Lorenz WW, Whetten RW, Sederoff R, Wheeler N, McGuire PE, Main D, Loopstra CA, Mockaitis K, deJong PJ, Yorke JA, Salzberg SL, Langley CH (2014) Decoding the massive genome of loblolly pine using haploid DNA and novel assembly strategies. Genome Biol 15:R59
Nuss DL (2005) Hypovirulence: mycoviruses at the fungal–plant interface. Nat Rev Microbiol 3:632–642
Onda Y, Takahagi K, Shimizu M, Inoue K, Mochida K (2018) Multiplex PCR targeted amplicon sequencing (MTA-Seq): simple, flexible, and versatile SNP genotyping by highly multiplexed PCR amplicon sequencing. Front Plant Sci 9:201
Pascale A, Proietti S, Pantelides IS, Stringlis IA (2020) Modulation of the root microbiome by plant molecules: the basis for targeted disease suppression and plant growth promotion. Front Plant Sci 10:1741
Patton RF (1961) The effect of age upon susceptibility of eastern white pine to infection by *cronartium ribicola*. Phytopathol 51(7):429–424
Pavan S, Jacobsen E, Visser RGF, Bai Y (2010) Loss of susceptibility as a novel breeding strategy for durable and broad-spectrum resistance. Mol Breed 25:1–12
Pavy N, Lamothe M, Pelgas B, Gagnon F, Birol I, Bohlmann J, Mackay J, Isabel N, Bousquet J (2017) A high-resolution reference genetic map positioning 8.8 K genes for the conifer white spruce: structural genomics implications and correspondence with physical distance. Plant J 90:189–203
Pham T, Chen H, Zhang RR, Dai LL, Vu T (2015) Changes of monoterpenes in stem of Chinese white pine (*Pinus armandi*) saplings following treatment with methyl jasmonate. For Stud 60(1):69–81
Pilet-Nayel M-L, Moury B, Caffier V, Montarry J, Kerlan M-C, Fournet S, Durel CE, Delourme R (2017) Quantitative resistance to plant pathogens in pyramiding strategies for durable crop protection. Front Plant Sci 8:1838
Quesada T, Resende MFRJ, Munoz P, Wegrzyn JL, Neale DB, Kirst M, Peter GF, Gezan SA, Nelson CD, Davis JM (2014) Mapping fusiform rust resistance genes within a complex mating design of loblolly pine. Forests 5:347–362
Richardson BA, Ekramoddoulah A, Liu J-J, Kim M-S, Klopfenstein NB (2010) Current and future molecular approaches to investigate the white pine blister rust pathosystem. For Path 40:314–331
Roden DM (2017) Phenome-wide association studies: a new method for functional genomics in humans. J Physiol 595(12):4109–4115
Roux F, Voisin D, Badet T, Balagué C, Barlet X, Huard-Chauveau C, Roby D, Raffaele S (2014) Resistance to phytopathogens e tutti quanti: placing plant quantitative disease resistance on the map. Mol Plant Pathol 15:427–432
Runyon JB, Gray CA, Jenkins MJ (2020) Volatiles of high-elevation five-needle pines: Chemical signatures through ratios and insight into insect and pathogen resistance. J Chem Ecol 46:264–274
Schoettle AW, Sniezko RA, Kegley A, Burns KS (2014) White pine blister rust resistance in limber pine: evidence for a major gene. Phytopathol 104:163–73
Sievers F, Wilm A, Dineen D, Gibson TJ, Karplus K, Li W, Lopez R, McWilliam H, Remmert M, Söding J, Thompson JD, Higgins DG (2011) Fast, scalable generation of high-quality protein multiple sequence alignments using Clustal Omega. Mol Syst Biol 7:539
Smith JA, Blanchette RA, Burnes TA, Jacobs JJ, Higgins L, Witthuhn BA, David AJ (2006a) Proteomic comparison of needles from blister rust-resistant and susceptible *Pinus strobus* seedlings reveals up regulation of putative disease resistance proteins. Mol Plant-Microbe Interact 19:150–160
Smith JA, Blanchette RA, Burnes TA, Gillman JH, David AJ (2006) Epicuticular wax and white pine blister rust resistance in resistant and susceptible selections of eastern white pine (*Pinus strobus*). Phytopathol 96(2):171–7
Sniezko RA, Koch J (2017) Breeding trees resistant to insects and diseases: putting theory into application. Biol Invas 19:3377–3400
Sniezko RA, Smith J, Liu J-J, Hamelin R (2014) Genetic resistance to fusiform rust in southern pines and white pine blister rust in white pines—a contrasting tale of two rust pathosystems–Current status and future prospects. Forests 5(9):2050–2083
Sniezko RA, Danchok R, Savin DP, Liu J-J, Kegley A (2016) Genetic resistance to white pine blister rust in limber pine (*Pinus flexilis*): major gene resistance in a northern population. Can J For Res 46:1173–1178

Sniezko RA, Johnson JS, Savin DP (2020) Assessing the durability, stability, and usability of genetic resistance to a non-native fungal pathogen in two pine species. Plants People Planet 2:57–68

Sniezko RA, Samman S, Schlarbaum SE, Kriebel HB (eds) (2004) Breeding and genetic resources of five-needle pines: growth, adaptability, and pest resistance. In: IUFRO Working Party 2.02.15. Medford, OR, USA. Proceedings RMRS-P-32. Fort Collins, CO, U. S. Department of Agriculture, Forest Service, Rocky Mountain Research Station. p 259

Sniezko RA, Kegley AJ, Danchok R (2008) White pine blister rust resistance in North America, Asian and European species—results from artificial inoculation trials in Oregon. Ann For Res 51:53–66

Stahl EA, Dwyer G, Mauricio R, Kreitman M, Bergelson J (1999) Dynamics of disease resistance polymorphism at the *Rpm1* locus of Arabidopsis. Nature 400:667–671

Stevens KA, Wegrzyn JL, Zimin A, Puiu D, Crepeau M, Cardeno C, Paul R, Gonzalez-Ibeas D, Koriabine M, Holtz-Morris AE, Martínez-García PJ, Sezen UU, Marçais G, Jermstad K, McGuire PE, Loopstra CA, Davis JM, Eckert A, de Jong P, Yorke JA, Salzberg SL, Neale DB, Langley CH (2016) Sequence of the sugar pine megagenome. Genetics 204:1613–1626

Sturrock RN, Frankel SJ, Brown AV, Hennon PE, Kliejunas JT, Lewis KJ, Worrall JJ, Woods AJ (2011) Climate change and forest diseases. Plant Pathol 60:133–149

Sweeney K, Stone J, Cook K, Sniezko RA, Kegley A, Schoettle AW (2012) Needle reactions in resistance to *Cronartium ribicola*: hypersensitivity response or not? In: Sniezko RA, Yanchuk AD, Kliejunas FT, Palmieri AM, Alexander JM, Frankel SJ (eds). 2012 Proceedings of the fourth international workshop on the genetics of host-parasite interactions in forestry: Disease and insect resistance in forest trees. Gen. Tech. Rep. PSW-GTR-240. Albany, CA, U.S. Department of Agriculture, Forest Service, Pacific Southwest Research Station, p. 368-371

Syring JV, Tennessen JA, Jennings TN, Wegrzyn J, Scelfo-Dalbey C, Cronn R (2016) Targeted capture sequencing in whitebark pine reveals range-wide demographic and adaptive patterns despite challenges of a large, repetitive genome. Front Plant Sci 7:484

Takai A, Eto M, Hirano K, Takeya K, Wakimoto T, Watanabe M (2018) Protein phosphatases 1 and 2A and their naturally occurring inhibitors: current topics in smooth muscle physiology and chemical biology. J Physiol Sci 68(1):1–17

Thomma BP, Nürnberger T, Joosten MH (2011) Of PAMPs and effectors: the blurred PTI-ETI dichotomy. Plant Cell 3(1):4–15

Thordal-Christensen H (2020) A holistic view on plant effector-triggered immunity presented as an iceberg model. Cell Mol Life Sci 77:3963–3976

Tomback DF, Achuff P (2010) Blister rust and western forest biodiversity: ecology, values and outlook for white pines. Forest Pathol 40(3–4):186–225

van Schie CCN, Takken FLW (2014) Susceptibility genes 101: how to be a good host. Annu Rev Phytopathol 52:551–5814

van Loon LC, Rep M, Pieterse CM (2006) Significance of inducible defense-related proteins in infected plants. Annu Rev Phytopathol 44:135–62

Vannier N, Agler M, Hacquard S (2019) Microbiota-mediated disease resistance in plants. PLoS Pathog 15 (6):e1007740

Vázquez-Lobo A, De La Torre AR, Martínez-García PJ, Vangestel C, Wegzryn JL, Ćalić I, Burton D, Davis D, Kinloch B, Vogler D, Neale DB (2017) Finding loci associated to partial resistance to white pine blister rust in sugar pine (*Pinus lambertiana* Dougl.). Tree Genet Genomes 13:108

Verma SS, Yajima W, Rahman MH, Shah S, Liu J-J, Ekramoddoullah AK, Kav NNV (2012) A cysteine-rich anti-microbial peptide from *Pinus monticola* (PmAMP1) confers multiple disease resistance in canola (*Brassica napus*). Plant Mol Biol 79:61–74

Vieira ML, Santini L, Diniz AL, Munhoz Cde F (2016) Microsatellite markers: what they mean and why they are so useful. Genet Mol Biol 39(3):312–328

Villari C, Sniezko RA, Rodriguez-Saona LE, Bonello P (2017) Accelerating dynamic genetic conservation efforts: Use of FT-IR spectroscopy for the rapid identification of trees resistant to destructive pathogens. In: Sniezko RA, Man G, Hipkins V, Woeste K, Gwaze D, Kliejunas JT, McTeague BA, (eds) tech. cords. 2017 Gene conservation of tree species—banking on the future. Proceedings of a workshop. Gen. Tech. Rep. PNW-GTR-963. Portland, OR, U.S. Department of Agriculture, Forest Service, Pacific Northwest Research Station, p 69

Vleeshouwers VG, Oliver RP (2014) Effectors as tools in disease resistance breeding against biotrophic, hemibiotrophic, and necrotrophic plant pathogens. Mol Plant Microbe Interact 27:196–206

Vorholt JA, Vogel C, Carlström CI, Müller DB (2017) Establishing causality: opportunities of synthetic communities for plant microbiome research. Cell Host Microbe 22:142–55

Wallis CM, Reich R (2010) Lodgepole pine provenances differ in chemical defense capacities against foliage and stem diseases. Can J Forest Res 40(12):2333–2344

Waszczak C, Carmody M, and Kangasjärvi J (2918) Reactive oxygen species in plant signaling. Annu Rev Plant Biol 69:209–236

Weiss M, Sniezko RA, Puiu D, Crepeau MW, Stevens K, Salzberg SL, Langley CH, Neale DB, De La Torre AR (2020) Genomic basis of white pine blister rust quantitative disease resistance and its relationship with qualitative resistance. Plant J 104(2):365–376

Westbrook JW, Chhatre VE, Wu L-S, Chamala S, Neves LG, Muñoz P, Martínez-García PJ, Neale DB, Kirst M, Mockaitis K, Nelson CD, Peter GF, Davis JM, Echt CS (2015) A consensus genetic map for *Pinus taeda* and *Pinus elliottii* and extent of linkage disequilibrium in two genotype-phenotype discovery populations of *Pinus taeda*. G3 Genes Genomes Genet 25:1685–1694

Wilcox PL, Amerson HV, Kuhlman EG, Liu BH, O'Malley DM, Sederoff RR (1996) Detection of a major gene for resistance to fusiform rust disease in loblolly pine by genomic mapping. Proc Natl Acad Sci USA 93:3859–3864

Zamany A, Liu J-J, Ekramoddoullah A, Sniezko R (2011) Antifungal activity of a *Pinus monticola*—antimicrobial peptide 1 (Pm-AMP1) and its accumulation in *Cronartium ribicola*-infected western white pine. Can J Microbiol 57:667–679

Zamany A, Liu J-J, Ekramoddoullah AK (2012) Comparative proteomic profiles of *Pinus monticola* needles during early compatible and incompatible interactions with *Cronartium ribicola*. Planta 236:1725–1746

Zhang M, Coaker G (2017) Harnessing effector-triggered immunity for durable disease resistance. Phytopathology 107(8):912–919

Functional Genomics of Mediterranean Pines

9

Concepción Ávila, Rafael A. Cañas, Fernando N. de la Torre, María Belén Pascual, Vanessa Castro-Rodríguez, Francisco R. Cantón, and Francisco M. Cánovas

Abstract

Studies of functional genomics in Mediterranean pines have been reviewed in this chapter. Only a few species are considered true representatives of the Mediterranean region, based on chloroplast DNA studies, distribution and climate adaptation. This chapter begins by analyzing the technological developments from a historical perspective. A large proportion of the studies has focused on the acquisition of nutrients and their metabolic interconversions with particular emphasis on nitrogen uptake, assimilation, storage and remobilization. At the end of this chapter, the analysis and discussion focus on the molecular and genomic approaches to study the responses of Mediterranean pines to environmental cues, and also on the studies that contributed to a better understanding of wood formation.

9.1 Introduction

The Mediterranean climate is a middle latitude temperate climate with warm or hot summers (Köppen 1900). It is also the climate of the geographic region surrounding the Mediterranean Sea located between the North of Africa, the South of Europe and the Middle East. The extension of this region is quite considerable, comprising more than 3,700 km in longitude and more than 1,600 km in latitude, and contrasting climates can be found in the different areas (Lionello et al. 2012). In fact, changes in the climatic conditions can be significant between the mountains surrounding the Mediterranean Sea and the coasts where the real temperate Mediterranean climate is found. This is the case of the Alps, Pyrenees, Balkan Mountains and Sierra Nevada in the South, where alpine environments can be found with cold winters under 0 ºC, snow and short frost-free periods. In this context, two main groups can be considered among the pine species distributed in the

C. Ávila (✉) · R. A. Cañas · F. N. de la Torre · M. B. Pascual · V. Castro-Rodríguez · F. R. Cantón · F. M. Cánovas (✉)
Facultad de Ciencias, Departamento de Biología Molecular y Bioquímica, Universidad de Málaga, Campus de Teatinos s/n, Málaga, Spain
e-mail: cavila@uma.es

F. M. Cánovas
e-mail: canovas@uma.es

R. A. Cañas
e-mail: racanas@uma.es

F. N. de la Torre
e-mail: fdelatorre@uma.es

M. B. Pascual
e-mail: bpascual@uma.es

V. Castro-Rodríguez
e-mail: vavicaro@uma.es

F. R. Cantón
e-mail: frcanton@uma.es

A. R. De La Torre (ed.), *The Pine Genomes*, Compendium of Plant Genomes,
https://doi.org/10.1007/978-3-030-93390-6_9

Mediterranean region: one adapted to highlands, and a second one adapted to the typical Mediterranean climate (Farjon 2017). An example of the first group is *Pinus sylvestris*. This pine species is one of the most widely distributed in the world, extending from Portugal to Russia, and from Sierra Nevada (Spain) to Finnmark in Norway (Farjon and Filer 2013). However, this pine needs cold temperatures around 0 °C in January, and consequently, is restricted to high mountains/alpine conditions in the Mediterranean area. For that reason, *P. sylvestris* functional genomics will not be considered in this chapter. Additionally, there are other pine species distributed around the world and adapted to grow in the Mediterranean climate, such as *Pinus coulteri* found in the California coast (USA) and Baja California (Mexico) (Farjon 2017). These species, however, cannot be strictly considered "Mediterranean pines".

Chloroplast DNA studies have divided Mediterranean pines into different sections: *Canarienses* (*P. canariensis*), *Halepenses* (*P. brutia* and *P. halepensis*) and *Pineae* (*P. pinea*), which correspond to almost all Mediterranean shore and island pines, and a separate group comprising mountain pines from areas surrounding the Mediterranean (part of subsection *Pinus*, e.g., *P. nigra*, *P. pinaster* and *P. sylvestris*). In fact, the chloroplast DNA analysis suggests that the *Canarienses*, *Halepenses* and *Pineae* sections form a clade separated from the subsection *Pinus* (Barbero et al. 2000).

Therefore, in this chapter we focus on only those strictly Mediterranean pines: *P. brutia*, *P. halepensis*, *P. pinea*, *P. nigra* and *P. pinaster* as evidenced by their chloroplast DNA studies, distribution and climate. *P. canariensis* has also been included because of its phylogenetic relationships with *P. pinea*, *P. pinaster*, *P. brutia* and *P. halepensis* in the *Pinaster* subsection, and the existence of plantations in the Mediterranean area (Parks et al. 2012; Farjon and Filer 2013).

Among the Mediterranean pines, maritime pine (*P. pinaster*) is the most extensively studied species, and a large collection of functional genomic resources is available (Cañas et al. 2019). Furthermore, the sequencing and assembly of the genome have been recently accomplished by a European consortium of researchers (see the chapter by Sterck et al. in this book). Maritime pine belongs to the subsection *Pinus*, and it is a native pine species widely distributed in the Mediterranean area. It extends across the Western Mediterranean basin from France to Morocco and from Portugal to Tunisia. It is classified into three groups: Atlantic, Mediterranean and Maghrebian (Morocco, Tunisia and Algeria and the populations of Southern Spain). It grows from sea level to 2,000 m and tolerates a wide range of changing characteristics in climate conditions, soil quality and water availability. These characteristics boost its genetic variability enabling a series of capabilities that probably explain why it is widely distributed in the Mediterranean region (Tapias et al. 2004), and why most studies on pines' adaptive response to changing scenarios have been performed in this species (Meijón et al. 2016).

9.2 Technical Approaches for Studies of Functional Genomics

Most conifers are evergreen trees with long lifespans and life cycles, big sizes, enormous genomes and great amounts of phenolic secondary metabolites (De la Torre et al. 2014; Cañas et al. 2019). The pines from the Mediterranean basin also have these characteristics that make the experimental work on biochemistry, molecular biology, genetics and omics sciences difficult. As a consequence of these adversities and inconveniences, it is extremely difficult to obtain mutant lines, although methods of genetic transformation have been recently developed (Cañas et al. 2019). As non-classical model species, most studies on the functional genomics of pines have been focused on the expression of genes and proteins. Pioneering studies were based on the application of classical molecular techniques of molecular biology such as northern, western blotting and determination of enzyme activities (Cánovas et al. 1991; Cantón et al. 1993; Allona et al. 1994; García-Gutiérrez et al. 1995; Nahalkova et al. 2001). Usually, these techniques were used to study only one or a few collections of genes and proteins in a limited number of samples and conditions (Hofmann and Clokie 2018). In a field of study involving just a limited number of researchers, it was extraordinarily difficult to obtain a general overview of fundamental pine processes in comparison to plant model species such as *Arabidopsis*, rice, or poplar (Cañas et al. 2019). But in recent years, the development of omics/high-throughput techniques has allowed overcoming this situation, and now with a single experiment it is possible to obtain a holistic vision of the functional genomics of any species in a determined situation. These recent developments in high-performance technologies enable approaches of functional genomics in non-model species, including Mediterranean pines, allowing in-depth studies that were inconceivable 20 years ago.

9.2.1 Transcriptomics

As mentioned before, the former gene expression studies were carried out mainly through northern blot analysis, a single gene-directed method (Cantón et al. 1993; García-Gutiérrez et al. 1995). Later, the initial characterizations of Mediterranean pine transcriptomes were made producing expression sequence tag (EST) libraries to elaborate gene catalogs, to identify single nucleotide polymorphisms (SNPs) and to develop association genetic studies (Cantón et al. 2003; Dubos et al. 2003; Le Dantec et al. 2004; Paiva et al. 2008a; Palme et al. 2008; Lepoittevin et al. 2010). These approximations yielded a limited amount of information biased to the more expressed genes and even to identify mainly their 3' region because cDNA synthesis is based on the use of oligo-dT primers (Nam et al. 2002; Skrypina et al. 2003). The costs of the results were high because they were based on the individual Sanger's sequencing of each clone, which also limited the depth of transcriptome data. A variant of this strategy was also used to determine differential gene expression in two contrasting samples using suppression-subtractive hybridization (SSH) (Alonso et al. 2007; Canales et al. 2010; Perdiguero et al. 2012; Ramos et al. 2012; Sancho dos Santos and de Vasconcelos 2012; Flores-Monterroso et al. 2013). Nevertheless, all these resources were extremely useful and allowed the construction of the first nucleotide databases for Mediterranean pines (Fernández-Pozo et al. 2011), and the design of new tools to evaluate the expression patterns of thousands of genes and ultimately, of the entire transcriptome.

As shown in Fig. 9.1, the initial approximation to characterize transcriptome-level expression in *P. pinaster* was to use custom microarrays constructed by printing in solid support the large cDNA collections already available (Canales et al. 2012b; Villalobos et al. 2012; Perdiguero et al. 2013; Cañas et al. 2015a) and even the use of related microarrays constructed for other conifer species such as loblolly pine and spruce (Perdiguero et al. 2012; de Vega-Bartol et al. 2013). This was costly and time-consuming due to the amount of work needed to manage (e.g., storage and setup) the EST and cDNA clone collections. The generation of complete transcriptome sequences was possible, thanks to the emergence of next generation sequencing (NGS), which allowed the design of new commercial oligonucleotide-microarrays with thousands of transcript sequences covering the entire transcriptome (Perdiguero et al. 2012, 2013; Cañas et al. 2015b; Chano et al. 2017a, b, 2020; Pascual et al. 2018; El-Azaz et al. 2020).

Alternatively, the direct use of NGS sequencing (RNA-seq) is becoming the most popular approach nowadays to quantify gene expression (Pinosio et al. 2014; Cañas et al. 2015b, 2016; Gaspar et al. 2017; Fox et al. 2018; Rodrigues et al. 2018; Hernández-Escribano et al. 2020; Olsson et al. 2020). In non-model species such as pines, the transcriptome assemblies from NGS data have built a global view over transcriptomes in the first round (Canales et al. 2014). However, the transcriptome is dynamic, and it is not possible to talk about a single transcriptome in a particular species. The fraction of the genome that is transcribed can be different in a cell type depending on

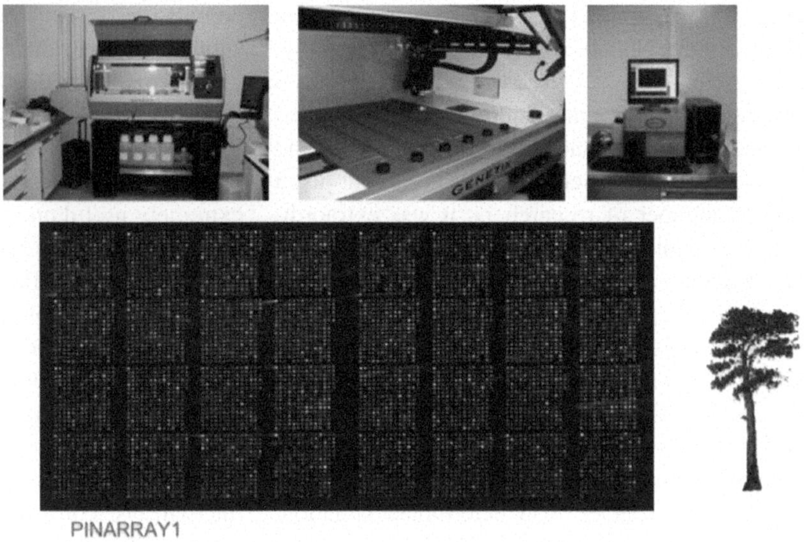

Fig. 9.1 Laboratory of functional genomics at University of Málaga for transcriptomics analysis of *Pinus pinaster*. Upper panel: Equipment for cDNA printing on glass slides. Lower panel: Image of the microarrray PINARRAY1

its anatomical position, developmental phase, environment and physiological conditions. RNA-seq determinations in samples from different organs, tissues or conditions are extending our vision about pine transcriptomes, allowing the identification of new transcripts and genes with very specific and low expressions. For example, the use of laser capture microdissection (LCM) to isolate different tissue samples of maritime pine seedlings has allowed the identification of transcripts not contained in a previous reference transcriptome (Cañas et al. 2017; Ortigosa et al. 2021). Even when the assembly of the pine genome is completed, its size and the high number of repeated sequences make it difficult to identify the functional elements it contains (e.g., protein-coding genes, sequences coding lncRNA and sncRNAs). This has occurred with the conifer genomes published to date, in which it is possible to find non-annotated genes by *blast* searches using transcript sequences from RNA-seq experiments. Thus, it will be necessary the use of read data from numerous RNA-seq to advance in the functional annotation of the maritime pine genome.

At the transcriptomic level, the NGS has also allowed the identification of miRNA in different developmental conditions of Mediterranean pines (Rodrigues et al. 2019; Perdiguero et al. 2021) although these types of studies have been limited. Unfortunately, there is no information about lncRNAs, epigenomics and epitranscriptomics traits. In the case of lncRNA, it would be possible to classify them from the information in the pine transcriptome databases (Waseem et al. 2020). However, to complete a good lncRNA study, a high-quality genome assembly would be necessary, and currently only a maritime pine assembly draft is available. For the same reason, massive epigenomic approaches based on ChIP-seq or DAP-seq have not been developed in coniferous species. With the progress in the genome assembly of maritime pine and new technical approaches, these constraints will likely be solved. Recent advances in genome sequencing, also called third generation NGS, have the potential of detecting epigenetic and epitranscriptomic marks in the bases of nucleic acids. These sequencing methods also allow the production of complete transcripts that gives direct information about splicing variants avoiding the assembly step in which different artifacts can be produced (Garalde et al. 2018; van Dijk et al. 2018).

9.2.2 Proteomics

A technical approach on functional genomics employed very often in Mediterranean pines is proteomics. The first studies used 2D-polyacrylamide gel electrophoresis (PAGE) to detect the presence/absence of protein spots between two contrasting samples (Gerber et al. 1993; Bahrman et al. 1994; Petit et al. 1995; Plomion et al. 1997; Costa and Plomion 1999). These analyses were mainly focused on the determination of genetic variability among different maritime pine provenances and the obtaining of genetic markers. A first step in the characterization of protein spots obtained in 2D-PAGE electrophoresis was the N-terminal amino acid sequencing of short internal peptides followed by the comparison with database sequences using *blast* search (Costa et al. 1998; Plomion et al. 2000). This methodology was limited in the number of peptides/proteins analyzed and in the identification process due to the limited number of sequences found in the public databases at this time.

The introduction of mass spectrometry (MS) in the technical pipeline revolutionized proteomics studies. MS was used to characterize protein spots isolated from a 2D-PAGE gel and proteins separated by liquid chromatography (LC) in shotgun methods (Gion et al. 2005; Paiva et al. 2008b; Garcés et al. 2014; Morel et al. 2014a, b; Taibi et al. 2017; Loewe et al. 2018; Salaj et al. 2019). One of the most important points in this type of technique is the comparison of the mass profiles from the digested peptides with the virtual mass profiles of a database that usually comes from a transcriptome. The best results in the identification of peptides are obtained when the protein databases come from the same species under study although it is possible to employ model organisms such as *Arabidopsis* or general public databases such as NCBI's plant-nr (Romero-Rodríguez et al. 2014). In the case of maritime pine, the number of transcriptomic sequences has allowed the use of protein databases from the same species. Nowadays, there are different reference databases that can be employed in peptide identification for Mediterranean pine studies. In fact, the low sequencing costs per sample allowed the obtaining of assemblies from single RNA-seq experiments in species without a reference transcriptome, improving peptide identification in proteomics analyses (Romero-Rodríguez et al. 2014; Cañas et al. 2017).

9.2.3 Metabolomics

High-throughput analysis of metabolites provides essential information on fundamental tree biology, including molecular insights of basic developmental processes and the response to a variety of environmental factors. Recent methodological advances in MS, nuclear magnetic resonance (NMR) and bioinformatics analyses have enabled the integration of metabolomics into the functional genomic approaches in trees. Several studies in Mediterranean pines have included metabolomics just as the main analytical tool and in combination with other omics such as transcriptomics and proteomics. For example, seasonal environmental changes in the levels of metabolites have been examined by ^{1}H-NMR analysis in the needles of 25-year-old *P. pinaster* trees growing in the forest under natural conditions (Cañas et al. 2015a). The same experimental approach was used to determine changes in the abundance of metabolites following the nutrition of young maritime pine trees with different inorganic nitrogen sources (Ortigosa et al. 2020). Moreover, the process of apical bud maturation has recently been explored by the determination of changes in proteins and metabolites using UPLC-LTQ-Orbitrap (Valledor et al. 2021). The natural variation of the metabolome was analyzed by GC–MS and LC-Orbitrap-MS in samples from ten provenances of *Pinus pinaster* (Meijón et al 2016). This elegant work illustrates the power of metabolomics to discriminate the secondary metabolism of Atlantic and Mediterranean provenances, providing new insights on the drought tolerance of Mediterranean pines depending on their geographic origin.

9.2.3.1 Genetic Transformation as a Biotechnological Tool for Studies of Functional Genomics

An important aspect of functional genomics is the analysis of the function of candidate genes within their biological context. Furthermore, climatic variations and an ever-increasing demand for forest products have led to the development of technologies assisting traditional tree breeding programs to produce phenotypically superior and genetically improved trees. This is of special interest in Mediterranean countries where rapid deforestation and the need for protection against drought-induced soil erosion have focused on improving forest trees. Reverse genetics and genetic transformation are reasonable approaches to reach that goal. One of the main drawbacks is plant regeneration through in vitro culture procedures after transformation that prolongs the process near a year or more. For this reason, in some studies the functional analyses of pine genes have been made through heterologous expression in model plant species with optimized and fast transformation methods (Pascual et al. 2015; Castro-Rodríguez et al. 2015; Rueda-López et al. 2015, 2017).

Nevertheless, different methods have been developed to determine gene function in Mediterranean pines. One successful approach is gene transient expression analysis. Although this method is not valid to obtain information about the gene's role in the whole plant, it gives important information about the molecular function of the genes in a cellular context. Two main techniques have been used for gene transient expression in Mediterranean pines: particle bombardment in excised cotyledons (Rey et al. 1996; Humara et al. 1998, 1999b; López et al. 2000) and electroporation in cotyledonary protoplasts (López et al. 2000; Gómez-Maldonado et al. 2001, 2004a, b; Rueda-López et al. 2008; Canales et al. 2012a; Craven-Bartle et al. 2013).

Despite the intrinsic problems for the permanent transformation of conifers, different genetic transformation methods have been developed in Mediterranean pines during the last decade. In this regard, somatic embryogenesis (SE) has become a powerful biotechnological tool to obtain large amounts of cloned material from elite genotypes, and for long-term storage by cryopreservation. In addition, SE enables functional genomics approaches in conifers via the silencing and overexpression of candidate genes determining important traits. The most relevant aspects of the cellular and molecular basis of conifer SE are examined in recently published reviews (Klimaszewska et al. 2016; Trontin et al. 2016a; von Arnold et al. 2019).

In Mediterranean pine species, plant regeneration via SE has been achieved in *Pinus pinea* (Carneros et al. 2009), *P. halepensis* (Montalbán et al. 2013; Pereira et al. 2017), *P. brutia* (Yildirim et al. 2006), *P. nigra* (Salajova and Salaj 2005; Salaj et al. 2019), *P. pinaster* (Lelu-Walter et al. 2006; Alvarez et al. 2013, 2019) and *P. canariensis* (Castander-Olarieta et al. 2020). The long generation time and long lifespan of conifers have been the main obstacles to performing functional studies in these woody plants. However, SE provides an excellent source of actively dividing and highly competent cells for genetic transformation. Efficient protocols are currently available for genetic transformation of embryogenic cell lines and subsequent plant regeneration (Klimaszewska et al. 2004; Trontin et al. 2007).

Two methodologies have been used for the stable genetic modification: biolistic/particle bombardment and *Agrobacterium*-mediated transformation. Transgenic plants through SE have been produced from *P. nigra* (Salaj et al. 2005) using particle bombardment, while the use of *Agrobacterium* has been reported for the regeneration of stably transformed *P. pinaster* (Trontin et al. 2007, 2016b; Tereso et al. 2006; Alvarez and Ordás 2013). Although SE offers the opportunity to produce transgenic plants in several forest tree species with important application in the forest industry, currently there are few reports of the regeneration of stably transformed pine species.

Genetic engineering provides the potential to add desirable characteristics of commercial importance, such as insect resistance, tolerance to herbicides and higher saccharification efficiency of wood. Pine species such as *P. taeda* (Tang and Tian 2003) and *P. radiata* (Grace et al. 2005) have been genetically modified with *Bt* toxin gene using biolistics. Tolerance to glufosinate has also been achieved in *P. radiata* using biolistics (Bishop-Hurley et al. 2001) and *Agrobacterium*-mediated transformation (Charity et al. 2005). Similar achievements have not yet been performed in Mediterranean pines, however, a large-scale platform for gene functional analysis in *P. pinaster* has been developed by FCBA and INRAE in the frame of the European project Sustainpine (see the section of Transgenesis in the chapter by Sterck et al.).

During the last decade, efficient protocols for the genetic transformation of conifers via somatic embryogenesis have been developed, however, given the complexity of the transformation process in pine, and the high number of factors that influence throughout the process, further research efforts are necessary to optimize some phases of the process.

9.3 Acquisition and Incorporation of Nutrients

A considerable number of molecular and genomic studies have been addressed to understand the acquisition of nutrients by Mediterranean pines. The nutrition of conifers depends to a great extent on the association of their roots with fungi to form the ectomycorrhizal mycelia that allow them to explore the soil and to acquire the mineral macro- and micronutrients that are necessary for tree growth. The ectomycorrhizal interaction provides the tree with mineral nutrients and metabolites, and the tree in return provides the fungus with metabolic products derived from photosynthesis. This allows a mineral nutrient-photosynthate interchange between the tree and the fungus. The current knowledge on the role of ectomycorrhizal fungi in the transport of nutrients from the soil to colonized roots has been recently reviewed (Becquer et al. 2019). The recent advances in high-throughput genomic technologies have enabled the identification of major families of transporters that are involved in the exchanges of nutrients within the mycorrhizal symbiosis and a wide range of membrane proteins have been identified and characterized (García et al. 2016). Many studies performed during the mycorrhization process of the Mediterranean pine *Pinus pinaster* have shed light on the molecular basis of how fungal transporters mediate nutrient acquisition (García et al. 2020). Little is known about how the progress of the symbiosis modifies the root transcriptome, although changes in the composition of the cell wall, defense, and carbon and nitrogen metabolism seem to be involved (Flores-Monterroso et al. 2013).

From the macronutrients, nitrogen has received special attention because it is a constituent of vitally important biomolecules such as nucleic acids, proteins, porphyrins, hormones and a range of secondary metabolites. For that reason, nitrogen availability largely determines the livelihood of pines and therefore this nutrient is often a major limiting factor for tree survival in Mediterranean forests. Due to the paramount importance of this essential nutrient, trees have developed a series of strategies for its acquisition, assimilation and storage to maintain sustained growth.

In the acidic soils of boreal forests, mineralization and nitrification are limited, therefore, ammonium and organic nitrogen in the form of amino acids are the predominant sources of nitrogen available for tree nutrition (Rennenberg et al. 2009; Näsholm et al. 2009). Accordingly, and unlike other plants, many conifer species prefer ammonium over nitrate as a source of nitrogen (Kronzucker et al. 1997). The acquisition of ammonium as a nitrogen source by conifers has been studied by Castro-Rodríguez et al. (2016). A search in the maritime pine transcriptome (Canales et al. 2014) revealed the existence of a small gene family encoding AMTs (ammonium transporters). The family is divided into two subfamilies named AMT1 and AMT2, as previously described in angiosperms (Castro-Rodríguez et al. 2017). The number and family composition are also conserved in other gymnosperms such as loblolly pine (*Pinus taeda*) and white spruce (*Picea glauca*) (Castro-Rodríguez et al. 2016). The individual members of the family are expressed at different times and in different cellular types in the tree, suggesting they play distinct and non-overlapping roles in ammonium transport. For example, PpAMT1.1 is highly represented at the early stages of tree development as well as in photosynthetic and reproductive tissues. However, PpAMT1.3 is mainly expressed in roots with up-regulation under nitrogen starvation and during the progress of mycorrhization. PpAMT1.3 has been proposed as a candidate to regulate the external ammonium uptake considering both its kinetic properties as a high-affinity transporter, and its location in the roots with increased expression in response to nitrogen starvation. Ammonium uptake in maritime pine roots is inhibited by a small cysteine-rich antimicrobial peptide (AMP1) which accumulates in the apoplast of root cells in response to increased ammonium availability (Canales et al. 2011). It has been suggested that AMP1 may act blocking specific AMT proteins regulating the incorporation of nitrogen from soil (Castro-Rodríguez et al. 2017).

In temperate climates, such as those in the Mediterranean basin, the availability of nitrate is much higher than in the soils of boreal forests. Therefore, nitrate represents a relevant source of nitrogen for the nutrition of Mediterranean pines. Genes putatively involved in the uptake of nitrate have been identified in *Pinus pinaster* (Castro-Rodríguez et al. 2017). Subsequent molecular analyses have shown that all subfamilies of the large *NITRATE TRANSPORTER 1/PEPTIDE TRANSPORTER* family (NPF) previously described in angiosperms (Léran et al. 2014) are present in the pine genome, suggesting that early diversification of this transporter family took place along the evolution of seed plants. Another gene family whose members are potentially involved in nitrate transport is *NRT2/NTR3*. NRT3 membrane proteins do not have transport activity but are required as a component for high-affinity nitrate uptake systems in *Arabidopsis* (Yong et al. 2010). Interestingly, the number of NRT3 members in the maritime pine NRT2/NRT3 family is unusually high in comparison to that described in angiosperms. These findings suggest that these membrane proteins may be involved in signaling nitrogen availability and in additional metabolic processes that are still unknown. Further research efforts are needed to elucidate the function of NRT3 proteins in conifers.

The growth of *P. pinaster* under the availability of different proportions of ammonium and nitrate has been recently evaluated (Ortigosa et al. 2020). In this work, it has been shown that the accumulation of biomass was higher in the seedlings grown in ammonium exhibiting higher nitrogen uptake efficiency than those grown in nitrate. In addition, the supplied nitrogen form determines nutrient storage site with ammonium accumulation mainly in the roots, and nitrate preferably in the stem. This difference in the distribution of both nutrients is also related to the metabolic response of pines, with the accumulation of arginine and soluble sugars in the roots of plants feed with ammonium, and asparagine in roots of plant feed with nitrate. These results suggest a limited ability of maritime pine to assimilate nitrate. The observed differences in the allocation of ammonium and nitrate suggest they can act as metabolic buffers to prevent interferences with the metabolism of photosynthetic organs. Taken together, the results indicate that ammonium is more efficiently incorporated and metabolically better utilized than nitrate in Mediterranean pines. However, in light of the above results, it would be interesting to determine whether ammonium is representing only a suitable nitrogen source for pine nutrition, or if it may also have a prominent role as a signaling molecule able to trigger developmental processes. This point deserves further attention and additional research efforts are necessary.

Besides inorganic sources such as ammonium and nitrate, organic nitrogen, mainly in the form of amino acids, represents a considerable proportion of the total nitrogen that is available in forest soils (Ganeteg et al. 2017). Consequently, the contribution of free amino acids to tree nutrition is much more important than initially thought, because they represent a source not only of nitrogen but also of carbon for plant growth and development. Overall, plant genomes encode a large number of transporters with different substrate specificities that facilitate the acquisition of amino acids from soil and internal traffic across different tissues, however, only a few of them have been characterized functionally. These transporters mainly belong to the permeases and LTH subfamilies (Dinkeloo et al. 2018). In maritime pine, sequences encoding 10 putative amino acid permeases have been identified using the available genomic resources. One of them, *PpAAP1*, has been found to be directly involved in the uptake of arginine by pine roots (Llebrés et al. unpublished). The absorption and intracellular transport of arginine are of paramount importance for pine nutrition because arginine is the amino acid with the highest nitrogen content, and therefore it represents a suitable source of nitrogen for immediate metabolic use and also for mid- and long-term storage. In fact, other results indicate that mycorrhizal fungi synthesize arginine that is translocated to the root mycelium, where it is degraded to provide ammonia that the plant assimilates, incorporating nitrogen into the metabolism (Chalot et al. 2006). An excess of nitrogen fertilization and the deposition of atmospheric nitrogen greatly increase the arginine content in conifer needles and wood, and it has been proposed that the arginine content may reflect the nitrogen status of conifers (Nordin et al. 2001).

Figure 9.2 summarizes the current knowledge on the major transporters potentially involved in

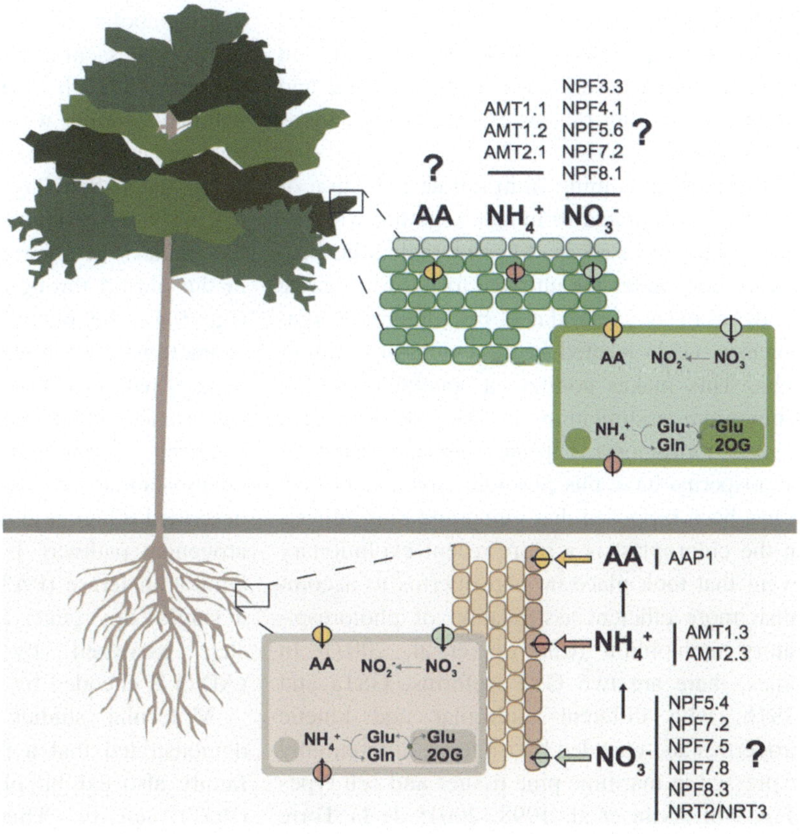

Fig. 9.2 Representative scheme of the uptake and transport of nitrate and ammonium in maritime pine and potential transporters involved. AA: Amino acid transporter; AMT: Ammonium transporter; Glu: Glutamate; Gln: Glutamine; NPF: Nitrate/peptide transporter; 2OG: 2-Oxoglutarate

the acquisition and distribution of inorganic and organic nitrogen in maritime pine (*P. pinaster*).

9.4 Nitrogen Metabolism

Special attention has also been paid to functional genomic approaches of nitrogen assimilation and metabolism (Xu et al. 2012; Krapp 2015; Castro-Rodríguez et al. 2017, 2020). Considering that the availability of nitrogen in forest soils is quite limited, several efficient mechanisms for the assimilation, storage, mobilization and recycling have been optimized by pines, in addition to nutrient uptake (reviewed by Cánovas et al. 2007). Nitrogen economy is especially relevant in fundamental tree biology because woody plants must cope with long lifespans that include many seasonal periods of growth and development. Efficient management of nitrogen resources is required for sustained growth, disease and pest resistance, response to climate change and adaptability. In fact, an appropriate N-status will result in higher biomass production, significant chance for reproduction and survival and will even determine, to a great extent, tree form and wood fiber quality (Cañas et al. 2016; Cánovas et al. 2018).

Ammonium coming from nitrate reduction or directly incorporated from soil, together with the ammonium released in different metabolic processes, constitutes the ultimate form of inorganic nitrogen to be assimilated by glutamine synthetases (GS1) located in the cytosol of conifer cells. This makes conifers a special model of ammonium assimilation, as they do not have GS2 isoforms located in the plastid, while most angiosperms have this plastidic enzyme. In fact, it has been proposed that glutamine biosynthesis in the chloroplast is a more recent evolutionary event that took place in angiosperms to accomplish more efficient assimilation of photorespiratory ammonium (Cánovas et al. 2007). In pines, there are two GS1 isoforms, GS1a and GS1b, with different molecular and kinetic properties and encoded by isogenes differentially expressed in maritime pine tissues and cell types (Fig. 9.3; Avila et al. 1998, 2001; de la Torre et al. 2002). GS1a is associated with chloroplast development and likely plays a similar role to GS2 in angiosperms (Cantón et al. 1999). However, *GS1b* is mainly expressed in roots and stems and is always associated with the vascular system (Fig. 9.3). Both *GS* genes in pine are differentially regulated by the transcription factor PpDof 5, and this regulatory mechanism suggests a transcriptional control of the expression of these genes to ensure the correct spatial distribution of the isoforms in the tree (Rueda-López et al. 2008). The role of *GS1b* as well as its coordinated co-regulation with genes involved in the phenylalanine pathway has also been studied (Craven-Bartle et al. 2013). This is not a minor topic since in trees, a large proportion of the carbon fixed in photosynthesis is immobilized as lignin during wood formation. Phenylalanine is deaminated for the biosynthesis of the phenylpropanoids and the released ammonium must be re-assimilated by the catalytic action of GS1b to keep the nitrogen homeostasis of the tree. In fact, large quantities of nitrogen should be recycled in this metabolic step, otherwise a continuous supply of this essential nutrient would be required to maintain a high metabolic flux through this pathway during wood formation. In addition, phenylalanine is the precursor of many other compounds that play vitally important roles in the growth and survival of land plants. The biosynthesis of phenylalanine takes place mainly in the plastid through two alternative pathways (Fig. 9.4), the phenylpyruvate pathway that is characteristic of microorganisms, and the arogenate pathway that is considered the most important in plants (Maeda and Dudareva 2012). Recently, a pathway for the biosynthesis of phenylalanine in the cytosol has also been described (Qian et al. 2019). The first step in the arogenate pathway is catalyzed by prephenate aminotransferase (PAT), an enzyme encoded by a single-copy gene. In contrast, the following step catalyzed by arogenate dehydratase (ADT) is encoded by a small gene family.

Molecular studies in maritime pine have demonstrated that a few enzymes of the ADT family also exhibit phenylpyruvate dehydratase (PDT) activity. This evidence supports the

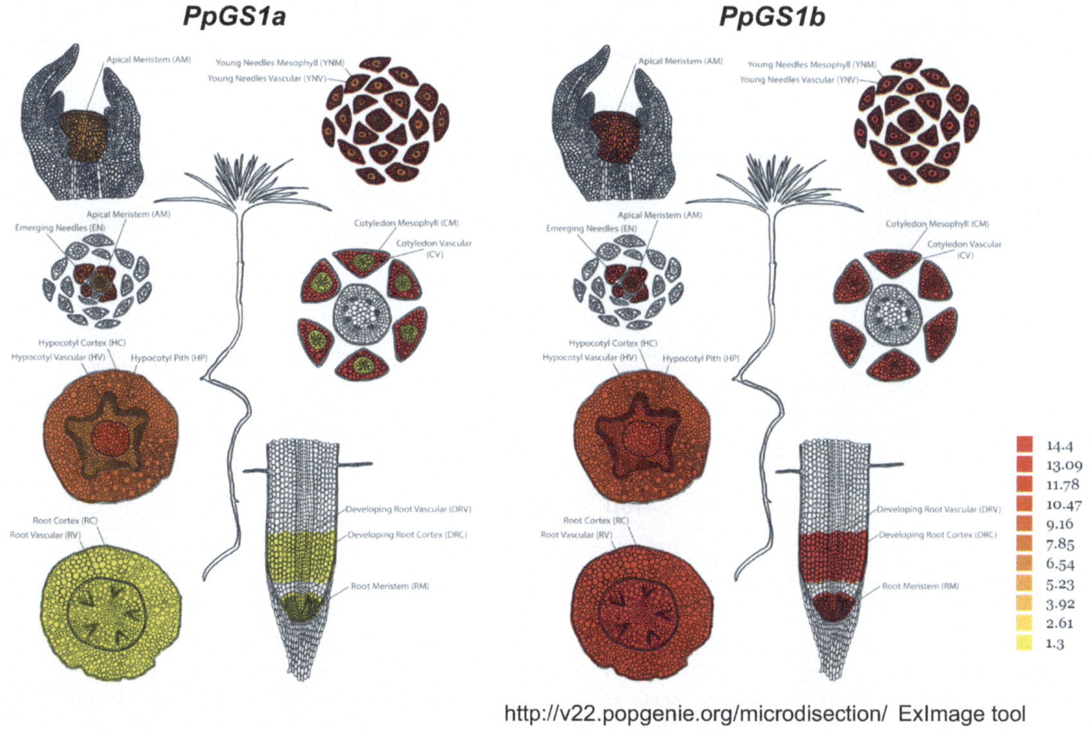

Fig. 9.3 Spatial gene expression of glutamine synthetase isogenes in different cell types of *P. pinaster*. Transcripts were identified using laser capture microdissection (LCM) combined with low-input RNA-seq. Image capture using the ExImage tool at ConGenIE.org (http://v22.popgenie.org/microdisection/). Gene expression is shown in relative units

phenylpyruvate pathway is also operative in pine (El-Azaz et al. 2016). Furthermore, it was found that a peculiarity of all ADT enzymes exhibiting PDT activity is that they contain a conserved small domain of 22 amino acid residues. An ortholog in *Arabidopsis*, ADT2, a bifunctional ADT/PDT enzyme is essential for embryo development and its deficiency causes embryo arrest and seed abortion (El-Azaz et al. 2018). Interestingly, pine ADT enzymes lacking this molecular feature are specifically involved in the production of phenylalanine to be used as a precursor of the lignin biosynthetic pathway (El-Azaz et al. 2020). Taken together, these results suggest an ancient origin of the phenylpyruvate pathway that is still fulfilling essential roles during development. Meanwhile, the arogenate pathway likely evolved more recently in plant evolution to provide phenylalanine for new metabolic fates. This is a very interesting research line that can provide new insights into the emergence of lignin biosynthesis, a key biochemical process for the expansion and adaptation of land plants (Weng and Chapple 2010).

With the aim of developing more in-depth studies about the regulation of genes involved in phenylalanine biosynthesis, transgenic plants of *P. pinaster* overexpressing or silenced for *PpMYB8* were obtained via somatic embryogenesis (El-Azaz et al. 2020). The overexpression of this TF-generated plants did not exhibit differences in growth after 2 years in the greenhouse, meanwhile, a significant reduction in growth and lignin content was observed in *RNAi-MYB8* pine. Furthermore, the transcriptomic results of these plants revealed a down-regulation of genes linked to phenylalanine metabolism, among which were two members of the arogenate dehydratase (ADT) gene family. The results obtained in this research suggest that

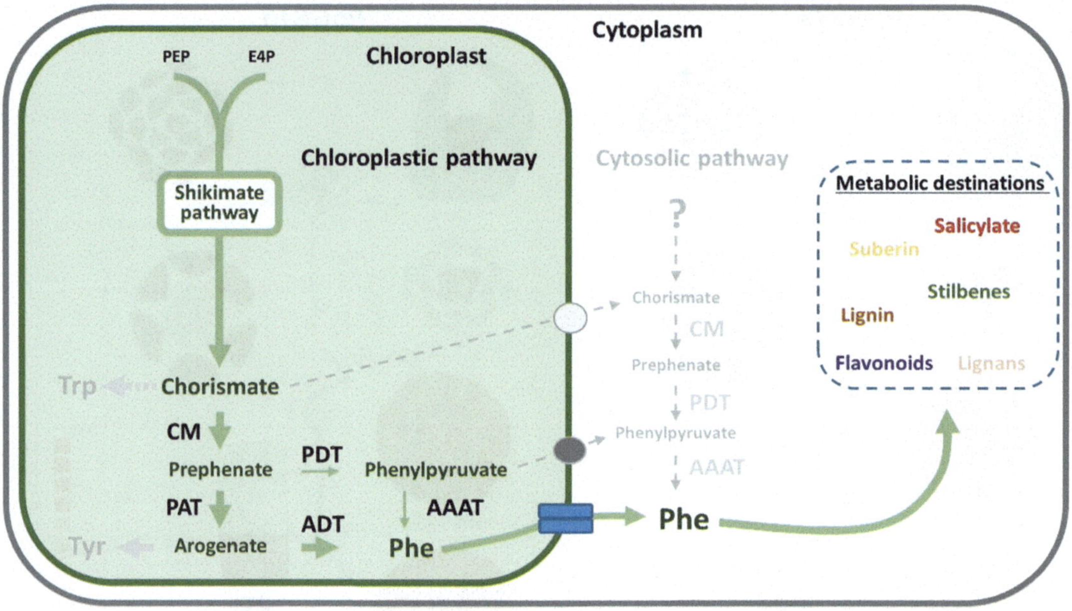

Fig. 9.4 Pathways for the biosynthesis of phenylalanine. Through the shikimate pathway, metabolites of intermediary carbon metabolism, erythrose-4-phosphate (E4P) and phosphoenolpyruvate (PEP) are channeled to the synthesis of aromatic amino acids phenylalanine, tyrosine and tryptophan. In plants, phenylalanine is mainly synthesized within plastids using two alternative pathways that differ in the respective use of arogenate or phenylpyruvate as an intermediary. The existence of a partial cytosolic phenylalanine biosynthetic pathway has also been described in some plants. CM: chorismate mutase; PAT: prephenate aminotransferase; PDT: prephenate dehydratase; ADT: arogenate dehydratase; AAAT: aromatic amino acid aminotransferase

MYB8 plays a pivotal role in coordinating the deposition of lignin with the biosynthesis of phenylalanine (El-Azaz et al. 2020).

The function of *PpNAC1*, a potential ortholog of *NST1* and *SDN1,* two transcription factors regulating secondary cell wall biosynthesis in *Arabidopsis*, has also been studied by RNAi interference analysis (Pascual et al. 2018). The silencing of PpNAC1 led to altered stem vascular structure and down-regulation of several genes associated with cell wall biogenesis and secondary metabolism. The results showed that PpNAC1 is a master regulator involved in the biosynthesis and metabolic utilization of phenylalanine in maritime pine, and that PpNAC1 mediates its activity through direct activation of *PpMyb8*, a TF associated with lignin and phenylpropanoid biosynthesis in conifers (Bomal et al. 2008; Bedon et al. 2010; Craven-Bartle et al. 2013). These recent advances pave the way for additional studies to unveil the regulatory mechanisms governing the biosynthesis and metabolic fate of phenylalanine in pine. As previously mentioned, functional studies in conifers have serious limitations. However, research efforts in this field should be encouraged to unravel the complexity of secondary metabolism in conifers.

In the long term, an excess of nitrogen is stored as vacuolar reserve proteins (VSP) that accumulate in seeds or vegetative organs. These proteins are particularly rich in arginine and amide amino acids, and the mobilization of these nitrogen reserves is vitally important for the growth and development of trees, both in the early stages and during the annual growth cycles (Cantón et al. 2005). In fact, it is well known that arginine is an important form of storage and transport of organic nitrogen that represents 40–50% of the total nitrogen reserves in pine seeds (Cañas et al. 2016). Figure 9.5 shows a schematic representation that illustrates the processes

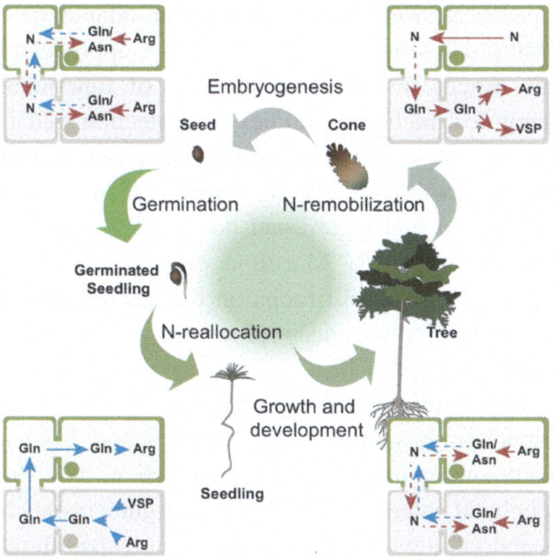

Fig. 9.5 General scheme showing the uptake, transport and storage of arginine in pine. Asn, asparagine; Arg, arginine; Gln, glutamine; N, nitrogen; VSP, vacuolar storage protein

of arginine uptake, transport and storage. Consequently, the study of arginine biosynthesis and molecular regulation is essential to understanding nitrogen homeostasis in Mediterranean pines.

Nevertheless, this is a relatively unexplored area of research, not only in conifers but also in plant metabolism in general. Recent studies highlight the important role of arginine metabolism during the embryogenesis and germination of maritime pine (Llebrés et al. 2018a, b). The subcellular localization of the enzymes involved in this metabolic pathway has been unambiguously determined (Llebrés et al. 2018b). These studies revealed that ornithine, a precursor of arginine biosynthesis in the chloroplasts, is synthesized not only in the plastid through the cyclic ornithine pathway but also in the cytosol through a compartmentalized linear pathway. Additional work has been conducted to further understand the molecular regulation of arginine biosynthesis. The enzyme N-acetylglutamate synthase (NAGK) is allosterically inhibited by arginine, the end product of the pathway. However, when nitrogen availability is high, the nitrogen sensor PII binds NAGK relieving the inhibition by arginine and favoring nitrogen storage. Two isoforms of PII (PIIa and PIIb), located in the chloroplast, have been characterized recently in *P. pinaster*. Expression studies have revealed a predominant role for PIIa during embryogenesis and seed germination (Llebrés et al. 2020). These advances in the understanding of the molecular basis of nitrogen metabolism during embryogenesis have great biotechnological potential to improve the efficiency of conifer regeneration via somatic embryogenesis and functional analysis of candidate genes in transgenic conifers (Trontin et al. 2013; Llebrés et al. 2018b).

The recombinant production in bacterial extracts and subsequent purification of key enzymes of the arginine pathway enabled their functional characterization and determination of their molecular and kinetic properties (Urbano-Gámez et al. 2020). Results derived from these functional studies indicate that, in addition to NAGK, the reaction catalyzed by ornithine transcarbamylase (OTC) is also an important point of regulation that controls the subcellular levels of ornithine. This amino acid is an intermediary precursor placed in the crossroads of several biosynthetic pathways such as arginine biosynthesis in the plastid, and polyamines in the cytosol.

Meanwhile, in herbaceous species, nitrogen assimilation and metabolism are highly coupled with photosynthetic activity in woody plants such as Mediterranean pines. These processes are much more dependent on the mobilization of nitrogen reserves (Cánovas et al. 2018). The photosynthetic capacity of Mediterranean pines (*Pinus halepensis, Pinus pinea* and *Pinus nigra*) has been examined comparing juvenile and adult needles (Kuusk et al. 2018a). There are major photosynthetic differences among young and mature needles besides nitrogen content and metabolic exchange between source and sink tissues. Differences in the morphology of juvenile and adult needles appear to be a consequence of the investment in two different tasks: the support of the tree and the photosynthetic capacity. Young needles maximize carbon gain for the establishment of new trees, meanwhile, adult needles drive higher investments in enhanced tolerance to environmental changes. Accordingly, a drought tolerance ranking of the species can be established, with *P. halepensis* being more tolerant, followed by *P. pinea* and *P. nigra* (Kuusk et al. 2018b). Transcriptome dynamics has been studied during the course of a complete year in the needles of *P. pinaster* (Cañas et al. 2015). This work provides an integrated view of the seasonal regulation of maritime pine growth, including changing photosynthetic capacities together with a view of the adaptive response of the trees. Gene co-expression analysis identified modules mainly correlated with the environment (temperature) except for one module which contained genes related to needle development. Overall, these results showed that environmental stimuli trigger major changes in the transcriptome which are buffered for a fine-tuning of metabolites during development.

9.5 Environmental Interactions

The ability of pine trees to adapt to changing environmental conditions, which in the Mediterranean basin can be extreme, represents a key factor for their survival. The existence of multiple natural populations, as well as extensive forest areas generated through reforestation, has allowed the identification of provenances differentially adapted to different stress-generating environmental factors. These natural populations are the main genetic resource to face studies targeting the identification of pine trees resistant to a wide set of biotic and abiotic stresses in a scenario of global warming and general environmental variations.

Multiple studies have specifically addressed the effect of different types of stress on pine species within the Mediterranean basin. Among these works, those carried out from a functional genomics perspective have addressed aspects related to abiotic (mainly drought) and biotic stress.

Drought is a stressful factor of enormous importance affecting Mediterranean pine species development, growth and viability. A high proportion of the research on this topic has consisted in the study of anatomical and biophysical aspects as well as environmental adaptations. Multiple research works have also approached this aspect from the perspective of functional genomics, including studies of molecular genetics as well as gene expression analyses. Given that an extensive and thorough review of the research in this regard can be found within this book in Chap. 5 "Maritime pine genomics in focus", here we will focus on the biotic stresses affecting Mediterranean pines using functional genomics approaches.

Interaction of Mediterranean pines with other organisms has been mainly approached from a plant/pathogen interaction perspective, but the beneficial interaction with other organisms such as endophytic bacteria and ectomycorrhizal fungi should not be forgotten (see the section on nutrient acquisition in this chapter). In this context, functional information about the interactions of Mediterranean pines with bacteria or viruses is currently limited and mainly linked to the detection of endophytic bacterial species (Paiva et al. 2014; Proença et al. 2017a, b) or associated with other pathogens, such as insects or nematodes, as described below.

Phytoplasmas are a group of obligate bacterial parasites of plant phloem transmitted by insect vectors and affecting most plant clades. *Candidatus* Phytoplasma pini causes malformations in multiple pine species including *P. halepensis* and results in a wide variety of symptoms including witch's broom, dwarfed leaves, stunting and yellowing (Schneider et al. 2005). Molecular information about this pest disease is very scarce despite its growing presence in the Mediterranean basin, where it is causing extensive forest damage (Fig. 9.6). However, the recent publication of a draft genome of this pathogen (Cai et al. 2020), in combination with the increased genomic resources available for functional molecular studies in pine (Cañas et al. 2019), represents a valuable tool for future research in this area. The genome of phytoplasmas lacks many genes involved in essential metabolic pathways and therefore, these cellular parasites depend on nutrients derived from the host (Tan et al. 2021). The molecular mechanisms underlying the

Fig. 9.6 A large witch's broom in a Mediterranean forest of pines in the province of Málaga, Spain

interchange of nutrients in this biotic interaction deserve further attention in future research work.

Different insect species are recurrent virulent pathogens of pines. Aleppo pine (*P. halepensis*) is frequently infected by the giant pine scale *Marchalina hellenica*. However, recent studies using ecophysiological approaches and metabolic analysis have shown that this pine species is highly resilient to infestation. Stomatal closure, photosynthetic inhibition, accumulation of carbohydrates, tricarboxilic acids, and production of phenolics seem to be the basis for the mechanism of resistance (Fotelli et al. 2020).

An important pathogenic insect species affecting pine trees is the processionary moth, *Thaumetopoea pityocampa,* which displays enormous damaging capacity in the Mediterranean basin. Processionary larvae growing in *P. nigra*, *P. pinaster* or *P. halepensis* differ in the microbial community present in their gut, suggesting alternative modulatory interactions between the host plant and this pathogen (Strano et al. 2018). The emission of specific volatile terpenoids, such as bornyl acetate, facing the processionary attack has been proposed as a possible mechanism of defense in *P. nigra* (Foti et al. 2020). In the same line, it has been reported that a strong increase in the concentration of polyphenolic compounds in the stems of *P. pinaster* infected with T. *pityocampa* (Moreira et al. 2013). Nevertheless, the available tools to combat this pest are currently quite limited. The recent whole genome sequencing of *T. pityocampa* represents an important starting point for the identification of potential targets for the action of pesticides (Shahraki et al. 2020). For example, it is well known that G-protein-coupled receptors play a major role in insect development. The identification of several G-protein-coupled receptors which are expressed in the larval stage paves the way for the design of specific ligand peptides able to act as blockers of insect development.

Pine wilt disease, one of the most severe diseases affecting pine species, is mediated by the pathogenic nematode *Bursaphelenchus xylophilus*. This pathogen is transmitted by the pine sawyer beetle *Monochamus galloprovincialis* and infects, among others, *P. pinaster* and *P. pinea* causing blockage of the host vascular elements and cavitation (Gonçalves et al. 2020). Different studies have shown that the virulence of the disease is related to the specific community of bacteria associated with the nematode (Han et al. 2003; Vicente et al. 2012; Proença et al. 2012). Metagenomic analyses of this nematode have allowed the identification of multiple bacterial species including *Serratia* spp., which not only is capable of degrading cellulose but also has enzymes participating in xenobiotic metabolism (α-pinene), thus suggesting an advantage for wood colonization (Vicente et al. 2013). Currently, there are no studies delving into the molecular regulation of this pathogenic process in Mediterranean pine species. In contrast, studies in other species such as *P. densiflora* and *P. thunbergii* suggest the existence of a conserved pattern of transcriptional regulation in response to this infection involving phenylpropanoid and flavonoid biosynthesis and oxidative stress (Lee et al. 2019; Park et al. 2020).

Mediterranean pine species are also affected by fungal pathogens such as *Fusarium circinatum*, the causal agent of pitch canker disease. *P. pinaster* and *P. pinea* have shown moderate resistance to this pathogen compared to other species such as *Pinus radiata* (Amaral et al. 2019). The molecular mechanism supporting the resistance in these species remains unclear but recent work in *P. pinaster* using RNA-seq analysis has shown that these plants are able to activate defense mechanisms involving complex signaling crosstalk between ethylene, jasmonic acid and salicylic acid and induction of *PR* (*P*athogenesis-*R*elated) genes (Hernández-Escribano et al. 2020). This work proposes that the pathogenicity in certain pine species triggered by this pathogen is based on its ability to manipulate host ethylene homeostasis, blocking of jasmonic acid signaling and preventing salicylic acid biosynthesis. Vivas et al. (2013) have also shown that the extension of damage caused by *F. circinatum* in *P. pinaster* is related to its maternal genotype and environment, suggesting an epigenetic regulatory mechanism.

Other fungal pathogens capable of infecting pines, such as the necrotroph *Botrytis cinerea*, the causal agent of gray mold disease, have also been detected in the Mediterranean basin. *P. pinaster* has shown to be able to display a non-host type II resistance mechanism under *B. cinerea* infection involving biphasic production of intracellular ROS, down-regulation of ROS-scavenging mechanisms and reduction of the antioxidant activity (Azevedo et al. 2008). *Diplodia sapinea* is the fungal pathogen causal of blight and canker disease affecting *P. nigra* plants subjected to drought stress. The pathogenic mechanism of *D. sapinea* involves the reduction of H_2O_2 generated during drought stress by increasing catalase and peroxidase activities (Sherwood et al. 2015).

Hemiparasitic plants such as *Viscum album* L. (mistletoe) frequently associated with Mediterranean pines may cause damage to species such as *P. pinaster*, *P. halepensis* or *P. nigra* (Gea-Izquierdo et al. 2019; Zamora and Mellado 2019). The presence of mistletoe reduces leaf area and correlates with tree decline, and its presence seems to be preceded by abiotic stresses that predispose to its infection (Gea-Izquierdo et al. 2019). The biotic interactions have been poorly studied and future studies are required to unveil the molecular and physiological mechanisms involved in the sequestration of pine metabolic resources.

9.6 Analysis of Wood Formation

Besides an important ecological role, some Mediterranean pine species have a significant economic impact as a major source of lumber and paper pulp. Transcriptomic analyses of developing secondary xylem made it possible to draw a general idea of the genes that encode the enzymes of the different biosynthetic processes involved, and to identify potential regulators of wood formation in conifers (Carvalho et al. 2013). Transcriptome profiling arose as a means to identify candidate genes associated with structural and compositional features associated with different types of wood in trees with different properties, which allow understanding the molecular basis of wood property variation and developing molecular markers for marker-assisted breeding programs. One advantage of working with trees is that up to six types of wood can be found in the same individual with different chemical, physical and anatomical properties (early wood, late wood, juvenile wood, mature wood, compression wood and opposite wood). The increasing demand for wood has led to a reduction of the tree rotation age and the breeding for fast-growing trees, aiming to increase wood yield in planted forests. However, as a direct consequence, the quality of the timber is reduced due to a higher proportion of juvenile and compression wood in young trees, both with lower properties for industrial uses compared to mature and normal wood. Therefore, molecular factors involved in juvenile and compression wood formation and properties must be considered in breeding programs for improving the quality of wood in short rotation forestry. Consequently, most wood development transcriptomic research in Mediterranean pines has focused on comparing transcriptome differences among developing secondary xylems. This is evident in *P. pinaster*, because of its importance to supply wood for processing industries in Southwestern Europe.

Proteomic analysis of developing secondary xylem in *P. pinaster* allowed the identification of proteins functionally associated with wood formation in this species. Two-dimensional polyacrylamide gel electrophoresis and microsequencing from developing xylem samples allowed identifying 26 compression wood-responsive proteins (Plomion et al. 2000). In a more advanced proteomic analysis, protein extracted from juvenile, mature, early, late, compression and opposite developing secondary xylems were analyzed by 2-DE. A total number of 1039 different spots were identified, and 300 proteins were excised from the polyacrylamide gel and analyzed by mass spectrometry or internal microsequencing, allowing to outline a functional distribution of the major proteins in maritime pine wood forming tissue (Gion et al. 2005). At the transcriptome level, an early report in *P. pinaster* used cDNA-amplified fragment length polymorphism (AFLP) analysis to screen several thousands of cDNA fragments from differentiating secondary xylem associated with the development of early wood, late wood, compression wood and opposite wood (Provost et al. 2003). About 100 transcript-derived fragments of differential expressed transcripts were identified, and among the genes with a known function, the results suggested a glycine-rich protein as a strong candidate gene for wood properties. At the same time, a large-scale cDNA sequencing effort was performed to identify genes related to wood formation in *P. pinaster*, producing a gene catalog (EST) from a composite library of developing xylem associated with early, late, opposite and compression wood from four different genotypes of the Corsican provenance (Cantón et al. 2003).

In a later study, a transcriptomic comparison between developing secondary xylem forming early and late wood in *P. pinaster* was carried out. This study differed from previous ones in that it analyzed a higher number of different transcripts by using cDNAs macroarrays, and performed an exhaustive phenotypic characterization of the samples collected along a growing season. As a result, 19% of 3512 analyzed genes showed a significant seasonal effect in their expression. Those genes were clustered into five distinct groups, with the largest two groups representing genes overexpressed in the early- or late-wood-forming tissues, respectively, whereas the other three clusters were characterized by responses to specific edaphoclimatic conditions (Paiva et al. 2008a).

Comparative analysis between developing secondary xylem producing crown/juvenil or base/mature has also been carried out in *P. pinaster*, following the same methodological approach (phenotypic characterization of samples and expression profiling by macroarray hybridization), together with proteomic analysis (Paiva et al. 2008b). This study identified 262 unigenes and 231 proteins that were differentially expressed along the stem. The authors found a good relationship between functional categories from transcriptomic and proteomic data, and between the molecular mechanisms represented by the differential expressed genes and the observed wood phenotypic differences.

The reprogramming of the transcriptome associated with the development of compression wood in *P. pinaster* was analyzed by comparing the transcriptomes of developing xylem-forming compression and opposite wood, sampled from the trunk of adult trees after 2 months of artificial mechanical bending. The analysis was carried out by using dual hybridization of microarrays constructed with cDNAs. These came from either a composite cDNA library of different developing secondary xylem or from SSH subtractive libraries (Villalobos et al. 2012). A total of 496 genes that changed their expression during the differentiation of compression wood were identified. Consistent with the deposition of a thicker secondary cell wall that characterizes compression wood development, the expression of a number of genes involved in the synthesis of cellulose, hemicellulose, lignin and lignans was up-regulated. The results suggested the coordinated modulation at the transcriptional level of genes involved in *S*-adenosylmethionine synthesis and ammonium assimilation, associated with increased demand for coniferyl alcohol for lignin and lignan synthesis.

The transcriptome of wood-forming tissue has also been analyzed in *P. canariensis*. A transcriptome of the cambial zone and differentiating xylem covering early- and late-wood differentiation from juvenile three-year-old trees was assembled by using Roche 454 Genome Sequencer FLX Titanium System technology. Transcriptome changes during the growing season were analyzed with microarrays (Chano et al. 2017a). This study identified a high number of previously undescribed genes, expressed during juvenile wood formation in pine. Wound xylem formation has also been analyzed in *P. canariensis* (Chano et al. 2017b) through microarray hybridization analysis. This study revealed alterations in the transcription profile of up to 1408 genes during the first period of healing, with a complete rearrangement of the transcriptional program in the cambial zone close to the injuries.

Overall, the different analyses of transcriptomes involved in wood formation in Mediterranean pine species have provided functional insights into the molecular machinery involved in wood formation. Transcriptomic studies have provided a better understanding of the molecular processes and metabolic requirements underlying secondary xylem differentiation. They have also revealed candidate genes potentially responsible for the intra-individual phenotypic differences found in different wood types, and have identified strong candidates for determining wood properties. Wood formation in Mediterranean pine species has been studied since the beginning of the development of early methodological and technical approaches to analyze the transcriptome. However, newer, more informative, methodologies of transcriptome and epitranscriptome analysis based on massive transcript sequencing (indirect or direct RNAseq) have not yet been systematically applied in the species. Opening the possibility of a more comprehensive and informative description of the transcriptome and transcriptional regulatory networks involved in wood formation, and their response to various endogenous and exogenous signals, is of great ecological and industrial interest in these species.

Acknowledgements Research in the authors' laboratory has been funded by Plant KBBE/MICINN SUSTAINPINE (PLE2009-0016), EU FP7-KBBE-2011-5 (ProCoGen), MINECO (BIO2012-33797, BIO2015-69285-R and BIO2015-73512-JIN), MICINN (RTI2018-094041-B-I00) and Junta de Andalucía grants (BIO-474 and BIO-114).

References

Allona I, Collada C, Casado R, Aragoncillo C (1994) Electrophoretic analysis of seed storage proteins from gymnosperms. Electrophoresis 15(8–9):1062–1067. https://doi.org/10.1002/elps.11501501158

Alonso P, Cortizo M, Cantón FR, Fernández B, Rodríguez A, Centeno ML, Cánovas FM, Ordás RJ (2007) Identification of genes differentially expressed during adventitious shoot induction in *Pinus pinea* cotyledons by subtractive hybridization and quantitative PCR. Tree Physiol 27:1721–1730. https://doi.org/10.1093/treephys/27.12.1721

Álvarez JM, Ordás RJ (2013) Stable *Agrobacterium*-mediated transformation of maritime pine based on kanamycin selection. Sci World J 681792.https://doi.org/10.1155/2013/681792

Álvarez JM, Cortizo M, Bueno N, Rodríguez A, Ordás RJ (2013) *CLAVATA1-LIKE*, a leucine-rich-repeat protein receptor kinase gene differentially expressed during adventitious caulogenesis in *Pinus pinaster* and *Pinus pinea*. Plant Cell Tiss Org 112:331–342. https://doi.org/10.1007/s11240-012-0240-8

Amaral J, Correia B, António C, Rodrigues AM, Gómez-Cadenas A, Valledor L, Hancock RD, Alves A, Pinto G (2019) *Pinus* susceptibility to Pitch Canker triggers specific physiological responses in symptomatic plants: an integrated approach. Front Plant Sci 10:509. https://doi.org/10.3389/fpls.2019.00509

Arrillaga I, Morcillo M, Zanon I, Lario F, Segura J, Sales E (2019) New approaches to optimize somatic embryogenesis in maritime pine. Front Plant Sci 10:138. https://doi.org/10.3389/fpls.2019.00138

Avila C, García-Gutiérrez A, Crespillo R, Cánovas FR (1998) Effects of phosphinothricin treatment on glutamine synthetase isoforms in Scots pine seedlings. Plant Physiol Biochem 36:857–863

Avila C, Suárez MF, Gómez-Maldonado J, Cánovas FM (2001) Spatial and temporal expression of two cytosolic glutamine synthetase genes in Scots pine: functional implications on nitrogen metabolism during early stages of conifer development. Plant J 25:93–102

Azevedo H, Lino-Neto T, Tavares RM (2008) The necrotroph *Botrytis cinerea* induces a non-host type II resistance mechanism in *Pinus pinaster* suspension-cultured cells. Plant Cell Physiol 49:386–395. https://doi.org/10.1093/pcp/pcn015

Barbero M, Loisiel R, Quezel P, Richardson DM, Romane F (2000) Pines of the Mediterrranean basin. In: Ecology and biogeography of *Pinus*. Cambridge University Press ed.: DM Richardson pp 153–170

Bahrman N, Zivy M, Baradat P, Damerval C (1994) Organization of the variability of abundant proteins in 7 geographical origins of maritime pine (*Pinus pinaster* Ait). Theor Applied Genet 88:407–411. https://doi.org/10.1007/bf00223652

Becquer A, Guerrero-Galán C, Eibensteiner JL, Houdinet G, Bücking H, Zimmermann SD, Garcia K (2019) The ectomycorrhizal contribution to tree nutrition. In: Cánovas F (ed) molecular physiology and biotechnology of trees, vol 89. Adv Bot Res pp 77–126

Bedon F, Bomal C, Caron S, Levasseur C, Boyle B, Mansfield SD, Schmidt A et al (2010) Subgroup 4R2R3-MYBs in conifer trees: gene gamily expansion and contribution to the isoprenoid- and flavonoid-oriented responses. J Exp Bot 61:3847–3864

Bishop-Hurley SL, Zabkiewiez RJ, Grace L, Gardner RC, Wagner A, Walter C (2001) Conifer genetic engineering: transgenic *Pinus radiata* (D. Don) and *Picea abies* (Karst) plants are resistant to the herbicide Buster. Plant Cell Rep 20:235–243

Bomal C, Bedon F, Caron S et al (2008) Involvement of *Pinus taeda* MYB1 and MYB8 in phenylpropanoid metabolism and secondary cell wall biogenesis: a comparative in planta analysis. J Exp Bot 59:3925–3939

Cai W, Shao J, Zhao Y, Davis RE, Costanzo S (2020) Draft genome sequence of 'Candidatus Phytoplasma pini'-related strain MDPP: a resource for comparative genomics of gymnosperm-infecting phytoplasmas. Plant Dis 104:1009–1010. https://doi.org/10.1094/PDIS-10-19-2127-A

Canales J, Bautista R, Label P, Gómez-Maldonado J, Lesur I, Fernández-Pozo N, Rueda-López M, Guerrero-Fernández D, Castro-Rodríguez V, Benzekri H, Cañas RA, Guevara M-Á, Rodrigues A, Seoane P, Teyssier C, Morel A, Ehrenmann F, Le Provost G, Lalanne C, Noirot C, Klopp C, Reymond I, García-Gutiérrez Á, Trontin J-F, Lelu-Walter M-A, Miguel C, Teresa Cervera M, Cantón FR, Plomion C, Harvengt L, Ávila C, Gonzalo Claros M, Cánovas FM (2014) De novo assembly of maritime pine transcriptome: implications for forest breeding and biotechnology. Plant Biotechnol J 12:286–299. https://doi.org/10.1111/pbi.12136

Canales J, Flores-Monterrosso A, Rueda-López M, Ávila C, Cánovas FM (2010) Identification of genes regulated by ammonium availability in the roots of maritime pine trees. Amino Acids 39:991–1001. https://doi.org/10.1007/s00726-010-0483-9

Canales J, Avila C, Cánovas FM (2011) A maritime pine antimicrobial peptide involved in ammonium nutrition. Plant Cell Environ 34:1443–1453

Canales J, Rueda-López M, Craven-Bartle B, Ávila C, Cánovas FM (2012a) Novel insights into regulation of asparagine synthetase in conifers. Front Plant Sci 3:100. https://doi.org/10.3389/fpls.2012.00100

Canales J, Ávila C, Cantón FR, Pacheco-Villalobos D, Díaz-Moreno S, Ariza D, Molina-Rueda JJ, Navarro-Cerrillo RM, Claros MG, Cánovas FM (2012b) Gene expression profiling in the stem of young maritime pine trees: detection of ammonium stress-responsive genes in the apex. Trees-Struct Funct 26:609–619. https://doi.org/10.1007/s00468-011-0625-z

Cañas RA, Canales J, Muñoz-Hernández C, Granados JM, Ávila C, García-Martín ML, Cánovas FM (2015) Understanding developmental and adaptive cues in pine through metabolite profiling and co-expression network analysis. J Exp Bot 66:3113–3127. https://doi.org/10.1093/jxb/erv118

Cañas RA, Feito I, Francisco Fuente-Maqueda J, Ávila C, Majada J, Cánovas FM (2015b) Transcriptome-wide analysis supports environmental adaptations of two *Pinus pinaster* populations from contrasting habitats. BMC Genomics 16.https://doi.org/10.1186/s12864-015-2177-x

Cañas RA, de la Torre F, Pascual MB, Ávila C, Cánovas FM (2016) Nitrogen economy and nitrogen environmental interactions in conifers. Agronomy 6:26

Cañas RA, Li Z, Pascual MB, Castro-Rodríguez V, Ávila C, Sterck L, Van de Peer Y, Cánovas FM (2017) The gene expression landscape of pine seedling tissues. Plant J 91:1064–1087. https://doi.org/10.1111/tpj.13617

Cañas RA, Pascual MB, de la Torre FN, Ávila C, Cánovas FM (2019) Resources for conifer functional genomics at the omics era. In: Cánovas F (ed) molecular physiology and biotechnology of trees, vol 89. Adv Bot Res pp 39–76

Cánovas FM, Cantón FR, Gallardo F, García-Gutiérrez A, De Vicente A (1991) Accumulation of glutamine-synthetase during early development of maritime pine (*Pinus pinaster*) seedlings. Planta 185:372–378. https://doi.org/10.1007/bf00201059

Cánovas FM, Ávila C, Cantón FR, Cañas RA, de la Torre F (2007) Ammonium assimilation and amino acid metabolism in conifers. J Exp Bot 58:2307–2318

Cánovas FM, Cañas RA, de la Torre FN, Pascual MB, Castro-Rodríguez V, Avila C (2018) Nitrogen metabolism and biomass production in forest trees. Front Plant Sci 9:1449

Cantón FR, García-Gutiérrez A, Gallardo F, De Vicente A, Cánovas FM (1993) Molecular characterization of a cDNA clone encoding glutamine synthetase from a gymnosperm, *Pinus sylvestris*. Plant Mol Biol 22:819–828. https://doi.org/10.1007/bf00027368

Cantón FR, Suárez M, Josè-Estanyol M, Cánovas FM (1999) Expression analysis of a cytosolic glutamine synthetase gene in cotyledons of Scots pine seedlings:

developmental, light regulation and spatial distribution of specific transcripts. Plant Mol Biol 40:623–634

Cantón FR, Le Provost G, García V, Barré A, Frigerio J-M, Fevereiro P, Ávila C, Mouret J-F, de Daruvar A, Cánovas FM, Plomion C (2003) Transcriptome analysis of wood formation in maritime pine. In: Espinal S, Barredo Y, Ritter E (eds) Sustainable forestry. DFA-AFA Press, Vitoria-Gasteiz, Woods products and biotecnology, pp 333–348

Cantón FR, Suárez MF, Cánovas FM (2005) Molecular aspects of nitrogen mobilisation and recycling in trees. Photosynth Res 83:265–278

Carneros E, Celestino C, Klimaszewska K, Park YS, Toribio M, Bonga JM (2009) Plant regeneration in Stone pine (*Pinus pinea* L.) by somatic embryogenesis. Plant Cell Tiss Org 98:165–178

Castander-Olarieta A, Moncalean P, Montalban AI (2020) *Pinus canariensis* plant regeneration through somatic embryogenesis. Forest Systems 29:eSC05

Carvalho A, Paiva J, Louzada J, Lima-Brito J (2013) The transcriptomics of secondary growth and wood formation in conifers. Mol Biol Int 2013:974324.https://doi.org/10.1155/2013/974324

Castro-Rodríguez V, García-Gutiérrez A, Canales J, Cañas RA, Kirby EG, Avila C, Cánovas FM (2015) Poplar trees for phytoremediation of high levels of nitrate and applications in bioenergy. Plant Biotechnol J. https://doi.org/10.1111/pbi.12384

Castro-Rodríguez V, Assaf-Casals I, López-Tienda J, Avila C, Miller AJ, Cánovas FM (2016) Deciphering the molecular basis of ammonium uptake and transport in maritime pine. Plant Cell Environ 39:1669–1682

Castro-Rodríguez V, Cañas RA, de la Torre F, Pascual MB, Ávila C, Cánovas FM (2017) Molecular fundamentals of nitrogen uptake and transport in trees. J Exp Bot 68:2489–2500

Castro-Rodríguez V, Avila C, Cánovas FM (2020) Get more bark for a buck: the economy of nitrogen in deciduous forest trees. J Exp Bot 71:4369–4372

Chalot M, Blaudez D, Braun A (2006) Ammonia: a candidate for nitrogen transfer at the mycorrhizal interface. Trends Plant Sci 11:263–266

Chano V, Collada C, Soto A (2017a) Transcriptomic analysis of wound xylem formation in *Pinus canariensis*. BMC Plant Biol 17.https://doi.org/10.1186/s12870-017-1183-3

Chano V, López de Heredia U, Collada C, Soto Á (2017b) Transcriptomic analysis of juvenile wood formation during the growing season in *Pinus canariensis*. Holzforschung 71:919–937. https://doi.org/10.1515/hf-2017-0014

Chano V, Sobrino-Plata J, Collada C, Soto A (2020) Wood development regulators involved in apical growth in *Pinus canariensis*. Plant Biol (Stuttgart, Germany). doi:https://doi.org/10.1111/plb.13228

Charity JA, Holland L, Grace LJ, Walter C (2005) Consistent and stable expression of the *nptII*, *uid*A and *bar* genes in transgenic *Pinus radiata* after *Agrobacterium tumefaciens*-mediated transformation using nurse cultures. Plant Cell Rep 23:606–616

Costa P, Bahrman N, Frigerio JM, Kremer A, Plomion C (1998) Water-deficit-responsive proteins in maritime pine. Plant Mol Biol 38:587–596. https://doi.org/10.1023/a:1006006132120

Costa P, Plomion C (1999) Genetic analysis of needle proteins in maritime pine 2. Variation of Protein Accumulation. Silvae Genet 48:146–150

Craven-Bartle B, Pascual BM, Cánovas FM, Ávila C (2013) A Myb transcription factor regulates genes of the phenylalanine pathway in maritime pine. Plant J 74:755–766. https://doi.org/10.1111/tpj.12158

de María N, Guevara MÁ, Perdiguero P, Vélez MD, Cabezas JA, López-Hinojosa M, Li Z, Díaz LM, Pizarro A, Mancha JA, Sterck L, Sánchez-Gómez D, Miguel C, Collada C, Díaz-Sala MC, Cervera MT (2020) Molecular study of drought response in the Mediterranean conifer *Pinus pinaster* Ait.: Differential transcriptomic profiling reveals constitutive water deficit-independent drought tolerance mechanisms. Ecol Evol 10:9788–9807. https://doi.org/10.1002/ece3.6613

de la Torre F, García-Gutiérrez A, Crespillo C, Cantón FR, Ávila C, Cánovas FM (2002) Functional expression of two pine glutamine synthetase genes in bacteria reveals that they encode cytosolic isoenzymes with different molecular and catalytic properties. Plant Cell Physiol 43:802–809

De la Torre AR, Birol I, Bousquet J, Ingvarsson PK, Jansson S, Jones SJM, Keeling CI, MacKay J, Nilsson O, Ritland K, Street N, Yanchuk A, Zerbe P, Bohlmann J (2014) Insights into conifer giga-Genomes. Plant Physiol 166:1724–1732. https://doi.org/10.1104/pp.114.248708

de Vega-Bartol JJ, Simões M, Lorenz WW, Rodrigues AS, Alba R, Dean JFD, Miguel CM (2013) Transcriptomic analysis highlights epigenetic and transcriptional regulation during zygotic embryo development of *Pinus pinaster*. BMC Plant Biol 13. https://doi.org/10.1186/1471-2229-13-123

Dinkeloo K, Boyd S, Pilot G (2018) Update on amino acid transporter functions and on possible amino acid sensing mechanisms in plants. Seminars Cell Dev Biol 74:105–113

Dubos C, Le Provost G, Pot D, Salin F, Lalane C, Madur D, Frigerio JM, Plomion C (2003) Identification and characterization of water-stress-responsive genes in hydroponically grown maritime pine (*Pinus pinaster*) seedlings. Tree Physiol 23:169–179. https://doi.org/10.1093/treephys/23.3.169

El-Azaz J, de la Torre F, Ávila C, Cánovas FM (2016) Identification of a small protein domain present in all plant lineages that confers high prephenate dehydratase activity. Plant J 87:215–229

El-Azaz J, Cánovas FM, Ávila C, de la Torre F (2018) The arogenate dehydratase ADT2 is essential for seed development in Arabidopsis. Plant Cell Physiol 59:2409–2420

El-Azaz J, de la Torre F, Pascual MB, Debille S, Canlet F, Harvengt L, Trontin J-F, Ávila C, Cánovas FM (2020) Transcriptional analysis of arogenate dehydratase genes identifies a link between phenylalanine biosynthesis and lignin biosynthesis. J Exp Bot 71:3080–3093. https://doi.org/10.1093/jxb/eraa099

Farjon A (2017) A handbook of the world's conifers 2nd edn. Brill, Leiden, The Netherlands

Farjon A, Filer D (2013) An atlas of the world's conifers. Brill, Leiden, The Netherlands

Fernández-Pozo N, Canales J, Guerrero-Fernández D, Villalobos DP, Diaz-Moreno SM, Bautista R, Flores-Monterroso A, Ángeles Guevara M, Perdiguero P, Collada C, Cervera MT, Soto Á, Ordás R, Cantón FR, Ávila C, Cánovas FM, Gonzalo Claros M (2011) EuroPineDB: a high-coverage web database for maritime pine transcriptome. BMC Genomics 12. https://doi.org/10.1186/1471-2164-12-366

Flores-Monterroso A, Canales J, de la Torre F, Avila C, Canovas FM (2013) Identification of genes differentially expressed in ectomycorrhizal roots during the *Pinus pinaster-Laccaria bicolor* interaction. Planta 237:1637–1650. https://doi.org/10.1007/s00425-013-1874-4

Fotelli MN, Lyrou FG, Avtzis DN, Maurer D, Rennenberg H, Spyroglou G, Polle A, Radoglou K (2020) Effective defense of aleppo pine against the giant scale *Marchalina hellenica* through ecophysiological and metabolic changes. Front Plant Sci 11:581693.https://doi.org/10.3389/fpls.2020.581693

Foti V, Araniti F, Manti F, Alicandri E, Giuffrè AM, Bonsignore CP, Castiglione E, Sorgonà A, Covino S, Paolacci AR, Ciaffi M, Badiani M (2020) Profiling volatile terpenoids from calabrian pine stands infested by the pine processionary moth. Plants (basel) 9:1362. https://doi.org/10.3390/plants9101362

Fox H, Doron-Faigenboim A, Kelly G, Bourstein R, Attia Z, Zhou J, Moshe Y, Moshelion M, David-Schwartz R (2018) Transcriptome analysis of *Pinus halepensis* under drought stress and during recovery. Tree Physiol 38:423–441. https://doi.org/10.1093/treephys/tpx137

Ganeteg U, Ahmad I, Jämtgård S, Aguetoni-Cambui C, Inselsbacher E, Svennerstam H, Schmidt S, Näsholm T (2017) Amino acid transporter mutants of *Arabidopsis* provides evidence that a non-mycorrhizal plant acquires organic nitrogen from agricultural soil. Plant Cell Environ 40:413–423

Garalde DR, Snell EA, Jachimowicz D, Sipos B, Lloyd JH, Bruce M, Pantic N, Admassu T, James P, Warland A, Jordan M, Ciccone J, Serra S, Keenan J, Martin S, McNeill L, Wallace EJ, Jayasinghe L, Wright C, Blasco J, Young S, Brocklebank D, Juul S, Clarke J, Heron AJ, Turner DJ (2018) Highly parallel direct RNA sequencing on an array of nanopores. Nat Methods 15:201–206. https://doi.org/10.1038/nmeth.4577

Garcés M, Le Provost G, Lalanne C, Claverol S, Barré A, Plomion C, Herrera R (2014) Proteomic analysis during ontogenesis of secondary xylem in maritime pine. Tree Physiol 34:1263–1277. https://doi.org/10.1093/treephys/tpt117

García K, Doidy J, Zimmermann SD, Wipf D, Courty PE (2016) Take a trip through the plant and fungal transportome of mycorrhiza. Trends Plant Sci 21:937–950

García K, Guerrero-Galán C, Frank HER, Haider MZ, Delteil A, Conéjéro G, et al. (2020) Fungal Shaker-like channels beyond cellular K+ homeostasis: a role in ectomycorrhizal simbiosis between *Hebeloma cylindrosporum* and *Pinus pinaster*. PLOS ONE 15 (11):e0242739

García-Gutiérrez A, Cantón FR, Gallardo F, Sánchez-Jiménez F, Cánovas FM (1995) Expression of ferredoxin-dependent glutamate synthase in dark-grown pine-seedlings. Plant Mol Biol 27:115–128. https://doi.org/10.1007/bf00019183

Gaspar D, Trindade C, Usie A, Meireles B, Barbosa P, Fortes AM, Pesquita C, Costa RL, Ramos AM (2017) Expression profiling in *Pinus pinaster* in response to infection with the pine wood nematode *Bursaphelenchus xylophilus*. Forests 8 (8). https://doi.org/10.3390/f8080279

Gea-Izquierdo G, Férriz M, García-Garrido S, Aguín O, Elvira-Recuenco M, Hernandez-Escribano L, Martin-Benito D, Raposo R (2019) Synergistic abiotic and biotic stressors explain widespread decline of *Pinus pinaster* in a mixed forest. Sci Total Environ 685:963–975. https://doi.org/10.1016/j.scitotenv.2019.05.378

Gerber S, Rodolphe F, Bahrman N, Baradat P (1993) Seed-protein variation in maritime pine (*Pinus pinaster* Ait) revealed by 2-dimensional electrophoresis—genetic determinism and construction of a linkage map. Theor Appl Genet 85:521–528. https://doi.org/10.1007/bf00220908

Gion JM, Lalanne C, Le Provost G, Ferry-Dumazet H, Paiva J, Chaumeil P, Frigerio JM, Brach J, Barré A, de Daruvar A, Claverol S, Bonneu M, Sommerer N, Negroni L, Plomion C (2005) The proteome of maritime pine wood forming tissue. Proteomics 5:3731–3751. https://doi.org/10.1002/pmic.200401197

Gómez-Maldonado J, Crespillo R, Ávila C, Cánovas FM (2001) Efficient preparation of maritime pine (*Pinus pinaster*) protoplasts suitable for transgene expression analysis. Plant Mol Biol Rep 19:361–366

Gómez-Maldonado J, Cánovas FM, Ávila C (2004) Molecular analysis of the 5'-upstream region of a gibberellin-inducible cytosolic glutamine synthetase gene (GS1b) expressed in pine vascular tissue. Planta 218:1036–1045. https://doi.org/10.1007/s00425-003-1185-2

Gómez-Maldonado J, Ávila C, de la Torre F, Cañas R, Cánovas FM, Campbell MM (2004) Functional interactions between a glutamine synthetase promoter and MYB proteins. Plant J 39:513–526. https://doi.org/10.1111/j.1365-313X.2004.02153.x

Gonçalves E, Figueiredo AC, Barroso JG, Henriques J, Sousa E, Bonifácio L (2020) Effect of *Monochamus galloprovincialis* feeding on *Pinus pinaster* and *Pinus pinea*, oleoresin and insect volatiles. Phytochemistry 169:112159

Grace LJ, Charity JA, Gresham B, Kay N, Walter C (2005) Insect-resistant transgenic *Pinus radiate*. Plant Cell Rep 24:103–111. https://doi.org/10.1007/s00299-004-0912-x

Han ZM, Hong YD, Zhao BG (2003) A study on pathogenicity of bacteria carried by pinewood nematode. J Phytopathol 151:683–689. https://doi.org/10.1046/j.1439-0434-2003-00790-x

Hernández-Escribano L, Visser EA, Iturritxa E, Raposo R, Naidoo S (2020) The transcriptome of *Pinus pinaster* under *Fusarium circinatum* challenge. BMC Genomics 21 (1). https://doi.org/10.1186/s12864-019-6444-0

Hofmann A, Clokie S (2018) Wilson and Walker's Principles and Techniques of Biochemistry and Molecular Biology, 8th edn. Cambridge University Press, Cambridge, UK

Humara JM, López M, Ordás RJ (1998) Modifying transient beta-glucuronidase expression in pine species using introns. Plant Cell Tiss Org 52:183–187. https://doi.org/10.1023/a:1006036704801

Humara JM, López M, Ordás RJ (1999) Transient expression of the uidA gene in *Pinus pinea* cotyledons: A study of heterologous promoter sequences. Plant Cell Tiss Org 56:69–78. https://doi.org/10.1023/a:1006250609262

Klimaszewska K, Rutledge RG, Séguin A (2004) Genetic transformation of conifers utilizing somatic embryogenesis. In: Peña L (ed) Methods in Molecular Biology. Humana Press, Totowa, NJ, pp 151–164

Klimaszewska K, Hargreaves C, Lelu-Walter MA, Trontin JF (2016) Advances in conifer somatic embryogenesis since year 2000. *In vitro* embryogenesis in higher plants. Ed. Germanà MA and Lambardi M 1359:131–166. https://doi.org/10.1007/978-1-4939-3061-6_8

Köppen W (1900) Versuch einer klassifikation der klimate, vorzugsweise nach ihren beziehungen zur pflanzenwelt. Geogr Z 6(593–611):657–679

Krapp A (2015) Plant nitrogen assimilation and its regulation: a complex puzzle with missing pieces. Curr Opin Plant Biol 25:115–122

Kronzucker HJ, Siddiqi MY, Glass ADM (1997) Conifer root discrimination against soil nitrate and the ecology of forest succesion. Nature 385:59–61

Kuusk V, Niinemets U, Valladares F (2018) Structural controls on photosynthetic capacity through juvenile-to-adult transition and needle aging in Mediterranean pines. Funct Ecol 32:1479–1491

Kuusk V, Niinemets U, Valladares F (2018) A major trade-off between structural and photosynthetic investments operative across plant and needle ages in three Mediterranean pines. Tree Physiol 38:543–557

Le Dantec L, Chagne D, Pot D, Cantin O, Garnier-Gere P, Bedon F, Frigerio JM, Chaumeil P, Leger P, García V, Laigret F, de Daruvar A, Plomion C (2004) Automated SNP detection in expressed sequence tags: statistical considerations and application to maritime pine sequences. Plant Mol Biol 54:461–470

Lee IH, Han H, Koh YH, Kim IS, Lee SW, Shim D (2019) Comparative transcriptome analysis of *Pinus densiflora* following inoculation with pathogenic (*Bursaphelenchus xylophilus*) or non-pathogenic nematodes (*B. thailandae*). Sci Rep. 9: 12180. https://doi.org/10.1038/s41598-019-48660-w

Lelu-Walter MA, Bernier-Cardou M, Klimaszewska K (2006) Simplified and improved somatic embryogenesis for clonal propagation of *Pinus pinaster* (Ait.). Plant Cell Tiss Org 92:31–45

Lepoittevin C, Frigerio J-M, Garnier-Gere P, Salin F, Cervera M-T, Vornam B, Harvengt L, Plomion C (2010) In vitro vs In silico detected SNPs for the development of a genotyping array: What can we learn from a non-model species? PLOS ONE 5: e11034

Léran S, Varala K, Boyer JC et al (2014) A unified nomenclature of nitrate transporter 1/peptide transporter family members in plants. Trends Plant Sci 19:5–9

Lionello P, Abrantes F, Congedi L, Dulac F, Gacic M, Gomis D, Goodess C, Hoff H, Kutiel H, Luterbacher J, Planton S, Reale M, Schroeder K, Struglia MV, Toreti A, Tsimplis M, Ulbrich U, Xoplaki E (2012) introduction: mediterranean climate-background information. Climate of the Mediterranean Region: from the Past to the Future, XXXV–XC. https://doi.org/10.1016/b978-0-12-416042-2.00012-4

Llebrés MT, Avila C, Cánovas FM, Klimaszewska K (2018) Root growth of hybrid *Pinus strobus* (L.) and *P. wallichiana* (A. B. Jacks.) somatic plants is affected by the nitrogen composition of the somatic embryo germination medium. Trees-Struct Funct 32:371–381. https://doi.org/10.1007/s00468-017-1635-2

Llebrés MT, Pascual MB, Debille S, Trontin JF, Avila C, Cánovas FM (2018) The role of arginine metabolic pathway during embryogenesis and germination in maritime pine (*Pinus pina*ster Ait.). Tree Physiol 38:471–484. https://doi.org/10.1093/treephys/tpx133

Llebrés MT, Pascual MB, de la Torre FN, Valderrama-Martin JM, Gómez L, Avila C, Cánovas FM (2020) Structural and functional characteristics of two molecular variants of the nitrogen sensor PII in maritime pine. Front Plant Sci 11:823

Loewe V, Navarro-Cerrillo RM, Sánchez Lucas R, Gómez FJR, Jorrín-Novo J (2018) Variability studies of allochthonous stone pine (*Pinus pinea* L.) plantations in Chile through nut protein profiling. J Proteomics 175:95–104

López M, Humara JM, Rodríguez R, Ordás RJ (2000) Transient *uidA* gene expression in electroporated cotyledonary protoplasts of *Pinus nigra* ssp salzmannii and in bombarded cotyledons. Can J Forest Res 30:448–455. https://doi.org/10.1139/cjfr-30-3-448

Maeda H, Dudareva N (2012) The shikimate pathway and aromatic amino acid biosynthesis in plants. Annu Rev Plant Biol 63:73–105

Meijón M, Feito I, Oravec M, Delatorre C, Weckwerth W, Majada J, Valledor L (2016) Exploring natural variation of *Pinus pinaster* Aiton using metabolomics:

Is it possible to identify the region of origin of a pine from its metabolites? Mol Ecol 25:959–976

Montalbán IA, Setien-Olarra A, Hargreaves CL, Moncalean P (2013) Somatic embryogenesis in *Pinus halepensis* Mill.: an important ecological species from the Mediterranean forest. Trees-Struct Funct 27:1339–1351

Moreira X, Lundborg L, Zas R, Carrillo-Gavilán A, Borg-Karlson AK, Sampedro L (2013) Inducibility of chemical defences by two chewing insect herbivores in pine trees is specific to targeted plant tissue, particular herbivore and defensive trait. Phytochemistry 94:113–122. https://doi.org/10.1016/j.phytochem.2013.05.008

Morel A, Teyssier C, Trontin JF, Eliasova K, Pesek B, Beaufour M, Morabito D, Boizot N, Le Mette C, Belal-Bessai L, Reymond I, Harvengt L, Cadene M, Corbineau F, Vagner M, Label P, Lelu-Walter MA (2014) Early molecular events involved in *Pinus pinaster* Ait. somatic embryo development under reduced water availability: transcriptomic and proteomic analyses. Physiol Plant 152:184–201. https://doi.org/10.1111/ppl.12158

Morel A, Trontin JF, Corbineau F, Lomenech AM, Beaufour M, Reymond I, Le Mette C, Ader K, Harvengt L, Cadene M, Label P, Teyssier C, Lelu-Walter MA (2014) Cotyledonary somatic embryos of *Pinus pinaster* Ait. most closely resemble fresh, maturing cotyledonary zygotic embryos: biological, carbohydrate and proteomic analyses. Planta 240:1075–1095. https://doi.org/10.1007/s00425-014-2125-z

Nahalkova J, Asiegbu FO, Daniel G, Hrib J, Vookova B, Pribulova B, Gemeiner P (2001) Isolation and immunolocalization of a *Pinus nigra* lectin (PNL) during interaction with the necrotrophs—*Heterobasidion annosum* and *Fusarium avenaceum*. Physiol Mol Plant Pathol 59:153–163. https://doi.org/10.1006/pmpp.2001.0341

Nam DK, Lee S, Zhou GL, Cao XH, Wang C, Clark T, Chen JJ, Rowley JD, Wang SM (2002) Oligo(dT) primer generates a high frequency of truncated cDNAs through internal poly(A) priming during reverse transcription. Proc Natl Acad Sci USA 99:6152–6156. https://doi.org/10.1073/pnas.092140899

Näsholm T, Kielland K, Ganeteg U (2009) Uptake of organic nitrogen by plants. New Phytol 182:31–48

Nordin A, Uggla C, Näsholm T (2001) Nitrogen forms in bark, wood and foliage of nitrogen-fertilized *Pinus sylvestris*. Tree Physiol 21:59–64

Olsson S, Grivet D, Cattonaro F, Vendramin V, Giovannelli G, Scotti-Saintagne C, Vendramin GG, Fady B (2020) Evolutionary relevance of lineages in the European black pine (*Pinus nigra*) in the transcriptomic era. Tree Genet Genomes 16:30. https://doi.org/10.1007/s11295-020-1424-8

Ortigosa F, Valderrama-Martín JM, Urbano-Gámez JA, García-Martín ML, Ávila C, Cánovas FM, Cañas RA (2020) Inorganic nitrogen form determines nutrient allocation and metabolic responses in maritime pine seedlings. Plants 9:481

Ortigosa F, Lobato-Fernández C, Hitomi Shikano, Concepción Ávila, Shu Taira, Cánovas FM, Cañas RA (2021) Ammonium regulates development and architecture of pine roots through hormonal cross-talk and differential expression of transcription factors in the apex. Plant Cell Environm. https://doi.org/10.1111/pce.14214

Paiva J, Garnier-Géré P, Rodrigues J, Alves A, Santos S, Graça J, Le Provost G, Chaumeil P, Da Silva-Perez D, Bosc A, Fevereiro P, Plomion C (2008a) Plasticity of maritime pine (*Pinus pinaster*) wood-forming tissues during a growing season. New Phytol 179:1180–1194. https://doi.org/10.1111/j.1469-8137.2008.02536.x

Paiva JAP, Garcés M, Alves A, Garnier-Gere P, Rodrigues JC, Lalanne C, Porcon S, Le Provost G, Pérez DD, Brach J, Frigerio JM, Claverol S, Barré A, Fevereiro P, Plomion C (2008b) Molecular and phenotypic profiling from the base to the crown in maritime pine wood-forming tissue. New Phytol 178:283–301. https://doi.org/10.1111/j.1469-8137.2008.02379.x

Paiva G, Abreu P, Proença DN, Santos S, Nobre MF, Morais PV (2014) *Mucilaginibacter pineti* sp. nov., isolated from *Pinus pinaster* wood from a mixed grove of pines trees. Int J Syst Evol Microbiol. 64(Pt 7): 2223–2228. https://doi.org/10.1099/ijs.0.057737-0

Palme AE, Wright M, Savolainen O (2008) Patterns of divergence among conifer ESTs and polymorphism in *Pinus sylvestris* identify putative selective sweeps. Mol Biol Evol 25:2567–2577. https://doi.org/10.1093/molbev/msn194

Park J, Jeon HW, Jung H, Lee HH, Kim J, Park AR, Kim N, Han G, Kim JC, Seo YS (2020) Comparative transcriptome analysis of pine trees treated with resistance-inducing substances against the nematode *Bursaphelenchus xylophilus*. Genes (basel) 11:1000. https://doi.org/10.3390/genes11091000

Parks M, Cronn R, Liston A (2012). Separating the wheat from the chaff: mitigating the effects of noise in a plastome phylogenomic data set from *Pinus* L. (Pinaceae). BMC Evol Biol 12:100. https://doi.org/10.1186/1471-2148-12-100

Pascual MB, Cánovas FM, Ávila C (2015) The NAC transcription factor family in maritime pine (*Pinus pinaster*): molecular regulation of two genes involved in stress responses. BMC Plant Biol 15.https://doi.org/10.1186/s12870-015-0640-0

Pascual MB, Llebrés M-T, Craven-Bartle B, Cañas RA, Cánovas FM, Ávila C (2018) *PpNAC1*, a main regulator of phenylalanine biosynthesis and utilization in maritime pine. Plant Biotechnol J.16:1094–1104. https://doi.org/10.1111/pbi.12854

Perdiguero P, Collada C, del Carmen BM, García Casado G, Teresa Cervera M, Soto Á (2012) Identification of water stress genes in *Pinus pinaster* Ait. by controlled progressive stress and suppression-subtractive hybridization. Plant Physiol Biochem 50:44–53. https://doi.org/10.1016/j.plaphy.2011.09.022

Perdiguero P, del Carmen BM, Teresa Cervera M, Collada C, Soto Á (2013) Molecular response to water stress in two contrasting Mediterranean pines (*Pinus pinaster* and *Pinus pinea*). Plant Physiol Biochem 67:199–208. https://doi.org/10.1016/j.plaphy.2013.03.008

Perdiguero P, Rodrigues AS, Chaves I, Costa B, Alves A, de María N, Vélez MD, Díaz-Sala C, Cervera MT, Miguel CM (2021) Comprehensive analysis of the isomiRome in the vegetative organs of the conifer *Pinus pinaster* under contrasting water availability. Plant Cell Environ 44:706–728. https://doi.org/10.1111/pce.13976

Pereira C, Montalbán IA, Goicoa T, Ugarte MD, Correia S, Canhoto JM, Moncalean P (2017) The effect of changing temperature and agar concentration at proliferation stage in the final success of Aleppo pine somatic embryogenesis. Forest Systems 26:Esc05

Petit RJ, Bahrman N, Baradat P (1995) Comparison of genetic differentiation in maritime pine (*Pinus pinaster* Ait) estimated using isozyme, total protein and terpenic loci. Heredity 75:382–389. https://doi.org/10.1038/hdy.1995.150

Pinosio S, González-Martínez SC, Bagnoli F, Cattonaro F, Grivet D, Marroni F, Lorenzo Z, Pausas JG, Verdú M, Vendramin GG (2014) First insights into the transcriptome and development of new genomic tools of a widespread circum-Mediterranean tree species, *Pinus halepensis* Mill. Mol Ecol Res 14:846–856. https://doi.org/10.1111/1755-0998.12232

Plomion C, Costa P, Bahrman N, Frigerio JM (1997) Genetic analysis of needle proteins in maritime pine .1. Mapping dominant and codominant protein markers assayed on diploid tissue, in a haploid-based genetic map. Silvae Genet 46 (2–3):161–165

Plomion C, Pionneau C, Brach J, Costa P, Bailleres H (2000) Compression wood-responsive proteins in developing xylem of maritime pine (*Pinus pinaster* Ait.). Plant Physiol 123:959–969. https://doi.org/10.1104/pp.123.3.959

Proença DN, Espírito Santo C, Grass G, Morais PV (2012) Draft genome sequence of *Pseudomonas* sp. strain M47T1, carried by *Bursaphelenchus xylophilus* isolated from *Pinus pinaster*. J Bacteriol 194:4789–4790. https://doi.org/10.1128/JB.01116-12

Proença DN, Francisco R, Kublik S, Schöler A, Vestergaard G, Schloter M, Morais PV (2017) The microbiome of endophytic, wood colonizing bacteria from pine trees as affected by pine wilt disease. Sci Rep 7:4205. https://doi.org/10.1038/s41598-017-04141-6

Proença DN, Whitman WB, Shapiro N, Woyke T, Kyrpides NC, Morais PV (2017) Draft genome sequence of the cellulolytic endophyte *Chitinophaga costaii* A37T2T. Stand Genomic Sci 12:53. https://doi.org/10.1186/s40793-017-0262-2

Provost GL, Paiva J, Pot D, Brach J, Plomion C (2003) Seasonal variation in transcript accumulation in wood-forming tissues of maritime pine (*Pinus pinaster* Ait.) with emphasis on a cell wall glycine-rich protein. Planta 217:820–830. https://doi.org/10.1007/s00425-003-1051-2

Qian Y, Lynch JH, Guo L, Rhodes D, Morgan JA, Dudareva N (2019) Completion of the cytosolic post-chorismate phenylalanine biosynthetic pathway in plants. Nat Commun 10:15

Ramos P, Le Provost G, Gantz C, Plomion C, Herrera R (2012) Transcriptional analysis of differentially expressed genes in response to stem inclination in young seedlings of pine. Plant Biol 14:923–933. https://doi.org/10.1111/j.1438-8677.2012.00572.x

Rennenberg H, Dannenmann M, Gessler A, Kreuzwieser J, Simon J, Papen H (2009) Nitrogen balance in forest soils: nutritional limitation of plants under climate change stresses. Plant Biol 11:4–23

Rey M, Humara JM, López M, González MV, Rodríguez R, Tavazza R, Ancora G, Ordás RJ (1996) Foreign gene expression in *Pinus nigra*, *P. radiata* and *P. pinea* following particle bombardment. In: Ahuja M.R., Boerjan W., Neale D.B. (eds), somatic cell genetics and molecular genetics of trees 49, pp113–117

Rodrigues AS, Chaves I, Costa BV, Lin YC, Lopes S, Ilhinhos AM, Van de Peer Y, Miguel CM (2019) Small RNA profiling in *Pinus pinaster* reveals the transcriptome of developing seeds and highlights differences between zygotic and somatic embryos. Sci Rep 9. https://doi.org/10.1038/s41598-019-47789-y

Rodrigues AS, De Vega JJ, Miguel CM (2018) Comprehensive assembly and analysis of the transcriptome of maritime pine developing embryos. BMC Plant Biol 18:379. https://doi.org/10.1186/s12870-018-1564-2

Romero-Rodríguez CM, Pascual J, Valledor L, Jorrín-Novo J (2014) Improving the quality of protein identification in non-model species. Characterization of *Quercus ilex* seed and *Pinus radiata* needle proteomes by using SEQUEST and custom databases. J Proteomics 105:85–91. https://doi.org/10.1016/j.jprot.2014.01.027

Rueda-López M, Crespillo R, Cánovas FM, Ávila C (2008) Differential regulation of two glutamine synthetase genes by a single Dof transcription factor. Plant J 56:73–85. https://doi.org/10.1111/j.1365-313X.2008.03573.x

Rueda-López M, Cañas R, Canales J, Cánovas F, Ávila C (2015) The over-expression of the pine transcription factor *PpDof5* in Arabidopsis leads to increased lignin content and affects carbon and nitrogen metabolism. Physiol Plant 155:369–383. https://doi.org/10.1111/ppl.12381

Rueda-López M, Pascual MB, Pallero M, Henao LM, Lasa B, Jáuregui I, Aparicio-Tejo PM, Cánovas FM, Ávila C (2017) Overexpression of a pine Dof transcription factor in hybrid poplars: A comparative study in trees growing under controlled and natural conditions. PLOS ONE 12 (4). https://doi.org/10.1371/journal.pone.0174748

Salaj T, Klubicova K, Matusova R, Salaj J (2019) Somatic embryogenesis in selected conifer trees *Pinus nigra* Arn. and *Abies* hybrids. Front Plant Sci 10:13. https://doi.org/10.3389/fpls.2019.00013

Salaj T, Moravcikova J, Grec-Niquet L, Salaj J (2005) Stable transformation of embryogenic tissues of *Pinus nigra* Arn. using a biolistic method. Biotechnol Lett 27:899–903. https://doi.org/10.1007/s10529-005-7178-4

Salajova T, Salaj J (2005) Somatic embryogenesis in *Pinus nigra*: maturation and regeneration ability of embryogenic tissue initiation, established cell lines. Biol Plant 49:333–339

Sancho dos Santos CS, de Vasconcelos MW (2012) Identification of genes differentially expressed in *Pinus pinaster* and *Pinus pinea* after infection with the pine wood nematode. Europ J Plant Pathol 132:407–418. https://doi.org/10.1007/s10658-011-9886-z

Schneider B, Torres E, Martín MP, Schröder M, Behnke HD, Seemüller E (2005) "Candidatus Phytoplasma pini", a novel taxon from *Pinus sylvestris* and *Pinus halepensis*. Int J Syst Evol Microbiol 55(Pt 1):303–307. https://doi.org/10.1099/ijs.0.63285-0

Shahraki A, Yu Y, Gul ZM, Liang C, Birgul Iyison N (2020) Whole genome sequencing of *Thaumetopoea pityocampa* revealed putative pesticide targets. Genomics 112:4203–4207. https://doi.org/10.1016/j.ygeno.2020.07.017

Sherwood P, Villari C, Capretti P, Bonello P (2015) Mechanisms of induced susceptibility to Diplodia tip blight in drought-stressed Austrian pine. Tree Physiol 35:549–562. https://doi.org/10.1093/treephys/tpv026

Skrypina NA, Timofeeva AV, Khaspekov GL, Savochkina LP, Beabealashvilli RS (2003) Total RNA suitable for molecular biology analysis. J Biotechnol 105(1–2):1–9. https://doi.org/10.1016/s0168-1656(03)00140-8

Strano CP, Malacrinò A, Campolo O, Palmeri V (2018) Influence of host plant on *Thaumetopoea pitycampa* gut bacterial community. Microb Ecol 75:487–494. https://doi.org/10.1007/s00248-017-1019-6

Taibi K, del Campo AD, Vilagrosa A, Belles JM, López-Gresa MP, Pla D, Calvete JJ, López-Nicolas JM, Mulet JM (2017) Drought Tolerance in *Pinus halepensis* Seed Sources As Identified by Distinctive Physiological and Molecular Markers. Front Plant Sci 8.https://doi.org/10.3389/fpls.2017.01202

Tang W, Tian Y (2003) Transgenic loblolly pine (*Pinus taeda* L.) plants expressing a modified δ-endotoxin gene of *Bacillus thuringiensis* with enhanced resistance to *Dendrolimus punctatus* Walker and *Crypyothelea formosicola* Staud. J Exp Bot 54:835–844

Tapias R, Climent J, Pardos JA, Gil J (2004) Life histories of Mediterranean pines. Plant Ecol 171:53–68

Tan Y, Li Q, Zhao Y, Wei H, Wang J, Baker CJ, Liu Q, Wei W (2021) Integration of metabolomics and existing omics data reveals new insights into phytoplasma-induced metabolic reprogramming in host plants. PLoS ONE 16:20246203

Tereso S, Miguel C, Zoglauer K, Valle-Piquera C, Oliveira MM (2006) Stable Agrobacterium-mediated transformation of embryogenic tissues from *Pinus pinaster* Portuguese genotypes. Plant Growth Regul 50:57–68. https://doi.org/10.1007/s10725-006-9126-2

Trontin J-F, Walter C, Klimaszewska K, Park YS, Lelu-Walter MA (2007) Recent progress on genetic transformation of four *Pinus* spp. Transgenic Plant J 1:314–329

Trontin J-F, Debille S, Canlet F, et al. (2013) Somatic embryogenesis as an effective regeneration support for reverse genetics in maritime pine: the Sustainpine collaborative project as a case study. In: Park Y-S, Bonga JM, eds. Proceeding of the IUFRO Working Party 2.09.02 conference on Integrating vegetative propagation, biotechnology and genetic improvement for tree production and sustainable forest management, 25–28/06/2012. Brno, Czech Republic, 184–187

Trontin JF, Klimaszewska K, Morel A, Hargreaves C, Lelu-Walter MA (2016a) Molecular aspects of conifer zygotic and somatic embryo development: A review of genome-wide approaches and recent Insights. In "In vitro embryogenesis in higher plants, methods in molecular biology vol. 1359" Germanà MA and Lambardi M (eds), Springer Science, New York, pp 167–207

Trontin JF, Ávila C, Debille S et al. (2016b) Somatic embryogenesis as an enabling technology for reverse genetics: achievements and prospects for breeding maritime pine (*Pinus pinaster* Ait.). Fourth international Conference of the IUFRO unit 2.09.02 (somatic embryogenesis and other vegetative propagation technologies) on 'Development and Application of Vegetative Propagation Technologies in Plantation Forestry to Cope with a Changing Climate and Environment', September 19–23 (La Plata, Argentina), p 57

Urbano-Gámez JA, El-Azaz J, Avila C, de la Torre FN, Cánovas FM (2020) Enzymes involved in the biosynthesis of arginine from ornithine in maritime pine (*Pinus pinaster* Ait.). Plants 9:1271

van Dijk EL, Jaszczyszyn Y, Naquin D, Thermes C (2018) The third revolution in sequencing technology. Trends Genet 34:666–681. https://doi.org/10.1016/j.tig.2018.05.008

Valledor L, Guerrero S, García-Campa L, Meijón M (2021) Proteometabolic characterization of apical bu maturation ion Pinus pinaster. Tree Physiol 41:508–521

Vicente CS, Nascimento FX, Espada M, Barbosa P, Hasegawa K, Mota M, Oliveira S (2013) Characterization of bacterial communities associated with the pine sawyer beetle *Monochamus galloprovincialis*, the insect vector of the pinewood nematode *Bursaphelenchus xylophilus*. FEMS Microbiol Lett 347:130–139. https://doi.org/10.1111/1574-6968.12232

Vicente CS, Nascimento F, Espada M, Barbosa P, Mota M, Glick BR, Oliveira S (2012) Characterization of bacteria associated with pinewood nematode *Bursaphelenchus xylophilus*. PLOS ONE 7: e46661. https://doi.org/10.1371/journal.pone.0046661

Villalobos DP, Díaz-Moreno SM, Said E-SS, Cañas RA, Osuna D, Van Kerckhoven SHE, Bautista R, Gonzalo Claros M, Cánovas FM, Cantón FR (2012)

Reprogramming of gene expression during compression wood formation in pine: Coordinated modulation of S-adenosylmethionine, lignin and lignan related genes. BMC Plant Biol 12. https://doi.org/10.1186/1471-2229-12-100

Vivas M, Zas R, Sampedro L, Solla A (2013) Environmental maternal effects mediate the resistance of maritime pine to biotic stress. PLOS ONE 8:e70148. https://doi.org/10.1371/journal.pone.0070148

von Arnold S, Clapham D, Abrahamsson M (2019) Embryology in conifers. In: Cánovas F (ed) molecular physiology and biotechnology of trees. Adv Bot Res 89:39–76

Waseem M, Liu Y, Xia R (2020) Long Non-Coding RNAs, the Dark Matter: An Emerging Regulatory Component in Plants. Int J Mol Sci 22 (1). https://doi.org/10.3390/ijms22010086.

Weng JK, Chapple C (2010) The origin and evolution of lignin biosynthesis. New Phytol 187:273–285

Xu G, Fan X, Miller AJ (2012) Plant nitrogen assimilation and use efficiency. Annu Rev Plant Biol 63:153–182

Yildirim T, Zeki K, Kani I (2006) Induction of embryogenic tissue and maturation of somatic embryos in *Pinus brutia* TEN. Plant Cell Tiss Org 87:67–76

Yong Z, Kotur Z, Glass AD (2010) Characterization of an intact two-component high-affinity nitrate transporter from *Arabidopsis* roots. Plant J 63:739–748

Zamora R, Mellado A (2019) Identifying the abiotic and biotic drivers behind the elevational distribution shift of a parasitic plant. Plant Biol (stuttg) 21:307–317. https://doi.org/10.1111/plb.12934

Pinus Sylvestris as a Reference Plant Species in Radiation Research: Transcriptomics of Trees from the Chernobyl Zone

10

Gustavo T. Duarte, Stanislav A. Geras'kin, and Polina Y. Volkova

Abstract

Pinus species are highly suitable for monitoring anthropogenic pollutants, including air contaminants, heavy metals and radionuclides. Scots pine (*Pinus sylvestris* L.) is one of the most widely distributed conifers in the world, and is characterized by a high sensitivity to environmental pollutants and an increased ability to accumulate pollutants in tissues. Because of its large genome size, this species is extremely sensitive to ionizing radiation, thus considered by the International Commission on Radiological Protection as a reference plant for radiation protection of biota. Current approaches in functional genomics and bioinformatics allow studying the transcriptional profiles of *P. sylvestris* in radioactively contaminated areas, revealing adaptive responses at a previously unattainable level. In the Chernobyl exclusion zone, where natural populations are under low-dose chronic radiation stress, such responses include direct structural protection of DNA integrity from radiation damage and modulation of reactive oxygen species production. Chronically irradiated pine populations have increased expression of genes encoding heat shock proteins and histones. However, no differential expression of genes related to DNA repair was revealed, while the transcriptional responses pointed to modulation of redox cellular homeostasis. These observations suggest that the response patterns to chronic low-dose and acute high-dose ionizing radiation exposure are partially different. Biomonitoring properties of *P. sylvestris* and a pipeline for assembling the transcriptome of this non-model species are also discussed.

10.1 *Pinus Sylvestris* in Biomonitoring of Anthropogenic Pollution

Trees are sessile long-living organisms with complex life cycles; therefore, environmental stressors often challenge their ability to survive and reproduce. Tree species belonging to the *Pinus* genus are widely used for monitoring of technogenic pollution, including air contaminants (Schulz et al. 1999; Lehndorff and Schwark 2004; Urbat et al. 2004; Ratola et al. 2014; Al-Alam et al. 2017), heavy metals (Mičieta and Murín 1998; Alaqouri et al. 2020) and radionuclides

G. T. Duarte
Max Plank Institute of Molecular Plant Physiology, Am Mühlenberg 1, 14476 Potsdam-Golm, Germany

Belgian Nuclear Research Centre (SCK CEN), Biosphere Impact Studies, Boeretang 200, B-2400 Mol, Belgium

S. A. Geras'kin · P. Y. Volkova (✉)
Russian Institute of Radiology and Agroecology, Kievskoe shosse 109 km, 249032 Obninsk, Russian Federation

A. R. De La Torre (ed.), *The Pine Genomes*, Compendium of Plant Genomes,
https://doi.org/10.1007/978-3-030-93390-6_10

(Geras'kin et al. 2019). Haploid megagametophyte of pines provides unique resources for genetic analysis (Lev-Yadun and Sederoff 2000), facilitating fast assessment of mutational rates and changes of the genetic structure under anthropogenic pollution. With a wide ecological niche, Scots pine (*Pinus sylvestris* L.) is one of the world's most widely distributed conifers, which is especially competitive in poor soils and in dry and extremely cold environments (Pyhäjärvi et al. 2019). The high sensitivity of *P. sylvestris* to environmental pollutants and its ability to accumulate a number of pollutants in bark and needle tissues make this species especially suitable for biomonitoring of environmental pollution.

10.1.1 Air and Soil Pollution

The large surfaces of forests provide high rates of deposition of pollutant gases, aerosols and cloud droplets (Fowler et al. 1999). Urban forests are also capable of removing air pollution by intercepting particulate matter on plant surfaces and absorbing gaseous pollutants through the leaf stomata (Nowak et al. 2018). This high retention ability allows passive sampling of natural vegetation in order to obtain well-defined samples at affordable costs for biomonitoring goals (Urbat et al. 2004). For example, ambient air monitoring in urban areas based on the persistent organic pollutant load on vegetation was suggested as a feasible and cost-effective way for controlling environmental quality (Lehndorff and Schwark 2004). In particular, pine needles stand out as effective biomonitors due to the excellent uptake properties of their waxy layer (Ratola et al. 2014).

Several gymnosperm species have been used for biomonitoring of multiple varieties of pollutants. For instance, *P. sylvestris* bark and *P. nigra* needles were used for assessing and characterizing deposition of airborne organic and inorganic pollutants in Germany (Schulz et al., 1999; Urbat et al., 2004). *P. pinea* needles were also used for monitoring the levels of 18 pesticides in Portugal (Ratola et al., 2014). Most recently, a multipollutant analysis of pesticides, polychlorinated biphenyls and polyaromatic hydrocarbons was performed using needles of *Cedrus atlantica* and *P. nigra* in France (Al-Alam et al. 2017). The diameter of *P. sylvestris* trees and the length of their needles were also used for the evaluation of soil contamination with diesel oil in Poland (Bęś et al. 2019). The fluctuation asymmetry (FA) indices of needle length, representing phenotypic deviations from the bilateral symmetry, were the highest in Scots pines grown on the most contaminated soil, whereas the opposite pattern was noted for the FA of needle weight (Bęś et al. 2019). Moreover, diesel oil contamination decreased the concentrations of photosynthetic pigments and influenced photosynthetic processes (Bęś et al. 2019). Interestingly, the evaluation of FA of *P. sylvestris* needles along a gradient of air pollution in the Karelian Isthmus (northwest Russia and southeast Finland) showed a direct link to the pollution load (Kozlov et al. 2002). However, the study based on traditional indices (growth, crown transparency and needle discoloration) failed to link the vitality of *P. sylvestris* stands with deposition of the main pollutants.

For using *P. sylvestris* morphology in biomonitoring, the pollution exposure must be recent, because pine trees are capable to recover even after long-term stress exposure. This was tested in the areas affected by a mineral fertilizer facility in Lithuania, after a distinct reduction of emissions of nitrogen and sulfur oxides and dust of mineral fertilizers. Recovery of more than a quarter of damaged trees was registered even in the case of 90% defoliation (Juknys et al. 2003).

10.1.2 Heavy Metals

Major anthropogenic sources of heavy metal (HM) contamination are mining and ore processing facilities, as well as road traffic (Vareda et al. 2019). Pieces of evidence indicate that *P. sylvestris* is a promising species for biomonitoring of HM pollution. For instance, Scots pine trees inhabiting heavily polluted plots in the immediate vicinity of zinc and iron smelters and a power plant in southern Poland had significantly higher concentrations of heavy metals in needles compared to the reference plots (Kandziora-Ciupa et al. 2016). These concentrations increased with the age of the needles, and probably caused elevated antioxidant responses. Because of that, protein and non-protein –SH groups in pine needles were suggested as suitable biomarkers for biological monitoring at heavy metal-polluted territories (Kandziora-Ciupa et al. 2016). The yearly comparison of *P. sylvestris* bark and needles along transects around the Stalowa Wola industry center in southeast Poland suggested that pine needles are suitable for atmospheric Cu and Zn pollution monitoring, while bark may be used for Cu contamination monitoring (Samecka-Cymerman et al. 2006). *P. sylvestris* and *P. nigra* from heavy metal polluted areas in Slovakia showed a visible effect of pollution in pollen grain morphology and needle-based meristematic cells (Mičieta and Murín 2018).

Interestingly, individuals of species belonging to the *Pinus* genus show a natural heterogeneity in sensitivity to heavy metal pollution. This observation was based on populations of *P. sylvestris* and *P. nigra* from southern Poland. Needles of sensitive trees had higher concentrations of Zn, Pb, Cd, and Cu compared with resistant ones (Chudzińska et al. 2014). For both species, resistant trees had a lower degree of genetic variation than the metal-sensitive trees, as assessed by several isoenzyme loci. Sensitive trees also had a smaller number of alleles identified and alleles per locus, while showing a significant excess of homozygotes in comparison to the expected values based on Hardy–Weinberg equilibrium (Chudzińska et al. 2014). The genetic changes induced by heavy metals suggested an important role of specific enzymes in the adaptation process, namely fluorescent esterase, shikimate dehydrogenase, glutamate–oxaloacetate transaminase and phosphoglucoisomerase (Chudzińska et al. 2014). Previously, isoenzymatic and cytogenetic analyses of the same populations of *P. sylvestris* allowed detecting differences in genetic composition between tolerant and sensitive trees, and the sensitive group also had a significant excess of homozygosity (Prus-Głowacki et al. 2006). Cytological analyses demonstrated numerous aberrations of chromosomes in meristematic root tissue of seedlings developed from seeds collected in the polluted area and a significant decrease of the mitotic index (Prus-Głowacki et al. 2006).

10.2 Effects of Radioactive Contamination on *P. sylvestris*

The high susceptibility of *Pinus* species to environmental pollutants is also reflected by an extremely high sensitivity to ionizing radiation exposure. In a dose-dependent manner, ionizing radiation impacts living organisms by two major pathways: (1) direct induction of various DNA damage; and (2) production of reactive oxygen species (ROS) by radiolysis of water molecules and disturbing of cellular redox homeostasis (Riley 1994). There are several factors affecting the radiation sensitivity of each species. It is known that larger genomes are more sensitive to DNA damage, which follows an approximated relationship to the cube root of the nuclear volume; e.g., a tenfold increase in genome size increases sensitivity to radiation by about twofold (Einset and Collins, 2018). Therefore, tree species with large genome sizes (e.g., conifers, including pines) are well-recognized radiosensitive species (ICRP 2008). The genome size of *P. sylvestris* comprises 22×10^9 bp (Bennett and Leitch 2012), not surprisingly being a radiosensitive species. Noteworthily, Scots pine is considered as one of the reference plants by the International Commission on Radiation Protection (ICRP).

10.2.1 Reference Plant of ICRP

Pinus species are especially important for assessing the consequences of radionuclide contamination in natural ecosystems. The high radiosensitivity of pines determined their recognition as a model-sensitive tree species in the List of Reference Animals and Plants developed by ICRP (2008). The creation of this list aimed to summarize the effects of radiation on several biota species, with regard to the effects on mortality, morbidity, reduced reproductive success and chromosomal damage. A set of derived consideration reference levels of irradiation was developed for each biotic type for optimizing the environmental protection of biota species and for suggesting the efficient instrument for decision-makers in the area of radiation protection (ICRP 2008).

The high radiosensitivity of *P. sylvestris* was observed by a well-known example of lethal pine forest damage, the so-called "Red Forest", as a consequence of the Chernobyl nuclear power plant (NPP) accident in 1986. Dead pine trees covered an area of about 10 km^2 located up to 5 km to the west of Unit 4 of the Chernobyl NPP (Geras'kin et al. 2008). In addition, a zone of sublethal damage (approximately 38 km^2) and a zone of weak impact, which covered the rest of the forests of the 30-km zone, were identified (Geras'kin et al. 2008). After the Chernobyl NPP disaster, the pine trees that survived showed inhibition of reproductive capacity and suppression of generative process, caused by the irradiation during the active growth period in April 1986 (Arkhipov et al. 1994). The beginning of the recovery processes in the next year after the Chernobyl accident, when the dose rates significantly decreased, was followed by a massive appearance of morphological abnormalities in survived pine trees: changes in the shape and size of needles, and development of "witch's brooms" and shoot bundles (Arkhipov et al. 1994). Another example of the high sensitivity of growth processes of pines to acute irradiation is the loss of apical dominance in young trees, which was observed in *P. sylvestris* in the severely contaminated Chernobyl exclusion zone (Yoschenko et al. 2011) and *P. densiflora* after the Fukushima Daiichi accident in 2011 (Yoschenko et al. 2016). Trees with morphological abnormalities had higher frequencies of aberrant cells in needle intercalary meristems than in the reference populations, and the concentration ratio of major classes of phytohormones in needles changed under radiation exposure (Geras'kin et al. 2021).

In the first year after the Chernobyl accident, those pine trees that received doses over 1 Gy had significantly decreased reproductive potential, measured as a decreased seed mass and number of seeds per cone and an increased portion of abortive seeds (Ipatyev et al. 1999). Reduction of reproductive potential was also observed after high-dose radiation exposure in the East Urals Radioactive Trail (EURT) (Kalchenko and Spirin 1989) and in the territories with enhanced concentrations of naturally occurring radionuclides in the north of Russia, where the dose rates were 11.3 μGy/h and higher (Evseeva et al. 2011).

However, chronic low-dose irradiation in the remote period after the Chernobyl accident did not lead to a significant reduction of the reproductive ability of Scots pine. Between 2007–2014, it was observed that the quality of seeds from chronically exposed populations mainly depended on the weather conditions rather than radiation (Geras'kin et al. 2017). At the same time, chronic low-dose exposure has been reported to cause morphological changes on *P. sylvestris*, such as an increase in the FA indices of needles and in the frequency of necrosis and morphoses (Makarenko et al. 2016). However, morphological effects may be phenotypically transient because of the gradual decline in dose rates at radioactively contaminated territories. A later study in the Chernobyl exclusion zone showed that FA indices of *P. sylvestris* needles were not significantly different from non-contaminated plots (Kashparova et al. 2020). Nevertheless, the transcriptome analysis of Scots pine populations indicated that chronic low-dose irradiation promotes changes in the gene expression profile of the trees (Duarte et al., 2019), meaning that they are continuously optimizing their growth and development to cope with the chronic stress condition.

Much attention has been given to the study of cytogenetic effects in irradiated pine populations. In the first years after the Chernobyl NPP accident, an increased level of cytogenetic abnormalities was observed at a relatively low dose (0.5 Gy, Fedotov et al. 2006). For instance, the frequency of cytogenetic abnormalities in the root meristem of seedlings from chronically exposed populations of *P. sylvestris* was significantly higher than the reference level even in the remote period after the Chernobyl accident (Geras'kin et al. 2019). In addition, long-term observations on chronically exposed Scots pine populations revealed the occurrence of qualitative differences in the temporal dynamics of cytogenetic effects. In reference populations, the frequencies of cytogenetic abnormalities changed with time in a cyclic manner, while in chronically exposed populations the cyclic patterns in temporal dynamics of cytogenetic abnormalities was disturbed (Geras'kin et al. 2018).

10.2.2 Mutation Rates

Response to radiation exposure occurs at different levels of biological organization: from DNA and chromosome modifications to population and community structure (Geras'kin 2016). Strong effects of radiation exposure on mutation rates suggest a significant impact of radioactive contamination on life-history traits. There is evidence that the accumulation of cellular alterations may afterward influence biological parameters that are important for populations, such as health and reproduction (Ipatyev et al. 1999; Theodorakis 2001; Evseeva et al. 2011; Lacaze et al. 2011).

Several studies revealed that *P. sylvestris* growing at radioactively contaminated areas showed increased frequency of mutations in isoenzymatic, amplified fragment length polymorphism (AFLP) and microsatellite loci (Kalchenko and Spirin 1989; Fedotov et al. 2006; Kuchma et al. 2011; Geras'kin and Volkova 2014; Volkova et al. 2017, 2018). For instance, the frequency of mutations in nine isoenzyme loci of *P. sylvestris* from the EURT comprised 7.8×10^{-4} per locus per generation, while an increase of chronic irradiation dose rates to 20, 30, and 50 times the reference levels led to the increase in total mutation frequency to 3.5, 6.7, and 8 times, respectively (Kalchenko and Spirin 1989). A study with Scots pine from areas contaminated by the Chernobyl accident showed that acute irradiation increased mutation frequencies up to 4–17 times in 20 studied isoenzymatic loci (Fedotov et al. 2006). Chronic radiation exposure also led to an increase of mutational frequencies in antioxidant and metabolic isoenzymatic loci (Volkova et al. 2017).

Somatic mutations were analyzed using DNA markers in *P. sylvestris* trees planted close to the Chernobyl NPP before and after the accident and in control trees of the same origin planted at plots with natural radiation background (Kuchma et al. 2011). The mutation rates for microsatellite loci were estimated as 2.8×10^{-4}–7.1×10^{-4} per locus for irradiated tree populations, while no mutations were detected in the controls. For AFLP loci, the observed mutation rates were three-fold higher than at control areas indicating that ionizing radiation causes strong DNA damage across the entire genome and that AFLPs may be an appropriate marker system for estimation of this damage (Kuchma et al. 2011). Indeed, a later study on chronically irradiated *P. sylvestris* trees showed that chronic radiation exposure changed the spectrum of mutations registered in AFLP loci (Volkova et al. 2018).

10.2.3 Changes of Population Genetic Structure

An increase in genetic and phenotypic variability is one of the most important trends observed in populations during adaptation to moderate stress (Slomka et al. 2011; Geras'kin et al. 2013). Changes in genetic structure due to the viability selection are likely to occur in populations exposed to rapidly and extremely changing environmental conditions after catastrophic events (Kuchma and Finkeldey 2011), and high mutation rates on radioactively contaminated territories may be partially responsible for the genetic structure modifications. The evaluation of isoenzyme markers in chronically irradiated *P. sylvestris* populations from areas contaminated

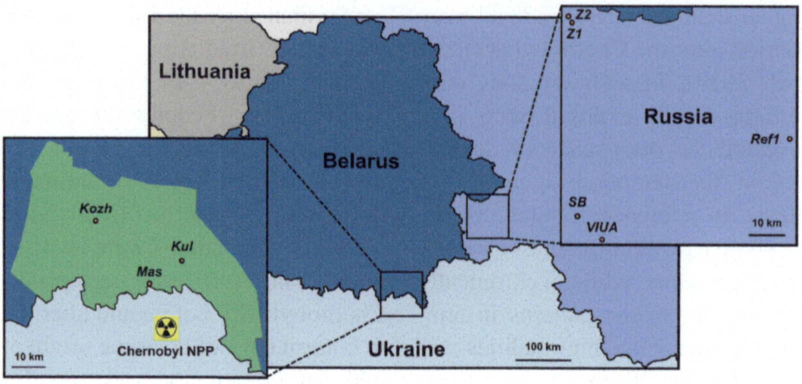

Fig. 10.1 Experimental plots used for studies of *P. sylvestris* responses to chronic radiation exposure (Geras'kin and Volkova 2014; Geras'kin et al. 2017, 2018, 2019; Volkova et al. 2017, 2018; Duarte et al. 2019). The green area in the magnified left box represents the Polesie State Radioecological Reserve, a part of the Chernobyl exclusion zone. The magnified right box represents relatively lower contaminated areas and the reference plot in the Bryansk region of Russia

by the Chernobyl accident (Fig. 10.1) suggested that the genotypic diversity of antioxidant enzymes on the contaminated areas increased faster than genotypic diversity of metabolic enzymes (Volkova et al. 2017). The effective number of alleles was higher in the group of metabolic enzymes, since their loci had a high allelic variability even in the control groups (Volkova et al. 2017). The estimation of genetic differences between *P. sylvestris* trees from the Chernobyl exclusion zone and reference plants using AFLP analysis showed that six percent of the investigated loci (15 out of 222) could be candidates for a selective response. Moderate differentiation was observed between groups of trees showing either weak or strong phenotypic responses to high radiation levels (Kuchma and Finkeldey 2011). A later study of AFLP loci in populations of *P. sylvestris* from the Chernobyl exclusion zone and the Bryansk region of Russia (Fig. 10.1) indicated that Nei's gene diversity significantly increased with the absorbed dose level. The increased values of Nei's gene diversity in radioactively contaminated areas may indirectly reflect the process of genomic adaptation to chronic stress (Volkova et al. 2018) and may also indicate an increase in deleterious alleles due to the increased mutation rates.

10.2.4 Epigenetic Responses

Epigenetic modifications underlie the ability of plants to tolerate long-term stress without changes in the DNA sequence (Meyer 2015). This is particularly important for long-living organisms such as trees. However, the epigenetic studies on trees represent only a small part of those carried out in plants and are limited to a few species (Amaral et al. 2020). The addition of a methyl group at the C_5 position of the cytosine ring (DNA CpG methylation) is the most common epigenetic modification in plant DNA and is a genome stabilization strategy (Flores et al. 2013). Methylated cytosines are distributed throughout the genome: all types of DNA sequences (genes, transposons and intergenic regions) can be methylated (Angers et al. 2010). Methylation of transposons inhibits their activity by blocking transposase binding sites and restricting transcription (Gehring and Henikoff 2007).

Hypermethylation can be a defense strategy aimed at preventing genomic instability and reducing the consequences of the mutagenic effect of ionizing radiation. Indeed, pine trees growing at the Chernobyl exclusion zone were found to be hypermethylated, while the hypermethylation level depended on the radiation exposure level (Kovalchuk et al. 2003). A later work demonstrated that the genomes of *P. sylvestris* trees growing on radioactively contaminated areas (Fig. 10.1) were still significantly hypermethylated compared to the reference plot, 30 years after the Chernobyl disaster (Volkova et al. 2018). The average level of genome methylation in the samples from radioactively contaminated plots was increased by 0.6–3%, however, without dependence on the absorbed dose. This suggests the presence of a threshold dose causing a sharp increase in the level of methylation in the genome, and indicates a possible feedback mechanism that controls the methylation status in order to avoid an excessive stress response. For instance, the increase in the percentage of methylated cytosines in the genome may reflect the general non-specific stress response of pines and may be related to the control of transposon activity (Volkova et al. 2018).

10.3 Transcriptional Profiling of *P. Sylvestris* Under Chronic Radiation Exposure

In spite of the long history of research, we are still far from the complete understanding of biological processes occurring in *P. sylvestris* populations under radiation exposure. Current genetic and molecular studies have uncovered complex regulatory systems that coordinate stress response and tolerance in conifers. Coordination of these responses is accomplished through multilevel regulatory mechanisms. The traditional biological approach has focused mostly on the identification of specific loci associated with the effect of radiation. However, the processes of radiation damage are complex, extended in time and usually nonlinear (Feinendegen et al. 2008). Therefore, it is meaningful to consider these processes in the context of broader regulatory networks. The unraveling of such networks responding to chronic stress exposure in natural populations inhabiting contaminated territories would alleviate the risk assessment and allow the development of new sensitive biomarkers of chronic radiation exposure (Volkova and Geras'kin 2018). In this context, hypothesis-generating "omic" technologies are highly suitable for radioecological field studies. However, up to now the high-throughput techniques remain underutilized in radioecology. One of the main obstacles is that ecotoxicology and radioecology usually deal with non-model organisms such as *P. sylvestris*, and high-throughput analysis of such species is still challenging (Volkova and Geras'kin 2018).

10.3.1 Transcriptome Analysis of Non-model Organisms

Gymnosperms have complex genomes, with up to 40 Gbp in size and harboring a high content of repetitive elements (De La Torre et al. 2014). For instance, while the model plant organism *Arabidopsis thaliana* is diploid and has a genome size of nearly 135 Mbp (The Arabidopsis Genome Initiative 2000), the coast redwood genome is hexaploid (Scott et al. 2016) and has 28.4 Gbp (Redwood Genome Project, https://treegenesdb.org/). Not surprisingly, gymnosperm species genomes are one of the last to be sequenced. Recently, draft conifer genomes have become available, as, for instance, Douglas fir (*Pseudotsuga menziesii*, Neale et al. 2017), giant sequoia (*Sequoiadendron giganteum*, Scott et al. 2020), Loblolly pine (*P. taeda*, Zimin et al. 2014, 2017), Norway spruce (*Picea abies*, Nystedt et al. 2013) and Sugar pine (*P. lambertiana*, Stevens et al. 2016). TreeGenes (https://treegenesdb.org/; Falk et al. 2018; Wegrzyn et al. 2019) and ConGenIE (https://congenie.org/, Sundell et al. 2015) databases host several gymnosperm genome projects. However, because of their complexity, gymnosperm genomes are yet fragmented and not fully annotated.

RNA sequencing (RNA-seq) has become an essential tool for high-throughput evaluation of gene expression, discovery of novel transcripts and identification of alternative splice forms. Moreover, it allows both coding and non-coding genes to be studied (Stark et al. 2019). The development of de novo assembly techniques,

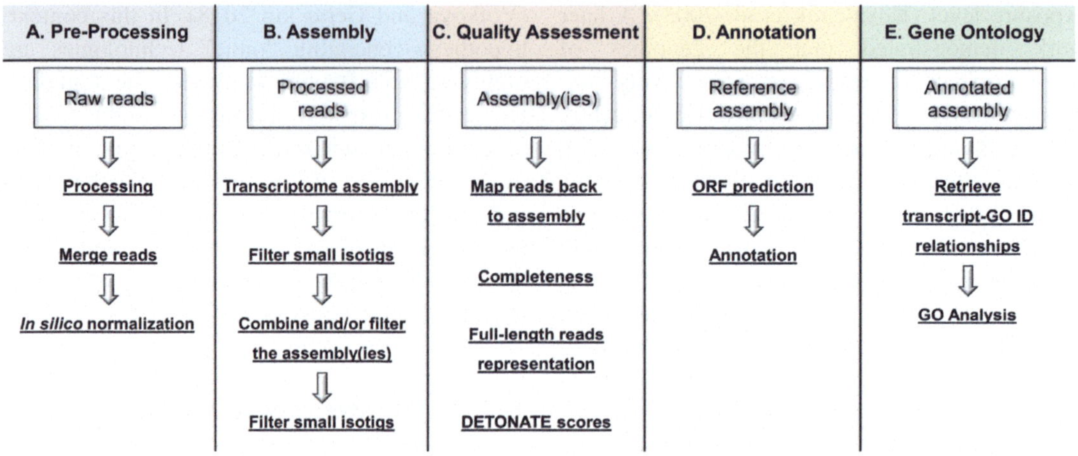

Fig. 10.2 Main steps on a transcriptomic study for non-model organisms. Adapted from (Duarte et al. 2021)

which use the redundancy and overlap between the reads for building the transcripts without requiring a reference for guiding the assembly, opened venues for high-throughput transcriptome studies involving non-model organisms. Indeed, the evaluation of conifer transcriptomes has flourished in recent years (Canales et al. 2014; Liu et al. 2017; Baker et al. 2018; Lee et al. 2018; Rodrigues et al. 2018; Duarte et al. 2019; Visser et al. 2019; Park et al. 2020), following the development of de novo transcriptome assemblers (Zerbino and Birney 2008; Robertson et al. 2010; Grabherr et al. 2011; Schulz et al. 2012; Xie et al. 2014; Liu et al. 2016). Nevertheless, a reliable transcriptome study depends on their selection in a case-specific way, following stringent quality evaluation criteria (Conesa et al. 2016; Hölzer and Marz 2019; Duarte et al. 2021). As suggested by a large-scale comparison between the performance of different de novo assemblers applied to different species (Hölzer and Marz 2019), each tool performs differentially depending on the dataset. Furthermore, a high-quality assembly is essential for proper understanding of the background elements controlling the physiological responses and adaptation processes to environmental stimuli. These patterns are revealed by the differentially expressed genes (DEGs), which quantification as component level rather than individual transcript abundance may help to reduce the impact of erroneously assembled contigs during de novo transcriptome analysis (Hsieh et al. 2019). Finally, the biological context of the transcriptome study is given by the functional annotation of the transcripts. It is possible to understand the context of the gene expression patterns by characterizing the biological processes, cellular components and molecular functions of the transcripts. Gene Ontology annotation (Ashburner et al. 2000), which provides a hierarchical representation of the gene functions, is often employed. The general overview of a transcriptomic study is represented in Fig. 10.2.

10.3.2 The Transcriptional Profile of Pine Populations at Radioactively Polluted Areas

Radioactive contamination is among the most long-lasting anthropogenic impacts on the environment, yet its long-term consequences to natural populations are poorly understood. As discussed in Sect. 10.2, Scots pine is a promising organism for understanding the effects of chronic low-dose radiation exposure (Volkova et al. 2017, 2018; Duarte et al. 2019; Geras'kin et al. 2019), especially due to its high radiosensitivity (ICRP 2008). The comprehensive evaluation of transcriptional changes promoted by chronic irradiation of Scots pine revealed a unique response profile (Duarte et al. 2019). The study was based on the transcriptome analysis of Scots pine samples from three contaminated plots, namely Kulazhin and Masany (Belarus Republic, Kul and Mas in Fig. 10.1) and Zabor'e (Russian Federation, Z1 in Fig. 10.1), affected by the Chernobyl NPP explosion in 1986, and from a reference plot in the Bryansk region of Russian Federation (Duarte et al. 2019). The trees, aged between 20 and 25 years, were not exposed to the high-dose acute irradiation just after the disaster and belonged to the first post-Chernobyl generation. They, therefore, have been exposed to low-dose chronic radiation during their lifetime. Inherently, a few of the DEGs identified in chronically irradiated populations have also been reported to respond to acute irradiation, but none of the DEGs was related to DNA repair processes. The Gene Ontology analysis revealed that processes related to the control of cell homeostasis were a common trend between the samples from the most contaminated areas (Duarte et al. 2019). Overall, these processes, directly or indirectly, are involved in the protection of DNA integrity from the effects of ionizing radiation. Most DEGs can be grouped into two general categories: modulation of ROS production and direct structural protection of DNA.

10.3.2.1 Modulation of Reactive Oxygen Species Production

ROS are by-products of water radiolysis caused by highly energetic electrons (Riley 1994). They are overproduced during radiation exposure and lead to oxidative damage to proteins, lipids and DNA (Riley 1994). During radiation-induced ROS burst, plants rely on their antioxidant system to contain ROS damage to the cells. That includes the activity of several enzymes (*e.g.*, catalases, superoxide dismutase, peroxidases and ascorbate–glutathione cycle enzymes) and on the production and accumulation of antioxidant molecules (e.g., α-tocopherol, ascorbic acid, carotenoids, glutathione and thioredoxin) (Sewelam et al. 2016). However, ROS are also important signaling molecules that are produced during different stress responses in plants (Sewelam et al. 2016; Zhu 2016). This observation has a major implication when considering adaptation to chronic radiation exposure.

Unlike the responses observed to acute irradiation, organisms exposed to chronic low-dose radiation may not continuously block ROS because it would have consequences for ROS-dependent signaling, and consequently to stress adaptation. Organisms facing chronic low-dose radiation stress likely adapt to higher ROS thresholds, and balance ROS signaling and buffering through the modulation of its biosynthesis and adjusting the steady-state levels of certain components of the antioxidant system (Duarte et al., 2019). Indeed, a study of Scots pine from radioactively contaminated regions revealed no significant changes in the activities of antioxidant enzymes in seed endosperms in comparison to the reference regions (Volkova et al. 2017). The transcriptional profiling of Scots pine populations in such conditions suggested that the balance is probably achieved by the fine-tuning of the gene expression of glutathione- and thioredoxin-related genes and of ROS-generating processes (Duarte et al. 2019). Accordingly, increased levels of oxidized glutathione have also been reported for pine trees from the same populations (Volkova et al. 2017). Interestingly, electron transport chains and photosynthesis are major ROS sources (Scheibe et al. 2005), and the transcriptomic evidence based on Scots pine suggests that they are under tight regulation (Duarte et al. 2019). In Scots pine, such a control seems to occur by the regulation of ion fluxes and may be linked to the control of the sensitivity to the hormone abscisic acid and gas exchanges via stomata, which would ultimately affect the photosynthesis rate (Duarte et al., 2019). A recent study involving different annual and perennial herbaceous plant species from the Chernobyl exclusion zone identified that photosynthetic parameters and in some cases ABA concentrations changed in a species-specific way depending on the chronic radiation dose (Volkova et al. 2021).

10.3.2.2 DNA Structural Protection

Ionizing radiation can directly damage the DNA, causing single- or double-strand breaks (Caplin and Willey 2018). Under acute radiation exposure, the DNA repair machinery is activated to prevent serious damages (Culligan et al. 2006; Wang et al. 2016). Conversely, such pattern was not observed on Scots pine under low-dose chronic radiation exposure (Duarte et al. 2019). It is plausible to hypothesize that the long-term maintenance of the DNA repair machinery must be demanding to the organism, while alternative solutions to adapt to the continuous stress can be more cost-effective. Interestingly, a common trend between chronically irradiated pine populations was the increased expression of genes encoding for heat shock proteins (HSPs) and histones (Duarte et al. 2019). HSPs are chaperones, molecular machines that assist on protein quality control and during recovery from stress conditions, thus being essential for cell integrity and longevity of an organism (Saibil 2013). Histones provide structural support for the chromosome, hence being mainly cell cycle-regulated (Jiang and Berger 2017). While oxidative stress induces the activity of HSPs (Ghosh et al. 2018), histones have been suggested to have a radioprotective role (Oleinick et al. 1999). The absence of activation of genes related to the repair machinery, and increased expression of histones and HSPs was also identified by the transcriptome analysis of shepherd's purse (*Capsella bursa-pastoris* L.) from areas within the Chernobyl exclusion zone (Volkova et al. 2021). Shepherd's purse is an annual plant from the *Brassicaceae* family. This common trend between short- and long-living plant species suggests that the modulation of both histones and HSPs without requiring increased expression of genes related to DNA repair machinery may be a transcriptional strategy for coping with chronic low-dose irradiation conditions.

DNA damage such as ionizing radiation-induced double-strand breaks can also trigger the activity of transposable elements (TEs) (Miousse et al. 2015; Wang et al. 2016). TEs are mobile DNA sequences, which may disrupt the genome integrity following an errant repair of double-strand breaks (Hedges and Deininger 2007). An overall analysis of TE families in chronically irradiated Scots pine revealed that the transcriptional control of certain TEs may be dose-dependent (Duarte et al. 2019). Indeed, genome hypermethylation identified on Scots pine trees from chronically irradiated areas (Volkova et al. 2018) indicates that the organisms living in such conditions must keep the activity of TEs under constant surveillance. Transcriptional regulation of TEs has also been identified in *C. bursa-pastoris* (Volkova et al. 2021), further supporting a common profile of plants under chronic low-dose radiation exposure.

10.4 Conclusions

Pinus sylvestris is a prominent species for biomonitoring research because of its high susceptibility to environmental stressors. The high radiosensitivity of the species of the *Pinus* genus made them reference tree species for the goals of radiation protection of the environment. Striking morphological and genetic effects were observed in *P. sylvestris* populations exposed to radionuclide contamination after the Chernobyl NPP accident in 1986, and responses to chronic low-dose irradiation are persisting even today, 35 years later. The identification of adaptation processes and improvement of biomonitoring techniques using sensitive edificator tree species at radioactively contaminated areas can help decision-makers to develop management programs in order to minimize biodiversity loss under chronic exposure to radionuclides.

References

Al-Alam J, Fajloun Z, Chbani A, Millet M (2017) The use of conifer needles as biomonitor candidates for the study of temporal air pollution variation in the Strasbourg region. Chemosphere 168:1411–1421

Alaqouri HAA, Genc CO, Aricak B, Kuzmina N, Menshikov S, Cetin M (2020) The possibility of using Scots pine needles as biomonitor in determination of heavy metal accumulation. Environ Sci Pollut Res 27:20273–20280

Amaral J, Ribeyre Z, Vigneaud J, Sow MD, Fichot R et al (2020) Advances and promises of epigenetics for forest trees. Forests 11:976

Angers B, Castonguay E, Massicotte R (2010) Environmentally induced phenotypes and DNA methylation: how to deal with unpredictable conditions until the next generation and after. Mol Ecol 19:1283–1295

Arkhipov NP, Kuchma ND, Askbrant S, Pasternak PS, Musica VV (1994) Acute and long-term effects of irradiation on pine (*Pinus sylvestris*) stands post-Chernobyl. Sci Total Environ 157:383–386

Ashburner M, Ball CA, Blake JA, Botstein D, Butler H et al (2000) Gene ontology: tool for the unification of biology. Gene Ontol Consort Nat Genet 25(1):25–29

Baker EAG, Wegrzyn JL, Sezen UU, Falk T, Maloney PE, et al (2018) Comparative transcriptomics among four white pine species. G3 (Bethesda) 8 (5):1461–1474

Bennett MD, Leitch IJ (2012) Plant DNA C-values database (release 6.0) http://data.kew.org/cvalues/

Bęś A, Warmiński K, Adomas B (2019) Long-term responses of Scots pine (*Pinus sylvestris* L.) and European beech (*Fagus sylvatica* L.) to the contamination of light soils with diesel oil. Environ Sci Pollut Res 26:10587–10608

Canales J, Bautista R, Label P, Gómez-Maldonado J, Lesur I et al (2014) *De novo* assembly of maritime pine transcriptome: implications for forest breeding and biotechnology. Plant Biotechnol J 12(3):286–299

Caplin N, Willey N (2018) Ionizing radiation, higher plants, and radioprotection: from acute high doses to chronic low doses. Front Plant Sci 9:847

Chudzińska E, Diatta JB, Wojnicka-Półtorak A (2014) Adaptation strategies and referencing trial of Scots and black pine populations subjected to heavy metal pollution. Environ Sci Pollut Res 21(3):2165–2177

Conesa A, Madrigal P, Tarazona S, Gomez-Cabrero D, Cervera A, et al (2016) A survey of best practices for RNA-seq data analysis. Genome Biol 17:13

Culligan KM, Robertson CE, Foreman J, Doerner P, Britt AB (2006) ATR and ATM play both distinct and additive roles in response to ionizing radiation. Plant J 48:947e961

De La Torre A, Birol I, Bousquet J, Ingvarsson P, Jansson S et al (2014) Insights into conifer giga-genomes. Plant Physiol 166:1724–1732

Duarte GT, Volkova PY, Geras'kin SA (2019) The response profile to chronic radiation exposure based on the transcriptome analysis of Scots pine from Chernobyl affected zone. Environ Pollut 250:618–626

Duarte GT, Volkova PY, Geras'kin SA (2021) A pipeline for non-model organisms for *de novo* transcriptome assembly, annotation, and Gene Ontology analysis using open tools: case study with Scots pine. Bio-protocol 11(3):e3912

Einset J, Collins AR (2018) Genome size and sensitivity to DNA damage by X-rays-plant comets tell the story. Mutagenesis 33(1):49–51

Evseeva TI, Belykh ES, Maistrenko TA, Geras'kin SA, Brown JE (2011) Assessment of the reproductive capacity of *Pinus sylvestris* trees growing under conditions of chronic exposure to radionuclides of uranium and thorium series. Russ J Ecol 42:382–387

Falk T, Herndon N, Grau E, Buehler S, Richter P et al (2018) Growing and cultivating the forest genomics database, TreeGenes. Database (oxford) 2018:1–11

Fedotov IS, Kalchenko VA, Igonina EV, Rubanovich AV (2006) Radiation and genetic consequences of ionizing irradiation on population of *Pinus sylvestris* L. within the zone of the Chernobyl NPP. Radiat Biol Radioecol 46(3):268–278 (*in Russian*)

Feinendegen L, Hahnfeldt P, Schadt EE, Stumpf M, Voit EO (2008) Systems biology and its potential role in radiobiology. Radiat Environ Biophys 47:5–23

Flores KB, Wolschin F, Amdam GV (2013) The role of methylation of DNA in environmental adaptation. Integr Comp Biol 52:359–372

Fowler D, Cape JN, Coyle M, Flechard C, Kuylenstierna J et al (1999) The global exposure of forests to air pollutants. Water Air Soil Pollut 116:5–32

Gehring M, Henikoff S (2007) DNA methylation dynamics in plant genomes. Biochim Biophys Acta 1769(5–6):276–286

Geras'kin SA, Volkova PY (2014) Genetic diversity in Scots pine populations along a radiation exposure gradient. Sci Total Environ 496:317–327

Geras'kin SA, Fesenko SV, Alexakhin RM (2008) Effects of non-human species irradiation after the Chernobyl NPP accident Environ Int 34:880–897

Geras'kin S, Evseeva T, Oudalova A (2013) Effects of long-term chronic exposure to radionuclides in plant populations. J Environ Radioact 121:22–32

Geras'kin S., Vasiliyev D, Makarenko E, Volkova P, Kuzmenkov A (2017) Influence of long-term chronic exposure and weather conditions on Scots pine populations. Environ Sci Pollut Res 24:11240–11253

Geras'kin SA (2016) Ecological effects of exposure to enhanced levels of ionizing radiation. J Environ Radioact 162–163:347–357

Geras'kin S, Oudalova A, Kuzmenkov A, Vasiliyev D (2018) Chronic radiation exposure modifies temporal dynamics of cytogenetic but not reproductive indicators in Scots pine populations. Environ Pollut 239:399–407

Geras'kin S, Volkova P, Vasiliyev D, Dikareva N, Oudalova A et al (2019) Scots pine as a promising indicator organism for biomonitoring of the polluted environment: A case study on chronically irradiated populations. Mutat Res Genet Toxicol Environ Mutagen 842:3–13

Geras'kin S, Yoschenko V, Bitarishvili S, Makarenko E, Vasiliev D et al (2021) Multifaceted effects of chronic radiation exposure in Japanese red pines from Fukushima prefecture. Sci Total Environ 763:142946

Ghosh S, Sarkar P, Basak P, Mahalanobish S, Sil PC (2018) Role of heat shock proteins in oxidative stress and stress tolerance. In: Asea A, Kaur P (eds) Heat Shock Proteins and Stress. Heat Shock Proteins 15. Springer, Cham

Grabherr MG, Haas BJ, Yassour M, Levin JZ, Thompson DA et al (2011) Full-length transcriptome assembly from RNA-seq data without a reference genome. Nat Biotechnol 29:644–652

Hedges DJ, Deininger PL (2007) Inviting instability: transposable elements, double-strand breaks, and the maintenance of genome integrity. Mutat Res-Fund Mol M 616(1–2):46–59

Hölzer M, Marz M (2019) *De novo* transcriptome assembly: A comprehensive cross-species comparison of short-read RNA-Seq assemblers. GigaScience 8 (5):giz039

Hsieh PH, Oyang YJ, Chen CY (2019) Effect of *de novo* transcriptome assembly on transcript quantification. Sci Rep 9:8304

ICRP (2008) Publication 108. Environmental protection —the concept and use of Reference Animals and Plants. Annals of the ICRP 38

Ipatyev V, Bulavik I, Braginsky V, Goncharenko G, Dvornik A (1999) Forest and Chernobyl: forest ecosystems after the Chernobyl nuclear power plant accident: 1986–1994. J Environ Radioact 42:9–38

Jiang D, Berger F (2017) Histone variants in plant transcriptional regulation. Biochim Biophys Acta 1860:123e130

Juknys R, Vencloviene J, Stravinskiene V, Augustaitis A, Bartkevicius E (2003) Scots pine (*Pinus sylvestris* L.) growth and condition in a polluted environment: from decline to recovery. Environ Pollut 125(2):205–212

Kalchenko VA, Spirin DA (1989) Genetics effects revealed in populations of *Pinus sylvestris* L. growing under exposure to small doses of chronic irradiation. Russ J Genet 25:1059–1064

Kandziora-Ciupa M, Ciepał R, Nadgórska-Socha A, Barczyk G (2016) Accumulation of heavy metals and antioxidant responses in *Pinus sylvestris* L. needles in polluted and non-polluted sites. Ecotoxicology 25:970–981

Kashparova E, Levchuk S, Morozova V, Kashparov V (2020) A dose rate causes no fluctuating asymmetry indexes changes in silver birch (*Betula pendula* (L.) Roth.) leaves and Scots pine (*Pinus sylvestris* L.) needles in the Chernobyl Exclusion Zone. J Environ Radioact 211:105731

Kovalchuk O, Burke P, Arkhipov A, Kuchma N, James SJ et al (2003) Genome hypermethylation in *Pinus sylvestris* of Chernobyl—a mechanism for radiation adaptation? Mutat Res 529:13–20

Kozlov MV, Niemela P, Junttila J (2002) Needle fluctuating asymmetry is a sensitive indicator of pollution impact on Scots pine (*Pinus sylvestris*). Ecol Indic 1:271–277

Kuchma O, Finkeldey R (2011) Evidence for selection in response to radiation exposure: *Pinus sylvestris* in the Chernobyl exclusion zone. Environ Pollut 159 (6):1606–1612

Kuchma O, Vornam B, Finkeldey R (2011) Mutation rates in Scots pine (*Pinus sylvestris* L.) from the Chernobyl exclusion zone evaluated with AFLP and microsatellite markers. Mutat Res 725:29–35

Lacaze E, Geffard O, Goyet D, Bony S, Devaux A (2011) Linking genotoxic responses in *Gammarus fossarum* germ cells with reproduction impairment, using the Comet assay. Environ Res 111:626–634

Lee IH, Kim J, Woo KS, Jang KH, Kim YH, Shim D (2018) *De novo* assembly and transcriptome analysis of the *Pinus densiflora* response to pine wilt disease in nature. Plant Biotechnol Rep 12:229–236

Lehndorff E, Schwark L (2004) Biomonitoring of air quality in the cologne conurbation using pine needles as a passive sampler—Part II: polycyclic aromatic hydrocarbons (PAH). Atmos Environ 38(23):3793–3808

Lev-Yadun S, Sederoff R (2000) Pines as model gymnosperms to study evolution, wood formation, and perennial growth. J Plant Growth Regul 19:290–305

Liu J, Li G, Chang Z, Yu T, Liu B et al. (2016) BinPacker: packing-based *de novo* transcriptome assembly from RNA-seq data. PLoS Comput Biol 12(2):e1004772

Liu Q, Wei Y, Xu L, Hao Y, Chen X, Zhou Z (2017) Transcriptomic profiling reveals differentially expressed genes associated with pine wood nematode resistance in Masson pine (*Pinus massoniana* Lamb.). Sci Rep 7:4693

Makarenko ES, Oudalova AA, Geras'kin SA (2016) Study of needle morphometric indices in Scots pine in the remote period after the Chernobyl accident. Radioprotection 51(1):19–23

Meyer P (2015) Epigenetic variation and environmental change. J Exp Bot 66:3541–3548

Mičieta K, Murín G (1998) Three species of genus *Pinus* suitable as bioindicators of polluted environment. Water Air Soil Pollut 104:413–422

Mičieta K, Murín G (2018) "*In situ*" biomonitoring of a polluted environment by wild plant and crop plant species. Nova Science Publishers Inc., NY

Miousse IR, Chalbot MC, Lumen A, Ferguson A, Kavouras IG, Koturbash I (2015) Response of transposable elements to environmental stressors. Mutat Res Rev Mutat Res 765:19–39

Neale DB, McGuire PE, Wheeler NC, Stevens KA, Crepeau MW, et al (2017) The Douglas-fir genome sequence reveals specialization of the photosynthetic apparatus in Pinaceae. G3 (Bethesda) 7(9):3157–3167

Nowak DJ, Hirabayashi S, Doyle M, McGovern M, Pasher J (2018) Air pollution removal by urban forests in Canada and its effect on air quality and human health. Urban for Urban Gree 29:40–48

Nystedt B, Street N, Wetterbom A, Zuccolo A, Lin YC et al (2013) The Norway spruce genome sequence and conifer genome evolution. Nature 497:579–584

Oleinick NL, Chiu S, Xue L, Mann KJ (1999) Modification of ionizing radiation damage to cellular DNA by factors affecting chromatin structure. In: Dizdaroglu M, Karakaya AE (eds.) Advances in DNA Damage and Repair. NATO ASI Series (Series A: Life Sciences). Springer, Boston, USA, p 302

Park J, Jeon HW, Jung H, Lee HH, Kim J et al (2020) Comparative transcriptome analysis of pine trees treated with resistance-inducing substances against the nematode *Bursaphelenchus xylophilus*. Genes 11 (9):1000

Prus-Głowacki W, Chudzińska E, Wojnicka-Półtorak AK, Kozacki L, Fagiewicz K (2006) Effects of heavy metal pollution on genetic variation and cytological disturbances in the *Pinus sylvestris* L. population. J Appl Genet 47(2):99–108

Pyhäjärvi T, Kujala ST, Savolainen O (2019) 275 years of forestry meets genomics in *Pinus sylvestris*. Evol Appl 13(1):11–30

Ratola N, Homem V, Silva JA, Araújo R, Amigo JM et al (2014) Biomonitoring of pesticides by pine needles—Chemical scoring, risk of exposure, levels and trends. Sci Total Environ 476–477:114–124

Riley PA (1994) Free radicals in biology: oxidative stress and the effects of ionizing radiation. Int J Radiat Biol 65:27e33

Robertson G, Schein J, Chiu R, Corbett R, Field M et al (2010) *De novo* assembly and analysis of RNA-seq data. Nat Methods 7:909–912

Rodrigues AS, De Vega JJ, Miguel CM (2018) Comprehensive assembly and analysis of the transcriptome of maritime pine developing embryos. BMC Plant Biol 18:379

Saibil H (2013) Chaperone machines for protein folding, unfolding and disaggregation. Nat Rev Mol Cell Biol 14:630–642

Samecka-Cymerman A, Kosior G, Kempers AJ (2006) Comparison of the moss *Pleurozium schreberi* with needles and bark of *Pinus sylvestris* as biomonitors of pollution by industry in Stalowa Wola (southeast Poland). Ecotox Environ Safe 65(1):108–117

Scheibe R, Backhausen JE, Emmerlich V, Holtgrefe S (2005) Strategies to maintain redox homeostasis during photosynthesis under changing conditions. J Exp Bot 56:e1481–e1489

Schulz MH, Zerbino DR, Vingron M, Birney E (2012) Oases: robust *de novo* RNA-seq assembly across the dynamic range of expression levels. Bioinformatics 28:1086–1092

Schulz H, Popp P, Huhn G, Stärk HJ, Schüürmann G (1999) Biomonitoring of airborne inorganic and organic pollutants by means of pine tree barks. I. Temporal and spatial variations. Sci Total Environ 232(1–2):49–58

Scott AD, Stenz NW, Ingvarsson PK, Baum DA (2016) Whole genome duplication in coast redwood (*Sequoia sempervirens*) and its implications for explaining the rarity of polyploidy in conifers. New Phytol 211 (1):186–193

Scott AD, Zimin AV, Puiu D, Workman R, Britton M, et al. (2020) A reference genome sequence for giant sequoia. G3 (Bethesda) 10(11):3907–3919

Sewelam N, Kazan K, Schenk PM (2016) Global plant stress signaling: reactive oxygen species at the crossroad. Front Plant Sci 7:187

Slomka A, Sutkowska A, Szczepaniak M, Malec P, Mitka J, Kuta E (2011) Increased genetic diversity of *Viola tricolor* L. in metal-polluted environments. Chemosphere 83:435–442

Stark R, Grzelak M, Hadfield J (2019) RNA sequencing: the teenage years. Nat Rev Genet 20:631–656

Stevens KA, Wegrzyn JL, Zimin A, Puiu D, Crepeau M et al (2016) Sequence of the sugar pine megagenome. Genetics 204(4):1613–1626

Sundell D, Mannapperuma C, Netotea S, Delhomme N, Lin YC, et al (2015) The plant genome integrative explorer resource: PlantGenIE.org. New Phytol 208 (4):1149–1156

The Arabidopsis Genome Initiative (2000) Analysis of the genome sequence of the flowering plant *Arabidopsis thaliana*. Nature 408:796–815

Theodorakis CW (2001) Integration of genotoxic and population genetic endpoints in biomonitoring and risk assessment. Ecotoxicology 10:245–256

Urbat M, Lehndorff E, Schwark L (2004) Biomonitoring of air quality in the Cologne conurbation using pine needles as a passive sampler—Part I: magnetic property. Atmos Environ 38(23):3781–3792

Vareda JP, Valente AJM, Durães L (2019) Assessment of heavy metal pollution from anthropogenic activities and remediation strategies: A review. J Environ Manage 246:101–118

Visser EA, Wegrzyn JL, Steenkamp ET, Myburg AA, Naidoo S (2019) Dual RNA-Seq analysis of the pine-*Fusarium circinatum* interaction in resistant (*Pinus tecunumanii*) and susceptible (*Pinus patula*) hosts. Microorganisms 7(9):315

Volkova PY, Geras'kin SA (2018) 'Omic' technologies as a helpful tool in radioecological research. J Environ Radioact 189:156–167

Volkova PY, Geras'kin SA, Kazakova EA (2017) Radiation exposure in the remote period after the Chernobyl accident caused oxidative stress and genetic effects in Scots pine populations. Sci Rep 7:43009

Volkova PY, Geras'kin SA, Horemans N, Makarenko E, Saenen E (2018) Chronic radiation exposure as an ecological factor: hypermethylation and genetic differentiation in irradiated Scots pine populations. Environ Pollut 232:105–112

Volkova PY, Duarte GT, Kazakova EA, Makarenko ES, Bitarishvili SV et al. (2021) Radiosensitivity of herbaceous plants to chronic radiation exposure: field study in the Chernobyl exclusion zone. Sci Total Environ 777:146206

Wang Z, Schwacke R, Kunze R (2016) DNA damage-induced transcription of transposable elements and long non-coding RNAs in *Arabidopsis* is rare and ATM-dependent. Mol Plant 9(8):1142–1155

Wegrzyn JL, Staton MA, Street NR, Main D, Grau E et al (2019) Cyberinfrastructure to improve forest health and productivity: the role of tree databases in connecting genomes, phenomes, and the environment. Front Plant Sci 10:813

Xie Y, Wu G, Tang J, Luo R, Patterson J et al (2014) SOAPdenovo-Trans: *de novo* transcriptome assembly with short RNA-seq reads. Bioinformatics 30:1660–1666

Yoschenko V, Kashparov V, Melnychuk M, Levchuk S, Bondar Y et al (2011) Chronic irradiation of Scots pine trees (*Pinus sylvestris*) in the Chernobyl exclusion zone: dosimetry and radiobiological effects. Health Phys 101:393–408

Yoschenko V, Nanba K, Yoshida S, Watanabe Y, Takase T et al (2016) Morphological abnormalities in Japanese red pine (*Pinus densiflora*) at the territories contaminated as a result of the accident at Fukushima Dai-Ichi Nuclear Power Plant. J Environ Radioact 165:60–67

Zerbino DR, Birney E (2008) Velvet: algorithms for *de novo* short read assembly using de bruijn graphs. Genome Res 18:821–829

Zhu JK (2016) Abiotic stress signalling and responses in plants. Cell 167:313e324

Zimin A, Stevens KA, Crepeau MW, Holtz-Morris A, Koriabine M et al (2014) Sequencing and assembly of the 22-gb loblolly pine genome. Genetics 196(3):875–890

Zimin AV, Stevens KA, Crepeau MW, Puiu D, Wegrzyn JL et al (2017) An improved assembly of the loblolly pine mega-genome using long-read single-molecule sequencing. GigaScience 6(1):1–4

11

Genomic Selection in Scots (*Pinus Sylvestris*) and Radiata (*Pinus Radiata*) Pines

Ainhoa Calleja-Rodríguez, Jaroslav Klápště, Heidi Dungey, Natalie Graham, Ahmed Ismael, Maria Rosario García-Gil, Sara Abrahamsson, and Mari Suontama

Abstract

Pines are economically important species widely planted across the globe. Therefore, the industry is interested in increasing the efficacy of their breeding programs, which explains the urge to implement new technological solutions to increase breeding efficiency for traditional and novel breeding objectives. Recently, the development of genotyping platforms accessible even for non-model organisms such as Scots and radiata pines has made it possible to evaluate genomic selection for their breeding programs. Genomic studies in both species are limited by the size and complexity of their genomes; therefore, genotyping platforms based on reduced genome representation have been implemented. Genomic selection studies performed within their breeding programs show viability and potential to increase the intensity of genetic progress compared to conventional (i.e., pedigree-based) strategies. Additionally, genomic prediction of traits with challenging or costly phenotyping, such as forest diseases or wood quality and cell wall chemistry, might allow for large-scale screening and selection of individuals with superior traits. This chapter presents a review of the recent research on genomic selection for Scots pine in Sweden, and radiata pine in New Zealand, as well as future perspectives for the implementation of this methodology in their breeding programs.

11.1 Introduction

11.1.1 Importance of Both Species

Pinus is the largest genus within the gymnosperms and *Pinaceae* Family, containing more than 100 species (Mullin et al. 2011). Pine species are predominantly distributed in the Northern Hemisphere, reaching the equator in Asia and being present in the Mediterranean areas of Africa, but a few species are also native to the Southern Hemisphere (Gernandt et al. 2011). They are adapted to very different ecological and environmental conditions, from tropical to boreal climates, and are able to grow in pure stands or mixed with other evergreen or deciduous trees (Plomion et al. 2007).

A. Calleja-Rodríguez (✉) · S. Abrahamsson · M. Suontama
SKOGFORSK (The Forestry Research Institute of Sweden), Box 3, 918 21 Sävar, SE, Sweden

J. Klápště · H. Dungey · N. Graham · A. Ismael
Scion (New Zealand Forest Research Institute), 49 Sala Street, Whakarewarewa, Rotorua 3020, New Zealand

M. R. García-Gil
Department of Forest Genetics and Plant Physiology, Umeå Plant Science Centre, Swedish University of Agricultural Sciences, 901 83 Umeå, SE, Sweden

A. R. De La Torre (ed.), *The Pine Genomes*, Compendium of Plant Genomes,
https://doi.org/10.1007/978-3-030-93390-6_11

The extensive distribution of Scots pine (*Pinus sylvestris* L.) across Eurasia is one of the main factors that contributes to the large genetic variation of the species (Mátyás et al. 2004). The species is distributed over an area of approximately 10,000–14,000 km (longitudinally) and 3,700 km (latitudinally); and naturally occurs from sea level up to 2,500 m (Boratyński 1991). The largest areas of Scots pine are mainly located in boreal forests of Fennoscandia, where there is a high demand of this species for commercial forest production. In Sweden, Scots pine and Norway spruce (*Picea abies* (L.) Karst.) are the two main forest tree species for commercial production, whose main uses are construction sawn wood, furniture, fiber- and chipboards, wood pulp and paper, among others Calleja-Rodriguez 2019; Krakau et al. 2013).

On the other hand, radiata pine (*Pinus radiata* D. Don) has a very small and fragmented native distribution, limited to five natural populations and three locations along the California and Baja California coasts (Burdon and Miller 1992). Within its natural distribution, the species can grow under very variable soil conditions, elevations and at different exposures, which ensures great genetic variation. Despite its reduced natural distribution, radiata pine is one of the most widely planted species worldwide and is used for production in New Zealand, Chile, Australia, South Africa, and Spain. It was introduced to New Zealand in 1859, and by 1881 there were already at least 3,284 hectares of plantations (Burdon and Miller 1992). Its main uses include construction wood, furniture, glue laminated beams, plywood, pulp, and paper (Mullin et al. 2011).

11.1.2 Scots Pine Breeding Program in Sweden

The Swedish Scots pine breeding program was developed and implemented by Skogforsk (the Forestry Research Institute of Sweden) and began during the 1950s. The base material that forms the founder populations of the Swedish Scots pine breeding program consists of 6,000 plus trees, selected in two stages. The first founders were selected during the 1950s–1970s for their superior phenotypes from naturally regenerated stands at full rotation age. As a result, 1,300 plus-trees were selected, many of which were directly established as the first-generation seed orchards (Wilhelmsson and Andersson 1993). During the 1980s, a second collection of founder trees resulted in 4,700 additional plus-trees. Most of these new selections were performed on trees with superior growth, better branch quality, and stem straightness, from planted or even-aged stands (Wilhelmsson and Andersson 1993).

The long-term Swedish Scots pine breeding program aims to integrate the improvement of economically important traits with the preservation of genetic diversity and adaptation to current and future biotic and abiotic factors, including climatic conditions (Andersson et al. 2011). To achieve those goals, the breeding program consists of a meta-population of 24 nucleus breeding populations, established with different adaptation outlines based on latitudinal range and temperature sums (Fig. 11.1). The nucleus populations have an effective population size of 50 to 100 individuals, with overlapping distributions according to photoperiod and temperature conditions (Rosvall et al. 2011; Wilhelmsson and Andersson 1993). The climatic range covered is broader than the temperature climate of Sweden, thus ensuring that suitable tree material will be available for adaptation to future climatic conditions (Calleja-Rodriguez 2019; Danell 1993).

The breeding objectives are fairly consistent across the whole country; however, the relative importance of the traits can vary from North to South (Wilhelmsson and Andersson 1993). For example, survival rates are very low in the harsh areas of northern Sweden compared to the southern areas; hence, survival is considered an important selection trait in northern Sweden (Abrahamsson et al. 2012; Calleja-Rodriguez et al. 2019a; Persson and Andersson 2003; Persson et al. 2010). When volume production per hectare was the breeding objective, overall tree growth, diameter, and survival ability were regarded as the most important traits throughout

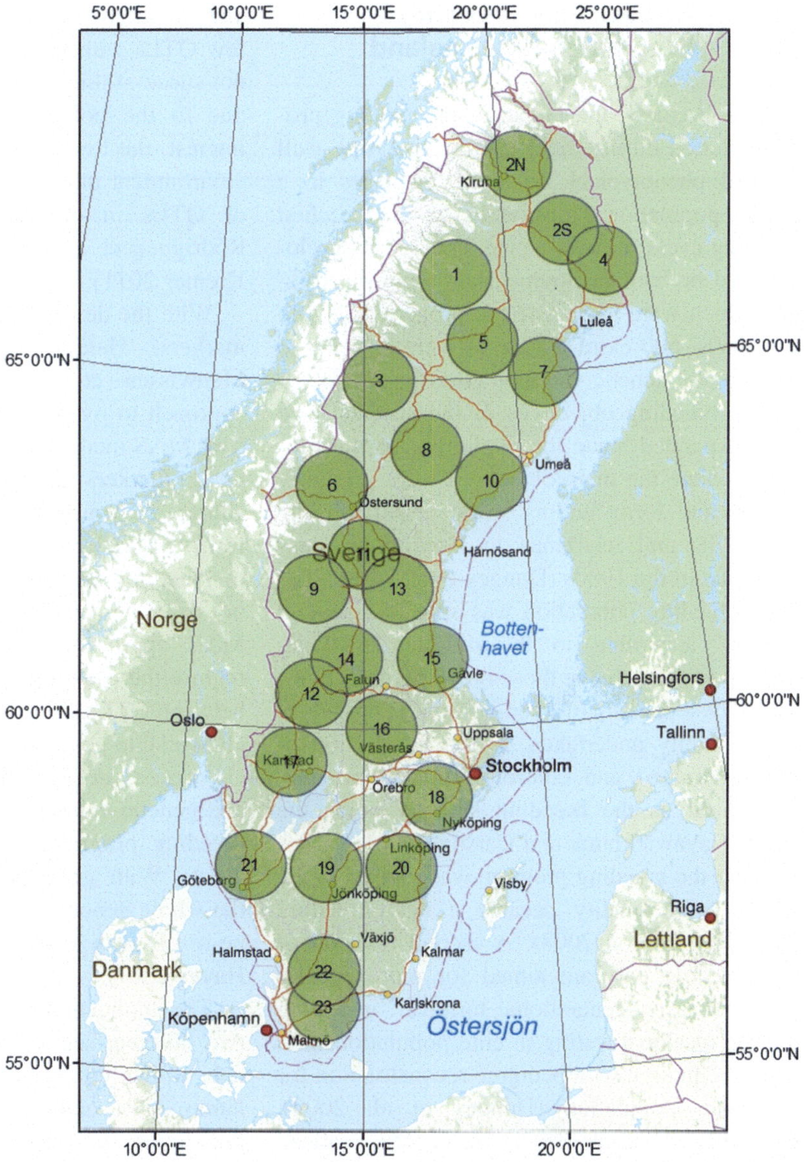

Fig. 11.1 Scots pine breeding populations in Sweden

the country. Wood quality traits were considered less important in harsh areas, according to the first review of the breeding program (Wilhelmsson and Andersson 1993). In the second review, new general breeding objectives were recommended in addition to those already in existence, such as more emphasis on resistance to biotic and abiotic stress, or basic wood density (Andersson et al. 2011).

Gains in stem volume production at full rotation age were 6–10% and 10–15% for the first and second round of seed orchards, respectively, but survival rates were not improved as a result of large environmental effects. However, from the recently finished third round of seed orchards, gains of 25% in volume production and between 5–15% in survival rates are expected (Andersson et al. 2007; Berlin et al. 2014; Rosvall et al. 2001).

11.1.3 Radiata Pine Breeding Program in New Zealand

In New Zealand, the radiata pine breeding program started during the 1950s by selecting well adapted phenotypical superior plus trees in a single population. Plus trees from the first breeding cycle of radiata pine were used for clonal seed orchard deployment and control crossing, which were later tested in progeny trials. In the 1960s, 600 extra plus trees were selected to increase the genetic variation (Burdon 2008).

The breeding objectives of the species have been expanded since the beginning of the program, where the main objectives were primarily growth and form, adding wood density during the 1970s and resistance to *Dothistroma septosporum* in the 1980s (Dungey et al. 2009). The main breeding population was split into several 'breeds' according to the different breeding objectives. To achieve those breeding objectives, recurrent selections of open-pollinated progenies were initially undertaken. Forward selection of open-pollinated and control-pollinated progeny were used as the breeding strategy since the 1980s (Jayawickrama and Carson 2000).

After the breeding program was revised in the 2000s wood quality became a new breeding objective (Burdon 2008). A second revision of the breeding program aimed for more genetic gain and faster generational turnover, including the use of clonal testing in elite populations and open-pollination with pedigree reconstruction for the main population (Dungey et al. 2009). According to Kimberley et al. (2015), realized genetic gains of the improved radiata pine in New Zealand were 25% in stand volume and 5.6% in height.

11.2 Genomic Selection

Prior to genomic selection, marker-assisted selection (MAS) was the first attempt to use molecular marker information to aid breeding programs and perform indirect selections (Neale and Williams 1991). MAS consisted of using linkage disequilibrium (LD) between markers and quantitative trait loci (QTL), i.e., detection of few QTLs with large effects. This approach was not successful in plant breeding (Bernardo 2016) due to the polygenic architecture of traits of interest, the low levels of LD, and the QTL-by-environment interactions, i.e., unique expression of QTLs in different environments (Calleja-Rodriguez et al. 2019b; Isik 2014; Neale and Kremer 2011).

With the development of genome-wide dense markers, Haley and Visscher (1998) and Meuwissen et al. (2001) suggested a new approach to overcome the obstacles of the standard MAS methods. This new approach involved using markers to predict phenotypes (genetic values) of young untested individuals and perform early selections to form the next generation of breeding populations, i.e., genomic selection or genomic predictions. Unlike the standard MAS approach, genomic selection does not require the preliminary detection of associations between QTLs and molecular markers, but rather all available genome-wide markers, regardless of effect size, are used for predictions of breeding and genetic values, and performing selections in breeding programs (Goddard et al. 2011; Isik 2014). With genomic selection, markers cover the whole genome, and all QTLs are assumed to be in LD with at least one marker (Goddard and Hayes 2007).

The application of genomic selection requires two different sets of genetically related individuals known as the training set or training population (TP) and the candidate or validation population (CP). The TP comprises individuals for which phenotypic data and marker information are available (Fig. 11.2). Genomic prediction models are developed with the information available in the TP and used to estimate genomic breeding values (GEBV) in the CP for which only marker information is available (i.e., no phenotypic information). After that process, selections are made accordingly (Goddard and Hayes 2009; Grattapaglia and Resende 2011).

Aided by genomic selection, forest tree breeding programs expect to achieve higher genetic gains primarily by shortening breeding cycles, for example, by reducing or entirely

Fig. 11.2 Genomic selection steps

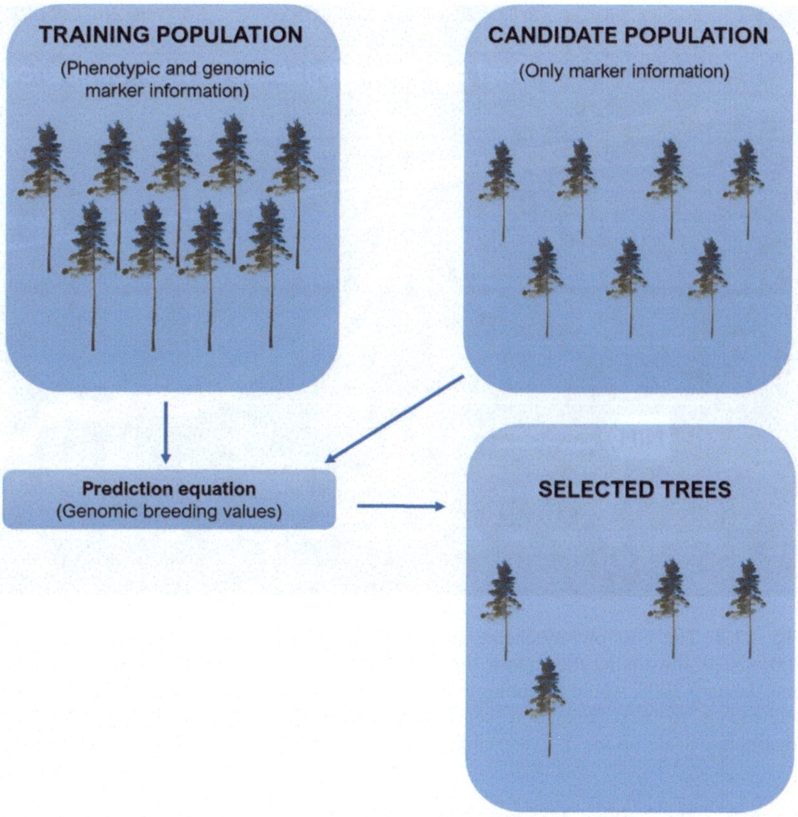

skipping progeny testing, and performing earlier selections (e.g., seedling stage) (Grattapaglia et al. 2018; Isik 2014). However, for most conifer species, reduction of breeding cycles implies that an earlier flowering age can also be reached (Almqvist 2001). If GEBV can be predicted for seedlings, more candidates could be genotyped at this stage than the number of candidates phenotyped in progeny tests, thereby increasing selection intensities and improving rates of genetic gain (Cobb et al. 2019). An additional advantage of genomic selection is its potential to capture Mendelian segregation, enabling the management of genetic diversity (Lorenz et al. 2011).

Application and research on genomic selection were only possible due to the advent of next-generation sequencing (NGS) (Elshire et al. 2011; Neves et al. 2013; Reis-Filho 2009) and the sudden drop in price for sequencing (Fig. 11.3) which allowed for the accelerated discovery of thousands of single nucleotide polymorphism (SNP) markers (Deschamps and Campbell 2010; Varshney et al. 2009). The drop in prices allowed, for the first time, genotyping at a scale that was large enough to be of interest to quantitative geneticists and tree breeders, i.e., at least a few thousand trees with 'known' parents. Although less reliable and with many challenges, this met the price point that opened the door for research. The release of the loblolly pine (*Pinus taeda* L.) genome was also an important milestone (Neale et al. 2014). While smaller genomes had been released for some time, this was the first large genome from *Pinus* and made genomic selection a possibility for this genus.

Applying genomics was an obvious next step for the forestry industry, following the lead of both crop and animal breeding. However, conifers have complex, highly-repetitive megagenomes, i.e., between 20 to 40 Gb (Neale et al. 2017), and other distinct characteristics that need consideration.

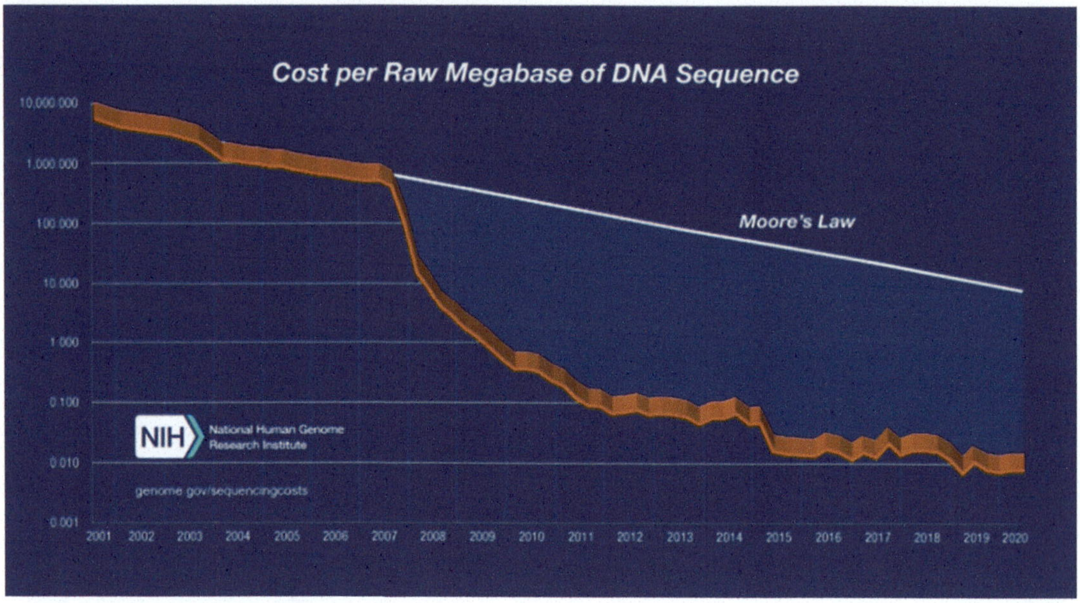

Fig. 11.3 The cost of sequencing per Megabase from 2001–2020, illustrating the sudden drop in prices, beginning January 2008, where technology moved from Sanger sequencing to next-generation sequencing. Note the logarithmic scale on the Y-axis. A hypothetical trend for Moore's law is shown for comparison. This law describes a long-term trend in the computer hardware industry that involves doubling 'compute power' every two years (See: Moore's Law [wikipedia.org]). *Source* DNA Sequencing Costs: Data (genome.gov) (Wetterstrand 2021)

11.2.1 Pedigree Versus Genomic Based Breeding Values and Genetic Parameter Estimations

For Scots and radiata pines, selection simultaneously takes several traits into account based on economic values and genetic parameters; this so-called index selection (Burdon and Klápště 2019; Rosvall and Mullin 2013) ensures that improvements are made for several traits at once, making breeding more effective.

Most of the traits targeted by improvement programs are polygenic, i.e., they are controlled by many genes, each of which has a very small effect on the phenotype, often less than 5% according to Neale and Kremer (2011). The phenotypic expression of polygenic traits is also affected by environmental factors (Goddard and Hayes 2009; White et al. 2007). Since both environment and multiple gene loci influence the phenotypic variance, breeding programs need to establish populations with large numbers of related individuals in many progeny trials located across many different environments, with the aim to estimate separately the genetic and environmental components of phenotypes (Lynch and Walsh 1998).

For breeding programs, accurate pedigree records are important as they connect the performance records of all tested individuals through populations, locations, and planting years. Kinship coefficients between the individuals are estimated with pedigrees by way of the numerator relationship matrix (NRM; Wright 1922), which is the expected proportion of the genome that individuals share (Goddard et al. 2011). To dissect genetic and environmental effects from phenotypic measurements, NRM is used in linear mixed models (Gilmour et al. 1995; Thomson et al. 2005) and pedigree-based Best Linear Unbiased Prediction (PBLUP) analysis through which breeding or genetic values, and genetic parameters such as heritability or genetic correlations, can be estimated (Henderson 1985) and selections made.

Haley and Visscher (1998) and Meuwissen et al. (2001) proposed to calculate the genomic realized relationship matrix (RRM or *G* matrix; VanRaden 2008) by estimating the kinship coefficients generated by the realized proportion of the genome (i.e., number of loci) shared between individuals (Goddard et al. 2011). The higher accuracy of RRM in estimating the relationships between individuals relies on the capacity of markers to account for the Mendelian sampling term. It also allows for the correction of possible errors in the pedigree, as long as the number of markers is sufficient to identify the identical-by-descent status across the genome (Meuwissen et al. 2013; Ødegård and Meuwissen 2014, 2015). The RRM is used in the genomic-based Best Linear Unbiased Prediction (GBLUP) analysis to estimate genetic parameter and genetic values, in the same way as PBLUP.

GBLUP methodology has been successfully evaluated in many forest species, including *Pinus pinaster* Ait. (Bartholome et al. 2016; Isik et al. 2016), *Eucalyptus nitens* (Suontama et al. 2018), Norway spruce (Chen et al. 2018; Zhou et al. 2020), *Picea engelmannii* × *glauca* (Ratcliffe et al. 2015) and *Pinus contorta* Douglas (Ukrainetz and Mansfield 2019). It requires all individuals involved in the estimation of RRM to be genotyped, which is easily achievable for simple breeding programs without a broad network of progeny or clonal trials.

Nevertheless, for advanced breeding programs with well-developed sets of trials and several cycles of evaluations and selections made, an alternative could be single-step Best Linear Unbiased Prediction (ssBLUP or HBLUP). Bartholome et al. (2016) reported that including previous generations of breeding populations into the TP improved the prediction accuracy of the genomic selection model's breeding values. In practice, to use a fully genotyped breeding program as the TP is unrealistic, not only due to cost but also because of the limited availability of some of the older generations. Therefore, forming a TP from both genotyped and non-genotyped individuals through a combination of phenotypes, pedigree and marker information is required (Legarra et al. 2009; Misztal et al. 2009). This involves generating a blended (hybrid) relationship matrix (HRM or H-matrix) where the RRM (VanRaden 2008) for genotyped individuals is blended into the NRM estimated for the whole population. This blending process consists of two critical steps upon which the model's resulting prediction accuracy depends (1) rescaling the marker-based relationship matrix according to the pedigree-based relationship matrix and (2) appropriate weighting of markers and pedigree-derived information. Proper attention to these two steps increases reliability and decreases breeding values bias (Gao et al. 2012). Promising results in *Picea glauca* (Moench) Voss (Cappa et al. 2017; Ratcliffe et al. 2017) showed that this alternative approach helped improve estimates from genetic parameters even when using only a reduced proportion of the marker information in the HBLUP model.

11.3 Genomic Selection in Scots Pine

The Scots pine genome has a size of about 22 Gb (Pyhäjärvi et al. 2020) but unlike other pine species with similar or larger genome sizes, for instance loblolly pine (Neale et al. 2014; Zimin et al. 2014) or *Pinus lambertiana* Dougl. (Stevens et al. 2016), its whole genome sequence has not been assembled yet.

The lack of a reference genome and well developed NGS technologies have slowed down the exploration of genomic selection for the species. Despite this, research on the topic has been undertaken using genotyping-by-sequencing (GBS) technology, which is an attractive alternative for this type of studies (Chen et al. 2013; Pan et al. 2015). GBS uses restriction enzymes to reduce sequencing of complex genomes and uses a barcoding system for multiplex sequencing, which increases its efficiency (He et al. 2014).

A complication of GBS data is the large amount of missing data due to the random distribution of coverage across samples, requiring a more tedious process of filtering, calling and imputation of SNP markers (Dodds et al. 2015; Hall et al. 2020). The first study of genomic selection for Scots pine in Sweden compared two different types of imputation methods, (i.e., random imputation and expectation maximization—EM), and evaluated their possible effect on genomic predictions (Calleja-Rodriguez 2019; Calleja-Rodriguez et al. 2020). Both methods retained similar number of SNPs and genomic predictions were slightly higher with EM imputation, yet no significant differences were found between both algorithms (Calleja-Rodriguez et al. 2020). However, for other plant species such as wheat and interior spruce, EM resulted in higher prediction efficiencies (El-Dien et al. 2015; Poland et al. 2012).

There are several factors that could influence the success of genomic selection and that can be controlled by breeders (Li and Dungey 2018), for instance, the composition and size of the TP. Larger sized TPs with a high genetic relationship with the CP will result in better predictions, as already reported for Norway spruce and *Eucalyptus urophylla* × *E. grandis* (Chen et al. 2018; Tan et al. 2017). Specifically, in Scots pine the composition of the TP and CP was evaluated within the same population and same full-sib families, verifying that as the number of individuals in the TP increased, the accuracy of predictions improved (Calleja-Rodriguez et al. 2020).

Another factor that influences the success of genomic selection is the extent of LD between SNP markers and QTL, which depends on both the effective population size and the number of SNP markers (Li and Dungey 2018). LD can be increased by reducing the effective population size or increasing the number of markers; however, decreasing the population size would also reduce the genetic diversity, thus favoring the latter approach of increasing the number of SNP markers (Grattapaglia and Resende 2011; Li and Dungey 2018). In addition to GBLUP and HBLUP models genomic predictions can also be evaluated with Bayesian models. Most of the Bayesian models presume a prior non-normal distribution of QTL allelic effects such that the variance at each locus can vary (Isik et al. 2017). Calleja-Rodriguez et al. (2020) stated that prediction accuracies increased for Bayesian LASSO (BL) and Bayesian Ridge Regression (BRR) models as the number of SNP markers increased, reaching the highest accuracies for growth and wood quality traits when the number of SNPs was between 6,000–8,719, the latter being the maximum number of SNPs available in the study (Calleja-Rodriguez et al. 2020).

The distribution of QTL effects and heritabilities of the traits of interest also influence the success of genomic selection. While this is beyond the control of breeders, different genomic selection models are available that accommodate different distributions of the allelic effects. Three genomic prediction models (i.e., GBLUP, BL and BRR) were used to evaluate the effect of the distribution of alleles for different growth and wood quality traits. Similar prediction efficiencies were obtained for all genomic models, which were comparable to those from pedigree-based models (Calleja-Rodriguez et al. 2020). Conversely, Tan et al. (2017) concluded that, based on the computational requirements, GBLUP was the most effective method, agreeing with the results for Scots pine (Calleja-Rodriguez 2019).

Beside the factors described above, there are others that also influence the success of genomic selection, such as genotype-by-environment interactions (G × E). Particularly in Northern Sweden, there is strong G × E among Scots pine progeny trials located in harsh and mild environments (Calleja-Rodriguez et al. 2019a; Kroon et al. 2011; Persson and Andersson 2003; Persson et al. 2010). Evaluation in both harsh and mild environments should be a priority to test the efficiency of genomic selection in the Swedish Scots pine breeding program, since the results for a single environment cannot be extrapolated to multiple environments (Chen et al. 2018, 2019).

Another advantage of using genomic selection is the possibility to discern between additive and non-additive effects, particularly when the mating design or the number of individuals available are insufficient to differentiate these effects using pedigree-based methods (Chen et al. 2019; El-Dien et al. 2018; Muñoz et al. 2014). This can lead to a long-term indirect benefit for the fundamental understanding of the genetic architectures of the traits of interest in breeding programs. PBLUP models with additive and additive plus dominance effects were unable to estimate non-additive effects for wood properties in Scots pine. On the contrary, a GBLUP model including epistatic effects was able to detect epistatic variance, ranging between 20 and 35% of the total variance, for wood density and dynamic modulus of elasticity (Calleja-Rodriguez et al. 2021). Furthermore, PBLUP models detected dominance variance for growth traits which became epistatic variance when GBLUP models were used. This indicates that epistatic variance may be confounded with dominance variance when models cannot allocate total genetic effects.

To date, genomic selection studies performed for Scots pine showed the potential of genomic selection to shorten the breeding cycle length by 50% (i.e., from 21 to 11 years), and in certain cases double the response of conventional phenotypic selection for the species (Calleja-Rodriguez 2019; Calleja-Rodriguez et al. 2020, 2021). Moreover, expected genetic gains of early genomic selection when considering total genetic effects were almost double when compared with traditional phenotypic selection, based on pedigrees and with a selection intensity of 1% (Calleja-Rodriguez et al. 2021).

11.4 Genomic Selection in Radiata Pine

Genomic selection in New Zealand radiata pine was initiated in the 2010s. There were several events and research published that stimulated this change. The key driver was interest from forest industry in speeding up the delivery of genetic gain after in-forest monitoring of productivity showed significant improvements with genetically improved planting stock. This was later verified from large-plot genetic gain trials, where it was demonstrated that genetic gain of up to 25% in stand volume with a value of over $NZ 3.5 billion had been delivered to the forest and that the increase in predicted production was real (Kimberley et al. 2015, 2016).

Further interest in genetic gain coincided with changes in forest valuations. Instead of valuations being based on models and inventory plots, valuations were now being based on real, albeit remotely sensed data using LiDAR (e.g., Bombrun et al. 2020; Dash et al. 2015; Watt et al. 2016). The interest and enthusiasm for developing genomic tools might not have existed if it were not for this change in industry outlook. Most of the applied genomics research is industry-driven and funded by the Radiata Pine Breeding Company (www.rpbc.nz). Funding is also provided by applied research grants from the NZ government, and from Scion, the New Zealand Forestry Research Institute.

Research advanced significantly in this time and also stimulated the New Zealand forest genomic revolution, a combination of theoretical and real advances in tree genomics. It is important to note that research into molecular genetics, MAS, linkage mapping, and quantitative genetics was active at the time, just at a smaller scale (Cato et al. 1999; Fisher et al. 1998; Kumar et al. 2004; Wilcox et al. 2001). The use of markers to validate pedigrees was also of real interest, led by Lambeth et al. (2001). The technology had been tested on several species, and while adopting NGS platforms posed some challenges, there were clear advantages to fully understanding the pedigree. The paper from Grattapaglia and Resende (2011) also provided some guidelines on the expected accuracy of genomic selection across a range of marker densities, numbers of quantitative trait loci, and heritabilities. This was the foundation paper behind designing New Zealand's first foray into genomic selection for radiata pine.

The New Zealand radiata pine breeding program is only three generations from wild populations. As the time to flowering is approximately 8 years for radiata pine, creating inbred lines is excessively time-consuming, unlike in some conifers (Russell and Ferguson 2008). Therefore, exploiting hybrid vigor through inbred lines is not a viable option. Because of long generation times in the breeding cycle, tree breeders take advantage of theoretical simulations to understand where and how to create the best outcomes and deliver the highest genetic gains. In New Zealand, simulations were based on times estimated for standard breeding operations and deployment (Li and Dungey 2018; Fig. 11.4). Using generation times of 19 (breeding) and 9 (genomics) years, simulations showed that, unsurprisingly, the amount of genetic gain attainable per generation was most significant with a larger training population or for traits with higher heritabilities. Large numbers of trees were ideal (>2000), and due to the size of the genome (~ 22 Gb), a marker assay of 60 k SNPs would also be ideal (Li and Dungey 2018). However, an interesting outcome of the simulation was the importance of the speed to flowering and the speed to deployment. Again, this is obvious in the context of delivering genetic gain in the fastest possible timeframe. Regardless, this result has also led to the development of a program of work to establish semi-automated mass production of radiata pine varieties (Burdon and Krajnakova 2020).

The next step toward implementing genomic selection was to develop the necessary genomic resources. Earlier research toward using association-based approaches to improve radiata pine (Li et al. 2016) had culminated in a catalogue of approximately 300 SNPs, around 100 of which were assayed using the Sequenom iPLEX® Gold MassARRAY® genotyping technology (Telfer et al. 2013). However, this level of SNP genotyping was not adequate, both in terms of the estimated number of SNPs and for genotyping populations of the size required for genomic selection (Grattapaglia and Resende 2011). Scion, in partnership with the Radiata Pine Breeding Company, thus embarked on simultaneous SNP discovery and genotyping platform development using exome capture GBS (Telfer et al. 2018, 2019). This platform had proven successful for loblolly pine (Neves et al. 2013) and other conifers with similar-sized genomes (Thistlethwaite et al. 2019; Vidalis et al. 2018), and access to the loblolly genome assembly meant that an assembled genome was not required for radiata pine.

Exome capture GBS has proven to be a highly successful workhorse for New Zealand's radiata pine genomic selection program. Now used to genotype over four thousand individuals across the New Zealand breeding program, this has well exceeded both the population size and the estimated number of SNPs (~ 80,000 with minor allele frequencies >0.03) that simulations indicated were required to implement genomic selection (Grattapaglia and Resende 2011). However, for genomic selection to be commercially viable, this still needed an order of magnitude reduction in the cost per test. For sequencing-based genotyping, cost reductions are often achieved through further multiplexing but this decreases the sequencing depth per SNP, and results in higher rates of missing data. Exome capture GBS already suffered from 5–10% missing data, with an over-representation of homozygote calls; thus, further sequencing depth reductions were not ideal.

Many groups interested in implementing genomic selection for conifer species joined an international Conifer SNP Consortium, with the shared goal to design robust and affordable SNP arrays using Thermo Fisher's Affymetrix technology (Bernhardsson et al. 2020). Similarly, Scion and the Radiata Pine Breeding Company developed a 36 k SNP array for radiata pine.

The degree of hidden relatedness in the radiata pine breeding population is currently unknown, as are the rates of mislabeling and misidentification, which in some instances can be high as 15–35% (Gömöry et al. 2000; Moriguchi et al. 2005). The correction of pedigree errors in other forest species such as *Eucalyptus* demonstrated substantial improvements to the accuracy of breeding values for growth and increased the response to selection by 50% (Klápště et al. 2017). Detecting and correcting pedigree errors in radiata pine could thus allow for similar improvements in the breeding program.

Parentage reconstruction has historically been achieved using a limited number of multi-allelic, highly variable microsatellite markers. However, ongoing upgrades to analysis software and laboratory equipment and consumables has impacted the cost-effectiveness and reproducibility of these markers both across labs and over time (Telfer et al. 2015). Nowadays, SNPs markers are far more efficient for parentage reconstruction due to the large numbers of SNPs that can be genotyped at a relatively low cost (Telfer et al. 2019).

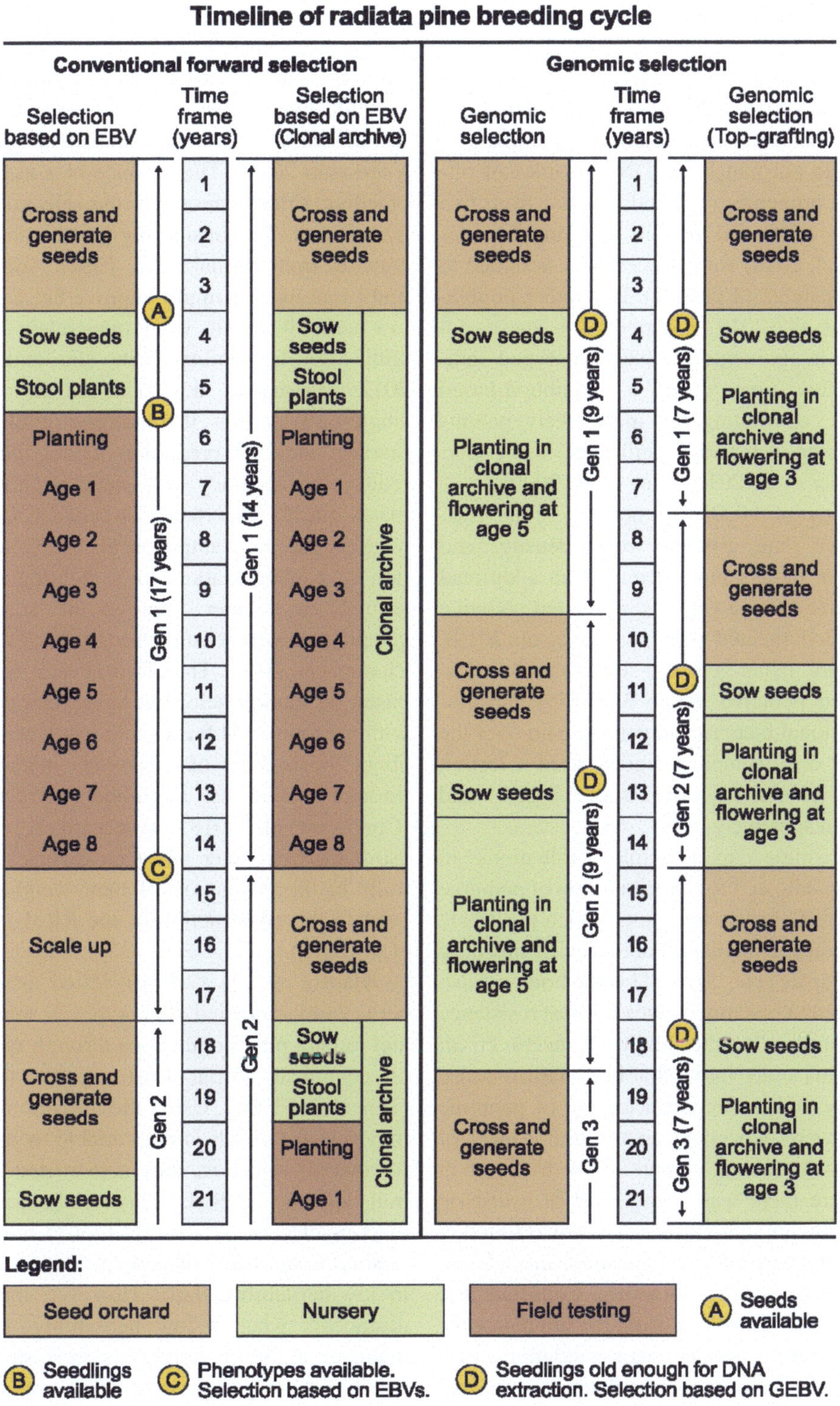

Fig. 11.4 Representation of the radiata pine breeding cycle assuming the timelines of forward selection and then putative timelines under genomic selection. A–B represents the start of a generation and growth in the nursery, B–C the field-testing phase, and C the beginning of a new generation. Total generation time in this instance is approximately 17 years, but in reality, this is often 1–5 years longer depending on the biology, timing, and availability of cones and pollen for crossing. The generation time under a genomic selection regime is optimistically between the D's, with a minimum generation time of 9 years

Parentage reconstruction in radiata pine has been successfully performed using a subset of 704 SNPs from the exome-capture GBS platform (Telfer et al. 2018, 2019). Using the genotypic exclusion method, i.e., excluding candidate parents that do not match the DNA profiles of offspring, all progenies (175 individuals) from two previously confirmed full-sib mapping families (Telfer et al. 2019) were successfully assigned to their true parents (4 parents). In another population, Li et al. (2019) reported 50% successful parentage assignments for 980 genotyped individuals, this time using a likelihood-based approach, i.e., finding the most likely parents and assessing statistical significance, using Cervus (Darrier et al. 2019; Slate et al. 2000).

Accuracy of GEBV of non-key traits (i.e., traits other than growth, wood density, and stiffness) in radiata pine increased an additional 7–11% after using a genomic-corrected pedigree over the documented pedigree (Li et al. 2019). Furthermore, using genomic selection as well as a corrected pedigree led to 37–115% and 13–77% additional increases in genetic gain over the traditional documented pedigree-based selection assuming generation intervals of 17 years and 14 years, respectively.

Using single-step genomic evaluations in radiata pine allowed for the correction of pedigree error before $\boldsymbol{G}$ matrix rescaling, which positively impacted on the prediction accuracy of breeding values (Klápště et al. 2018). This method was also used to predict breeding values for host resistance to Dothistroma needle blight (Klápště et al. 2020a). This study found that both narrow-sense heritability and prediction accuracy of genomic selection were dependent on the trait's level of expression. In the early years of the trials, or in years where there was a low level of a disease outbreak, phenotyping for resistance was found to be relatively unreliable and resulted in underestimated narrow-sense heritability. Consequently, breeding values estimated using the full dataset or predicted during model cross-validation were both shrunk towards the population mean, and thus both shared errors which upwardly biased prediction accuracy in terms of their correlation. Therefore, genomic prediction models should ideally use disease phenotypic data obtained when there are at least moderate levels of disease.

Multi-trait models are another way to improve the accuracy of genetic parameters (Calus and Veerkamp 2011). They reduce bias in estimated breeding values caused by the selection of correlated traits by considering genetic correlations between traits (Pollak et al. 1984). Additionally, multi-trait models might improve the accuracy of low heritability traits when moderately correlated with high heritability traits (Jia and Jannink 2012; Marchal et al. 2015). As discovered in single-trait analyses, the inclusion of all available markers into the prediction model might even reduce the accuracy of genetic parameter estimates due to imperfect LD between QTLs and markers (de los Campos et al. 2015) and selection of genetic markers or weighting might be required to further improve the accuracy of genetic parameters estimates (Lippert et al. 2013; Zhang et al. 2010). Therefore, several approaches based on marker selection have been developed within a multi-trait framework using assumptions about the number of traits each marker affects through pleiotropy to model marker effects (Cheng et al. 2018). Alternatively, posterior estimates of marker effect covariances between traits in the estimation of their weights can be used in the construction of the RRM (Karaman et al. 2018).

Klápště et al. (2020b) tested associations between markers and phenotypes in shining gum and radiata pine populations through two blocks of canonical partial least squares (PLS-CA) (Tenenhaus 1998). Using the measure of marker importance to each latent variable as a criterion for marker selection, they constructed a "multi-trait specific" RRM. This study found that marker-selection is a useful tool to improve the prediction accuracy of genomic breeding values in low-heritability traits. However, this benefit disappears when a low heritability trait (e.g., diameter at breast height) is in a strong relationship (high genetic correlation), regardless of whether this is positive or negative, with a high heritability trait (e.g., wood density).

Additionally, the study found that capturing a broader genetic diversity in the training population had a positive impact on marker selection efficiency, which might be a result of the higher genetic diversity required for precise estimation of genetic correlations (Bijma and Bastiaansen 2014).

11.5 Future Perspectives and Implementation of GS in Pine Breeding Programs

To date the genomic selection studies performed in Scots and radiata pines have shown the potential of GBLUP and HBLUP to be used as evaluation tools within the existing breeding programs. The implementation of genomic selection in the radiata pine breeding program has already started, whereas for Scots pine training of genomic selection models in breeding populations is currently under development.

Generally, in many conifer species there are biological constraints such as late onset of flowering which may become a challenge in terms of shortening generation intervals and should be tackled by developing the tools for flowering stimulation such as top-grafting and light treatment high tunnels. The Scots pine breeding cycle takes 21–30 years to complete, including a progeny testing phase up to 15 years (Rosvall 2011), and 15–18 years to flowering of breeding archive grafts (Mátyás et al. 2004). Thus, reducing the breeding cycle length for Scots pine will only be possible if genomic selection is applied simultaneously with stimulation to produce female flowers at earlier ages and/or if somatic embryogenesis for the species can be developed (Almqvist 2018; Li and Dungey 2018). The possibility of skipping the progeny testing phase altogether when using genomic selection is a highly attractive option both in terms of reducing costs as well as for more rapid turnover of breeding cycles. Nevertheless, establishment of progeny trials in each breeding cycle is still important in accounting for G × E as well as monitoring the impact of future climate change (Calleja-Rodriguez et al. 2019a; Li et al. 2017) and to follow phenotypic growth rates in the improved tree material as a mean of demonstrating the prediction ability of genomic selection models between breeding cycles (Ukrainetz and Mansfield 2020).

When considering genomic selection as part of the genetic evaluation in Scots pine, the cost of genotyping as well as the lack of a reference genome and well-developed genomic marker resources have been hindering factors. For Scots pine, the reference genome v1.0 and 50 K SNP chip are currently being developed in Sweden (Nilsson and Lundmark 2020). The expected improvements in prediction ability as well as potentially decreased genotyping costs would facilitate implementation of this methodology in future. Development of genomic marker resources will be followed by the breeding population specific analysis of genomic selection models. The breeding program is divided into 24 different populations across Sweden, and selection is conducted carefully within each population. Accounting for G × E across progeny testing sites within a breeding population as well as improving the prediction accuracy of genomic selection across generations will be the initial research focus. To date, Skogforsk Scots pine breeding program database has more than 1.2 million trees and almost 27 million measurements. Therefore, using methodology that only requires a subset of the breeding populations to be genotyped will be key to the implementation of genomic selection in the Swedish Scots pine breeding program.

The Radiata Pine Breeding Company (RPBC), as the dominant player in radiata pine breeding, is fully committed to implementing genomic resources in operational evaluation systems and research and development. From the operational evaluation side, the development of models integrating phenotypic, pedigree, genomic, and environmental data, which are computationally feasible, will be critical for the evaluation of G × E and the establishment of deployment strategies that reflect future climates. Additionally, the expansion of the radiata pine reference genome with whole-genome resequencing of genetically diverse individuals will allow for better genome representation on future marker arrays. It will also allow for the implementation of other advanced genomic strategies such as genome-wide association studies (GWAS) for the discovery of loci affecting economically important traits or exploration of LD patterns for detection of selection signatures.

Acknowledgements The authors are grateful to Gancho Slavov (Scion), Curt Almqvist (Skogforsk), Torgny Persson (Skogforsk), and Ulfstand Wennström (Skogforsk) for their comments. Skogforsk authors acknowledge T4F program in Sweden for financial support to write this chapter.

References

Abrahamsson S, Nilsson JE, Wu H et al (2012) Inheritance of height growth and autumn cold hardiness based on two generations of full-sib and half-sib families of *Pinus sylvestris*. Scand J for Res 27 (5):405–413. https://doi.org/10.1080/02827581.2012.663403

Almqvist C (2001) Improvement of flowering competence and capacity with reference to Swedish conifer breeding. Dissertation. Swedish University of Agricultural Sciences

Almqvist C (2018) Improving floral initiation in potted *Picea abies* by supplemental light treatment. Silva Fenn 52 (2):1:11. https://doi.org/10.14214/sf.7772

Andersson B, Karlson B, Ståhl P et al (2011) Breeding objectives under changing environmental and market conditions. In: Rosvall O (ed) Review of the swedish tree breeding programme. Skogforsk, Uppsala, Sweden, pp 15–19

Andersson B, Elfving B, Persson T et al (2007) Characteristics and development of improved *Pinus sylvestris* in northern Sweden. Can J for Res 37(1):84–92. https://doi.org/10.1139/X06-224

Bartholome J, Van Heerwaarden J, Isik F et al (2016) Performance of genomic prediction within and across generations in maritime pine. BMC Genomics 17 (604):1–14. https://doi.org/10.1186/s12864-016-2879-8

Berlin M, Ericsson T, Andersson Gull B (2014) Plantval, manual and background to technical implementation. In: Hallberg I (ed), Arbetsrapport, vol 851. Skogforsk, Uppsala, pp 1–66

Bernardo R (2016) Bandwagons I, too, have known. Theor Appl Genet 129(12):2323–2332. https://doi.org/10.1007/s00122-016-2772-5

Bernhardsson C, Zan Y, Chen Z et al (2020) Development of a highly efficient 50K single nucleotide polymorphism genotyping array for the large and complex genome of Norway spruce (*Picea abies* L. Karst) by whole genome resequencing and its transferability to other spruce species. Mol Ecol Resour 00:1–17. https://doi.org/10.1111/1755-0998.13292

Bijma P, Bastiaansen JWM (2014) Standard error of the genetic correlation: how much data do we need to estimate a purebred-crossbred genetic correlation? Genet Sel Evol 46(1):79. https://doi.org/10.1186/s12711-014-0079-z

Bombrun M, Dash JP, Pont D et al (2020) Forest-scale phenotyping: productivity characterisation through machine learning. Front Plant Sci 11(99):1–14. https://doi.org/10.3389/fpls.2020.00099

Boratyński A (1991) Range of natural distribution. In: Giertych M (ed) Developments in plant genetics and breeding, vol 3. Elsevier, pp 19–30

Burdon RD (2008) Breeding radiata pine-historical overview. N Z J for 52(4):4

Burdon RD, Miller JT (1992) Introduced forest trees in New Zealand: recognition, role, and seed source. 12. Radiata pine (*Pinus radiata* D. Don). FRI bulletin-Forest Research Institute, New Zealand Forest Service

Burdon RD, Klápště J (2019) Alternative selection methods and explicit or implied economic-worth functions for different traits in tree breeding. Tree Genet Genomes 15(79):1–15. https://doi.org/10.1007/s11295-019-1384-z

Burdon RD, Krajnakova J (2020). 21st Century Tissue Culture Partnership. 8395

Calleja-Rodriguez A (2019) Quantitative genetics and genomic selection of Scots pine. Dissertation. Faculty of Forest Sciences, Department of Forest Genetics and Plant Physiology. Swedish University of Agricultural Science

Calleja-Rodriguez A, Andersson Gull B, Wu HX et al (2019a) Genotype-by-environment interactions and the dynamic relationship between tree vitality and height in northern *Pinus sylvestris*. Tree Genet Genomes 15(36):1–15. https://doi.org/10.1007/s11295-019-1343-8

Calleja-Rodriguez A, Li Z, Hallingbäck HR et al (2019b) Analysis of phenotypic- and Estimated Breeding Values (EBV) to dissect the genetic architecture of complex traits in a Scots pine three-generation pedigree design. J Theor Biol 462:283–292. https://doi.org/10.1016/j.jtbi.2018.11.007

Calleja-Rodriguez A, Chen Z, Suontama M et al (2021) Genomic predictions with non-additive effects improved estimates of additive effects and predictions of total genetic values in *Pinus sylvestris*. Front Plant Sci 12(666820):1–16. https://doi.org/10.3389/fpls.2021.666820

Calleja-Rodriguez A, Pan J, Funda T et al (2020) Evaluation of the efficiency of genomic versus pedigree predictions for growth and wood quality traits in Scots pine. BMC Genomics 21(796):1–17. https://doi.org/10.1186/s12864-020-07188-4

Calus MPL, Veerkamp RF (2011) Accuracy of multi-trait genomic selection using different methods. Genet Sel Evol 43(1):26. https://doi.org/10.1186/1297-9686-43-26

Cappa EP, El-Kassaby YA, Munoz F et al (2017) Improving accuracy of breeding values by incorporating genomic information in spatial-competition mixed models. Mol Breed 37(10):125. https://doi.org/10.1007/s11032-017-0725-6

Cato SA, Corbett GE, Richardson TE (1999) Evaluation of AFLP for genetic mapping in *Pinus radiata* D. Don. Mol Breed 5:275–281. https://doi.org/10.1023/A:1009612709760

Chen C, Mitchell SE, Elshire RJ et al (2013) Mining conifers' mega-genome using rapid and efficient multiplexed high-throughput genotyping-by-sequencing (GBS) SNP discovery platform. Tree Genet Genomes 9:1537–1544. https://doi.org/10.1007/s11295-013-0657-1

Chen ZQ, Baison J, Pan J et al (2019) Increased prediction ability in Norway spruce trials using a marker x environment interaction and non-additive genomic selection model. J Hered 110(7):830–843. https://doi.org/10.1093/jhered/esz061

Chen ZQ, Baison J, Pan J et al (2018) Accuracy of genomic selection for growth and wood quality traits in two control-pollinated progeny trials using exome capture as the genotyping platform in Norway spruce. BMC Genomics 19(946):1–16. https://doi.org/10.1186/s12864-018-5256-y

Cheng H, Kizilkaya K, Zeng J et al (2018) Genomic prediction from multiple-trait bayesian regression methods using mixture priors. Genetics 209:89–103. https://doi.org/10.1534/genetics.118.300650

Cobb JN, Juma RU, Biswas PS et al (2019) Enhancing the rate of genetic gain in public-sector plant breeding programs: lessons from the breeder's equation. Theor Appl Genet 132:627–645. https://doi.org/10.1007/s00122-019-03317-0

Danell O (1993) Breeding programmes in Sweden: General approach. In: Lee SJ (ed), Corrected reprint from: progeny testing and breeding strategies, proceedings of the Nordic group of tree breeding, vol 302. Skogforsk. Forestry Commission, Edinburgh, UK, pp 1–4

Darrier B, Russell J, Milner SG et al (2019) A comparison of mainstream genotyping platforms for the evaluation and use of barley genetic resources. Front Plant Sci 10 (544):1–14. https://doi.org/10.3389/fpls.2019.00544

Dash JP, Marshall HM, Rawley B (2015) Methods for estimating multivariate stand yields and errors using k-NN and aerial laser scanning. Forestry: Int J Forest Res 88:237–247. https://doi.org/10.1093/forestry/cpu054

de los Campos G, Sorensen D, Gianola D, , 2015.de los Campos G, Sorensen D, Gianola D (2015) Genomic heritability: what is it? PLoS Genet 11(5):e1005048. https://doi.org/10.1371/journal.pgen.1005048

Deschamps S, Campbell MA (2010) Utilization of next-generation sequencing platforms in plant genomics and genetic variant discovery. Mol Breed 25:553–570. https://doi.org/10.1007/s11032-009-9357-9

Dodds KG, McEwan JC, Brauning R et al (2015) Construction of relatedness matrices using genotyping-by-sequencing data. BMC Genomics 16 (1047):1–15. https://doi.org/10.1186/s12864-015-2252-3

Dungey HS, Brawner JT, Burger F et al (2009) A new breeding strategy for *Pinus radiata* in New Zealand and New South Wales. Silvae Genet 58:28–38. https://doi.org/10.1515/sg-2009-0004

El-Dien OG, Ratcliffe B, Klápště J et al (2015) Prediction accuracies for growth and wood attributes of interior spruce in space using genotyping-by-sequencing. BMC Genomics 16(370):1–16. https://doi.org/10.1186/s12864-015-1597-y

El-Dien OG, Ratcliffe B, Klápště J et al (2018) Multienvironment genomic variance decomposition analysis of open-pollinated Interior spruce (*Picea glauca* x *engelmannii*). Mol Breed 38(26):1–16. https://doi.org/10.1007/s11032-018-0784-3

lshire RJ, Glaubitz JC, Sun Q et al (2011) A robust, simple genotyping-by-sequencing (GBS) approach for high diversity species. PLoS One 6(5):e19379.https://doi.org/10.1371/journal.pone.0019379

Fisher PJ, Richardson TE, Gardner RC (1998) Characteristics of single- and multi-copy microsatellites from *Pinus radiata*. Theor Appl Genet 96:969–979. https://doi.org/10.1007/s001220050828

Gao H, Christensen OF, Madsen P et al (2012) Comparison on genomic predictions using three GBLUP methods and two single-step blending methods in the Nordic Holstein population. Genet Sel Evol 44(8):1–8. https://doi.org/10.1186/1297-9686-44-8

Gernandt DS, Willyard A, Syring JV et al (2011) The conifers (Pinophyta). In: Plomion C et al (eds) Genetics, genomics and breeding of conifers. CRC Press and Edenbridge Science Publishers, New York, pp 1–39

Gilmour AR, Thompson R, Cullis BR (1995) Average information REML: an efficient algorithm for variance parameter estimation in linear mixed models. Biometrics 51(4):1440–1450. https://doi.org/10.2307/2533274

Goddard ME, Hayes BJ (2007) Genomic selection. J Anim Breed Genet 124:323–330

Goddard ME, Hayes BJ (2009) Mapping genes for complex traits in domestic animals and their use in breeding programmes. Nat Rev Genet 10:381–391. https://doi.org/10.1038/Nrg2575

Goddard ME, Hayes BJ, Meuwissen THE (2011) Using the genomic relationship matrix to predict the accuracy of genomic selection. J Anim Breed Genet 128:409–421. https://doi.org/10.1111/j.1439-0388.2011.00964.x

Grattapaglia D, Resende MDV (2011) Genomic selection in forest tree breeding. Tree Genet Genomes 7:241–255. https://doi.org/10.1007/s11295-010-0328-4

Grattapaglia D, Silva-Junior OB, Resende RT et al (2018) Quantitative genetics and genomics converge to accelerate forest tree breeding. Front Plant Sci 9 (1693):1–10. https://doi.org/10.3389/fpls.2018.01693

Gömöry D, Bruchánik R, Paule L (2000) Effective population number estimation of three Scots pine

(*Pinus sylvestris* L.) seed orchards based on an integrated assessment of flowering, floral phenology, and seed orchard design. For Genet 7:66–75

Haley CS, Visscher PM (1998) Strategies to utilize marker-quantitative trait loci associations. J Dairy Sci 81:85–97. https://doi.org/10.3168/jds.S0022-0302(98)70157-2

Hall D, Zhao W, Wennström U et al (2020) Parentage and relatedness reconstruction in *Pinus sylvestris* using genotyping-by-sequencing. Heredity 124:633–646. https://doi.org/10.1038/s41437-020-0302-3

He J, Zhao X, Laroche A et al (2014) Genotyping-by-sequencing (GBS), an ultimate marker-assisted selection (MAS) tool to accelerate plant breeding. Front Plant Sci 5(484):1–8. https://doi.org/10.3389/fpls.2014.00484

Henderson CR (1985) Best linear unbiased prediction of nonadditive genetic merits in noninbred populations. J Anim Sci 60:111–117

Isik F (2014) Genomic selection in forest tree breeding: the concept and an outlook to the future. New for 45:379–401. https://doi.org/10.1007/s11056-014-9422-z

Isik F, Bartholome J, Farjat A et al (2016) Genomic selection in maritime pine. Plant Sci 242:108–119. https://doi.org/10.1016/j.plantsci.2015.08.006

Isik F, Holland J, Maltecca C (2017) Genetic data analysis for plant and animal breeding, vol 400. Springer, New York

Jayawickrama KJS, Carson J (2000) A breeding strategy for the New Zealand Radiata Pine Breeding Cooperative. Silvae Genet 49:82–90

Jia Y, Jannink JL (2012) Multiple-trait genomic selection methods increase genetic value prediction accuracy. Genetics 192:1513–1522. https://doi.org/10.1534/genetics.112.144246

Karaman E, Lund MS, Anche MT et al (2018) Genomic prediction using multi-trait weighted GBLUP accounting for heterogeneous variances and covariances across the genome. G3 (Bethesda) 8:3549–3558. https://doi.org/10.1534/g3.118.200673

Kimberley MO, Moore JR, Dungey HS (2015) Quantification of realised genetic gain in radiata pine and its incorporation into growth and yield modelling systems. Can J Forest Res 45:1676–1687. https://doi.org/10.1139/cjfr-2015-0191

Kimberley MO, Moore JR, Dungey HS (2016) Modelling the effects of genetic improvement on radiata pine wood density. N Z J for 46(1):8. https://doi.org/10.1186/s40490-016-0064-0

Klápště J, Dungey HS, Graham NJ et al (2020) Effect of trait's expression level on single-step genomic evaluation of resistance to Dothistroma needle blight. BMC Plant Biol 20(205):1–13. https://doi.org/10.1186/s12870-020-02403-6

Klápště J, Suontama M, Dungey HS et al (2018) Effect of hidden relatedness on single-step genetic evaluation in an advanced open-oollinated breeding program. J Hered 109:802–810. https://doi.org/10.1093/jhered/esy051

Klápště J, Dungey HS, Telfer EJ et al (2020) Marker selection in multivariate genomic prediction improves accuracy of low heritability traits. Front Genet 11 (1240):1–15. https://doi.org/10.3389/fgene.2020.499094

Klápště J, Suontama M, Telfer E et al (2017) Exploration of genetic architecture through sib-ship reconstruction in advanced breeding population of *Eucalyptus nitens*. PLoS One 12(9):e0185137. https://doi.org/10.1371/journal.pone.0185137

Krakau UK, Liesebach M, Aronen T et al (2013) Scots Pine (*Pinus sylvestris* L.). In: Pâques LE (ed) Forest tree breeding in Europe: current state-of-the-srt and perspectives. Springer, Netherlands, Dordrecht, pp 267–323

Kroon J, Ericsson T, Jansson G et al (2011) Patterns of genetic parameters for height in field genetic tests of *Picea abies* and *Pinus sylvestris* in Sweden. Tree Genet Genomes 7:1099–1111. https://doi.org/10.1007/s11295-011-0398-y

Kumar S, Echt C, Wilcox PL et al (2004) Testing for linkage disequilibrium in the New Zealand radiata pine breeding population. Theor Appl Genet 108:292–298. https://doi.org/10.1007/s00122-003-1352-7

Lambeth C, Lee BC, O'Malley D et al (2001) Polymix breeding with parental analysis of progeny: an alternative to full-sib breeding and testing. Theor Appl Genet 103:930–943. https://doi.org/10.1007/s001220100627

Legarra A, Aguilar I, Misztal I (2009) A relationship matrix including full pedigree and genomic information. J Dairy Sci 92:4656–4663. https://doi.org/10.3168/jds.2009-2061

Li Y, Dungey HS (2018) Expected benefit of genomic selection over forward selection in conifer breeding and deployment. PLoS ONE 13(12):e0208232. https://doi.org/10.1371/journal.pone.0208232

Li Y, Wilcox P, Telfer E et al (2016) Association of single nucleotide polymorphisms with form traits in three New Zealand populations of radiata pine in the presence of genotype by environment interactions. Tree Genet Genomes 12(63):1–12. https://doi.org/10.1007/s11295-016-1019-6

Li Y, Klápště J, Telfer E et al (2019) Genomic selection for non-key traits in radiata pine when the documented pedigree is corrected using DNA marker information. BMC Genomics 20(1026):1–10. https://doi.org/10.1186/s12864-019-6420-8

Li YJ, Suontama M, Burdon RD et al (2017) Genotype by environment interactions in forest tree breeding: review of methodology and perspectives on research and application. Tree Genet Genomes 13(60):1–18. https://doi.org/10.1007/s11295-017-1144-x

Lippert C, Quon G, Kang EY et al (2013) The benefits of selecting phenotype-specific variants for applications of mixed models in genomics. Sci Rep 3(1815):1–9. https://doi.org/10.1038/srep01815

Lorenz AJ, Chao SM, Asoro FG et al (2011) Genomic selection in plant breeding: knowledge and prospects.

Adv Agron 110:77–123. https://doi.org/10.1016/B978-0-12-385531-2.00002-5

Lynch M, Walsh B (1998) Genetics and analysis of quantitative traits. Sinauer Sunderland, MA

Marchal A, Legarra A, Tisné S et al (2015) Multivariate genomic model improves analysis of oil palm (*Elaeis guineensis* Jacq.) progeny tests. Mol Breed 36(2):1–13. https://doi.org/10.1007/s11032-015-0423-1

Mátyás C, Ackzell L, Samuel C (2004) EUFORGEN technical guidelines for genetic conservation and use for Scots pine (*Pinus sylvestris*). Bioversity International

Meuwissen T, Hayes B, Goddard M (2013) Accelerating improvement of livestock with genomic selection. Annu Rev Anim Biosci 1(1):221–237. https://doi.org/10.1146/annurev-animal-031412-103705

Meuwissen THE, Hayes BJ, Goddard ME (2001) Prediction of total genetic value using genome-wide dense marker maps. Genetics 157:1819–1829

Misztal I, Legarra A, Aguilar I (2009) Computing procedures for genetic evaluation including phenotypic, full pedigree, and genomic information. J Dairy Sci 92:4648–4655. https://doi.org/10.3168/jds.2009-2064

Moriguchi Y, Goto S, Takahashi M (2005) Genetic management of seed orchards based on information revealed by molecular markers. J Japanese for Soc 87:161–169

Mullin TJ, Andersson Gull B, Bastien C et al (2011) Economic importance, breeding objectives and achievements. In: Plomion C et al (eds) Genetics, genomics and breeding of conifers. CRC Press and Edenbridge Science Publishers, New York, pp 40–127

Muñoz PR, Resende MFR, Gezan SA et al (2014) Unraveling additive from nonadditive effects using genomic relationship matrices. Genetics 198:1759–1768. https://doi.org/10.1534/genetics.114.171322

Neale DB, Williams CG (1991) Restriction-fragment-length-polymorphism mapping in conifers and applications to forest genetics and tree improvement. Can J for Res 21:545–554. https://doi.org/10.1139/x91-076

Neale DB, Kremer A (2011) Forest tree genomics: growing resources and applications. Nat Rev Genet 12:111–122. https://doi.org/10.1038/nrg2931

Neale DB, Martínez-García PJ, Torre ARDL et al (2017) Novel insights into tree biology and genome evolution as revealed through genomics. Annu Rev Plant Biol 68:457–483. https://doi.org/10.1146/annurev-arplant-042916-041049

Neale DB, Wegrzyn JL, Stevens KA et al (2014) Decoding the massive genome of loblolly pine using haploid DNA and novel assembly strategies. Genome Biol 15(R59):1–13. https://doi.org/10.1186/gb-2014-15-3-r59

Neves LG, Davis JM, Barbazuk WB et al (2013) Whole-exome targeted sequencing of the uncharacterized pine genome. Plant J 75:146–156. https://doi.org/10.1111/Tpj.12193

Nilsson O, Lundmark T (2020) SLU receives major grants for forest research. https://www.slu.se/en/ew-news/2019/1/slu-receives-major-grants-for-forest-research. Accesed 17 Dec 2020

Pan J, Wang B, Pei Z et al (2015) Optimization of the genotyping-by-sequencing strategy for population genomic analysis in conifers. Mol Ecol Resour 15:711–722. https://doi.org/10.1111/1755-0998.12342

Persson T, Andersson B (2003) Genetic variance and covariance patterns of growth and survival in northern *Pinus sylvestris*. Scand J for Res 18:332–343. https://doi.org/10.1080/02827580310003993

Persson T, Andersson B, Ericsson T (2010) Relationship between autumn cold hardiness and field performance in northern *Pinus sylvestris*. Silva Fenn 44:255–266. https://doi.org/10.14214/sf.152

Plomion C, Chagne D, Pot D et al (2007) Pines. In: Kole C (ed), Forest trees. Genome mapping and molecular breeding in plants, vol 7. Springer, Berlin, pp 29–92

Poland J, Endelman J, Dawson J et al (2012) Genomic selection in wheat breeding using genotyping-by-sequencing. Plant Genome 5:103–113. https://doi.org/10.3835/plantgenome2012.06.0006

Pollak EJ, van der Werf J, Quaas RL (1984) Selection bias and multiple trait evaluation. J Dairy Sci 67:1590–1595. https://doi.org/10.3168/jds.S0022-0302(84)81481-2

Pyhäjärvi T, Kujala ST, Savolainen O (2020) 275 years of forestry meets genomics in *Pinus sylvestris*. Evol Appl 13:11–30. https://doi.org/10.1111/eva.12809

Ratcliffe B, El-Dien OG, Klápště J et al (2015) A comparison of genomic selection models across time in interior spruce (*Picea engelmannii* x *glauca*) using unordered SNP imputation methods. Heredity 115:547–555. https://doi.org/10.1038/hdy.2015.57

Ratcliffe B, El-Dien OG, Cappa EP et al (2017) Single-step BLUP with varying genotyping effort in open-pollinated *Picea glauca*. G3 (Bethesda) 7:935–942. https://doi.org/10.1534/g3.116.037895

Reis-Filho JS (2009) Next-generation sequencing. Breast Cancer Res 11(S12):1–7. https://doi.org/10.1186/bcr2431

Rosvall O (2011) Review of the Swedish tree breeding program. Skogforsk, Uppsala, Sweden

Rosvall O, Mullin TJ (2013) Introduction to breeding strategies and evaluation of alternatives. In: Mullin TJ, Lee SJ (eds) Best practice for tree breeding in Europe. Skogforsk, Gävle, Sweden

Rosvall O, Almqvist C, Lindgren D et al (2011) Updates from research on selection and mating strategies. In: Rosvall O (ed) Review of the Swedish tree breeding programme. Skogforsk, Uppsala, Sweden, pp 28–38

Rosvall O, Jansson G, Andersson B et al (2001) Genetic gains from present and future seed orchards and clone mixes. SkogForsk, Uppsala, ISSN 1103

Russell JH, Ferguson DC (2008) Preliminary results from five generations of a western redcedar (*Thuja plicata*) selection study with self-mating. Tree Genet Genomes 4:509–518. https://doi.org/10.1007/s11295-007-0127-8

Slate J, Marshall T, Pemberton J (2000) A retrospective assessment of the accuracy of the paternity inference program cervus. Mol Ecol 9:801–808. https://doi.org/10.1046/j.1365-294x.2000.00930.x

Stevens KA, Wegrzyn JL, Zimin A et al (2016) Sequence of the sugar pine megagenome. Genetics 204:1613–1626. https://doi.org/10.1534/genetics.116.193227

Suontama M, Klápště J, Telfer E et al (2018) Efficiency of genomic prediction across two *Eucalyptus nitens* seed orchards with different selection histories. Heredity 122:370–379. https://doi.org/10.1038/s41437-018-0119-5

Tan BY, Grattapaglia D, Martins GS et al (2017) Evaluating the accuracy of genomic prediction of growth and wood traits in two Eucalyptus species and their F-1 hybrids. BMC Plant Biol 17(110):1–15. https://doi.org/10.1186/s12870-017-1059-6

Telfer E, Graham N, Stanbra L et al (2013) Extraction of high purity genomic DNA from pine for use in a high-throughput genotyping platform. N Z J for 43(3):1–8. https://doi.org/10.1186/1179-5395-43-3

Telfer E, Graham N, Macdonald L et al (2018) Approaches to variant discovery for conifer transcriptome sequencing. PLoS ONE 13(11):e0205835. https://doi.org/10.1371/journal.pone.0205835

Telfer E, Graham N, Macdonald L et al (2019) A high-density exome capture genotype-by-sequencing panel for forestry breeding in *Pinus radiata*. PLoS ONE 14 (09):e0222640.https://doi.org/10.1371/journal.pone.0222640

Telfer EJ, Stovold GT, Li Y et al (2015) Parentage reconstruction in *Eucalyptus nitens* using SNPs and microsatellite markers: a comparative analysis of marker data power and robustness. PLoS ONE 10 (7):e0130601.https://doi.org/10.1371/journal.pone.0130601

Tenenhaus M (1998) La régression PLS: théorie et pratique. Editions technip

Thistlethwaite FR, Ratcliffe B, Klápště J et al (2019) Genomic selection of juvenile height across a single-generational gap in Douglas-fir. Heredity 122:848–863. https://doi.org/10.1038/s41437-018-0172-0

Thomson R, Brotherstone S, White IMS (2005) Estimation of quantitative genetic parameters. Philos Trans R Soc Biol Sci 360:1469–1477

Ukrainetz NK, Mansfield SD (2019) Assessing the sensitivities of genomic selection for growth and wood quality traits in lodgepole pine using Bayesian models. Tree Genet Genomes 16(14):1–19. https://doi.org/10.1007/s11295-019-1404-z

Ukrainetz NK, Mansfield SD (2020) Prediction accuracy of single-step BLUP for growth and wood quality traits in the lodgepole pine breeding program in British Columbia. Tree Genet Genomes 16(64):1–13. https://doi.org/10.1007/s11295-020-01456-w

VanRaden PM (2008) Efficient methods to compute genomic predictions. J Dairy Sci 91:4414–4423. https://doi.org/10.3168/jds.2007-0980

Varshney RK, Nayak SN, May GD et al (2009) Next-generation sequencing technologies and their implications for crop genetics and breeding. Trends Biotechnol 27:522–530. https://doi.org/10.1016/j.tibtech.2009.05.006

Vidalis A, Scofield DG, Neves LG et al (2018) Design and evaluation of a large sequence-capture probe set and associated SNPs for diploid and haploid samples of Norway spruce (*Picea abies*). bioRxiv 291716. doi: https://doi.org/10.1101/291716

Watt MS, Dash JP, Watt P et al (2016) Multi-sensor modelling of a forest productivity index for radiata pine plantations. N Z J for Sci 46(9):1–14. https://doi.org/10.1186/s40490-016-0065-z

Wetterstrand KA (2021) DNA sequencing costs: data from the NHGRI Genome Sequencing Program (GSP) https://www.genome.gov/about-genomics/fact-sheets/DNA-Sequencing-Costs-Data. Accesed 7 Dec 2020

White TL, Adams WT, Neale DB (2007) Forest Genetics. CABI, Cambridge, MA

Wilcox PL, Richardson TE, Corbett GE et al (2001) Framework linkage maps of *Pinus radiata* D. Don based on pseudotestcross markers. For Genet 8:109–117

Wilhelmsson L, Andersson B (1993) Breeding of Scots pine (*Pinus sylvestris*) and lodgepole pine (*Pinus contorta* ssp. *latifolia*). In: Lee SJ (ed) Corrected repring from: progeny testing and breeding strategies, proceedings of the Nordic group of tree breeding, vol 302Skogforsk. Forestry Commission, Edinburgh, UK, pp 5–15

Wright S (1922) Coefficients of inbreeding and relationship. Am Nat 56:330–338

Zhang Z, Liu J, Ding X et al (2010) Best linear unbiased prediction of genomic breeding values using a trait-specific marker-derived relationship matrix. PLoS ONE 5(9):e12648. https://doi.org/10.1371/journal.pone.0012648

Zhou L, Chen Z, Olsson L et al (2020) Effect of number of annual rings and tree ages on genomic predictive ability for solid wood properties of Norway spruce. BMC Genomics 21(323):1–12. https://doi.org/10.1186/s12864-020-6737-3

Zimin A, Stevens KA, Crepeau MW et al (2014) Sequencing and assembly of the 22-Gb loblolly pine genome. Genetics 196:875–890. https://doi.org/10.1534/genetics.113.159715

Ødegård J, Meuwissen THE (2014) Identity-by-descent genomic selection using selective and sparse genotyping. Genet Sel Evol 46(3):1–8. https://doi.org/10.1186/1297-9686-46-3

Ødegård J, Meuwissen THE (2015) Identity-by-descent genomic selection using selective and sparse genotyping for binary traits. Genet Sel Evol 47(8):1–4. https://doi.org/10.1186/s12711-015-0090-z

12 Community-Based Genome Resource Needs in Pines

David B. Neale

Abstract

Community-based genome resources will be fundamentally important if pine genomics research is going to advance at a rate needed to genetically improve pine species for commodity production and also to facilitate conservation and restoration programs. The paramount resource is the reference genome sequence; yet as of 2021, only three genome sequences have been completed, published and released into the public domain; *P. taeda, P. lambertiana and P.pinaster.* Next generation sequencing technologies and computational methods for assembly and annotation are now well-developed for pines. A strategic plan must be developed so that a global and well-coordinated effort begins to sequence many, if not all, pine genomes.

12.1 A Short History of Pine Genetics and Genomics

The study of the genetics of pines, conifers and essentially all forest trees has historically been conducted using some kind of common garden approach where variation in the environment is held constant and thereby phenotypic variation can be equated with genetic variation (White et al. 2007). Polygenic traits such as height and volume growth, survival and disease resistance are commonly measured and components of variation (provenance, source, family, within family, etc.) are measured using quantitative genetic methods (White et al. 2007; Neale and Wheeler 2019). This approach has been used, and continues to be used, most efficiently in tree breeding programs. The common garden approach has also been used in geographical studies to estimate components of variation for a host of adaptive and physiological traits such as phenology, cold-hardiness and drought tolerance. (Neale and Wheeler 2019).

Beginning in the 1970s, the first monogenic molecular genetic markers were developed, called isozymes or allozymes (White et al. 2007). These markers were widely used in dozens of pine species to estimate a suite of population genetic parameters such as within and between population variation, gene flow, genetic drift, mating systems and hybridization (White et al. 2007; Neale and Wheeler 2019). Allozymes were followed by a suite of new DNA-based molecular markers such as RFLPs, RAPDs, AFLPs and SSRs (White et al. 2007), nearly all of which interrogated neutral genetic variation and were thus informative for demographic population genetic parameters (drift, gene flow, hybridization and mating system) (Neale and Wheeler 2019). The study of non-neutral and adaptive

D. B. Neale (✉)
University of California-Davis, One Shields Ave, Davis, CA, USA
e-mail: dbneale@ucdavis.edu

A. R. De La Torre (ed.), *The Pine Genomes*, Compendium of Plant Genomes,
https://doi.org/10.1007/978-3-030-93390-6_12

patterns of genetic variation using molecular markers did not advance during this period because markers interrogating non-neutral DNA sequences generally did not exist.

Beginning in the 1980s, approaches were developed to identify the genes underlying polygenic traits and then develop monogenic molecular markers to estimate non-neutral and adaptive genetic variation. The first approach was **quantitative trait locus (QTL)** mapping followed shortly thereafter by **genome-wide association studies (GWAS)** (White et al. 2007; Chaps. 6–8 of this volume). This approach has been used in a large number of pine species in the last 30 years (De La Torre et al. 2019a; Weiss et al. 2020; Neale and Wheeler 2019). The molecular genetic marker used is the **single nucleotide polymorphism (SNP)** found in well-annotated protein-coding genes. What soon followed were landscape genomics or **environmental association analyses (EAA)** where genetic variation in SNPs is associated with variation in the environment (generally moisture and temperature variables) (De La Torre et al. 2019b; Neale and Wheeler 2019). SNPs were first discovered from mining transcriptome databases (Neale and Wheeler 2019, Chap. 1 of this volume). Searching for SNPs in just expressed genes was done simply because whole-genome sequences in pines and other conifers did not exist due to the cost and challenges associated with sequencing very large and complex genomes.

Beginning in 2010, the first projects to sequence conifer genomes began (Neale and Wheeler 2019). To date, three pine genomes have been sequenced, published and released to the public domain; *P. taeda* (Neale et al. 2014; Wegrzyn et al. 2014; Zimin et al. 2014), *P. lambertiana* (Stevens et al. 2016; Crepeau et al. 2017) and *P. pinaster* (Chap. 5 of this volume). Thus, the first necessary modern genomic resources to facilitate genetic and genomic studies in pines have just really begun.

12.2 Future Needs and Applications in Pine Genomics

The paramount need for advancing pine genomics is reference genome sequences for a large number of species. All other important needs will rapidly fall into place once reference genome sequences are available. Ever since the sequencing of the human genome in 2000, it has become very clear that the power and efficiency of genetic and genomic research in any organism are greatly facilitated by having a well-annotated reference genome sequence. As we noted above, these critically important resources were delayed in development in pines and other conifers due to the size and complexity of these genomes. With the advent of next generation sequencing technologies (NGS) and advanced computational methods for genome assembly, this limitation is no longer true. However, as of this writing, only three of the 116 pine species have been sequenced and these resources released into the public domain. Why does this situation exist? Is it a lack of interest, lack of technical skill in the community or lack of funding? It seems to be a combination of all three. So how can this problem be overcome? I submit that the need is leadership and a well-developed strategic plan to get these resources developed. Currently, a very *ad hoc* approach is used to get pine genomes sequenced. An international advisory committee should be assembled, much like the human genome project in the early days where the sequencing of individual chromosomes was assigned to different countries, but in the case of pines, species priorities need to be determined and specific countries and funding agencies identified. Many pine species cross international borders so bi- or multi-national efforts might be established. The new North American Forest Genetics Society (https://treegenesdb.org/nafgs) could take the lead on assembling this international committee. A priority system would need to be developed and agreed upon. A simple classification system of high, medium and low might be sufficient. The economic value and the IUCN status of individual species would be the primary determinants of priority classification. Examples of high priority economically valuable species would be *P. radiata* and *P. sylvestris* where genome sequencing projects were begun but have not yet been completed. Other high-value species such as *P. contorta*, *P. elliottii*, *P. ponderosa and P. caribaea* might be early target species. However, species with an IUCN status of threatened or endangered (T&E) would be of extremely high value. There are many species in this category and nearly all have absolutely no genomic resources that would facilitate conservation and restoration. These might be called the "forgotten" species. The one pine species with T&E status and a genome sequence under development is *P. albicaulis* (https://whitebarkfound.org/our-work/genome-project/). Medium priority species might be as much as 80% of the species of the genus. These species are neither of very high economic value nor have T&E status. Genomic resources for many of these species might develop regionally where they have some specific regional importance and a public agency or private foundation would fund genome sequencing. A recent example of this is *P. longeva* (https://whitebarkfound.org/our-work/). Finally, the low priority species would simply be the ones where publicly released genome sequences already exist, just as *P. taeda, P. lambertiana and P. pinaster*.

There are many community genomic resources that would rapidly develop if reference genomes were in place. Because these are genomic resources, and not necessarily hypothesis-driven research topics, the development of these resources would also be well-served by an international advisory committee. The first would be **Transcriptome Databases.** These resources have developed in pines in an *ad hoc* and hypothesis-driven manner over the last 20 years using microarray technology and then RNA-seq technology (Neale and Wheeler 2019). These resources are reasonably well-curated at the TreeGenes Database (Falk et al. 2018; Wegrzyn et al. 2019; https://treegenesdb.org); however, because of large differences in technologies and the spatial and temporal variation in gene expression studies, comparative genomic opportunities are challenging. Coordinated transcriptome databases among a large number of species would be an important community resource. The next genetic resource would be **SNP Databases**. These resources also exist in a small number of species but certainly have a significant amount of ascertainment basis due to technologies used and nucleic acid source (DNA or mRNA).

Once reference genomes sequences, transcriptome databases and polymorphism databases community resources are developed and reside in well-curated relational databases or well-federated databases, then the modern genomic resources are in place to support a broad spectrum of hypothesis-driven research. Two research areas where some work has begun and will greatly facilitate tree breeding and conservation and restoration programs are genome-wide association studies (GWAS) and environmental association analysis (EAA) (Neale and Wheeler 2019). These studies are much less likely to have a high degree of standardization because questions and needs among species vary greatly. Nevertheless, a curated relational database for these resources would be quite valuable (see CartograTree, Vasquez-Gross et al. 2013, https://treegenesdb.org).

The health and economy of much of the world's population depend on the health of pine forests. Genomic tools have much to contribute toward that health and economy. Pine genetics and genomics have lagged for a very long time, that day is now over!

References

Crepeau MW, Langley CH, Stevens KA (2017) From pine cones to read clouds: rescaffolding the mega-genome of sugar pine (*Pinus lambertiana*), G3 Genes|Genomes|Genetics 7(5):1563–1568. https://doi.org/10.1534/g3.117.040055

De La Torre AR, Puiu D, Crepeau MW, Stevens K, Salzberg SL, Langley CH, Neale DB (2019a) Genomic architecture of complex traits in Loblolly

pine (*Pinus taeda*). New Phytol 221:1789–1801. https://doi.org/10.1111/nph.15535
De La Torre AR, Wilhite B, Neale DB (2019b) Environmental genome-wide association reveals climate adaptation is shaped by subtle to moderate allele frequency shifts in loblolly pine. Genome Biol Evol 11 (10):2976–2989. https://doi.org/10.1093/gbe/evz220
Falk T, Herndon N, Grau E, Buehler S, Richter P, Zaman S, Baker EM, Ramnath R, Ficklin S, Staton M, Feltus FA, Jung S, Main D, Wegrzyn JL (2018) Growing and cultivating the forest genomics database, TreeGenes, *Database*, bay084, https://doi.org/10.1093/database/bay084
Neale DB, Wegrzyn JL, Stevens KA, Zimin AV, Puiu D, Crepeau MW, Cardeno C, Koriabine M, Holtz-Morris AE, Liechty JD, Martínez-García PJ, Vasquez-Gross HA, Lin BY, Zieve JJ, Dougherty WM, Fuentes-Soriano S, Wu L, Gilbert D, Marçais G, Roberts M, Holt C, Yandell M, Davis JM, Smith K, Dean JFD, Lorenz WW, Whetten RW, Sederoff R, Wheeler N, McGuire PE, Main D, Loopstra CA, Mockaitis K, de Jong PJ, Yorke JA, Salzberg SL, Langley CH (2014) Decoding the massive genome of loblolly pine using haploid DNA and novel assembly strategies. Genome Biol 15:R59
Neale DB, Wheeler NC (2019) The conifers: genomes, variation and evolution. Springer Nature Switzerland AG
Stevens KA, Wegrzyn JL, Zimin A, Puiu D, Crepeau M, Cardeno C, Paul R, Gonzalez-Ibeas D, Koriabine M, Holtz-Morris AE, Martinez-Garcia PJ, Sezen UU, Marçais G, Jermstad K, McGuire PE, Loopstra CA, Davis JM, Eckert A, de Jong P, Yorke JA, Salzbert SL, Neale DB, Langley CH (2016) Sequence of the sugar pine megagnome. Genetics 204:1613–1626. https://doi.org/10.1534/genetics.116.193227
Vasquez-Gross HA, Yu JJ, Figueroa B, Gessler DDG, Neale DB, Wegrzyn JL (2013) CartograTree: connecting tree genomes, phenotypes and environment. Mol Ecol Resour 13:528–537. https://doi.org/10.1111/1755-0998.12067
Wegrzyn JL, Liechty JD, Stevens KA, Wu LS, Loopstra CA, Vasquez-Gross AH, Dougherty WM, Lin BY, Zieve JJ, Martínez-García PJ, Holt C, Yandell M, Zimin AV, Yorke YA, Crepeau MW, Puiu D, Salzberg SL, de Jong PJ, Mockaitis K, Main D, Langley CH, Neale DB (2014) unique features of the loblolly pine (*Pinus taeda* L.) megagenome revealed through sequence annotation Genetics 196(3):891–909. https://doi.org/10.1534/genetics.113.159996
Wegrzyn JL, Staton MA, Street NR., Main D, Grau E, Herndon N, Buehler S, Falk T, Zaman S, Ramnath R, Richter P, Sun L, Condon B, Almsaeed A, Chen M, Mannapperuma C, Jung S, Ficklin S (2019) cyberinfrastructure to improve forest health and productivity: the role of tree databases in connecting genomes, phenomes, and the environment. Front Plant Sci 10:813. https://doi.org/10.3389/fpls.2019.00813
Weiss M, Sniezko R, Puiu D, Crepeau MW, Stevens K, Salzberg SL, Langley CH, Neale DB, De La Torre AR (2020) Genomic basis of white pine blister rust quantitative disease resistance and its relationship with qualitative resistance. Plant J. https://doi.org/10.1111/tpj.14928
White TL, Adams WT, Neale DB (2007) Forest Genetics. CABI Publishing, Cambridge, MA
Zimin A, Stevens KA, Crepeau MW, Holtz-Morris A, Koriabine M, Marçais G, Puiu D, Roberts M, Wegrzyn JL, de Jong PJ, Neale DB, Salzberg SL, Yorke JA, Langley CH (2014) Sequencing and Assembly of the 22-Gb Loblolly Pine Genome. Genetics 196(3):875–890. https://doi.org/10.1534/genetics.113.159715

The manufacturer's authorised representative in the EU is Springer Nature Customer Service Centre GmbH, Europaplatz 3, 69115 Heidelberg, Germany. If you have any concerns regarding our products, please contact ProductSafety@springernature.com

Printed and bound by CPI Group (UK) Ltd, Croydon, CR0 4YY
30/09/2026
02225699-0019